Atlas of Crabs of the Persian Gulf

Reza Naderloo

Atlas of Crabs of the Persian Gulf

Springer

Reza Naderloo
Center of Excellence in Phylogeny
of Living Organisms
School of Biology University of Tehran
Tehran
Iran

ISBN 978-3-319-49372-5 ISBN 978-3-319-49374-9 (eBook)
DOI 10.1007/978-3-319-49374-9

Library of Congress Control Number: 2017945213

Printed on acid-free paper

This Springer imprint is published by Springer Nature
The registered company is Springer International Publishing AG
The registered company address is: Gewerbestrasse 11, 6330 Cham, Switzerland

Foreword I

It is with great pleasure that I introduce Dr. Reza Naderloo's *Atlas of Crabs of the Persian Gulf*. It will prove to be a valuable reference not only for specialists in the taxonomy of Brachyura, but for marine biologists, ecologists, conservationists, and anyone interested in the biodiversity of the marine environment. For a long time, those of us involved in the taxonomy and ecology of Brachyura tended to think of the Persian Gulf as a simple extension of the Indo-West Pacific region. Dr. Naderloo's *Atlas* shows that this is not so. Faunistically and ecologically speaking, the Gulf is a lot more than an appendix of the big ocean. The marine life of the Gulf has evolved to adapt to a unique environment subject to particular physical, chemical, and biological factors. It is unfortunately also critically affected by the impact of us humans. This *Atlas* will therefore validate the appreciation and protection of the rich and valuable marine biodiversity of the Persian Gulf, not only for the inhabitants of the countries bordering the Gulf, but for the rest of the world as well.

Rancho Palos Verdes, CA, USA
June 2017

Peter Castro

Foreword II

A good atlas of the fauna of an interesting region is always an asset to scientists studying biodiversity – be it for systematics, ecology or conservation. It allows any new researcher entering a field to grasp the diversity of an area and quickly introduces him or her to the taxa present. Even for experienced scientists, a good atlas is a useful tool to help him plan for interesting research questions for the future. The present atlas of the Persian Gulf brachyuran fauna by Reza Naderloo is exactly that. The brachyuran crab fauna of this major body of sea is defined by the unique physiochemical properties and biogeographical history, and while the area shares faunistic elements with the Arabian Sea and western Indian Ocean, it also has many species known only from its waters. The present atlas is the result of many years of painstaking work, put together by someone with a deep knowledge of the region and a great passion for his animals. It is scientifically valuable in its up-to-date taxonomy, concise literature, excellent figures and detailed maps, documenting a total of 37 families, 150 genera and almost 256 species. But it is a list that will certainly grow in the years ahead. I was amused to read that the author actually captured a passing remark I made to him some time back about one of the pioneers of carcinological research in the Persian Gulf, the well-known Danish scientist Knud Stephensen. I had commented that while Stephensen's 1946 tome was still a major reference for the region, his work was but a "superficial scratch to this group, with many more discoveries to be expected". In the years since 1946, some 50 species have been added to the area, many of which are not just new records but new species as well. As Stephensen's work set the stage for subsequent generations of researchers on Indian Ocean carcinology, I am certain Naderloo's will as well!

National University of Singapore, Singapore
June 2017,

Peter Ng

Preface

The importance of biodiversity and its conservation have been subject of serious debate since "the Rio Earth Summit" in 1992. For years, scientists, environmental activists, and even politicians have addressed different aspects of biodiversity, have highlighted the services provided by biodiversity, and are concerned about its loss. Therefore, biodiversity conservation looked to be an immediate action which had to be taken into account. Wilson (1988) once cited a celestial statement in conserving the biodiversity "... that fauna and flora of country will be thought part of the natural heritage as important as its art, language, and ...." But here, the main concern is the amount of barrels of crude oil extracted daily in the Persian Gulf. We follow the oil price fluctuations by surfing different media every day, but have no interest to hear even a single word about the diversity change and its consequences on the environment and our lives. We do not know that the second largest population of an interesting marine mammal *Dugong dugong* lives in the seagrass meadows of the southeastern Gulf. Regional politicians discuss about their own share from a common oil field for hours, but do not pay their share to conserve the biodiversity of the environment. The Persian Gulf is surrounded by eight rich oil--producing countries including Bahrain, Iran, Iraq, Kuwait, Oman, Qatar, Saudi Arabia, and the UAE. Oil and gas-based industries of such ever-developing countries in the Gulf are the main threat for such a fragile environment. Marine pollutions originate mainly from these industries, particularly when taking in account that 60% of the exported oil and its derivatives are shipped using tankers through the Strait of Hormuz. Scientists working in the Gulf will find tar balls and trace metals distributed everywhere in the different coastal habitats. In particular, the coastal environments of the Gulf are severely subjected to such anthropologic effects, mainly resulting from coastal developments. Habitat destruction by using dredging and filling in shallow coastal regions for economical and recreational purposes and overfishing, when accompanied with mismanagement, are possibly two more important stresses. Supposedly, desalination plants are silent but severely affect pollution source in the Gulf. The desalination plants are mostly based along the Kuwait, Saudi Arabia, and UAE coasts and discharge hot and hypersaline water into the coastal environment (Lattemann and Höpner 2008). Reportedly, other countries around the Gulf are planning to establish desalination plants along their coasts, without examining the capacity of the small-scaled marine basin with limited freshwater input. For these, we should add stresses coming from sewage discharges, recently increasing air dust, and alluvial sediments. I have witnessed all such environmental corruptions whenever I went for sampling in the last 13 years. Furthermore, the effects of these stresses will undoubtedly be doubled under the direct effect of the recently alarming global warming, and they will be multiplied by silently entering alien and invasive species. A regional conceptual framework is necessarily needed in order to document and regularly monitor the marine biodiversity in the region. However, the success of such regional framework is unlikely as the function of ROPME (Regional Organization for the Protection of the Marine Environment) in the last four decades since its foundation in 1979 is a good evidence for this fact. The main impediment for local scientists to perform regional collaboration in documenting and conserving biodiversity is the political turbulence between the countries bordering the Persian Gulf.

There is no doubt that for any biologist, in particular for those quantifying and evaluating the biodiversity, "species" is the main concern and a basic entity. The species must be identified, named, and classified using the most updated classification hierarchy. Taxonomists take this fundamental but cumbersome responsibility. Unfortunately, few comprehensive taxonomic researches have been conducted on the different marine taxa in the region so far, mostly because of the paucity of a regional taxonomist. Expertise in the taxonomy of the different marine groups is being developed among a new generation of regional scientists. Most of them have studied abroad and returned to the region to pursue their field of expertise. I am among the scientists of the generation and started to study the brachyuran crabs since 2002 when I started my masters in animal biosystematics at the University of Tehran. Since then, I have been collecting material and notes during my several field trips over the past 13 years. I did substantial works during my PhD at the Senckenberg Museum of Frankfurt where I performed extensive taxonomic studies on the material I had collected myself and the materials collected from the region, in particular those of Stephensen (1946), Titgen (1982), and Apel (2001).

An atlas like this is a basic and an important step in understanding and reconstructing the biodiversity of life in this geographical region. The book provides baseline information on the brachyuran fauna of the Persian Gulf and the Gulf of Oman and, at the same time, reflects the huge gap of required fundamental knowledge in other groups of marine animals. Within crustaceans, for example, there is a lack of such a data for caridean shrimps, amphipods, isopods, cumaceans, mysids, ostracodes, tanaids, and many others. I hope that young scientists will fill the gaps in the near future.

Tehran, Iran Reza Naderloo

Acknowledgment

I am most grateful to Alireza Sari (University of Tehran, Iran), who is the fundamental carcinologist in Iran with wide information of Iranian crustaceans. Alireza Sari encouraged me and first got me interested in the brachyuran crabs of the Persian Gulf when I was a master student under his supervision at the University of Tehran between 2002 and 2005. I am especially thankful and extremely honored working under the supervision of the late Michael Türkay (Senckenberg Museum, Frankfurt am Main, Germany) during my PhD in Frankfurt am Main. Michael Türkay helped me in the course of preparation of this atlas in any possible occasions by providing working space, material, and funding during the last 8 years. My special thanks is due to Michael Apel (Museum Mensch und Natur, Munich, Germany), who did the comprehensive ecological and taxonomic works on the brachyuran crabs in the region, particularly, along the Arabian coasts during the years 1991–2001, and compiled a huge amount of information on this group. Michael Apel has always helped me from my first visit to Senckenberg Museum in 2005 up to now and further encouraged me to continue the study of the brachyuran crabs of the region. He allowed me to use the data from his unpublished PhD thesis in this atlas which is greatly appreciated.

This publication was really impossible without contributions of many other individuals. These are Peter Ng and Christopher Mendoza (National University of Singapore), Paul Clark (Natural History Museum, London), Peter Castro (California State Polytechnic University, USA), Peter Dworschak (Natural History Museum, Vienna), Danièle Guinot and Laure Corbari (Muséum national d'Histoire naturelle, Paris, MNHN), Mortiz Sonnewald (Senckenberg Museum, Frankfurt am Main, Germany), and Venkataraman Krishnamoorthy and Kandasamy Valarmathi (Zoological Survey of India, Kolkata, India). James Bishop (Kuwait Institute for Scientific Research) and Alireza Keikhosravi (Shahid Hakim University of Sabzevar, Iran) helped me in editing the introduction and improved the text.

I have been helped by a number of my students in different issues. These are Sepideh Barzgar, Rashed Abdollahi, Mehrnaz Afkhami, Abolfazl Arab, Arash Reeisbahrami, Saeed Ebrahimnejad, Samaneh Pazoki, Elahe Parvizi, and Zahra Rajaie who are greatly acknowledged. The photographs presented in this book have been mostly taken by Sven Träkner (Senckenberg Museum, Frankfurt am Main, Germany) and Rashed Abdollahi (Zoology Museum, University of Tehran) and some by Harry Taylor (Natural History Museum, London) and Soheil Shahbazi. All photographs are very much appreciated [all help by all photographers].

Contents

Introduction

1

Abbreviations

CL	Carapace length
CB	Carapace breadth
G1	Male first pleopod
G2	Male second pleopod
MNHN	Muséum National d'Histoire Naturelle, Paris, France
NHM	Natural History Museum, London
NHMW	Naturhistorisches Museum Wien, Vienna, Austria
RMNH	Naturalis, formerly Nationaal Natuurhistorisch Museum, Leiden, Netherlands
SMF	Research Institute and Natural Museum Senckenberg, Frankfurt am Main, Germany
SNMNH	Saudi National Museum of Natural History, Riyadh, Saudi Arabia
ZMG	Zoological Museum of Gottingen, Germany
ZMUC	Zoological Museum, University of Copenhagen, Denmark
ZRC	Zoological Reference Collection, Raffles Museum of Biodiversity Research, National University of Singapore
ZUTC	Zoology Museum, University of Tehran, Iran

The Persian Gulf

The Persian Gulf is a semi-enclosed sea which is located in a subtropical, hyper-arid region of the north-western Indian Ocean. The Persian Gulf is a shallow sedimentary basin stretching from southeast to north-west, about 1000 km long and between 200 to 300 km wide, and covers an area of about 251,000 km^2. The basin is characterized with generally eastward dipping seafloor, with its deepest regions along the Iranian coast. A maximum depth of 100 m is recorded in the Strait of Hormuz, and the average depth of the basin is 35 m (Barth and Khan 2008). The low-salinity oceanic water enters to the Persian Gulf through the Strait of Hormuz, flowing along the Iranian coast in a counter-clockwise direction, remarkably increasing in salinity along the Arabian side and discharging into the Gulf of Oman (Sale et al. 2010). Harsh environmental conditions result mostly from high evaporation and low fresh water input (Fig. 1.1).

The present structure of the Persian Gulf primarily formed due to tectonic movements in the Tertiary period. It was basically established in the Miocene (25–5 Mya), when the orogenic process resulted in the

R. Naderloo, *Atlas of Crabs of the Persian Gulf*, DOI 10.1007/978-3-319-49374-9_1

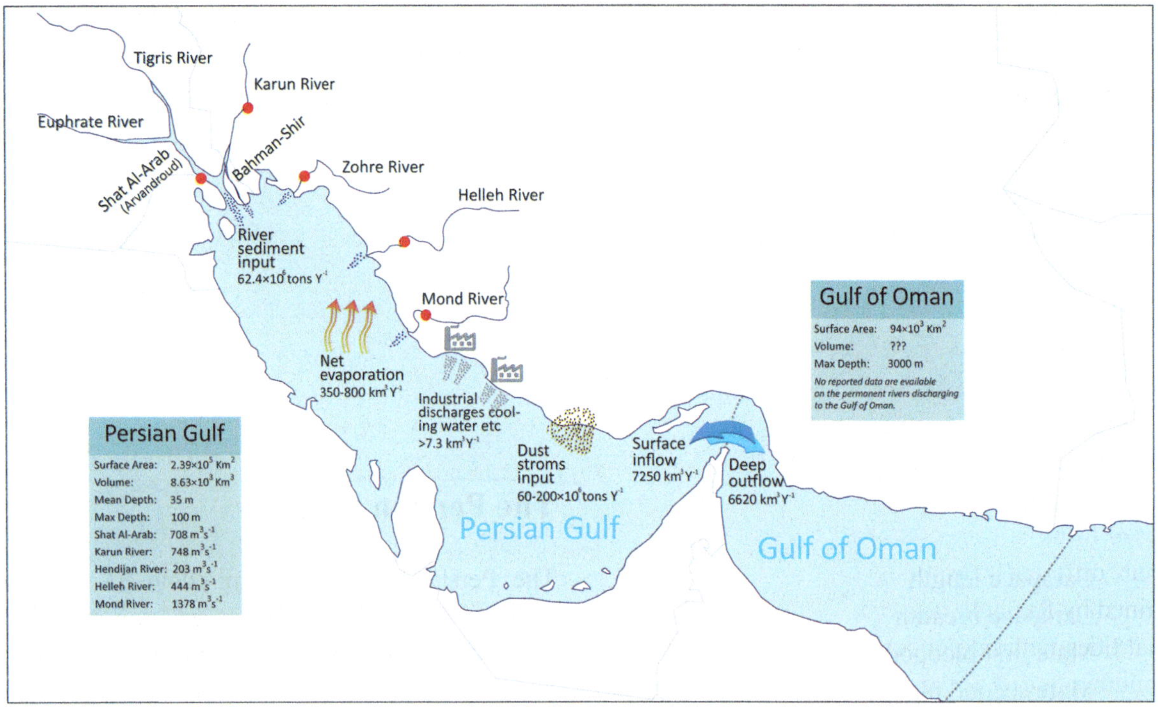

Fig. 1.1 Map of the Persian Gulf and the Gulf of Oman showing their boundaries. Estimates of the water exchange, dust and sediment input were drived from Sheppard et al. (2010). Estimates of the discharge amount of the major rivers flowing into the Persian Gulf are shown in the Persian Gulf box (Data from Reynolds 1993). *Red circles* indicate the recording stations for the estimates

formation of the Taurus, the Zagros Mountains and the Oman mountains (Barth and Khan 2008). Considerable sea level fluctuations occurred during the period of 110 ka BP and 17 ka BP, when the sea level was between 120 and 150 m below its present level, leading to the complete evaporation of the Persian Gulf (e.g. Sheppard et al. 1992; Barth and Khan 2008). The global temperature increased during the Holocene resulting in the sea level transgression that started about 15 ka BP and continued till 6 ka BP. At that time, the Persian Gulf sea level was about 2.5–3.5 m higher than today's level (for detailed see Barth and Khan 2008).

There are four main rivers discharging into the Persian Gulf along the northeastern coast, namely, Shat Al-Arab, Hendijan, Helleh, and Mond. Shat Al-Arab River (known as Arvandroud River in Iran) is formed by combination of Euphrates, Tigris and Karun rivers, and flows down along the border of Iran and Iraq. The Shat Al-Arab River is the largest river with length of 195 km and average width of 500 m and average discharge of 1456 m^3/s (Reynolds 1993). It constructs a large estuarine system with considerable productive marshlands and leads to formation of a large river plume of 30–40 km in the northern Persian Gulf (Al-Yamani 2008). Unfortunately, in recent decades intensive damming in quickly developing countries like Turkey, Syria, Iraq and Iran has reduced the amount of water discharge through this river to the Gulf. This subsequently severely affected the estuary ecosystem, hydrological system and sedimentation in the northern Gulf. Other three rivers, namely Hendijan (203 m^3/s), Helleh (444 m^3/s) and Mond (1387 m^3/s) which feed the Gulf mainly originate from Zagros Mountains in Iran (Fig. 1.1).

The substrate of the Persian Gulf Basin is mostly characterized with sedimentary structures (Sheppard et al. 2010; Sale et al. 2010; Naderloo and Türkay 2012). Muddy flat is the dominant substrate along the Iranian and Arabian coast of the Gulf, and usually stretching to deeper zones in most parts. The muddy flats are widely distributed on the northern part of the Gulf where it supports a highly valuable primary production, and provides nursery and spawning ground for shrimps and a grazing ground for migratory birds. In contrast, rocky structures are patchily distributed within the Gulf and generally are more seen along the Iranian coast than the Arabian side (Naderloo and Türkay 2012). Other ecologically important habitats, e. g. mangroves, corals, and sandy beaches, all with their own diversity, can be patchily found within the Gulf. Despite this ecologically diverse environment, the Persian Gulf is generally considered biologically impoverished, partly because of its young age, but mainly because of its harsh environment (Sheppard et al. 2010). Recently, Price and Iszak (Price and Izsak 2005) addressed the importance of the biodiversity in the Persian Gulf by highlighting the high value of taxonomic distinctness and high beta diversity within the Gulf. This makes the environment worthy of conservation attention.

The tidal system in the Persian Gulf, due to its semi-enclosed form, is very complex (Jones 1986a) leading to four different variations of the tidal system as explained by Raffaelli and Hawkins (1996). But, semidiurnal tides with two unequal tides occur daily with a maximum range of 5 m in the northern part, and a minimum of 2 m in middle part of the Gulf. The most ecologically important fact is that the lowest tides generally occur during the cold months along the Iranian coast (Naderloo and Türkay 2012) and Kuwait coast (Jones 1986a), and the coast experienced low spring tides during the day in cold seasons and during the night in warm seasons. This is essential for intertidal organisms because it protects them from freezing in winter and from desiccation in summer. It is also important for the coastal biologists to perform sampling during the long days of summer. The dominant wind in the Persian Gulf is known as "Shemal", a wind blowing northwesterly throughout the year. Winter Shemal is predominantly stronger, bringing denser air from higher latitudes over to the Persianw Gulf leading to higher temperature fluctuations ranging between 15° and 30 ° C (Reynolds 1993). More detailed information on the different aspects of the Persian Gulf is recently provided by Riegl and Purkis (2012).

The Gulf of Oman

The Gulf of Oman connects the Persian Gulf to the Indian Ocean via narrow Strait of Hormuz. The Gulf of Oman with about 94,000 m^2 surface area is westerly defined by an imaginary line running from Minab on the Iranian coast to Qabr al-Hindi on the Oman coast, and its eastern side is delineated by an imaginary line stretching from Ras Jiwani on the border of Iran and Pakistan to Ras al-Hadd on the Oman coast (Fig. 1.1). In contrast to the shallow Persian Gulf, the Gulf of Oman is quite deep, about 200 m near Strait of Hormuz, getting immediately even deeper up to 2000 m in its middle part and then to maximum depth of 3000 m in eastern part (Reynolds 1993). The Gulf of Oman is markedly different from the Persian Gulf as it is affected by the monsoon systems in the northern Arabian Sea. The monsoons bring southward blowing cool winds in summer and northwardly blowing winds in winter. Therefore, the temperature fluctuation of water is balanced in the Gulf of Oman ranging from 22° to 31 ° C (Reynolds 1993; Piontkovski et al. 2012). Salinity is the second most important physical factor limiting the biodiversity. While it is always more than 37 psu and reaching up to 40 psu in most parts of the Persian Gulf, the maximum summer salinity recorded for the Gulf of Oman is 37 psu (Reynolds 1993). Intertidal regions support wide variety of habitats from sedimentary muddy-sandy in sheltered environments to exposed sandy beaches and active rocky habitats. Comparatively, the Gulf of Oman has more exposed rocky habitats than the coast of the Persian Gulf.

Tidal system of the Gulf of Oman is semidiurnal as seen in the Persian Gulf with two unequal tides, but it is affected by tidal regime of the Arabian Sea to a greater extent. Unfortunately, geomorphology and biology of the Gulf of Oman has not received much attention compared to that of the Persian Gulf, as the oil exploration in the Persian Gulf accelerated the research in the Gulf by financial supports of the oil companies.

Infraorder Brachyura Linnaeus, 1758

History of Brachyuran Research in the Region

Biodiversity of the Persian Gulf is generally poorly documented. Nevertheless, the brachyuran crabs have received better attention compared to other marine taxonomic groups, and other infraorders of the class Decapoda. The study of this group in the Gulf was started by Heller (1861a) who recorded just one species, *Epixanthus frontalis* from the northern Persian Gulf based on a collection made by T. Kotschy 1843/1844.

Heller's (1861a) work was basically on the infraorders Anomura and Brachyura of the Red Sea and the material was deposited in the Natural History Museum of Vienna. Alcock (1895, 1896, 1898, 1899a, 1899b, 1900, 1901, 1905) has recorded 51 brachyuran species from the Persian Gulf in his works on the crabs collected by scientific vessel "Investigator". Unfortunately, the Persian Gulf sampling localities are not mentioned by Alcock and it is thought that some localities were outside the Gulf, or even "Investigator" had not ever been in inner the Gulf (Apel 2001). Further, Investigator's collection is mostly deposited in the Indian Museum of Calcutta, where, nowadays, it is not easily accessible to scientists, therefore re-examination of the Alcock's material has not been easily possible for long time. The second major contribution to the Persian Gulf decapod crustaceans was made by G. Nobili in his 1906 work entitled "Crustacés décapodes et stomatopodes. In: Mission J. Bonnier et Ch. Perèz - Golfe Persique, 1901". He examined the materials collected by Belgian yacht "Selinka" in the Persian Gulf in 1901 and recorded 50 brachyuran species from the Gulf, the majority of which were from the Arabian side, particularly from the UAE coast. After these works in the early twentieth century, there was no important investigation on the Gulf's crabs for several decades; only few scattered records of these animals from the area followed the earlier studies. MacGilchrist (1905) recorded seven brachyuran species and Klunzinger (1913) and Chopra and Das (1930) each recorded one.

N.V. Bogoyavlensky from "Moscow Society of Natural Science, Ethnography, and Anthropology" made a good sampling from the intertidal coasts of Kuwait, Bahrain and Iran. This collection was deposited in the Zoological Museum of the Moscow University, and unfortunately has not received much attention. Fauvel (1911) studied the polychaetes of this collection and provided basic information of the sampling sites. The brachyuran crabs of the collection was examined by Apel (2001).

The largest collection from the Persian Gulf, so far, was performed by "Danish Scientific Expedition in 1937/38". Iranian government invited Danish fisheries scientists to conduct a preliminary survey in the Gulf in order to improve the Iranian fisheries industry. For this purpose, a 35 ton fishing boat named "Rashgoo", was sold to the Iranian government in 1936 and was sailed to the Persian Gulf by Danish sailors. They made a good collection from the Iranian waters, but mainly from the subtidal zones. The collection is now deposited in Zoological Museum of the Copenhagen University, Denmark. Different taxa of the collection were examined by specific taxonomic experts and provided key literatures which are broadly used nowadays. Brachyuran crabs were examined by K. Stephensen (Fig. 1.2). In his landmark work "Brachyuran Crabs of the Iranian Gulf", he identified 124 species (Stephensen 1946). Thus, he updated the number of brachyuran species known to that time to162. His work is a good reference devoted to Persian Gulf's crabs, but taxonomically speaking, it is thought to be just a "superficial scratch to this group" (Peter Ng, pers comm). Stephensen (1946) described three new species, namely *Dentoxanthus iranicus* Stephensen, 1946, *Typhlocarcinus dentatus* Stephensen, 1946, and *Thalamita iranica* Stephensen, 1946, from the collection.

Fig. 1.2 Knud Stephensen (1882–1947), curator in ZMUC of Copehenhagen from 1910 to 1947

In the early 1970s, an Austrian carcinologist, Gerhard Pretzmann from the Natural History Museum of Vienna, who was an expert on the fresh water crabs, made a small collection from the coast of Bandar Abbas along the Iranian side. Pretzmann (1971) reported 12 intertidal crab species, five of which were new records, and two were new to science, namely *Uca iranica* Pretzmann, 1971, and *Macrophthamus* (*Macrophthamus*) *ressli* Pretzmann, 1971. The latter has been later considered as a junior synonym of *Macrophthalmus laevis* A. Milne-Edwards, 1867 (see Naderloo et al. 2011). The next large faunistic study in the Persian Gulf was conducted along the Saudi Arabian coasts by the Saudi Arabian-American Oil Company (ARAMCO) in the 1970s. Basson et al. (1977) published the result of this investigation in a referable book entitled "Biotopes of the Western Arabian Gulf". This was the first ecological study in the Persian Gulf, however, it failed to identify the majority of the specimens up to species level, and many of the specimens were even misidentified, thus it suffered severely from taxonomic problems (Apel 2001). Titgen (1982) from Texas University did his PhD project on the systematic and ecology of Decapoda in Dubai and listed 196 species from the Gulf, most of which were not properly studied and the identification of some species is highly questionable. Nevertheless, he was the first who performed a zoogeographical analysis based on the decapod crustaceans in the Persian Gulf. Jones and Clayton (1983) described two new camptandriid species from muddy shores of Kuwait, namely *Cleistostoma kuwaitense* Jones and Clayton, 1983 [currently known as *Leptochryseus kuwaitensis* (Jones and Clayton 1983)] and *Paracleistostoma arabicum* Jones and Clayton, 1983 [currently known as *Manningis arabicum* (Jones and Clayton 1983)]. Jones (1986a, b) are two further literatures dealing with marine intertidal organisms of Kuwait including brachyuran crabs.

A huge oil spill following the Gulf War of 1991 triggered for some important ecological and faunistic surveys in the coastal zone of the Arabian coast, mainly on the muddy shore along Saudi Arabia and Kuwait. Michael Apel (Fig. 1.3), a German carcinologist from Senckenberg Museum of Frankfurt, participated in these projects and devoted over a decade of his life to collecting and examining the Persian Gulf's crabs. He published some important papers scrutinizing the taxonomy and distribution of different intertidal Decapoda, with special emphasis on the brachyuran crabs (e.g. Apel 1994a, b, 1996a; Apel and Spiridonov 1998; Apel and Türkay 1992, 1999). Finally, in his PhD dissertation, Apel (2001) thoroughly reviewed all faunistic and taxonomic works on the Brachyura of the Gulf and undertook intensive sampling along the Saudi Arabian and UAE coasts. He raised the valid species number of the Brachyura occurring in the Gulf to188 species.

Fig. 1.3 Michael Apel (born in 1964), currently Director of Museum Mensch und Natur, Munich, Germany

Al-Ghais and Cooper (1996) and Cooper (1997) investigated the mangal associated Brachyura of Abu Dhabi (UAE) and Al-Khayat and Jones (1999) compared macrofauna in natural and planted mangroves in Qatar.

In the recent years, Iranian scientists conducted worthy contributions to the marine organisms of the Persian Gulf. In which, brachyuran crabs received well attention by the local scientists, particularly by the author. Two important intertidal groups, superfamily Grapsoidea and family Macrophthlamidae, have been thoroughly revised and several new species have been described (Naderloo 2011; Naderloo et al. 2011). These new species are *Nanosesarma sarii* Naderloo and Türkay, 2009, *Parasesarma persicum* Naderloo and Schubart, 2010, *Chaenostoma sinuspersici* (Naderloo and Türkay 2011), *Hiplyra ramli* Naderloo and Apel, 2012, *Paratymolus apeli* Naderloo and Türkaym, 2015, *Menaethiops abumusa* Naderloo, 2015, *Palapedia apeli* Naderloo, 2015, and *Palapedia persica* Naderloo, 2015. Naderloo and Türkay (2012) provided an updated checklist of the decapod crustaceans presenting along the Iranian coast of the Gulf and recorded 150 Decapoda. Of which, 83 (55 %) belong to the brachyuran crabs. Comparatively, brachyuran fauna of Iraqi shores has been received less attention. Only recently, few studies have documented brachyuran crabs of Iraq (Ng et al. 2009; Naser et al. 2010, 2012; Naser 2011), but Brachyura of Iraq have still remained largely underrepresented. In addition to the above mentioned studies which merely focused on the Persian Gulf's fauna, there are several new species and new records of Brachyura described or reported from the Persian Gulf in the literatures (Galil 2001, 2005a, 2009 ; Castro and Ng 2010; Galil et al. 2012).

In comparison with the Persian Gulf, general biodiversity of the Gulf of Oman is markedly underestimated. This is mainly because of scarcity of scientific samplings. The brachyuran fauna of the Gulf of Oman, principally initiated by "Danish Expedition 1937/38". As already stated, this expedition has comprehensively sampled along the Iranian coast of the Persian Gulf, with having only 10 sampling stations along northern coast of the Gulf of Oman. In his report, Stephensen (1946) recorded 24 species from the Gulf of Oman, of which 22 were new records to the region. This revealed the fact that fauna in the region was fully unknown. Hogarth (1988) recorded five species of hermit crabs from Muscat, Oman and a year later 19 decapods from the same region (Hogarth 1989). Ismail and Ahmed (1993) performed a sampling in the Khor Kalba in the east coast of UAE and recorded seven brachyuran species. Hornby (1997) made a small collection of decapod crustaceans from the east coast of UAE with reporting 11 crabs. In 1994, Hywel-Davies, in his M.Sc. thesis on biodiversity of mangroves of Muscat, recorded four brachyuran crabs (Hywel-Davies 1994). Moreover, five species of ghost crabs of the genus *Ocypode* were recorded by Clayton (1996) from Oman. In a comprehensive study of Portunidae, Apel and Spiridonov (1998) recorded 29 species from the Gulf of Oman. In his thesis, Apel (2001) made a limited sampling along the east coast of UAE together with reviewing the literatures, he reported 106 species of Brachyura in Gulf of Oman. Ghotbeddin et al. (2012a) collected 34 species of brachyuran crabs from Chabahar Bay in Iranian coast of the Gulf of Oman. Most recently, there are some few records of the brachyuran crabs from the Iranian coast of the Gulf: Fatemi et al. (2011) recorded three species of the family Dotillidae, Ng et al. (2011a) recorded *Calappa exanthematosa* Alcock and Anderson, 1894, Ng et al. (2011b) recorded *Calappa dumortieri* Guinot, 1964, Fatemi et al. (2012) recorded *Cryptopodia echinosa* Chion and Ng, 1998, Valinasab et al. (2012) recorded *Galene bispinosa* (Herbst, 1783), Ghotbeddin and Naderloo (2014) recorded four xanthid crabs, and Naderloo and Fatemi (2015) recorded an alien species, *Percnon affine* (H. Milne Edwards 1853). Finally, Naderloo et al. (2015) provided an annotated checklist of the Decapoda of the Gulf of Oman by examining freshly collected samples and reviewing already published literatures. They recorded a total of 176 brachyuran crabs from the Gulf of Oman.

Taxonomy

The Brachyura Linnaeus, 1758, is an infraorder of the order Decapoda Latreille, 1802. Decapoda is an incredibly diverse group and decapod taxonomy is an active field with researchers making constantly new discoveries (De Grave et al. 2009). According to De Grave et al. (2009) the order Decapoda contains 233 families, 2725 genera and 17, 635 species distributed worldwide, finding in any possible habitat. Ng et al. (2008) published "Systema Brachyurum", a valuable catalogue on the extant brachyuran species, with invaluable notes on the taxonomically problematic taxa. According to Ng et al. (2008), Brachyura with 8616 species is the most species-rich infraorder within the Decapoda. Of this number, 6559 are extant, whereas 1781 are exclusively fossils, and 276 have both living and fossil records. Currently there are 256 species of the brachyuran crabs belonging to 144 genera and 37 families from the Persian Gulf and Gulf of Oman. Of which, 223 are recorded from the Persian Gulf and 176 are from the Gulf of Oman. The low number of species recorded from the Gulf of Oman is mainly due to fewer sampling efforts.

This atlas provided the basic taxonomic list for the brachyuran species recorded from the region, but detailed taxonomic revisionary studies are urgently needed in different groups. Examples are the diverse families Leucosiidae, Epialtidae, Majidae, Xanthidae and Pilumindae. Morphologic differences are clearly seen in populations of some taxa occurring in the region. Further morphological and molecular data would be useful to clarify the taxonomic position of the different species in the region. *Pilumnopeus convexus* (Maccagno, 1936) (Pilumnidae) is an outstanding example which has frequently been confused with *Pilumnopeus vaquelini* (Audouin, 1826). Species within several xanthid genera (e.g. *Medaeops*, *Macromedaeus*, *Etisus*) are morphologically poorly differentiated and the members of the family Pilumnidae are taxonomically difficult taxon which all need urgent taxonomic revision.

There are two species recently recorded as aliens to the region. These are *Eriochier hepuensis* Dai, 1991 (Varunidae) from the northern estuarine system of the Persian Gulf (Naderloo 2014) and *Percnon affine* (H. Milne Edwards, 1843) (Plagusiidae) from the Iranian coast of the Gulf of Oman (Naderloo and Fatemi 2015). The brachyuran crabs live in variety of habitats ranging from marshlands, through different intertidal substrates to shallow subtidal coral reefs and seaweeds down to maximum depths in the Persian Gulf and the Gulf of Oman.

Main Characteristics of the Brachyuran Crabs

Carapace (Fig. 1.4a) large, prominent, comprising of 5 cephalic and 3 thoracic somites, with lateral linea brachyura. **Front** prominant, usually separated from orbit. **Eyes** always with distinct stalk, eyestalk with two segments, situated in complete or sometimes incomplete orbits. **First antennae** (antennules) with three-segmented peduncle, with short flagella, longitudinally or transversely folded in distinct hiatus. **Second antennae** (antennae) with three-segmented peduncle, first segment (basal antennal segment) usually immovable, flagella short or long. **Mandibles** intensely calcified, large, with prominant molar and incisor processes. Two biramous **maxillae**. Three **maxillipeds**, third maxillipeds (Fig. 1.4b) well developed, with distinct plate-form ischium and merus, functioning as an operculum in covering buccal cavern. **Pereiopods** (walking legs) uniramous, with seven segments including coxa, basis, ischium, merus, carpus, propodus, and dactylus (Fig. 1.5b); basis and coxa are tightly fused, while ischium is tightly fused to merus; first pereiopod chelate. **Chelipeds** (Fig. 1.5a) with propodus elongate, forming palm and immovable finger, dactylus forms movable finger. Last four pairs of walking legs getting smaller distally; last two or just last ones are sometimes modified for doing specific functions, like swimming (e.g. Portunidaae), object carrying (e.g. Dromiidae), markedly reduced (e.g. Palicidae) or completely missing (e. g. Hexapodidae). **Abdomen** (Fig. 1.6a, b) seven segmented (with telson), some somites fused together in some groups, particularly in males; abdomen of males relatively narrow, tightly folded against sternum, with or without holding mechanism; abdomen of females usually broad, free, without holding mechanism. **Male gonopore** (Fig. 1.6c) on coxa of pereiopod 5 (e.g. Xanthoidea, Pilumnoidea, Potamoidea) or thoracic sternite 8 (e.g. Grapsoidea, Ocypodoidea), supplemented with penis for transferring sperms to G1. **Female gonopore** (Fig. 1.6d) on coxa of pereiopod 3 (section Podothremata) or on sternite 6 (section Eubrachyura). **Pleopods** 1 and 2 (Fig. 1.6c) modified into uniramous gonopods in males, together functioning as sperm transporter to female gonopore, G1 is normally long and robust, while G2 is short in most groups but in some narrow and long (e.g. Menippidae, Oziidae) and used to pump sperms into sperm channel along the G1, pleopods 3–5 are completely missing in males; pleopods of females exist, pleopod 1 uniramous, generally reduced, rarely absent, pleopods 2–5 biramous, setose, are used for carrying eggs.

The infraorders Brachyura is primarily divided into two sections based on the position of the genital opening of males and females; section Podothremata or "primitive" crabs (genital opening of males and females coxal) and Eubrachyura or "advanced" crabs (genital opening of females sternal and males coxal or sternal). The section Eubrachyura is subsequently divided into two subsections Heterotremata (genital openings of males on coxa of fifth pereiopod) and Thoracotremata (genital opening of males on eighth sternum) (Guinot 1968;

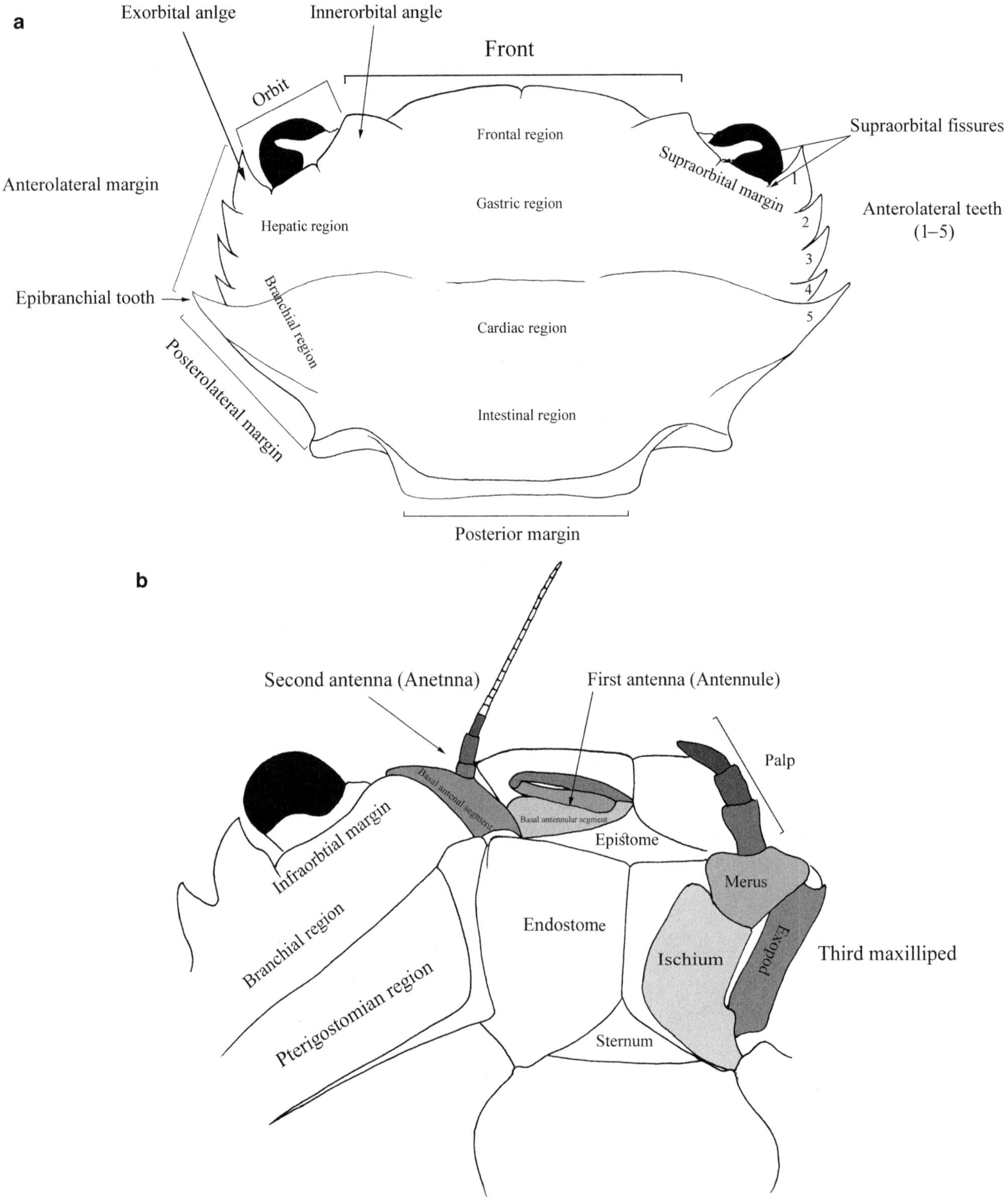

Fig. 1.4 Schematic drawing of the carapace indicating important diagnostic characteritics and the terms used in the text. (**a**) dorsal view, (**b**) anteroventral view. The illustration is from *Thalamita iranica* (Portunidae)

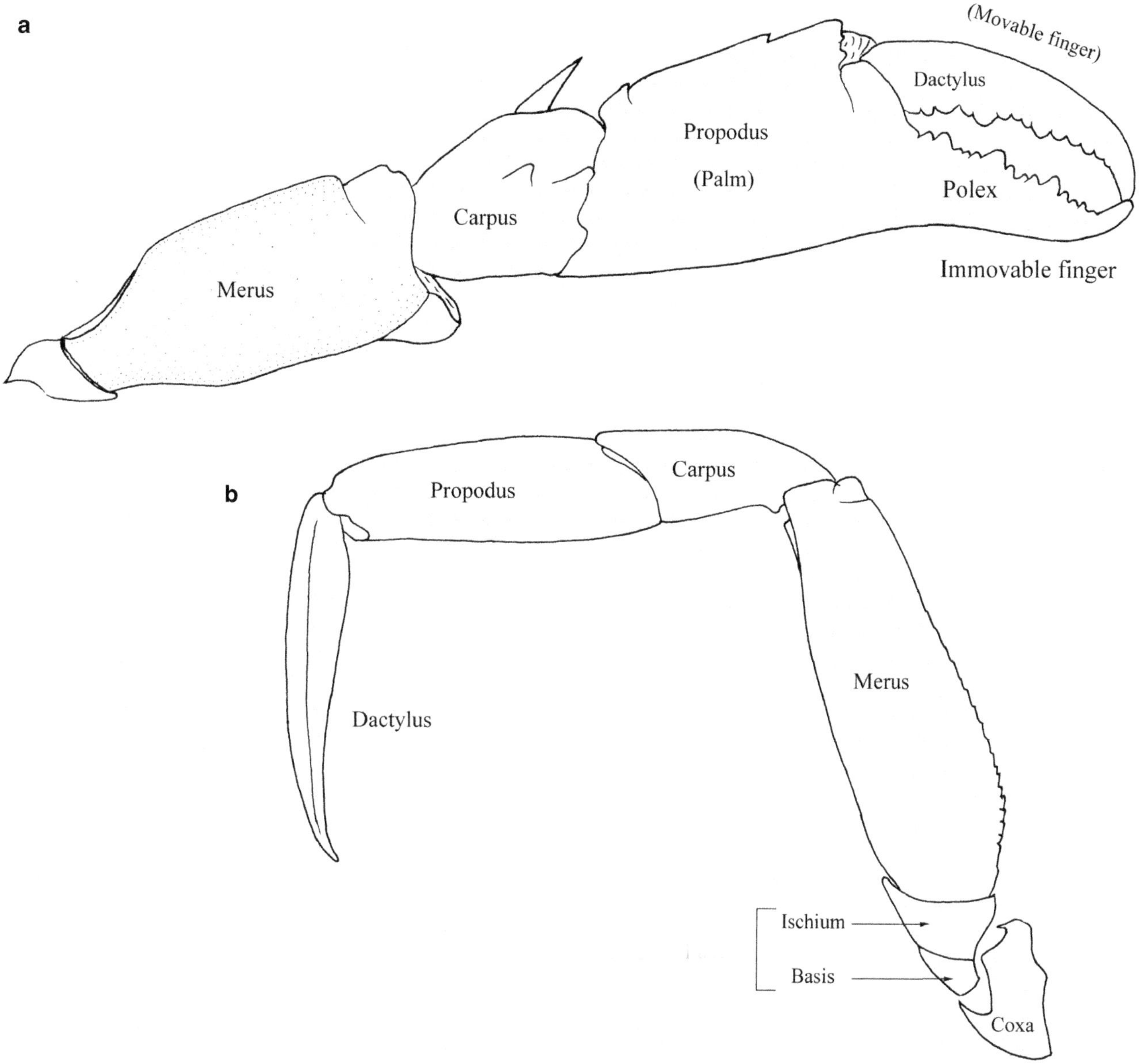

Fig. 1.5 Schematic drawing of the pereiopods indicating important diagnostic characteritics and the terms used in the text. (**a**) cheliped, (**b**) walking leg

De Saint Laurent 1980a, 1980b; Ng et al. 2008). The unity and monophyly of the sections and subsections have frequently been questioned by authors (e.g. Brösing et al. 2007; Ahyong et al. 2007; Ng et al. 2008). I prefer to use the high level classification in the presently accepted form, although it does not convey phylogenetic history, it shows only distinguishable shared characters of the members of the groups. I found out that it is not necessary to use superfamilies, due to the instability of most of the groups like Ocypodoidea, Grapsoidea and Goneplacoidea.

Ng's (1998) work is a good starting point for students beginning to basically understand brachyuran crabs. This work is thought to largely deal with commercially

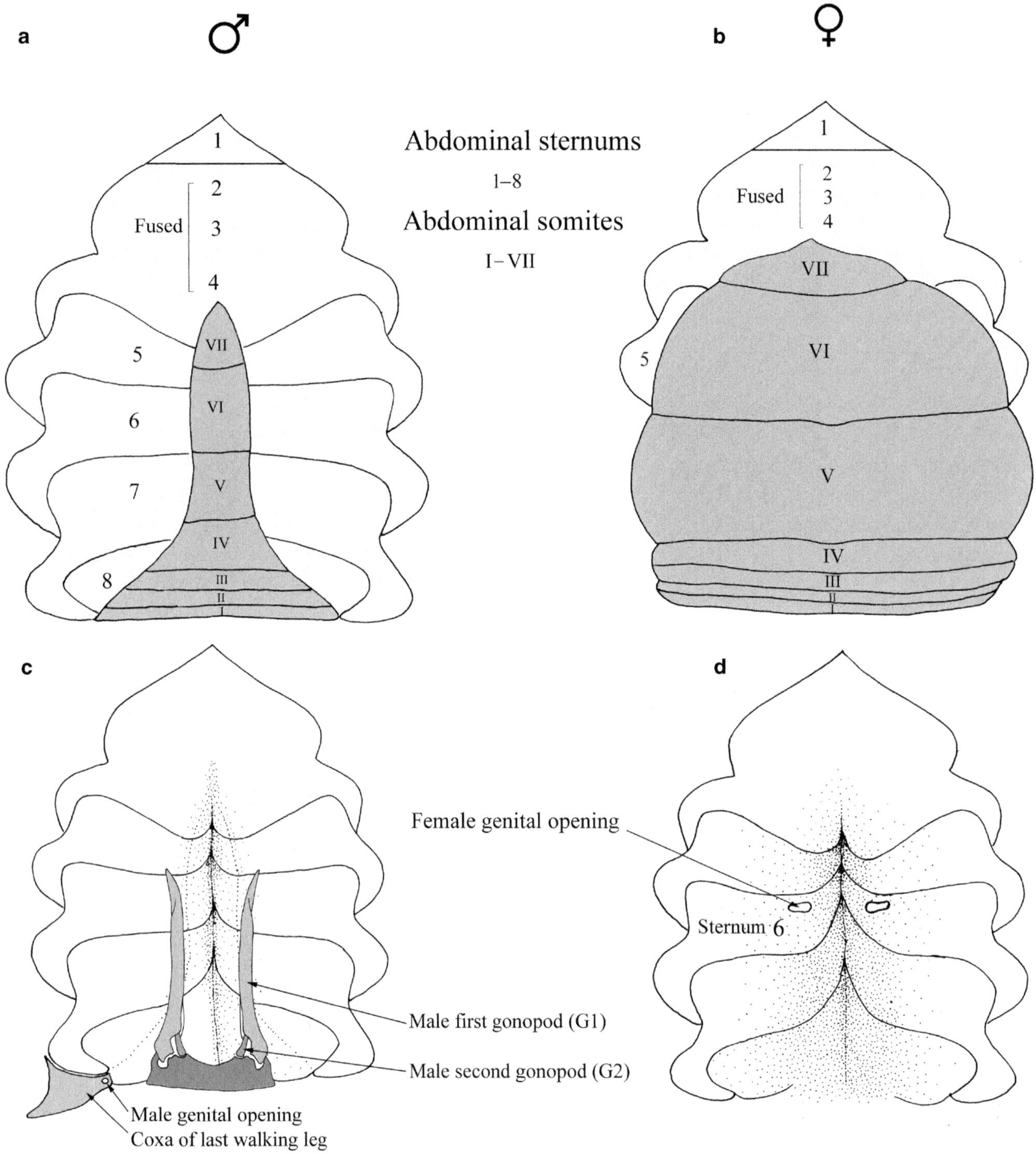

Fig. 1.6 Schematic drawing of the abdominal part with important diagnostic characteritics and the terms used in the text. (**a**) abdomen of male, (**b**) abdomen of female, (**c**) gonopod and gonopore of male, abdomen removed, (**d**) gonopore of female, abdomen removed

important species, nevertheless it provides a basic descriptions and illustrations of the crab's external morphology, accompanying with landmark identification keys for known extant families. Ng et al. (2008) is a key catalogue of extant brachyuran species, which is indispensable to any scientists working on the taxonomy of this group. Any scientist has to have a hardcopy of this book on his/her desk or a PDF file on his/her desktop!

About the Atlas

The Atlas is first provided for crustacean taxonomists working on the taxonomy and biodiversity of the group in the region, with providing a valid list of the species and their synonymies. Secondly, the Atlas with identification guide and discriminative photos can also help general marine scientists, ecologists and students to readily identify the species in the fields and in the labs as well.

Classification systems and hierarchical nomenclatures used in this atlas are mainly based on Ng et al. (2008) and De Grave et al. (2009). Sections and families are arranged according to their evolutionary relationships and superfamilies are not used in this work in order to avoid the complexity of the taxonomic hierarchies. Genera and species are arranged in alphabetical order. Identification keys are provided for all families, genera and species using main synapomophies. Short but precise descriptions containing the main characteristics are also provided for every family. Descriptions and key of the families are mainly based on the three main literatures including Ng (1998), Davie (2002) and Števčić (2005). Wherever available, references specifically contributed for a certain family are used, e. g. family Palicidae (Castro 2000a), Euryplacidae (Castro and Ng 2010), Portunidae (Apel and Spiridonov 1998), and Trapeziidae and Tetraliidae (Castro et al. 2004). The descriptions are supplemented with remarks, mainly with taxonomic notes if there is a necessity.

"Synonymies" encompass the regional works, including those from the Persian Gulf, Gulf of Oman and Arabian Sea, and sometimes references from the Red Sea and Gulf of Aden are included. Further, original descriptions, overall revision of given taxa and monographs are also listed under synonymies in order to provide additional detailed taxonomic accounts. "Taxonomic remarks" for species are just given when are necessary.

Measurements for the specimens in the figures are included in the legends. These are carapace length (CL) which was measured from the middle of the frontal margin to the middle of the posterior margin of the carapace, and carapace breadth (CB) which was measured across the widest breadth of the carapace, in most cases between the longest anterolateral or epibranchial teeth.

Color photograph from dorsal view of the carapace is provided for all species. Some species are provided with additional photos, particularly ventral view, showing discriminative characters. As in the Brachyura, the male first gonopod is most important discriminant character, line drawing of the character was made for the majority of the species. Drawings of abdomen, carapace, cheliped and third maxilliped were made for few species where the photos do not properly help for readily distinguishing the species.

Distribution within the Persian Gulf and the Gulf of Oman are given based on the records from localities defined in the Fig. 1.1. Unfortunately, because of paucity of the sampling, many species have been recorded from limited localities, therefore point distribution are provided for all the species. Some species have already been recorded from the Persian Gulf and the Gulf of Oman, by non-taxonomists. These are mostly from Basson et al. (1977), Titgen (1982), and Hogarth (1989). I neither could re-examine the material nor can easily rely on the records, therefore the 28 species with doubtful and questioned records are not in taxonomic list but listed in a table (Table 1.1). In addition, Alcock's (1895–1905) records are based on the material collected by "Investigator" which possibly were not in the inner the Persian Gulf. Therefore, if there was no further finding of the Alcock's (1895–1905) records, the records are also included in the list of doubtful species (Table 1.1).

Table 1.1 The list of brachyuran species with doubtful records from the region that their occurrence could not be convincingly verified by the author

Species	Locality/ References
Achaeus spinosus Miers, 1879	Persian Gulf, unknown (Alcock 1895)
Benthopanope indica (De Man, 1888)	Persian Gulf, unknown (Titgen 1982)
Calappa gallus (Herbst, 1803)	Persian Gulf, unknown (Alcock 1895)
Camptoplax coppingeri Miers, 1884	Persian Gulf, UAE (Nobili 1906a)
Carpilius maculatus (Linnaeus, 1758)	Persian Gulf, UAE (Guinot 1967a; Titgen 1982)
Cryptochirus coralliodytes Heller, 1861	Gulf of Oman, Oman (Hogarth 1989)
Doclea canalifera Stimpson, 1857	Persian Gulf, Qatar (Kuronuma 1974)
Echinoecus pentagonus (A. Milne-Edwards, 1879)	Persian Gulf, unknown (Guinot 1967a, b)
Euxanthus sculptilis Dana, 1852	Persian Gulf, unknown (Alcock 1898)
Hyastenus diacanthus (De Haan 1839)	Persian Gulf, UAE (Evans et al. 1973)
Leptodius sanguineus (H. Milne Edwards, 1834)	Persian Gulf, unknown (Alcock 1898)
Liomera margaritata (A. Milne-Edwards, 1873)	Persian Gulf, unknown (Guinot 1967a)
Leucosia elata A. Milne-Edwards, 1879	Persian Gulf, unknown (Alcock 1896)
Medaeops granulosus (Haswell, 1882)	Persian Gulf (Alcock 1898; Nobili 1906a, b; Apel 2001), Gulf of Oman (Stephensen 1946)
Notonyx nitidus (A. Milne-Edwards, 1879)	Persian Gulf, unknown (Alcock 1900)
Pilumnopeus laevis (Dana, 1852)	Persian Gulf, UAE (Titgen 1982)
Pilumnopeus vaquelini (Audouin, 1826)	Persian Gulf, Iran (Naderloo and Türkay 2012)
Pilumnus scabriusculus Adams and White, 1849	Persian Gulf, UAE (Titgen 1982)
Portunus stephensoni Moosa, 1981	Persian Gulf, Saudi Arabia [Basson et al. (1977) as P. *emarginatus*]
Portunus xantusii (Stimpson, 1860)	Persian Gulf, Saudi Arabia (Mohammed and Al-Ssadh 1996)
Pseudoliomera speciosa (Dana, 1852)	Persian Gulf, unknown (Alcock 1898)
Pseudopalicus investigatoris (Alcock, 1900)	Persian Gulf, unknown (MacGilchrist 1905)
Pseudophilyra melita De Man, 1888	Persian Gulf, unknown (Stephensen 1946)
Systroplax angusta (Rathbun, 1914)	Persian Gulf, Iran [Naderloo and Sari (2007a) as *Carcinoplax* cf. *angusta*]
Thalamita danae Stimpson, 1858	Gulf of Oman, Oman (Hogarth 1989)
Thalamita sima H. Milne Edwards, 1834	Persian Gulf, unknown (Alcock 1899a)
Xaiva biguttata (Risso, 1816)	Persian Gulf, unknown (Apel 2001)
Xenophthalmodes moebii Richters, 1880	Persian Gulf, unknown (Alcock 1900)

Key to families of the infraorder Brachyura in the Persian Gulf and the Gulf of Oman

1. Three pairs of walking legs, last pairs absent .... Hexapodidae
- Four pairs of walking legs 2
2. Orbits extraordinary longitudinal .... Xenophthalmidae
- Orbits normal, transverse 3
3. Last pairs of walking legs subchelate or chelate 4
- Last pairs of walking legs not subchelate or chelate 5
4. Third maxillipeds with triangular merus, not covering whole buccal cavern; second maxillipeds and sometimes mandibles are exposed .... Dromiidae
- Third maxillipeds with quadrate merus, completely covering buccal cavern; a plate-like structure between sixth abdominal somite and telson .... Dromiidae
5. Third maxillipeds elongate, with distinct triangular merus 6
- Third maxillipeds not remarkably elongate; with merus quadrate or subquadrate 9

6. Opening of afferent respiratory canal below orbits, adjacent to endostome, distinct canal along sides of buccal cavern; chelipeds without strong cutting edge 7
- Opening of afferent respiratory canal at base of chelipeds, no canal along sides of buccal cavern; carapace expanded laterally or with one spine; chelipeds with strong cutting edge, or with strong tooth 8
7. Abdomen of males and females with somites mostly fused, in females strongly arched, forming a dome-like plate over the sternum; in ovigerous females when abdomen is closed eggs are not visible; fingers of chelipeds in normal size; carapace and appendages glabrous .... Leucosiidae
- Abdomen of males and females with somites freely moving, flat, without dome-like structure; in ovigerous females when abdomen is closed eggs are visible externally; fingers of chelipeds long, narrow, serrated with spine-like teeth; carapace and appendages setose .... Iphiculidae
8. Walking legs with propodus and dactylus paddle-shaped .... Matutidae
- Walking legs with propodus and dactylus not paddle-shaped .... Calappidae
9. Dactylus of last walking legs paddle-shaped .... Portunidae
- Dactylus of last walking legs not paddle-shaped 10
10. Last leg strongly reduced in size in comparison with other legs .... Palicidae
- Last leg not strongly reduced in size, equal or subequal to others 11
11. Socket of antennules longer than broad, antennules folded longitudinally 12
- Socket of antennules broader than long, antennules folded transversely or obliquely 16
12. Carapace distinctly longer than broad; flagella of antenna long, intensely setose .... Corystidae
- Carapace not distinctly longer than broad; flagella of antenna short 13
13. Carapace poorly calcified, dorsal surface normally flat, without spines and hooked setae; abdomen usually with 5 somites .... Hymenosomatidae
- Carapace well calcified, dorsal surface gently to strongly convex, usually covering with spines and hooked setae, abdomen usually with 6 somites 14
14. Eyes with nearly complete orbits; basal antennal segment moderately to very broad .... Majidae
- Eyes with incomplete orbits; basal antennal segment relatively narrow 15
15. Eyestalks generally long, retractile; basal antennal segment extremely narrow, always distally free; merus of third maxillipeds not as broad as ischium .... Inachidae
- Eyestalks short or long, when long, then basal antennal segment not extremely narrow, always fused distally; merus of third maxillipeds as broad as ischium .... Epialtidae
16. Carapace generally triangular, sometimes hexagonal, always extremely uneven; front produced, triangular; chelipeds markedly larger than walking legs .... Parthenopidae
- Carapace not triangular, in different shapes; front not distinctly produced; chelipeds relatively short in comparison to walking legs 17
17. Merus of third maxillipeds completely fused to ischium; males usually much smaller than females; symbionts in bivalves .... Pinnotheridae
- Merus of third maxillipeds well developed; males subequal or larger than females; free-living crabs 18
18. Distinct gap between third maxillipeds 19
- No gap between third maxillipeds 26

19. Gap between third maxillipeds rhomboidal; mandibles are visible when maxillipeds are closed 20

- Gap between third maxillipeds relatively narrow, not rhomboidal; mandibles are not visible when maxillipeds are closed 21

20. Merus and ischium of third maxillipeds with oblique ridge besets with setae; pterygostomian regions with network-like pattern of short setae .... Sesarmidae

- Merus and ischium of third maxillipeds without oblique setal ridge; pterygostomian regions maybe with short setae, but without any pattern .... Grapsidae

21. Front distinctly narrow in comparison to breadth of carapace, simple, eyestalks elongate 22

- Front not distinctly narrow in comparison to breadth of carapace, truncate, lobate or dentate 24

22. Carapace sligtly convex; interantennular septum broad; setal tufts between margin of carapace and walking legs or abdomen 23

- Carapace relatively flat, usually transversely rectangular; interantennular septum remarkably narrow; no setal tufts between branchial opening chamber and walking legs or abdomen .... Macrophthalmidae

23. Carapace subglobular, squarish, transversely rectangular; setal tufts between first and second walking legs; G1 with curved apical part; small-sized crabs .... Dotillidae

- Carapace squarish to transversely rectangular; eyestalks extremely long; setal tufts between second and third walking legs; G1 straight .... Ocypodidae

24. Carapace relatively rhomboidal; front strongly dentate; abdomen somites 3–5 of males and females tightly fused together .... Plagusiidae

- Carapace subquadrate to transversely rectangular; front entire; abdomen somites 3–5 not fused together 25

25. Anterolateral margins of carapace entire, without teeth; abdomen somites 2–3 fused together; G1 long, apically strongly bent .... Camptandriidae

- Anterolateral margins of carapace prominently dentate, with large teeth; abdomen somites 2–3 not fused; G1 normal .... Varunidae

26. Male abdominal somites 3–5 or 3–4 tightly fused, immovable, sutures sometimes visible 27

- Male abdominal somites not fused, freely moving 29

27. Anterolateral margins of carapace with four to six teeth, tubercles or spines, posterior surface of carapace uneven, regions well-defined .... Xanthidae

- Anterolateral margins of carapace with only one tubercle or tooth on epibranchial region; posterior surface of carapace smooth, regions poorly defined 28

28. Carapace not broader than long; front distinctly broader than posterior margin; chelipeds remarkably large; outer surface of palm with dense patch of setae; male abdominal somites 3–5 tightly fused; small-sized crabs; symbionts of soft corals, black corals, scleractinian corals (Acroporidae) .... Trapeziidae

- Carapace distinctly broader than long; front as broad as or slightly broader than posterior margin; cheliped palm without patch of setae; male abdominal somites 3–4 tightly fused, large-sized crabs; free living .... Carpiliidae

29. Front distinctly broader than posterior margin of carapace; anterolateral margins without any dentation; small-sized crabs; obligate symbionts of scleractinian corals and other colonial, in particular anthozoans .... Tetraliidae

- Front not broader than posterior margin of carapace; anterolateral margins with dentation 30

30. G1 stout, straight; G2 very elongate, usually longer than G1 31
- G1 slender to stout, slightly sinuous-shaped; G2 distinctly shorter than G1 33

31. Front granulate; anterolateral margins dentate, with numerous denticles .... Eriphiidae
- Front dentate, or lobate; anterolateral margins with distinct lobes or teeth 32

32. Carapace with anterolateral teeth truncated; large chelipeds with large curved tooth on proximal part of cutting edge .... Oziidae
- Carapace with anterolateral teeth pointed; large chelipeds with truncated large tooth on proximal part of cutting edge .... Menippidae

33. G1 slender, distinctly S-shaped; G2 very short, stout, comma-shaped .... Pilumnidae
- G1 moderately stout, slightly sinuous; G2 about 1/2 to 1/4 of G1 34

34. G2 about 1/4 as long as G1; sternum 8 of males completely covered by abdominal somites 2 and 3 when abdomen closed; telson much longer than somite .... Euryplacidae
- G2 about 0.3 times as long as G1 or slightly longer; sternum 8 of males exposed when abdomen closed 35

35. Anterolateral margins of carapace entire, without any denticle; first three abdominal somites are seen from upper view of carapace; carapace oval .... Pseudoziidae
- Anterolateral margins of carapace with denticles or tubercles; first three abdominal somites are hardly seen from upper view of carapace; carapace xanthoid, hexagonal or subqaudrate 36

36. Carapace subquadrate, with epibranchial spine, posterolateral margins nearly straight; front entire, straight .... Goneplacidae
- Carapace xanthoid-shaped to hexagonal, posterolateral margins converging distally; front with distinct lobes or denticles .... Galenidae

Part I

Section Podotremata Guinot, 1977

Family Dromiidae De Haan, 1833 (Sponge Crabs)

2

Diagnosis Carapace ovoid, circular, or hexagonal, distinctly convex. Front narrow, usually entire. Eyes distinctly small in comparison to front. Third maxillipeds operculiform. Chelipeds medium-sized, fingers usually shorter than manus, cutting edges with rounded t eeth. Walking legs with last two smallest, in the same size, obliquely turned over the carapace, terminally subchelate, used for carrying sponges or truncates. Abdomen of both sexes with six movable somites plus telson; small plate-like segment usually between sixth somite and telson. G1 short, stout; G2 long, slender. Genital opening of males and females coxal.

Key to species of the family Dromiidae in the region

1. Carapace flattened, subpentagonal, dactylus of third walking legs large, claw-shaped which is used for carrying bivalve shell .... *Conchoecetes artificiosus*
- Carapace convex, in different shapes, dactylus of third walking legs not claw-shaped, usually surrounded by spines of merus 2
2. Chelipeds with epipode; abdominal somites 5–6 completely fused; uropods large, not included in the locking mechanism of male abdomen, large-sized species .... *Lauridromia dehaani*
- Chelipeds without epipode; all abdominal somites freely movable; uropods small, included in the locking mechanism of male abdomen 3
3. Dorsal surface of carapace granulate, uropods small, elongate, small-sized species .... *Epigodromia granulata*
- Dorsal surface of carapace smooth, uropods very small, hardly visible 4
4. Lateral margins of carapace smooth, without any denticle or tubercle; front bidentate; sternal grooves of females ending close together, medium-sized species .... *Lewindromia unidentata*
- Lateral margins of carapace with teeth; front distinctly tridentate; sternal grooves of females ending apart from each other 5
5. Anterolateral margins of carapace with three tubercle-like teeth, first one largest; front with three short teeth, not deeply separated, small-sized species .... *Cryptodromia fallax*
- Anterolateral margins of carapace with one acute teeth; front with three long teeth, deeply separated from each other; small-sized species .... *Cryptodromia hilgendorfi*

R. Naderloo, *Atlas of Crabs of the Persian Gulf*, DOI 10.1007/978-3-319-49374-9_2

Genus *Conchoecetes* Stimpson, 1858

***Conchoecetes artificiosus* (Fabricius, 1798)**
(Figs. 2.1 and 2.9)

Dromia artificiosa Fabricius 1798: 92.
Cancer artificiosus — Herbst 1803: 54, pl 58, fig 7.
Conchoecetes artificiosus — Stimpson 1858c: 240; Nobili 1906a: 94; Stimpson 1907: 80, pl 21, fig 5; Titgen 1982: 247; Tirmizi and Kazmi 1986: 15, figs 3a–g, 4a–e; McLay 1993: 175 (in key); Apel 2001: 43; Ng et al. 2008: 33 (in list).

Type locality: "In Oceano Indico, Dom. Daldorff", possibly Tharangambadi (south coast of India).

Distribution

Persian Gulf: Iran (Nobili 1906a), based on a female (MNHN B.6950) collected by Bonnier and Peréz from Bushehr in Iran in 1901.

Gulf of Oman: No record.

World distribution: Indian Ocean: South Africa, Madagascar, Mozambique, Persian Gulf, Pakistan, India, Sri Lanka.

Habitat: Subtidal, 30–100 m, substrates sandy and muddy. This species bears one valve of lamellibranch over its carapace.

Fig. 2.1 *Conchoecetes artificiosus* (Fabricius, 1798). Male (MNHN B6950), CL = 14.6, CB = 14.8 mm, between Al-Qatif and Bushehr, Iran, Persian Gulf, 1902, coll. Bonnier and Peréz (Photo: S. Träkner)

Genus *Cryptodromia* Stimpson, 1858

Cryptodromia fallax (Latreille, in Milbert, 1812)
(Figs. 2.2, 2.8a and 2.9)

Dromia fallax Lamarck 1818: 264.
Cryptodromia canaliculata Stimpson 1858c: 240.
Cryptodromia canaliculata — Alcock 1901: 50, pl 2, fig 8; Nobili 1906b: 145; Stephensen 1946: 62; Guinot 1967a: 240 (in list); Titgen 1982: 247 (in list); Hogarth 1989: 104.
Dromia (*Cryptodromia*) *canaliculata* — Alcock 1899b: 142, 143.
Cryptodromia fallax — Alcock 1901: 77 (in list); McLay 1993: 198 (in key), 206, fig 18e; Apel 2001: 44; Naderloo and Türkay 2012: 28; Naderloo et al. 2013: 449, tab 1; Ng et al. 2008: 33 (in list).

Type locality: *Dromia fallax*: Réunion (Ile Bourbon). — *Cryptodromia canaliculata*: Japan.

Distribution

Persian Gulf: Iran [Stephensen (1946) as *Cryptodromia canaliculata*; Naderloo and Türkay 2012; Naderloo et al. 2013]. Alcock (1899b, 1901) recorded this species as *Cryptodromia canaliculata* Simpson, 1858, from an unknown locality.

Gulf of Oman: No record.

World distribution: Indo-West Pacific: South Africa, East Africa, Madagascar, Mauritius, Seychelles, Red Sea, Gulf of Aden, southern Oman, Persian Gulf, Maldives, Japan, Philippines, Indonesia, New Caledonia, Gilbert and Ellice Islands, Marshall Islands and French Polynesia.

Habitat: Intertidal to 10 m, substrate rocky/cobble.

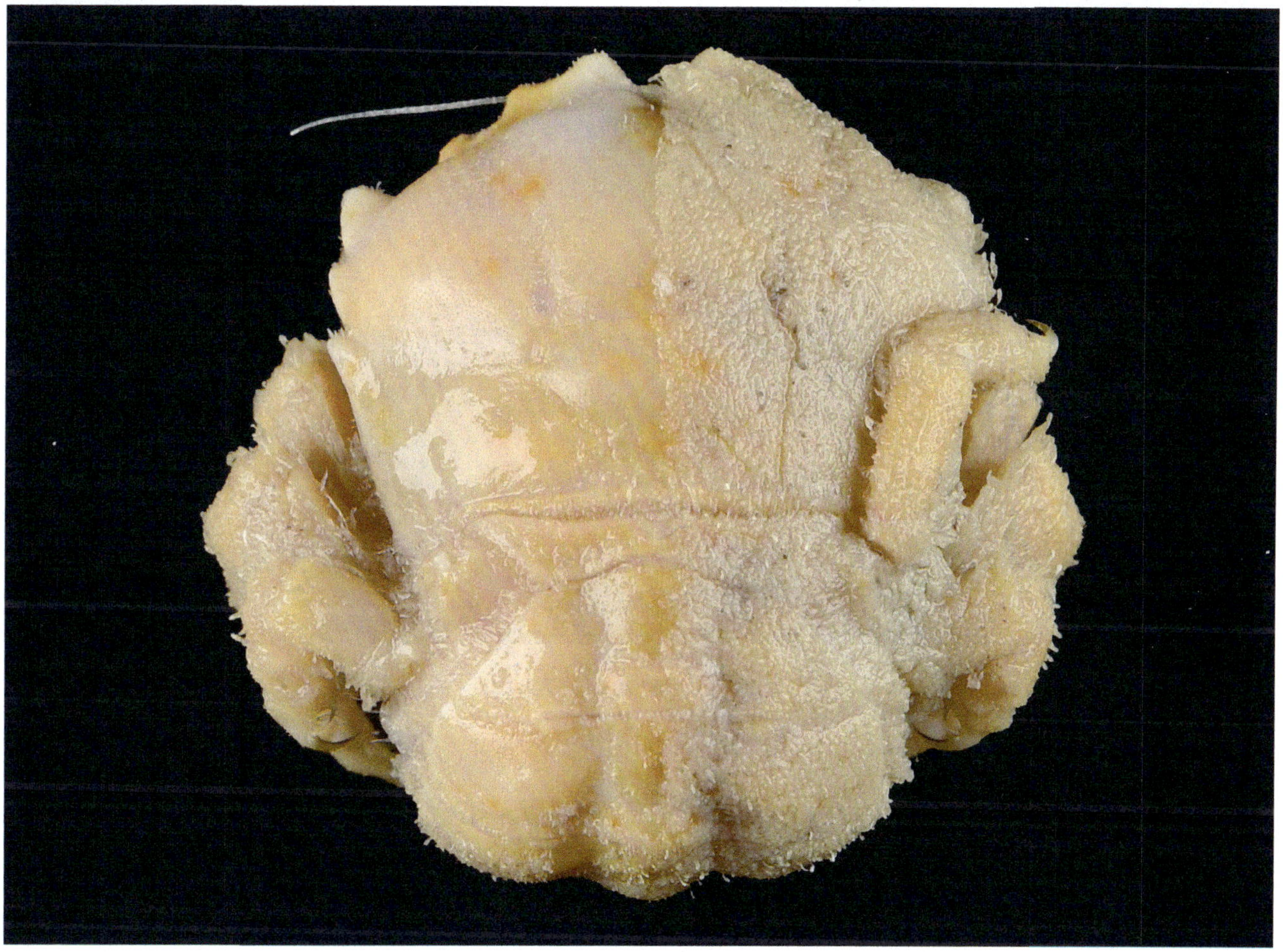

Fig. 2.2 *Cryptodromia fallax* (Latreille, in Milbert, 1812). Female (ZUTC Brach1266), CL = 11.5, CB = 13.5 mm, Qeshm Island, Iran, Persian Gulf, 11.05.2008, coll. R. Naderloo (Photo: S. Träkner)

Cryptodromia hilgendorfi **De Man, 1888**
(Figs. 2.3, 2.8c and 2.9)

Cryptodromia hilgendorfi De Man 1888a: 404, pl 18, fig 3.
Cryptodromia hilgendorfi — Alcock 1899b: 145; Alcock 1901: 52, pl 3, fig 11; Nobili 1906a: 93; Nobili 1906b: 146; Guinot 1967a: 240 (in list); Titgen 1982: 247; McLay 1993: 198 (in key), 205, fig 18d; Apel 2001: 44; Ng et al. 2008: 33 (in list).
Dromia (Cryptodromia) hilgendorfi — Alcock 1899b: 145.

Type locality: Indonesia (Edam Island).
Distribution
Persian Gulf: UAE (Nobili 1906a). Alcock (1899b, 1901) recorded this species from an unknown locality.
Gulf of Oman: No record.
World distribution: Indo-West Pacific: Madagascar, Comoros, Gulf of Aden, Red Sea, Persian Gulf, Philippines, China, Queensland, Chesterfield Islands, New Caledonia and Funafuti Atoll.
Habitat: Subtidal, 35–109 m, substrate rocky.

Fig. 2.3 *Cryptodromia hilgendorfi* De Man, 1888. Male (SMF 44551), CL = 6.2, CB = 6.8 mm, Djibouti, Gulf of Aden, collected using Baum trawl by Meteor, 06.03.1987, coll. H. Thiel (Photo: S. Träkner)

Genus *Epigodromia* McLay, 1993

Epigodromia granulata **(Kossmann, 1878)**
(Figs. 2.4, 2.8e and 2.9)

Epidromia granulata Kossmann 1878: 256.
Cryptodromia granulata — Nobili 1906b: 147, pl 9, fig 5; Gordon 1950: 208, figs 2a, b; Guinot 1967a: 240 (in list); Lewinsohn 1977a: 22, fig 5; Titgen 1982: 247 (in list); Hogarth 1989: 104.
Epigodromia granulata — McLay 1993: 216; Apel 2001: 44; Ng et al. 2008: 34 (in list).
Type locality: Red Sea.

Distribution

Persian Gulf: Saudi Arabia (Lewinsohn 1977).

Gulf of Oman: No record.

World distribution: Northwestern Indian Ocean: Red Sea, Gulf of Aden, southern Oman and Persian Gulf.

Habitat: Shallow subtidal to 10 m.

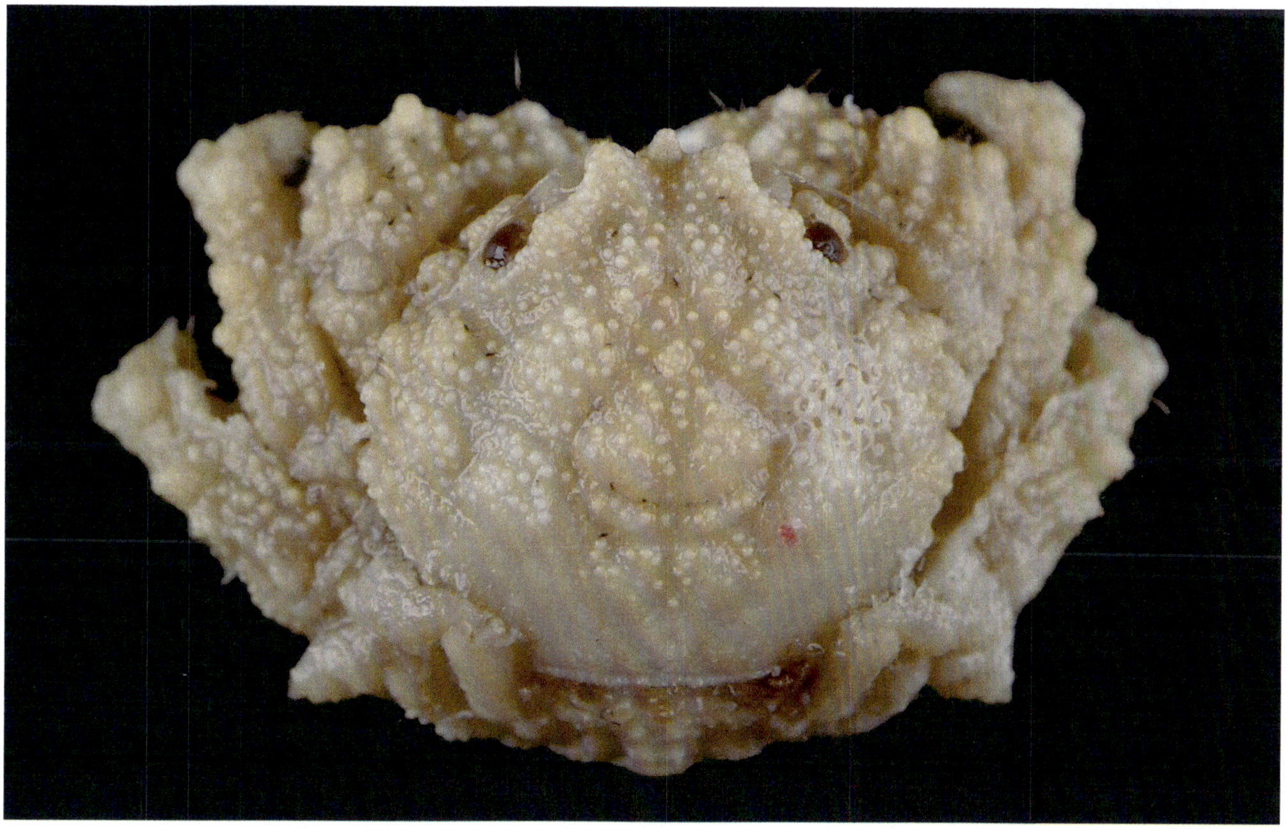

Fig. 2.4 *Epigodromia granulata* (Kossmann, 1878). Male (RMNH 16593), CL = 6.2, CB = 7.1 mm, Juraid Island, Saudi Arabia, Persian Gulf, depth 1–4 m, 04.10.1956, coll. C.E. Dawson (Photo: S. Träkner)

Genus *Lauridromia* McLay, 1993

Lauridromia dehaani (Rathbun, 1923)

(Figs. 2.5, 2.6, 2.8b and 2.9)

Dromia dehaani Rathbun 1923: 68.

Dromia rumphii — Alcock 1899b: 137 (partly); Alcock 1901: 44 (partly).

Dromia dehaani — Tirmizi and Kazmi 1986: 20–23, figs 5, 6.

Dromia dormia — Stephensen 1946: 61, figs 3a, b.

Dromidiopsis dormia — Guinot 1967a: 239 (material from the Persian Gulf).

Lauridromia dehaani — McLay 1993: 145; Apel 2001: 45; Naderloo and Sari 2007a: 342, tab 1; Ng et al. 2008: 35 (in list).

Type locality: Japan.

Distribution

Persian Gulf: Iran [Stephensen (1946) as *D. dromia*; Naderloo and Sari 2007a], Bahrain [Stephensen (1946) as *D. dromia*]. Alcock (1899b) recorded this species as *D. rumphii* from an unknown locality.

Gulf of Oman: Iran [Stephensen (1946) as *D. dromia*; Naderloo et al. 2015].

World distribution: Indo-West Pacific: Gulf of Aden, Persian Gulf, Gulf of Oman, Pakistan, India, Java, Hong-Kong, Taiwan and Japan.

Habitat: Subtidal to 160 m.

Fig. 2.5 *Lauridromia dehaani* (Rathbun ,1923). Male (ZMUC CRU845), CL = 58.6, CB = 68.4 mm, W Bushehr, Iran, Persian Gulf, 13.03.1937, coll. G. Thorson (Photo: S. Träkner)

Fig. 2.6 *Lauridromia dehaani* (Rathbun, 1923). Female (ZUTC Brach1051), CL = 71.0, CB = 82.1 mm, Iran, Persian Gulf, depth 79 m, 19.12.2003, coll. R. Naderloo (Photo: H. Rajaie)

Genus *Lewindromia* Guinot and Tavares, 2003

Lewindromia unidentata (Rüppell, 1830)

(Figs. 2.7, 2.8d and 2.9)

Dromia unidentata Rüppell, 1830: 16, pl 4, figs 2, 2a, pl 6, fig 9.
Dromia unidentata — Alcock 1899b: 139; Alcock 1901: 47, pl 2, fig 6.
Dromidia unidentata — Kossmann 1880: 67; Nobili 1906a: 92; Nobili 1906b: 145; Stephensen 1946: 63; Guinot 1967a: 240 (in list); Basson et al. 1977: 243, 261; Tirmizi and Kazmi 1986: 281, fig 8a–k.
Cryptodromia incisa — Zarenkov 1971: 169.
Cryptodromiopsis unidentata — McLay 1993: 192, figs 7a–k, 18a; Apel 2001: 44.
Lewindromia unidentata — Guinot and Tavares 2003: 74–78, fig 11; Ng et al. 2008: 35 (in list).

Type locality: Red Sea.

Distribution

Persian Gulf: Bahrain (Stephensen 1946), Saudi Arabia (Basson et al. 1977), UAE (Nobili 1906a). Alcock (1899b) recorded this species from an unknown locality.

Gulf of Oman: No record.

World distribution: Indo-West Pacific: Madagascar, East Africa, Gulf of Aden, Red Sea, southern Oman, Persian Gulf, Pakistan, India, Sri Lanka, Andaman, Mergui Archipelago, Thailand, Singapore, Indonesia, Australia, New Caledonia, Japan, Hawaii and Eastern Islands.

Habitat: Subtidal, 4–112 m.

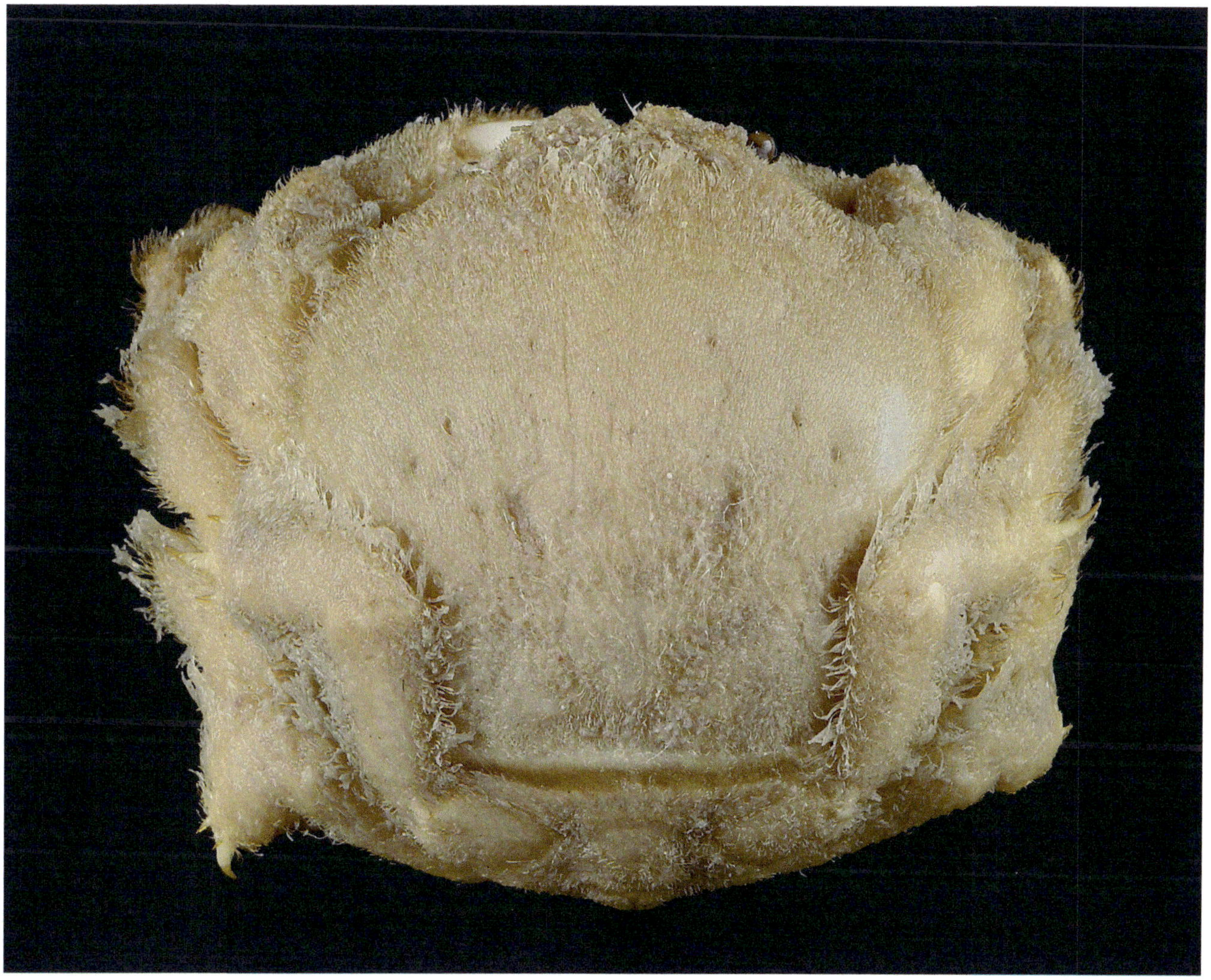

Fig. 2.7 *Lewindromia unidentata* (Rüppell, 1830). Male (ZMUC CRU849), CL = 29.5, CB = 29.0 mm, 100 km NNE of Bahrain, Persian Gulf, depth 27–30 m., 12.02.1990, coll. Abramusseun and Anderson (Photo: S. Träkner)

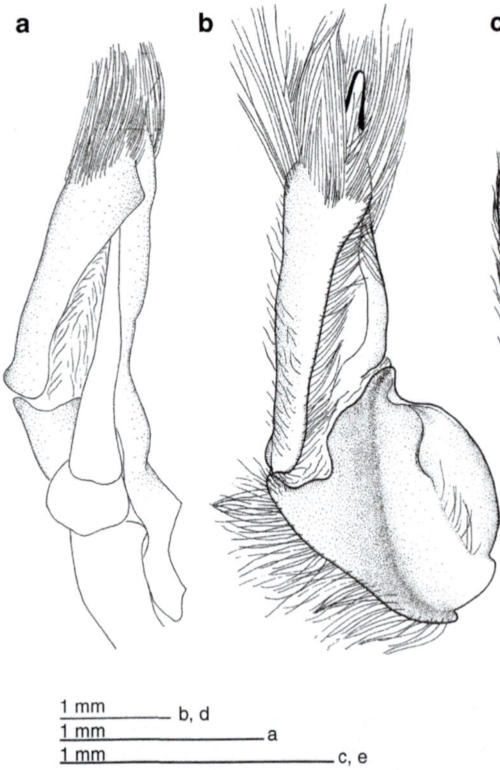

Fig. 2.8 Male G1, ventral view: (**a**) *Cryptodromia fallax* (Latreille, in Milbert, 1812) (SMF 1517); (**b**) *Lauridromia dehaani* (Rathbun, 1923) (ZMUC CRU845); (**c**) *Cryptodromia hilgendorfi* De Man, 1888 (SMF 44551); (**d**) *Lewindromia unidentata* (Rüppell, 1830) (ZMUC CRU849); (**e**) *Epigodromia granulata* (Kossmann, 1878) (RMNH CRUS.D.11.593)

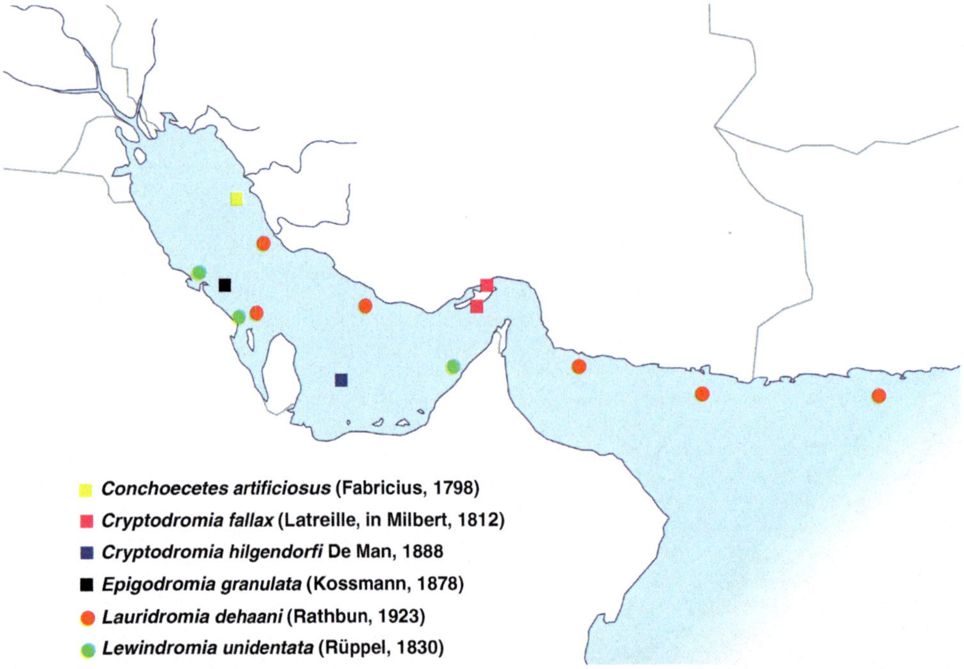

Fig. 2.9 Map showing the recorded localities of *Conchoecetes artificiosus* (Fabricius, 1798); *Cryptodromia fallax* (Latreille, in Milbert, 1812), *Cryptodromia hilgendorfi* De Man, 1888, *Epigodromia granulata* (Kossmann, 1878), *Lauridromia dehaani* (Rathbun, 1923), and *Lewindromia unidentata* (Rüppell, 1830)

Part II

Section Eubrachyura, Subsection Heterotremata

Family Calappidae De Haan, 1833 (Box Crabs, Shame-Faced Crabs)

3

Diagnosis Carapace subcircular, transversely ovate or subovate, posterolaterally expanded. Front distinctly narrow in relation to carapace breadth, slightly dentate. Eyes markedly small in comparison to front. Third maxillipeds with merus distinctly triangular. Chelipeds large, fingers usually shorter than manus, cutting edges with rounded teeth. Walking legs normal-formed, distinctly small in relation to chelipeds. Abdomen of males with somites 3–5 completely fused. G1 short; G2 long, slender. Genital opening of males usually coxal.

Key to species of the genus *Calappa* in the region

1. Carapace nearly triangular, slightly convex, with posterolateral margins expanded, without spines; transverse granular ridges over carapace; no prominant spines on lateral and posterior margins of carapace .... *Calappa hepatica*
- Carpace round, markedly convex, scattered granules over carapace, spines or teeth on lateral or posterior margins of carapace 2
2. Outer surface of chelipeds granulate, without longitudinal ridge running subparallel to lower margin; G1 not markedly curved distally .... *Calappa exanthematosa*
- Outer surface of chelipeds nearly smooth, with ridge running subparallel to lower margin; G1 markedly curved distally 3
3. Posterior margin of carapace without median tooth .... *Calappa gueirini*
- Posterior margin of carapace with median tooth, spine-like, conical, larger than adjacent teeth; small marron spots over posterior surface of carapce; posterolateral teeth pale yellow, basically maroon coloured .... *Calappa dumortieri*

Genus *Calappa* Weber, 1795

Calappa dumortieri Guinot, 1964
(Figs. 3.1, 3.2, 3.3a and 3.4)

Calappa dumortieri Guinot 1964: 21, figs 7–10, 16, 19a, b, 20, pl 1, fig 2, pl 2, fig 2.
Calappa philargius—Nobili 1906b: 148; Holthuis 1958: 45.
Calappa dumortieri — Guinot 1967a: 245 (in list); Serène 1968: 41 (in list); Galil 1997: 290, figs 10b, 11, 13b, 30, 35b; Ng et al. 2008: 48 (in list); Ng et al. 2011b: 1781, fig 1.
Type locality: Red Sea.
Distribution
Persian Gulf: Iran (present study).
Gulf of Oman: Iran (Ng et al. 2011b; Naderloo et al. 2015).
World distribution: Northwestern Indian Ocean: Red Sea, Somalia, Persian Gulf and Gulf of Oman.
Habitat: Subtidal, 5–76 m.

R. Naderloo, *Atlas of Crabs of the Persian Gulf*, DOI 10.1007/978-3-319-49374-9_3

Fig. 3.1 *Calappa dumortieri* Guinot, 1964. Female (ZUTC 5036), CL = 76.4, CB = 97.21 mm, Ras-Meydani, Iran, Gulf of Oman, depth 40 m, 09.2010, coll. N. Ghotbeddin (Photo: R. Abdollahi)

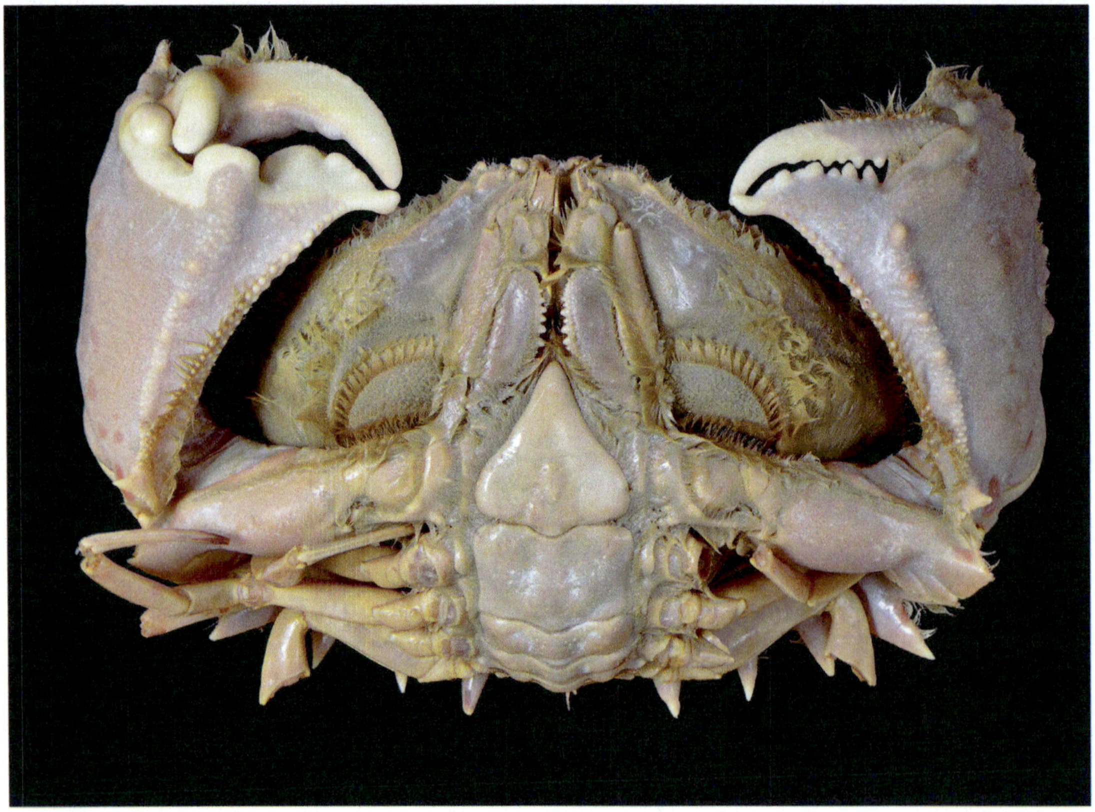

Fig. 3.2 *Calappa dumortieri* Guinot, 1964. The same specimen, ventral view (Photo: R. Abdollahi)

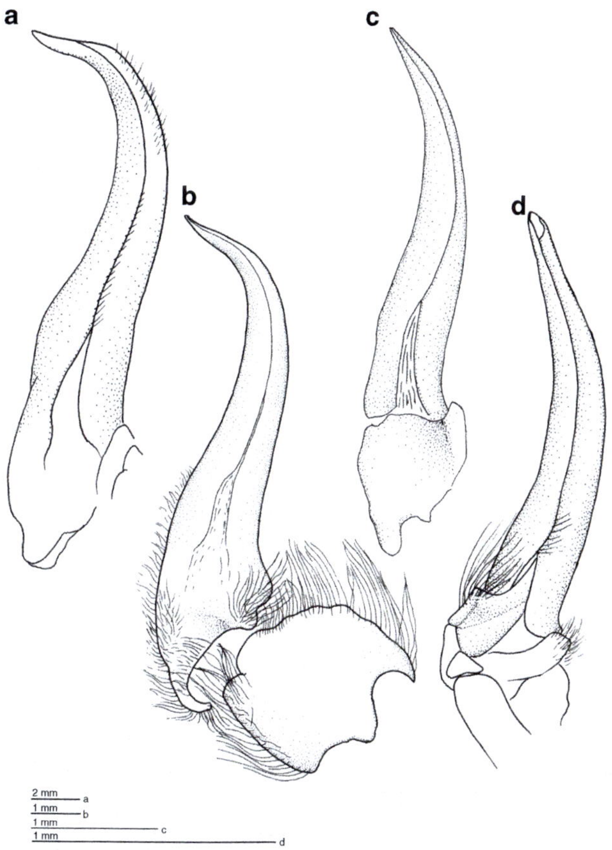

Fig. 3.3 Male G1, ventral view: (**a**) *Calappa dumortieri* Guinot, 1964 (ZUTC 5058); (**b**) *Calappa guerini* Brito Capello, 1871 (ZMUC CRU971); (**c**) *Calappa hepatica* (Linnaeus, 1758) (SMF: Saudi Arabia); (**d**) *Calappa exanthematosa* Alcock and Anderson, 1894 (ZUTC 5083)

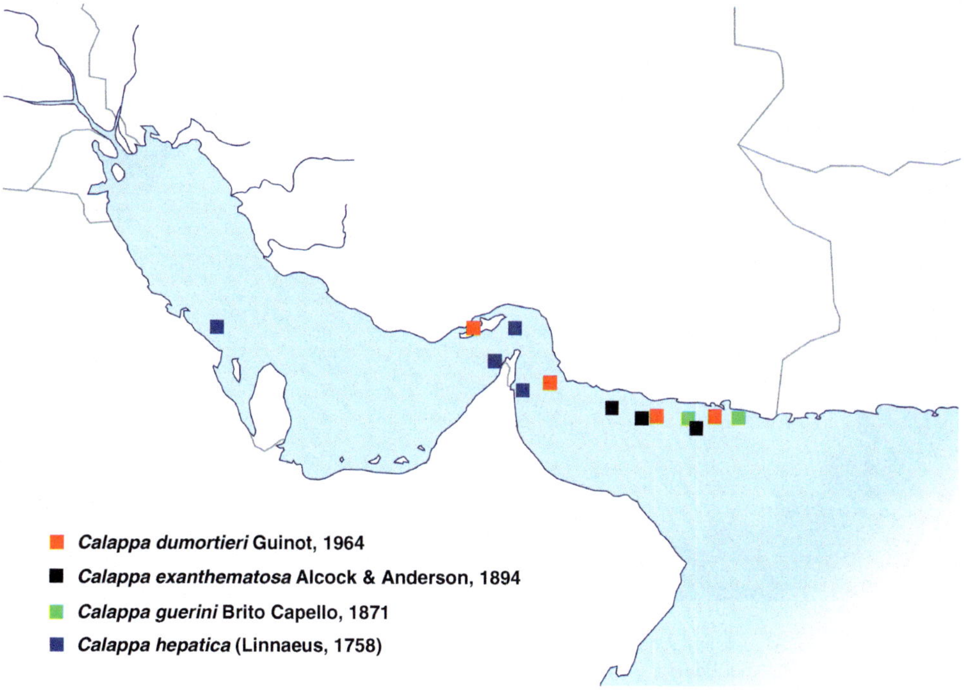

Fig. 3.4 Map showing the recorded localities of *Calappa dumortieri* Guinot, 1964, *Calappa exanthematosa* Alcock and Anderson, 1894, *Calappa guerini* Brito Capello, 1871, and *Calappa hepatica* (Linnaeus, 1758)

***Calappa exanthematosa* Alcock and Anderson, 1894**
(Figs. 3.3d, 3.4, 3.5, and 3.6)

Calappa exanthematosa Alcock and Anderson 1894a: 177.
Calappa exanthematosa — Alcock and Anderson 1895: pl 15, figs 1, 1a; Alcock 1896: 146; Ng et al. 2011a: 1–14, figs 1, 2, 4, 5b, 6b, 7b, 8c, d, 9d–f.
Calappa japonica — Sankarankutty and Subramanian 1976: 21; Galil 1997: 301, figs 10d, 13d, 15; Spiridonov and Apel 2007: 2853, figs 1a, b; Khvorov et al. 2012: 1737 (in list).
Type locality: Bay of Bengal.
Distribution
Persian Gulf: No record.
Gulf of Oman: Iran (Ng et al. 2011a; Naderloo et al. 2015).
World distribution: Indian Ocean: Red Sea, Gulf of Aden, southern Oman, Gulf of Oman, Bay of Bengal and western Australia.
Habitat: Subtidal, 10–196 m, substrate sandy.

Remarks *Calappa exanthematosa* Alcock and Anderson, 1894, has been considered to be synonym with *C. japonica* Ortmann, 1892, for decades. But, Ng et al. (2011a) recently thoroughly examined huge number of specimens from the Indian Ocean and recognized the identity of the species. Therefore, all records of *C. japonica* from the northwestern Indian Ocean must be regarded as *C. exanthematosa*.

Fig. 3.5 *Calappa exanthematosa* Alcock and Anderson, 1894. Male (ZUTC 5083), CL = 77.71, CB = 120.57 mm, Iran, Gulf of Oman, 05.07.2009, coll. Rashidi (Photo: R. Abdollahi)

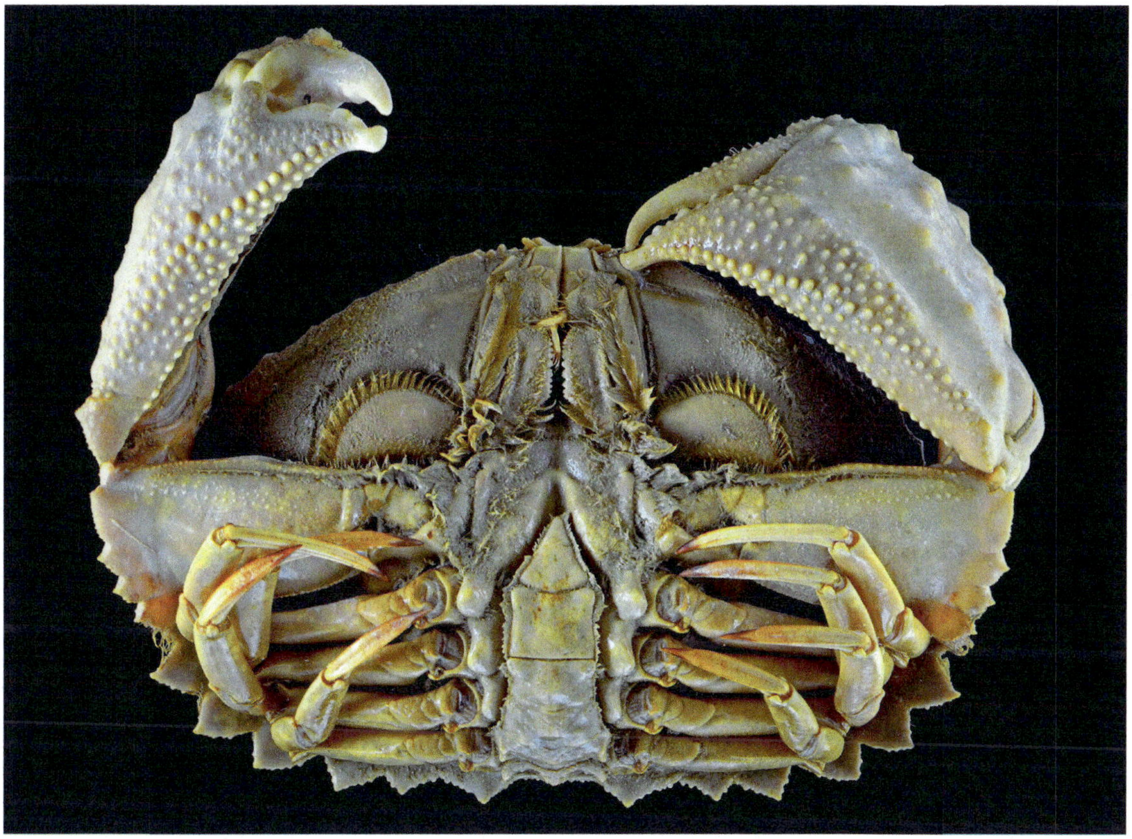

Fig. 3.6 *Calappa exanthematosa* Alcock and Anderson, 1894. The same specimen, ventral view (Photo: R. Abdollahi)

Calappa guerini Brito Capello, 1871

(Figs. 3.3b, 3.4, 3.7, and 3.8)

Calappa guerini Brito Capello 1871: 128, 133, pl 2, fig 2.
Calappa lophos — Alcock 1896: 144 (partly: material from the Persian Gulf); Stephensen 1946: 65, figs 5a, b; Guinot 1967a: 245 (in list; partly); Titgen 1982: 247 (in list); Tirmizi and Kazmi 1986: 53, fig 16a–g; Galil 1997: 302 (partly: material from the western Indian Ocean), fig 16.
Calappa guerini — Apel 2001: 46, 47; Ng et al. 2008: 48 (in list).

Type locality: East coast of India (Yanam).

Distribution

Persian Gulf: Alcock (1896) recorded this species as *Calappa lophos* from an unknown locality.

Gulf of Oman: Iran [Stephensen (1946) as *Calappa lophos*; Naderloo et al. 2015].

World distribution: Western Indian Ocean: Madagascar, South Africa, Mozambique, Reunion, Seychelles, Persian Gulf, eastern Gulf of Oman, Pakistan, Mumbai, Madras, Sri Lanka and eastern India.

Habitat: Shallow subtidal to 10 m, substrate sandy.

Calappa hepatica (Linnaeus, 1758)

(Figs. 3.3c, 3.4, 3.9, and 3.10)

Cancer hepaticus Linnaeus 1758: 630.
Cancer tuberculatus Herbst 1785: 204, pl 13, fig 78.
Calappa hepatica — White 1847a: 44; Alcock and Anderson 1894b: 203; Alcock 1896: 142; Nobili 1906b: 148; Stephensen 1946: 65; Guinot 1967a: 245 (in list); Titgen 1982: 247 (in list); Galil 1997: 296, figs 10e–f, 13e–f, 14, 31; Apel 2001: 47; Ng et al. 2008: 48 (in list).

Type locality: *Cancer hepaticus*: Unknown. — *Cancer tuberculatus*: Pacific Ocean.

Distribution

Persian Gulf: Iran (Stephensen 1946), Saudi Arabia (Apel 2001), UAE (Apel 2001). Alcock (1896) recorded this species as *Calappa lophos* from an unknown locality.

Gulf of Oman: UAE (Apel 2001).

World distribution: Indo-West Pacific: South Africa, East Africa, Madagascar, Seychelles, Somalia, Red Sea, Gulf of Aden, southern Oman, Persian Gulf, Gulf of Oman, Pakistan, India (Mumbai), Andaman Sea, Thailand, China, Japan, Tonga and Hawaii.

Habitat: Shallow intertidal to 100 m, substrates muddy to gravel.

Fig. 3.7 *Calappa guerini* Brito Capello, 1871. Female (ZUTC 5032), CL = 51.72, CB = 81.57 mm, Pasabandar, Iran, Gulf of Oman, depth 22 m, 08.07.2009, coll. N. Ghotbeddin (Photo: R. Abdollahi)

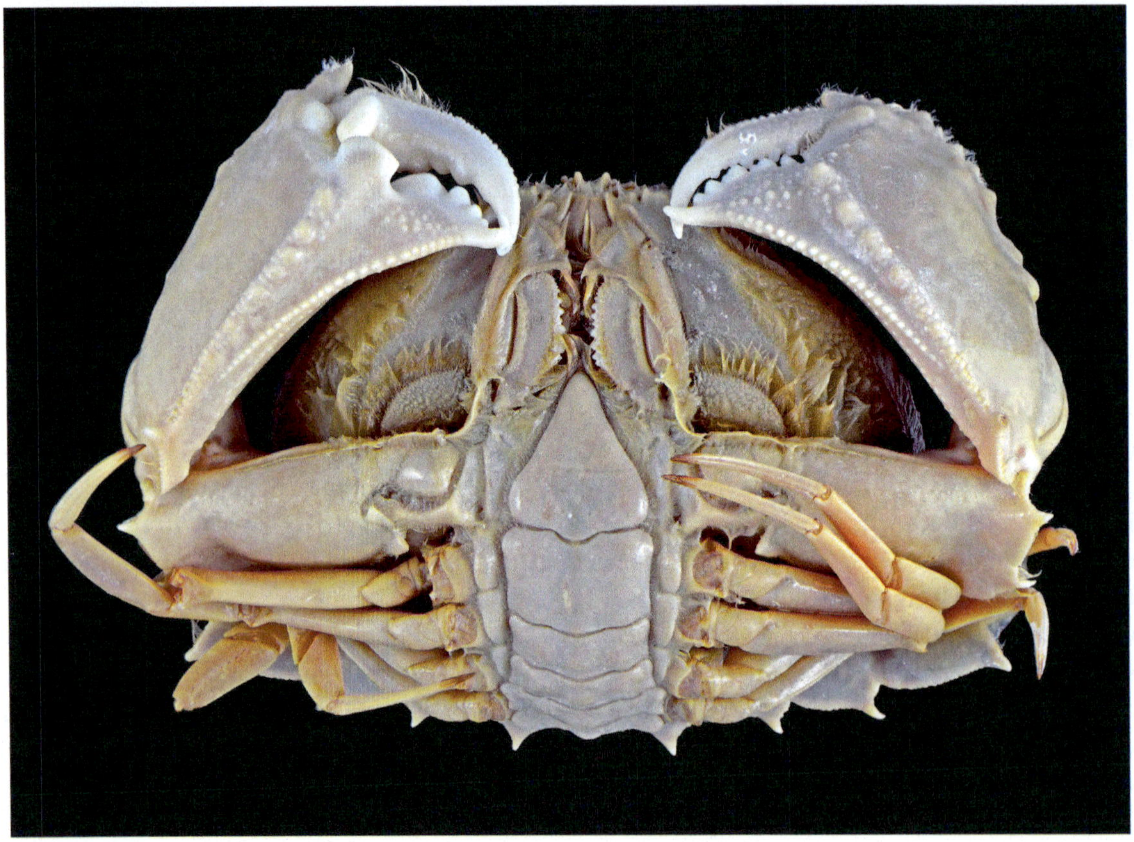

Fig. 3.8 *Calappa guerini* Brito Capello, 1871. The same specimen, ventral view (Photo: R. Abdollahi)

Fig. 3.9 *Calappa hepatica* (Linnaeus, 1758). Male (ZMUC CRU848), CL = 6.2, CB = 4.5 mm, Larak Island, Iran, Persian Gulf, 06.04.1938, coll. B. Loppenthin (Photo: S. Träkner)

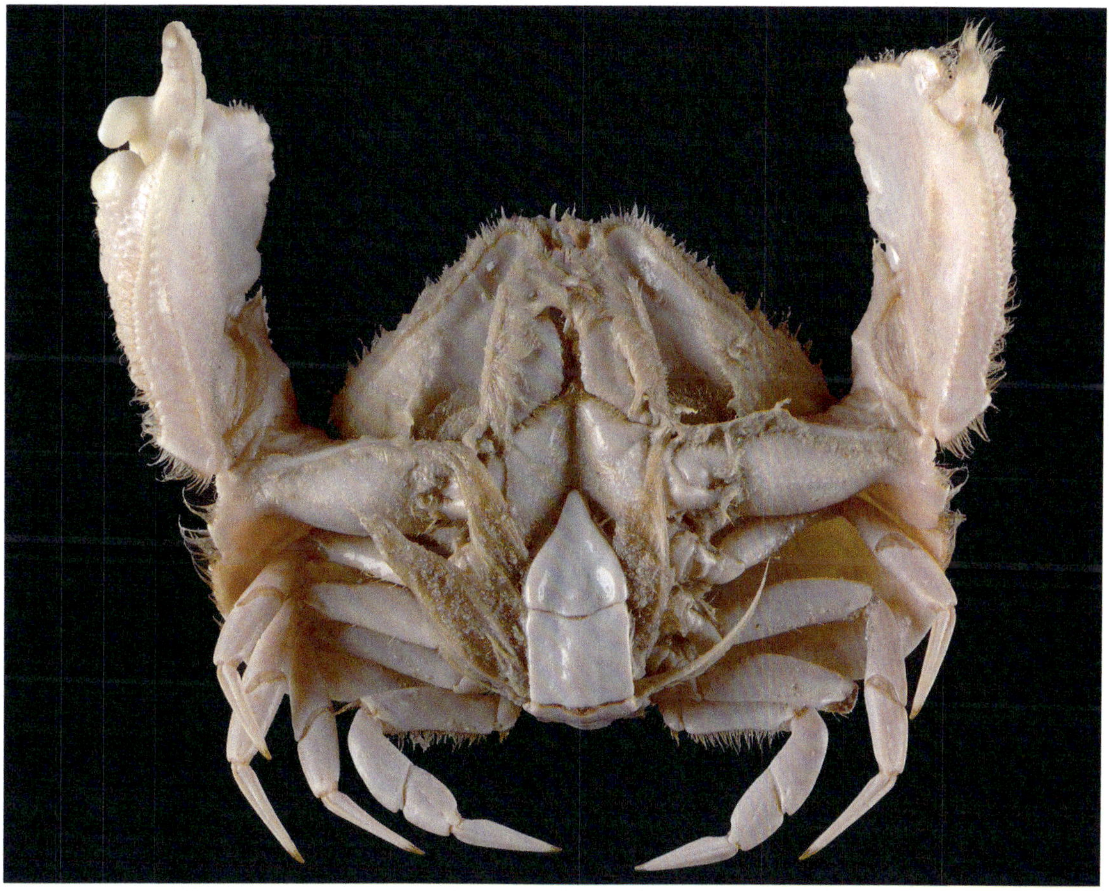

Fig. 3.10 *Calappa hepatica* (Linnaeus, 1758). The same specimen, ventral view (Photo: S. Träkner)

Family Matutidae De Haan, 1835 (Moon Crabs)

4

Diagnosis Carapace subcircular, transversely ovate or subovate, with laterally markedly elongated spines. Front distinctly narrow in relation to carapace breadth, slightly dentate. Eyestalks strongly short in comparison to front. Third maxillipeds with merus distinctly triangular. Chelipeds medium-sized, cutting edges without large tooth. Walking legs with propodus and dactylus paddle-shaped. Abdomen of males with sternums 3–5 completely fused. G1 short; G2 long, slender. Genital opening of males usually coxal.

Key to species of the genus *Matuta* in the region

- Upper surface of carapace with fine red spots; lower margin of cheliped palm smooth, not serrated .... *Matuta victor*
- Upper surface of carapace with fine red lines, forming mesh-like pattern; lower margin of cheliped palm serrated .... *Matuta planipes*

Genus *Matuta* Weber, 1795

Matuta planipes Fabricius, 1798

(Figs. 4.1, 4.2a, b and 4.3)

Matuta planipes Fabricius 1798: 369.
Matuta planipes — Stephensen 1946: 67, figs 5c, d; Guinot 1967a: 246 (in list); Basson et al. 1977: 80, 224, 243; Titgen 1982: 247 (in list); Galil and Clark 1994: 35, 49, pl 12a, b, fig 6c, d; Apel 2001: 48; Ng et al. 2008: 50 (in list).
Cancer lunaris — Herbst 1799: 43, pl 48 (6).
Matuta victor — Latreille 1818: 273, figs 3, 4.
Matuta lunaris — Alcock 1896: 161; Jones 1986a: 156, pl 45.

Type locality: Indian Ocean (exact locality unknown).

Distribution

Persian Gulf: Iran (Stephensen 1946; Naderloo and Sari 2007a), Kuwait [Jones (1986a) as *Matuta lunaris*], Saudi Arabia (Basson et al. 1977; Apel 2001), UAE (Titgen 1982).

Gulf of Oman: Iran (Naderloo et al. 2015).

World distribution: Indo-West Pacific: Persian Gulf, Gulf of Oman, Pakistan, India, Sri Lanka, Malaysia, Indonesia, Singapore, China, Japan, Australia and New Caledonia.

Habitat: Shallow subtidal, substrate sandy.

R. Naderloo, *Atlas of Crabs of the Persian Gulf*, DOI 10.1007/978-3-319-49374-9_4

Fig. 4.1 *Matuta planipes* Fabricius, 1798. Female (ZUTC Brach1053), CL = 29.80, CB = 49.06 mm, between Bandar-Abbas and Qeshm Island, Iran, Persian Gulf, depth 20 m, 06.11.2003, coll. R. Naderloo (Photo: R. Abdollahi)

Matuta victor (Fabricius, 1781)

(Figs. 4.2c, d, 4.3, and 4.4)

Cancer victor Fabricius 1781: 502.
Cancer lunaris — Forskål 1775: 91 (partly).
Matuta victor — Fabricius 1798: 369; H. Milne Edwards 1837a: 115, pl 20 (3–6); Alcock 1896: 160; Nobili 1906b: 149; Galil and Clark 1994: 395, 48, pl 13a, b, fig 7a, b; Apel 2001: 48; Ng et al. 2008: 50 (in list); Naderloo and Türkay 2012: 29; Naderloo et al. 2013: 449, tab 1, 458, tab 3.
Matuta lunaris — Pretzmann 1971: 484; Titgen 1982: 247 (in list); Jones 1986a: 156, pl 45; Tirmizi and Kazmi 1986: 60–64, fig 18a–d.
?*Matuta lunaris* — Hogarth 1989: 114.
Type locality: India (Malabar).

Distribution

Persian Gulf: Iran (Pretzmann 1971; Naderloo and Türkay 2012; Naderloo et al. 2013), Kuwait (Jones 1986a), Bahrain (Hornby 1997), UAE (Apel 2001).

Gulf of Oman: Iran (Ghotbeddin et al. 2012a; Naderloo et al. 2015).

World distribution: Indo-West Pacific: East Africa, Red Sea, Gulf of Aden, Socotra, Oman, Persian Gulf, Gulf of Oman, Pakistan, India, Sri Lanka, Andaman Sea, Malaysia, Thailand, China, Hong Kong, Japan, Philippines, Singapore, Indonesia, Australia, New Caledonia, New Hebrides and Fiji.

Habitat: Shallow subtidal, substrate sandy.

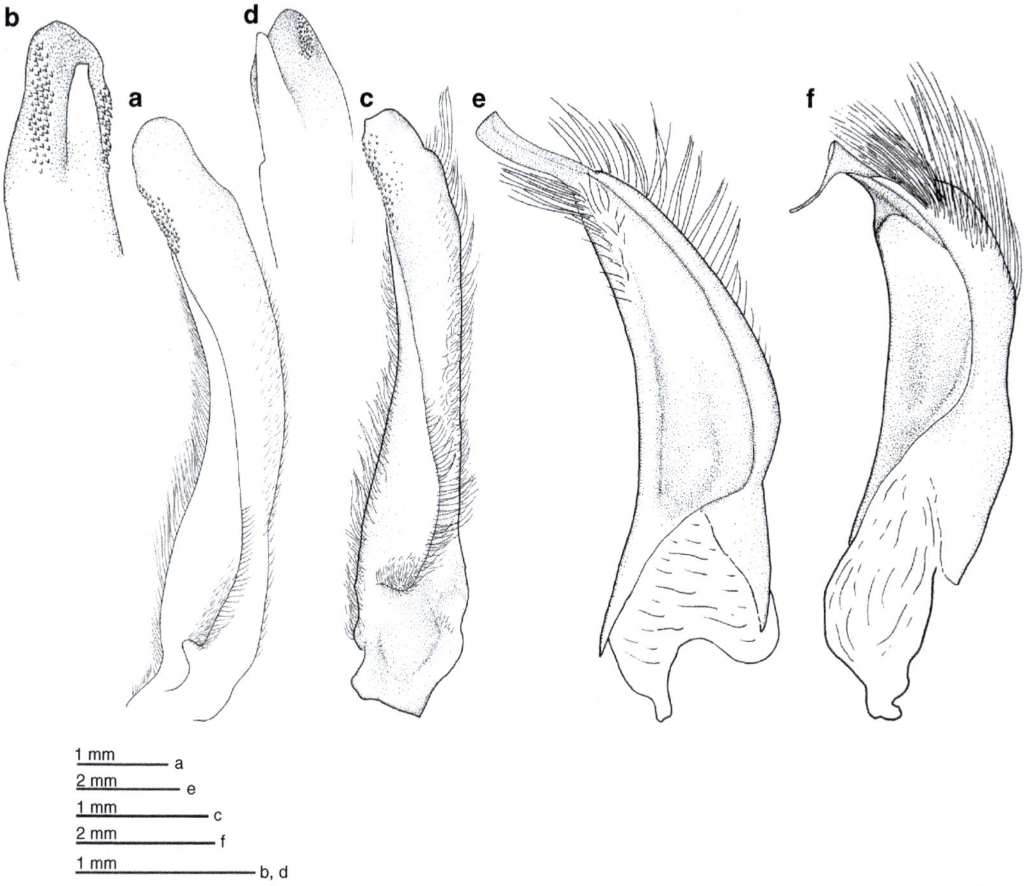

Fig. 4.2 Male G1, ventral view: (**a**) *Matuta planipes* Fabricius, 1798 (ZUTC Brach1844); (**b**) dorsal view of apical part of (**a**); (**c**) *Matuta victor* (Fabricius, 1781) (ZUTC Brach1296); (**d**) dorsal view of apical part of (**c**); (**e**) *Dorippe quadridens* (Fabricius, 1793) (ZMUC CRU844); (**f**) *Dorippoides nudipes* Manning and Holthuis, 1986 (ZMUC CRU972)

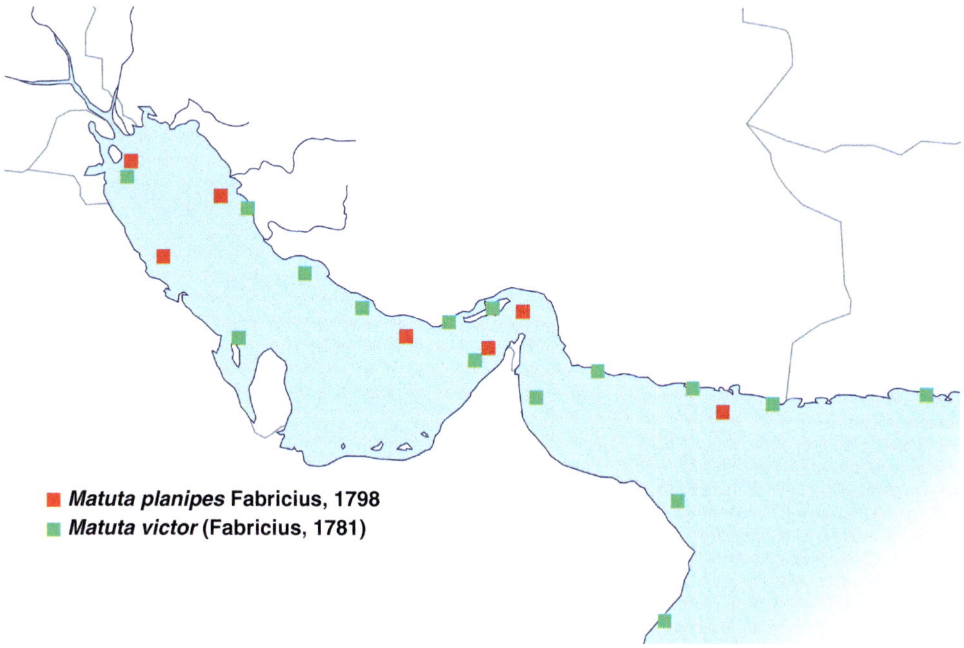

Fig. 4.3 Map showing the recorded localities of *Matuta planipes* Fabricius, 1798, and *Matuta victor* (Fabricius, 1781)

Fig. 4.4 *Matuta victor* (Fabricius, 1781). Male (ZUTC Brach1296), CL = 40.05, CB = 60.20 mm, Qeshm Island, Iran. Persian Gulf, 07.05.2008, coll. R. Naderloo (Photo: S. Träkner)

Family Carpiliidae Ortmann, 1893 (Coral Crabs)

5

Diagnosis Carapace transversly ovate, distincly broader than long, posterior surface smooth, without tubercles, regions not defined, anterolateral margins entire, without lobes or tubercles, with distinct epibranchial lobe. Front short, medially projected. Orbits about one third of front, eyes small. Third maxillipeds without gap between, merus relatively quadrate. Chelipeds unequal, fingers apically pointed, immovable finger of larger cheliped with large blunt tooth proximally on cutting edge. Walking legs medium-sized, smooth, without setae. Abdomen of males with somite 3–4 tightly fused, suture between somites 3 and 4 absent, suture between somites 4 and 5 distinct, abdomen of females with free-moving somites. G1 stout, nearly straight; G2 longer than G1. Genital opening of males coxal. Genital opening of females sternal.

Genus *Carpilius* Desmarest, 1823

Carpilius convexus (Forskål, 1775)

(Figs. 5.1, 5.2, 5.3 and 8.4a)

Cancer convexus Forskål 1775: 88.
Carpilius convexus — Rüppell 1830: 13, pl 3, fig 2, pl 6, fig 6; Stephensen 1946: 156, fig 40; Guinot 1967a: 262; Basson et al. 1977: 261; Crosnier in Serène 1984: 302, figs 210, 211, pl 44f; Carpenter et al. 1997: 39; Apel 2001: 91; Ng et al. 2008: 54 (in list).
Type locality: Red Sea.

Distribution

Persian Gulf: Iran (Stephensen 1946; present study), Saudi Arabia (Basson et al. 1977; Apel 2001).
Gulf of Oman: UAE (Apel 2001).
World distribution: Indo-West Pacific: South Africa, East Africa, Mauritius, Seychelles, Red Sea, Gulf of Oman, Persian Gulf, Pakistan, India, Maldives, Lakshadweep Islands, Andaman Sea, Nicobar, Tores Strait, Japan and Australia.

Habitat: Shallow subtidal to 20 m, substrate coral reef with sand.

R. Naderloo, *Atlas of Crabs of the Persian Gulf*, DOI 10.1007/978-3-319-49374-9_5

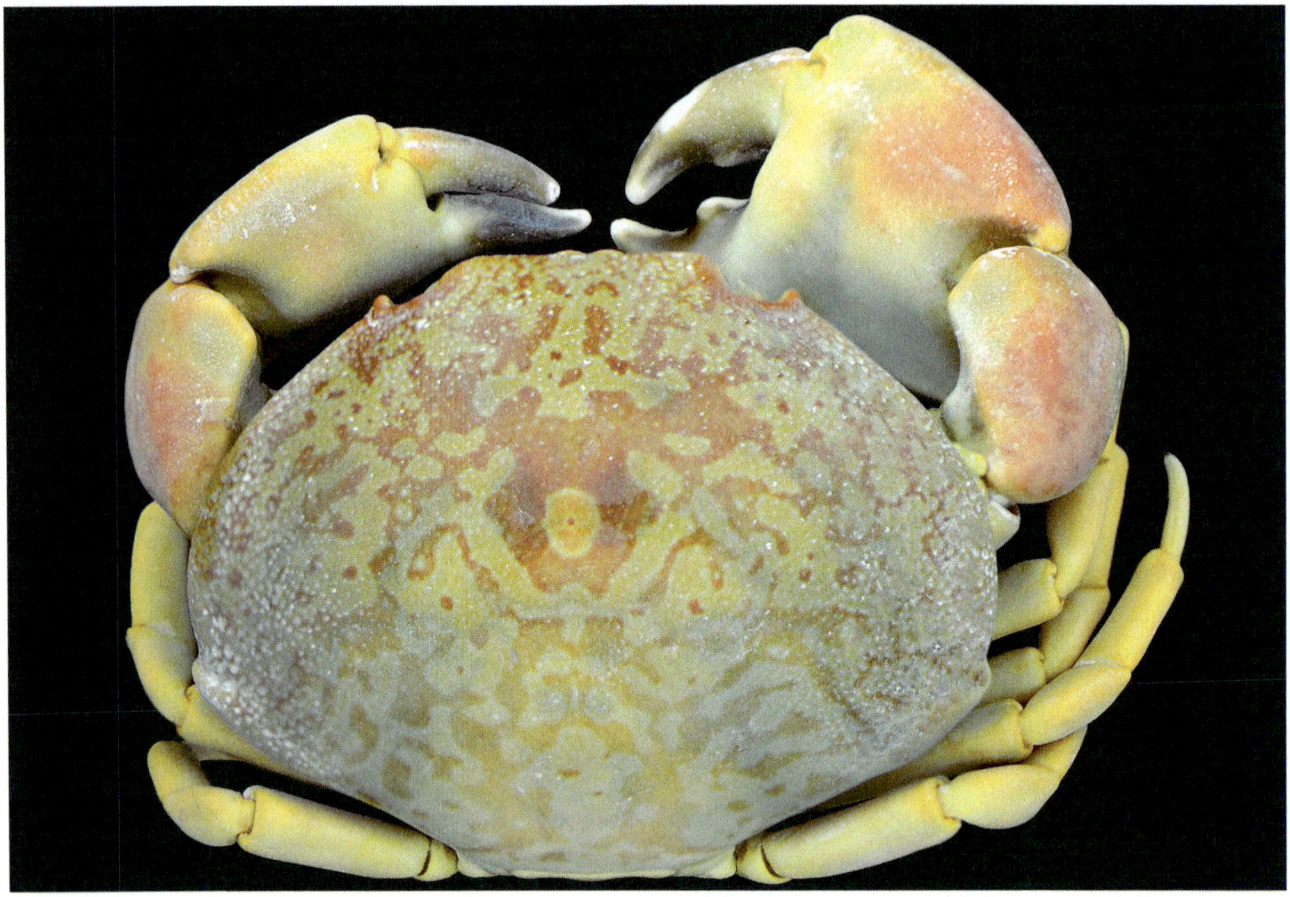

Fig. 5.1 *Carpilius convexus* (Forskål, 1775). Female (ZUTC 5316), CL = 74.93, CB = 98.99 mm, Park Dowlat, Abu Musa Island, Iran, Persian Gulf, 01.05.2014, coll. R. Naderloo and H. Rahimian (Photo: R. Abdollahi)

Fig. 5.2 *Carpilius convexus* (Forskål, 1775). The same specimen in its habitat, Abu Musa Island, Iran, Persian Gulf, 03.05.2014 (Photo: R. Abdollahi)

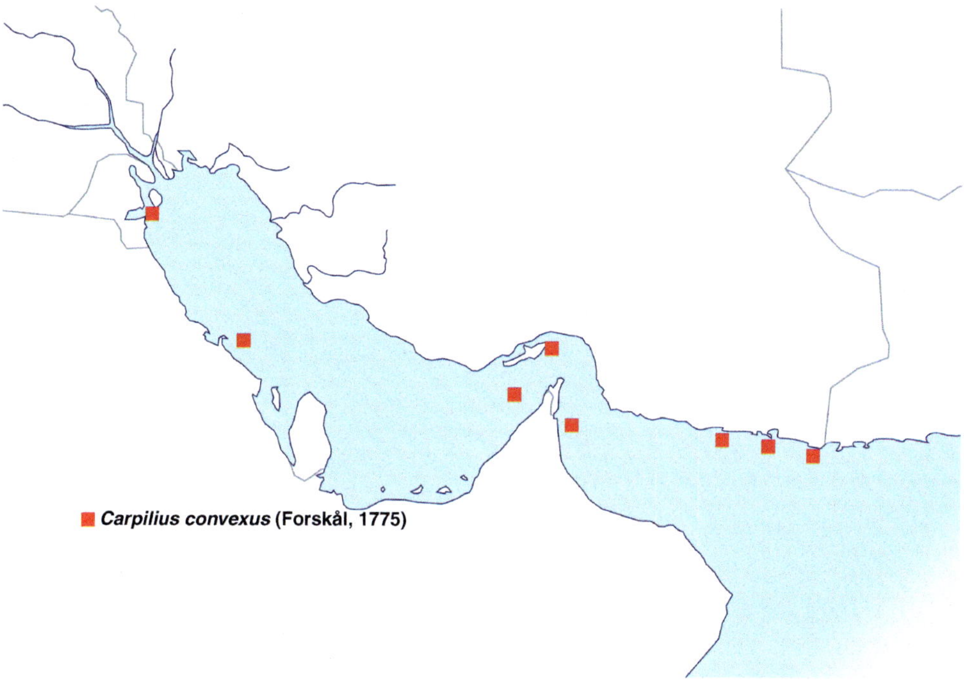

Fig. 5.3 Map showing the recorded localities of *Carpilius convexus* (Forskål, 1775)

Family Corystidae Samoulle, 1819

6

Diagnosis Carapace longitudinally oval, distinctly longer than broad, strongly convex, regions weakly defined; lateral margins often with teeth or spines along entire length, no distinct junction between anterolateral and posterolaterl margins of carapace. Front distinct, narrow, produced, with two or three teeth or lobes. Orbits incomplete. Socket of antennules broader than long, antennules folded longitudinally. Basal antennal segment broad, flagella of antenna long, intensely setose. Third maxillipeds elongate, reaching to antennules, ischium longer than merus, merus quadrate or subquadrate. Epistome absent. Chelipeds markedly sexually dimorphic. Walking legs medium-sized, last leg not reduced in size, slighly broadened. Abdomen of males with five somites, first somite of male and female abdomen visible from dorsal view, abdomen of females not fully covering gential opening.

Genus *Jonas* Hombron and Jacquinot, 1846

Jonas indicus (Chopra, 1935)

(Figs. 6.1 and 6.2)

Gomeza distincta var. *indica* Chopra 1935: 500–505, pl 9, figs 4, 5.
not *Corystes* (*Oeidea*) *distincta* De Haan 1950: 45, pl 13, fig 2.
Gomeza distincta var. — MacGilchrist 1905: 263; Stephensen 1946: 114.
Jonas indicus — Apel 2001: 69; Ng et al. 2008: 56 (in list).

Type locality: *Gomeza distincta*: Japan. — *Gomeza distincta* var. *indica*: Bay of Bengal.

Distribution

Persian Gulf: MacGilchrist (1905) recorded as *Gomeza distincta* var. *indica* from the Strait of Hormuz.

Gulf of Oman: No record.

World distribution: Northern Indian Ocean: Persian Gulf and Bay of Bengal.

Habitat: Subtidal, depth 36–95.5 m, substrates muddy and sandy.

R. Naderloo, *Atlas of Crabs of the Persian Gulf*, DOI 10.1007/978-3-319-49374-9_6

Fig. 6.1 *Jonas indicus* (Chopra, 1935). Male holotype (C 1766/1, Indian Museum, Calcutta, unfortunately the specimen was not found in the Calcutta Museum), CL = 6.1, CB = 4.6 mm, mouth of Hooghly River, Bay of Bengal, 02–03.1928, coll. Pilot vessel "Lady Fraser", the photo was taken from the original figure provided in Chopra (1935) (Photo: S. Träkner)

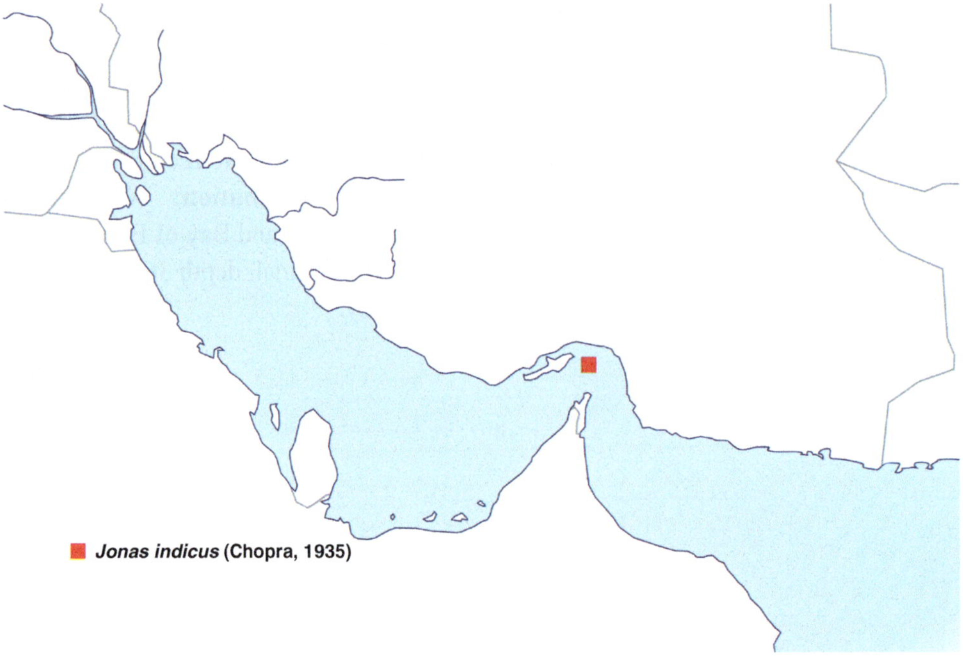

Fig. 6.2 Map showing the recorded (type) locality of *Jonas indicus* (Chopra, 1935)

Family Dorippidae MacLeay, 1838 (Porter Crabs)

7

Diagnosis Carapace quadrate to subquadrate, lateral margins relatively straight, parallel, slightly diverging posteriorly, no clear junction between anterolateral and posterolateral margins. Front relatively broad. Eyes small. Antennules folded obliquely, antennae and antennules large. Third maxillipeds with triangular merus, not completely covering anterior part of buccal cavern. Chelipeds relatively small, robust, subequal. Walking legs relatively long, with dense setae, last two legs markedly reduced, directed upward, subcheliform for carrying shells and sea anemones. Abdomen of males and females with seven somites, all freely moving, first and second somites are seen from dorsal view. G1 stout; G2 narrow, shorter than G1.

Key to species of the family Dorippidae in the region

- Carapace distinctly longer than broad, dorsal surface with large granules; inner orbital angle extending beyond of exorbital angle; G1 apical process narrow, twisted .... *Dorippoides nudipes*
- Carapace broader than long, no large granules on dorsal surface, small granules on posterolateral region of carapace; inner orbital angle not extending beyond exorbital angle; G1 with long spade-shaped apical process.. *Dorippe quadridens*

Genus *Dorippe* Weber, 1795

***Dorippe quadridens* (Fabricius, 1793)**
(Figs. 4.2e, 7.1, and 7.2)

Cancer quadridens Fabricius 1793: 464.
Cancer dorsipes — Linnaeus 1764: 452.
Cancer quadridens — Manning and Holthuis 1981: 30.
Dorippe quadridens — Alcock and Anderson 1894b: 204; Holthuis and Manning 1990: 18, figs 5–12; Apel 2001: 45; Naderloo and Sari 2007a: 342, tab 1; Ng et al. 2008: 59 (in list).
Dorippe dorsipes — Miers 1884a: 185, 257; Alcock 1896: 277; Nobili 1906a: 95; Nobili 1906b: 172; Stephensen 1946: 63, 200, 215, fig 4a, b; Basson et al. 1977: 81, 256.
Dorippe frascone — De Man 1896: 371; Guinot 1967a: 244 (in list); Titgen 1982: 247 (in list).

Type locality: India (Tharangambadi).
Distribution
Persian Gulf: Iran (Nobili 1906a; Stephensen 1946; Naderloo and Sari 2007a), Kuwait (Apel 2001), Saudi Arabia [Basson et al. (1977) as *Dorippe dorsipes*; Apel 2001], Bahrain (Stephensen 1946), UAE (Apel 2001). Alcock (1896) and Holthuis and Manning (1990) recorded this species from an unknown locality.
Gulf of Oman: Iran (Naderloo et al. 2015).
World distribution: Indo-West Pacific: Madagascar, Saint Brandon, Seychelles, Gulf of Aden, Red Sea, Persian Gulf, Gulf of Oman, Pakistan, India, Tharangambadi, Madras, Sri Lanka, Andaman Islands, Mergui Archipelago, Malaysia, Singapore, Indonesia, Thailand, Vietnam, China, Taiwan and Australia.
Habitat: Subtidal, 15–45 m, substrate muddy.

R. Naderloo, *Atlas of Crabs of the Persian Gulf*, DOI 10.1007/978-3-319-49374-9_7

Fig. 7.1 *Dorippe quadridens* (Fabricius, 1793). Male (ZMUC CRU844), CL = 26.5, CB = 28.3 mm, Iran, Strait of Hormuz, 22.04.1937, coll. G. Thorson (Photo: S. Träkner)

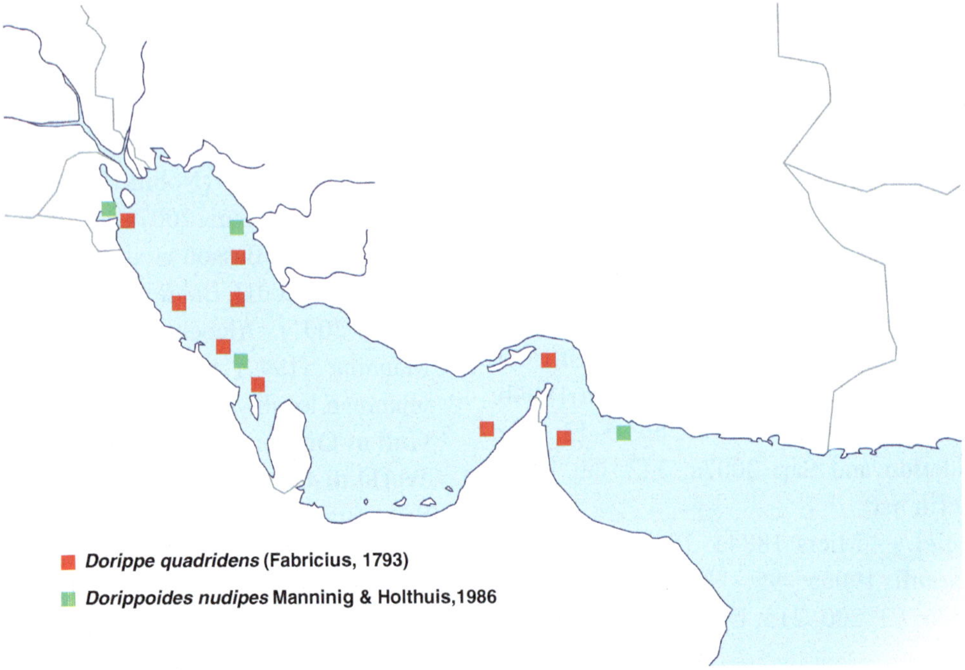

Fig. 7.2 Map showing the recorded localities of *Dorippe quadridens* (Fabricius, 1793), and *Dorippoides nudipes* Manning and Holthuis, 1986

Genus *Dorippoides* Serène and Romimohtarto, 1969

Dorippoides nudipes Manning and Holthuis, 1986
(Figs. 4.2f, 7.2, and 7.3)

Dorippoides nudipes Manning and Holthuis 1986: 364, fig 1c.
Dorippe (*facchino*?) — Stephensen 1946: 64, 200, 215, fig 4c, d.
Dorippe? *facchino* — Basson et al. 1977: 243.
Dorippe (*Dorippoides*) *facchino* — Tirmizi and Kazmi 1993: 367.
Dorippoides facchino — Tirmizi and Kazmi 1986: 43, fig 12.
Dorippoides nudipes — Holthuis and Manning 1990: 66, fig 26; Apel 2001: 46; Ng et al. 2008: 59 (in list).

Distribution

Persian Gulf: Iran [Stephensen (1946) as *Dorippoides* (*facchino* De Haan?); Apel 2001], Kuwait (Apel 2001; Holthuis and Manning 1990), Saudi Arabia [Basson et al. (1977) as *Dorippoides? facchino*].

Gulf of Oman: Iran [Stephensen (1946) as *Dorippoides* (*facchino* De Haan?)].

World distribution: Western Indian Ocean: Madagascar, Gulf of Aden, Red Sea, Persian Gulf, Gulf of Oman and Pakistan.

Habitat: Subtidal, 15–40 m, substrate muddy.

Fig. 7.3 *Dorippoides nudipes* Manning and Holthuis, 1986. Male (ZMUC CRU972), CL = 26.3, CB = 30.1 mm, Jask, Iran, Gulf of Oman, depth 12 m, 24.04.1937, coll. G. Thorson (Photo: S. Träkner)

Family Eriphiidae MacLeay, 1838 (Stone Crabs)

8

Diagnosis Carapace distinctly broader than long, anterolateral margins with several denticles. Front straight, broad, broader than posterior margin. Orbits shorter than front, eyestalks small. Third maxillipeds without gap between, completely covering buccal cavern, merus subquadrate. Sutures between sternums 4–5 and 5–6 interrupted. Chelipeds unequal, larger cheliped usually with proximal large blunt tooth. Walking legs with dense stout setae. Abdomen of males and females with free-moving somites. G1 stout; G2 narrow, longer than G1. Genital opening of males coxal. Genital opening of females sternal.

Genus *Eriphia* Latreille, 1817

Eriphia smithii MacLeay, 1838
(Figs. 8.1, 8.2, 8.3, and 8.4b)

Eriphia smithii MacLeay 1838: 60.
Eriphia laevimana var. *smithii* — Alcock 1898: 216.
Eriphia laevimana var. *smithi* — Nobili 1906a: 291.
Eriphia sebana smithii — Stephensen 1946: 140, fig 34c, d; Guinot 1967a: 271 (in list); Basson et al. 1977: 47, 145, 227, 261, fig 22; Titgen 1982: 129, 251 (in list).
Eriphia smithii — Crosnier in Serène 1984: 311, figs 236, 237, pl 47c; Hogarth 1989: 106, 113; Hogarth 1994: 102; Tirmizi and Ghani 1996: 24–27, fig 8; Hornby 1997: 16; Apel 2001: 92; Koh and Ng 2008: 342–345, figs 13–15, 18c; Ng et al. 2008: 63 (in list); Naderloo and Türkay 2012: 29; Khvorov et al. 2012: 1738; Naderloo et al. 2013: 449, tab 1, tab 4.

Type locality: South Africa.

Distribution

Persian Gulf: Iran (Stephensen 1946; Naderloo and Türkay 2012; Naderloo et al. 2013), Saudi Arabia (Basson et al. 1977), UAE (Nobili 1906a; Titgen 1982; Hornby 1997; Apel 2001).

Gulf of Oman: Iran (Ghotbeddin et al. 2012a; Naderloo et al. 2015).

World distribution: Indo-West Pacific: South Africa, East Africa, Madagascar, Seychelles, Somalia, Red Sea, Gulf of Aden, southern Oman, Persian Gulf, Gulf of Oman, Pakistan, India, Andaman Sea, Thailand, China, Taiwan, Japan, Tonga and Hawaii.

Habitat: Intertidal, substrate rocky/cobble.

R. Naderloo, *Atlas of Crabs of the Persian Gulf*, DOI 10.1007/978-3-319-49374-9_8

Fig. 8.1 *Eriphia smithii* MacLeay, 1838. Female (ZUTC 5126), CL = 32.71, CB = 42.94 mm, Tis, Chabahar, Iran, Gulf of Oman, 05.19.2013, coll. R. Naderloo (Photo: R. Abdollahi)

Fig. 8.2 *Eriphia smithii* MacLeay, 1838. Abd Village, Iran, Gulf of Oman, 29.12.2013, the specimen is not preserved (Photo: R. Naderloo)

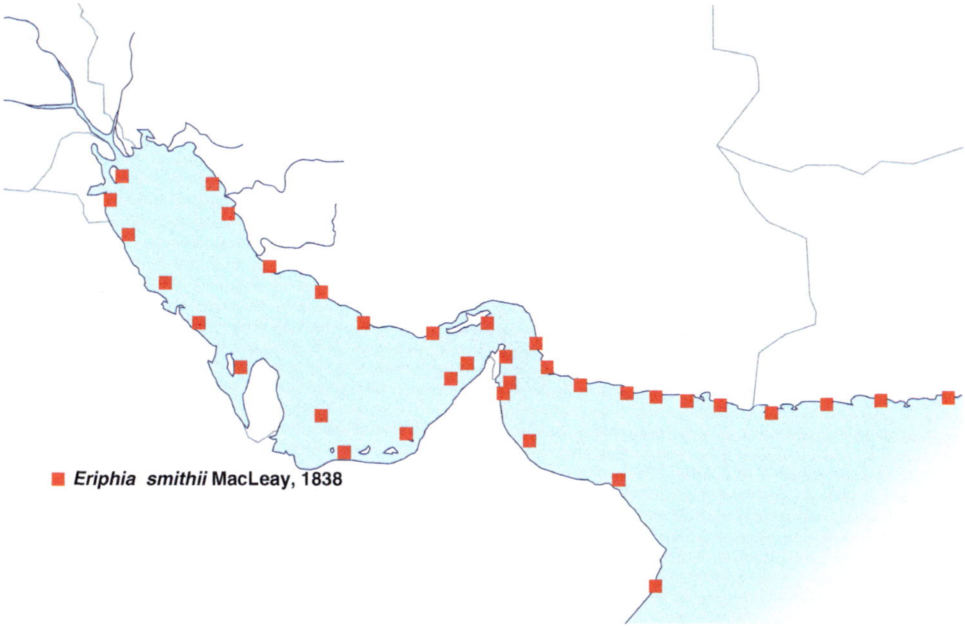

Fig. 8.3 Map showing the recorded localities of *Eriphia smithii* MacLeay, 1838

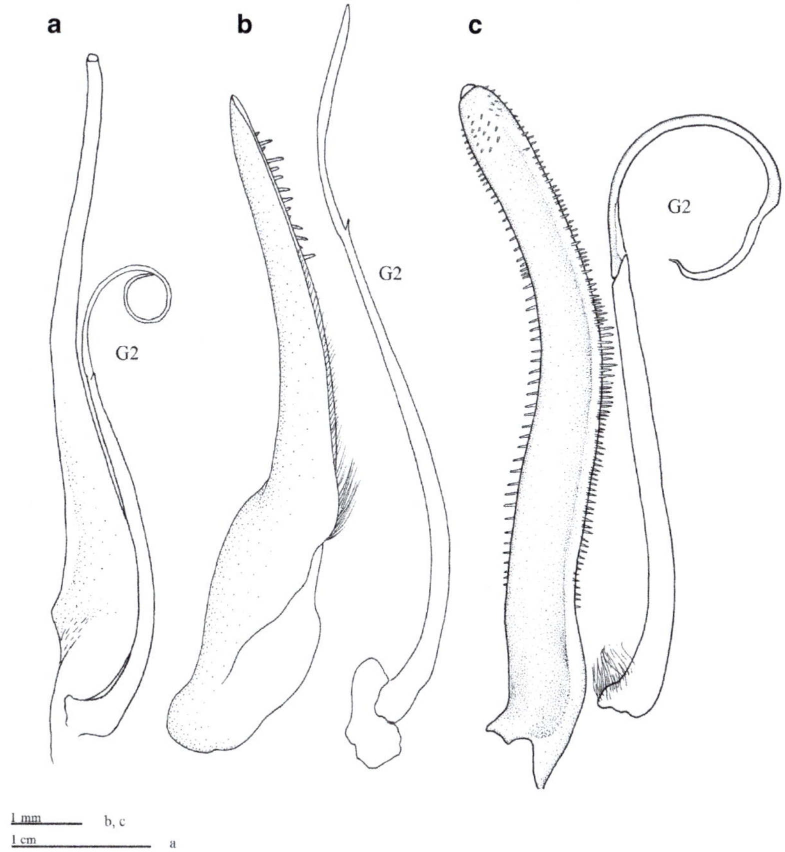

Fig. 8.4 Male G1 and G2, ventral view: (**a**) *Carpilius convexus* (Forskål, 1775) (SMF 38483); (**b**) *Eriphia smithii* MacLeay, 1838 (ZUTC 5095); (**c**) *Menippe rumphii* (Fabricius, 1798) (ZUTC Brach1280)

Family Menippidae Ortmann, 1893 (Stone Crabs)

9

Diagnosis Carapace distinctly broader than long, anterolateral margins with three broad lobes and tubercles. Front with four lobes, medians larger, frontal margin shorter than posterior margin of carapace. Orbits half as broad as front, eyestalks short. Third maxillipeds without gap between, completely covering buccal cavern, merus subquadrate. Sutures between abdominal sternums 4–5 and 7–8 entire. Chelipeds unequal, larger cheliped usually with proximal large blunt tooth. Walking legs medium-sized, with scarce setae on distal appendages. Abdomen of males and females with free-moving somites. G1 stout; G2 narrow, longer than G1. Genital opening of males coxal, genital opening of females sternal.

Genus *Menippe* De Haan, 1833

Menippe rumphii (Fabricius, 1798)

(Figs. 8.4c, 9.1, and 9.2)

Cancer rumphii Fabricius 1798: 336.
Cancer rumphii — Herbst 1799: 63, pl 49, fig 2.
Menippe rumphii — Alcock 1898: 178 (?partly); Stephensen 1946: 139; Guinot 1967a: 271 (in list); Titgen 1982: 252 (in list); Sakai 1999: 31, pl 16d; Apel 2001: 93; Naderloo and Türkay 2012: 30; Naderloo et al. 2013: 449, tab 1.
Ozius rugulosus — Chhapgar 1957a: 433, pl 10, figs d–f; Tirmizi and Ghani 1996: 65, fig 25.

Type locality: "in Indica orientali Dom. Daldorff", more probably Tharangambadi on the south coast of India.

Distribution

Persian Gulf: Iran (Naderloo and Türkay 2012; Naderloo et al. 2013).

Gulf of Oman: Iran (Ghotbeddin et al. 2012a; Naderloo et al. 2015).

World distribution: Indian Ocean: East Africa, Madagascar, Red Sea, Gulf of Aden, Socotra, Persian Gulf, Gulf of Oman, Pakistan, India, Lakshadweep Islands, Sri Lanka, Mergui Archipelago, Nicobar, Andaman Sea, Thailand, Taiwan and Indonesia.

Habitat: Intertidal, substrate rocky/cobble.

R. Naderloo, *Atlas of Crabs of the Persian Gulf*, DOI 10.1007/978-3-319-49374-9_9

Fig. 9.1 *Menippe rumphii* (Fabricius, 1798). Male (ZUTC 1280), CL = 37.6, CB = 54.9 mm, Zeytun Park, Qeshm Island, Iran, Persian Gulf, 08.12.2011, coll. R. Naderloo (Photo: S. Träkner)

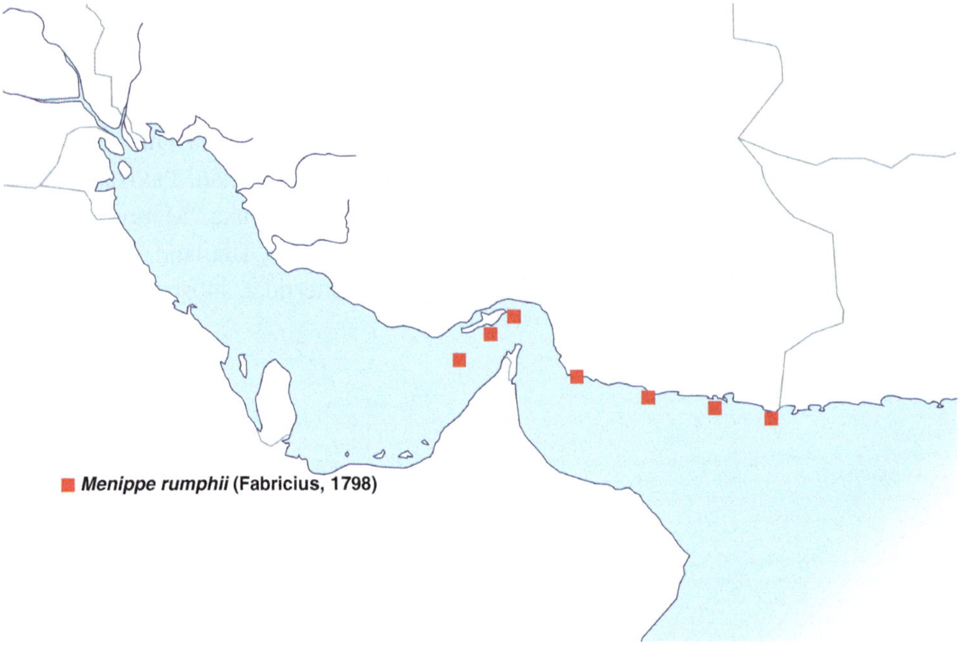

Fig. 9.2 Map showing the recorded localities of *Menippe rumphii* (Fabricius, 1798)

Family Oziidae Dana, 1851 (Forceps Crabs)

10

Diagnosis Carapace markedly broader than long, ovate, posterior surface slightly uneven, regions weakly defined. Front broad, with four small tubercles. Eyestalks short, about one third of frontal margin. Third maxillipeds with quadrate-shaped merus, completely covering buccal cavern. Chelipeds unequal, movable finger of larger chelipeds with backwardly curved tooth on proximal part of cutting edge; fingers of smaller cheliped markedly slender. Walking legs relatively small-sized, without setae. Abdomen of males and females with free-moving somites. G1 stout, relatively straight; G2 narrow, longer than G1. Genital opening of males coxal, genital opening of females sternal.

Key to species of the family Oziidae in the region

1. Supraorbital and infraorbital angles in close contact with each other, without orbital hiatus ..... *Lydia tenax*
- Supraorbital and infraorbital angles not in close contact with each other, with orbital hiatus 2

2. Carapace nearly as long as broad, relatively convex, anterolateral margins spinose; small chelipeds with normal fingers; long setae on carapace and appendages .... *Eupilumnus calmani*
- Carapace markedly broader than long, distinctly flat, anterolateral margins lobate or dentate; small chelipeds with markedly elongated finger 3

3. Dorsal surface of carapace finely granular, anterolateral margins with irregular lobes; upper surface of cheliped merus and palm uneven .... *Epixanthus corrosus*
- Dorsal surface of carapace smooth, anterolateral margins with four distinct lobes; upper surface of cheliped merus and palm smooth .... *Epixanthus frontalis*

Genus *Epixanthus* Heller, 1861

Epixanthus corrosus A. Milne-Edwards, 1873
(Figs. 10.1, 10.2a, and 10.3)

Epixanthus corrosus A. Milne-Edwards 1873: 241, 242, pl 9, figs. 1–1a.
Epixanthus corrosus — Serène 1984: 307, figs. 224, 225, pl 66a; Apel 2001: 92; Ng et al. 2008: 65 (in list).

Type locality: New Caledonia.
Distribution
Persian Gulf: Saudi Arabia (Apel 2001).
Gulf of Oman: Iran (Ghotbeddin et al. 2012a; Naderloo et al. 2015).
World distribution: Indo-West Pacific: Madagascar, Comoros, Mauritius, Red Sea, Persian Gulf, Gulf of Oman, Taiwan and New Caledonia.
Habitat: Intertidal to shallow subtidal, substrate rocky/cobble.

R. Naderloo, *Atlas of Crabs of the Persian Gulf*, DOI 10.1007/978-3-319-49374-9_10

Fig. 10.1 *Epixanthus corrosus* A. Milne-Edwards, 1873. Female (ZUTC 5522), CL = 11.96, CB = 20.50 mm, Iran, Gulf of Oman, 2010, coll. N. Ghotbeddin (Photo: S. Träkner)

Epixanthus frontalis (H. Milne Edwards, 1834)

(Figs. 10.2b, 10.3, 10.4, and 10.5)

Ozius frontalis H. Milne Edwards 1834: 406.
Epixanthus kotschii Heller 1861a: 325, pl 2, figs. 14, 15.
Epixanthus frontalis — Stephensen 1946: 139; Guinot 1967a: 270 (in list); Titgen 1982: 251 (in list); Serène 1984: 307 (in key), figs. 222, 223, pl 45f; Jones 1986a: 162, pl 48; Hogarth 1989: 106, 113, 1994: 101; Tirmizi and Ghani 1996: 22, fig. 7; Hornby 1997: 16; Al-Khayat and Jones 1999: 5; Apel 2001: 92; Ng et al. 2008: 65 (in list); Naderloo and Türkay 2012: 30; Naderloo et al. 2013: 449, tab 1, 458, tab 4; Khvorov et al. 2012: 1739 (in list).

Type locality: *Ozius frontalis*: Tharangambadi, south coast of India. — *Epixanthus kotschii*: Persian Gulf (Khark Island).

Distribution

Persian Gulf: Iran [Heller (1861a) as *Epixanthus kotschii*; Stephensen 1946; Apel 2001; Naderloo and Türkay 2012; Naderloo et al. 2013], Kuwait (Jones

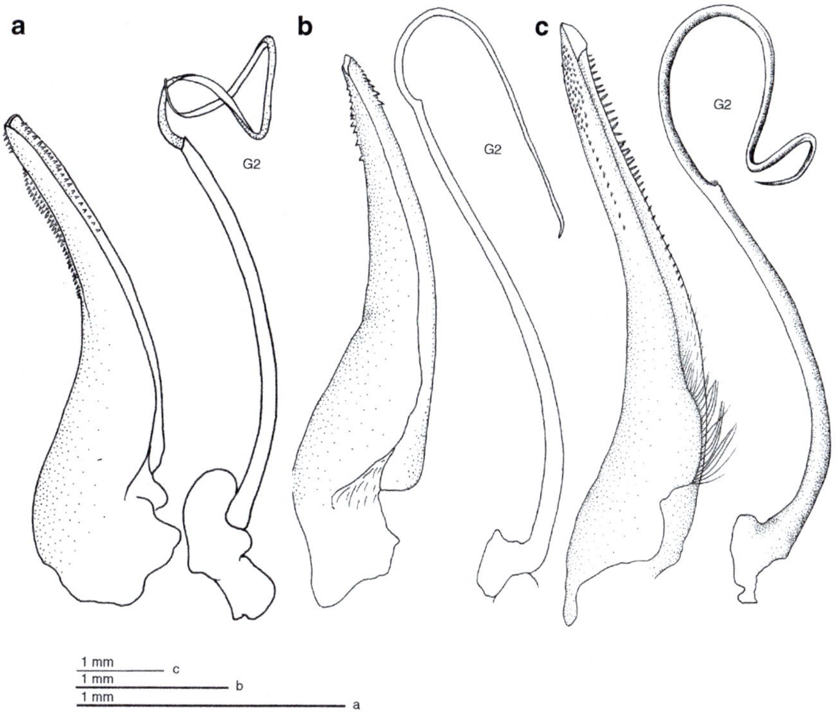

Fig. 10.2 Male G1 and G2, ventral view: (**a**) *Epixanthus corrosus* A. Milne-Edwards, 1873 (SMF: Socortra); (**b**) *Epixanthus frontalis* (H. Milne Edwards, 1834) (ZUTC 5015); (**c**) *Lydia tenax* (Rüppell, 1830) (ZUTC 5517)

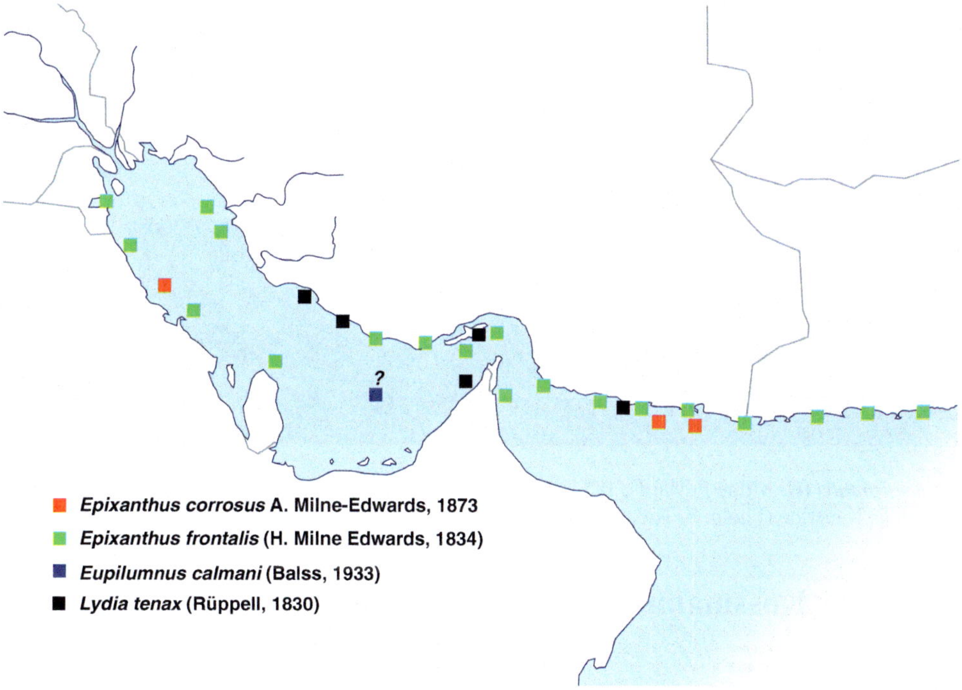

Fig. 10.3 Map showing the recorded localities of *Epixanthus corrosus* A. Milne-Edwards, 1873, *Epixanthus frontalis* (H. Milne Edwards, 1834), *Eupilumnus calmani* (Balss, 1933), and *Lydia tenax* (Rüppell, 1830)

1986a), Saudi Arabia (Apel 2001), Qatar (Al-Khayat and Jones 1999).

Gulf of Oman: Iran (Ghotbeddin et al. 2012a; Naderloo et al. 2015), Oman (Hogarth 1989), UAE (Hornby 1997).

World distribution: Indo-West Pacific: South Africa, East Africa, Madagascar, Seychelles, Somalia, Red Sea, southern Oman, Persian Gulf, Gulf of Oman, Pakistan, India, Sri Lanka, Bay of Bengal, Andaman Sea, Mergui Archipelago, Indo-Malayan Archipelago, Thailand, China, Taiwan, Japan, Philippines and Australia.

Habitat: Intertidal, substrate rocky/cobble.

Fig. 10.4 *Epixanthus frontalis* (H. Milne Edwards, 1834). Male (SMF 40807), 5 km E to Chiruyeh, Iran, Persian Gulf, 26° 43′N, 53° 46′E, 23.10.2010, coll. R. Naderloo (Photo: S. Träkner)

Genus *Eupilumnus* Kossmann, 1877

Eupilumnus calmani (Balss, 1933)

(Figs. 10.3 and 10.6)

Globopilumnus calmani Balss 1933: 9, pl 1, figs 6, 7.

Globopilumnus calmani — Guinot 1960: 110, figs. 9–13; Guinot 1964: 87, 1967a: 271 (in List); Titgen 1982: 251 (in List); Apel 2001: 93.

Eupilumnus calmani — Ng et al. 2008: 65 (in list).

Type locality: Red Sea.

Distribution

Persian Gulf: Guinot (1960, 1964) recorded this species from an unknown locality.

Fig. 10.5 *Epixanthus frontalis* (H. Milne Edwards, 1834). 5 km E to Chiruyeh, Iran, Persian Gulf, 26° 43′N, 53° 46′E, 23.10.2010 (Photo: R. Naderloo. The specimen is not preserved)

Gulf of Oman: No record.

World distribution: Northwestern Indian Ocean: Red Sea, Aldabra, Persian Gulf and Macclesfield Bank.

Habitat: Unknown.

Genus *Lydia* Gistel, 1848

Lydia tenax (Rüppell, 1830)

(Figs. 10.2c, 10.3, and 10.7)

Cancer tenax Rüppell 1830: 11, pl 3, fig. 1, pl 6, fig. 5.
Ozius (*Euruppellia*) *tenax* — Alcock 1898: 187.
Lydia tenax — Crosnier in Serène 1984: 309, pl 46f; Hogarth 1989: 106, 113, 1994: 102; Apel 2001: 93; Ng et al. 2008: 65 (in list); Naderloo and Türkay 2012: 30; Naderloo et al. 2013: 449, tab 1; Khvorov et al. 2012: 1739 (in list).
Lydia (*Ozius*) *tenax* — Hornby 1997: 16.

Type locality: Red Sea (Massaua).

Distribution

Persian Gulf: Iran (Stephensen 1946; Naderloo and Türkay 2012; Naderloo et al. 2013), UAE (Hornby 1997).

Gulf of Oman: Iran (Naderloo et al. 2015).

World distribution: Western Indian Ocean: East Africa, Seychelles, Red Sea, Gulf of Aden, Socotra, southern Oman, Persian Gulf, Gulf of Oman and Pakistan.

Habitat: Intertidal to shallow subtidal, substrate rocky/cobble.

Fig. 10.6 *Eupilumnus calmani* (Balss, 1933). Male (MNHN-IU-2014-12,929), CL = 5.05, CB = 5.71 mm, Persian Gulf, 18.04.1954 (Photo: S. Träkner)

Fig. 10.7 *Lydia tenax* (Rüppell, 1830). Male (ZUTC 5517), CL = 32.58, CB = 48.28 mm, Park-Qadir, Abu Musa Island, Iran, Persian Gulf, 13.05.2015, coll. R. Naderloo (Photo: R. Abdollahi)

Family Euryplacidae Stimpson, 1871

11

Diagnosis Carapace transversely rectangular, quadrate or trapezoid, dorsal surface smooth, region not well defined, anterolateral margins arched, nearly straight, with one to three teeth in addition to exorbital tooth. Eyestalks short to long, orbital hiatus usually closed by basal antennal segment. Third maxillipeds completely covering buccal cavern, merus quadrate with anterolateral angle angular. Chelipeds relatively large, fingers usually narrow, longer than palm. Walking legs slender and long, without spines, serrated with setae. Abdomen of males and females with seven free-moving somites. G1 long, slender, straight or slightly curved, tapering distally; G2 short, less than one third length of G1, apical part with two unequal acute or obtuse processes. Genital opening of males coxo-sternal, large with prolonged penis. Genital opening of females sternal.

Key to species of the family Euryplacidae in the region

1. Carapace transeversly hexagonal, markedly broader than long (CB/CL about 1.7), anterolateral margins of carapace with one short tooth (except exorbital tooth) .... *Henicoplax eriochir*
- Carapace nearly subquadrate to hexagonal, slighly broader than long (CB/CL about 1.2), anterolateral margins of carapace with two or three distinct teeth (except exorbital tooth) 2
2. Eyestalks long, markedly longer than cornea; dorsal surface of carapace with fine brown spots, without particular flakes; anterolateral margins with three distinct teeth, second one largest, last one obtuse .... *Trissoplax dentata*
- Eyestalks short, not longer than cornea 3
3. Last tooth of anterolateral margins of carapace indistinct; dorsal surface of carapace with two or more large reddish brown flakes in central portion .... *Eucrate indica*
- Last tooth of anterolateral margins of carapace distinct; carapace and chelipeds covered with small red spots; usually two larger distinct red spots in bronchial region of males, the spots usually not distinct in females .... *Eucrate crenata*

Genus *Eucrate* De Haan, 1835

Eucrate crenata (De Haan, 1835)

(Figs. 11.1, 11.2a, and 11.3)

Cancer (*Eucrate*) *crenatus* De Haan 1835: 51, pl 15, fig 1.
Pilumnoplax sulcatifrons Stimpson 1858b: 93 (39).
Eucrate crenata — Nobili 1906a: 145, 1906b: 296.
Not *Eucrate crenata* — Tirmizi and Ghani 1982: 107, fig 3, 1996: 77 (in key), 80, fig 31.

Type locality: Japan.

Distribution

Persian Gulf: Kuwait (Pers Obser), UAE (Nobili 1906a).

Gulf of Oman: No record.

World distribution: Indo-West Pacific and Mediterranean: Madagascar, Seychelles, Red Sea, Arabian Sea, Persian Gulf, Gulf of Thailand, Korea, Japan, Indonesia, northern Australia and Mediterranean.

Habitat: Intertidal to 36 m, substrate rocky/cobble.

R. Naderloo, *Atlas of Crabs of the Persian Gulf*, DOI 10.1007/978-3-319-49374-9_11

Fig. 11.1 *Eucrate crenata* (De Haan, 1835). Male specimen, Ras Al-Ajuza, Kuwait, Persian Gulf, coll. R. Naderloo (Photo: R. Abdollahi. The specimen is not preserved)

Eucrate indica Castro and Ng, 2010

(Figs. 11.3 and 11.4)

Eucrate indica Castro and Ng 2010: 32–35, figs 2d, 4d–f, 9a–f, 15a–c.

Eucrate alcocki — Ng and Davie 2002: 378; Rajkumar and Wasagam 2009: 832, fig 1.

(?) *Eucrate crenata* var. *dentata* — Alcock 1900: 299 (in key), 301.

Type locality: Thailand (Phuket).

Distribution

Persian Gulf: Castro and Ng (2010) recorded this species from an unknown locality.

Gulf of Oman: No record.

World distribution: Indian Ocean: Persian Gulf, India, Thailand and Peninsular Malaysia.

Habitat: Shallow subtidal, substrate sandy.

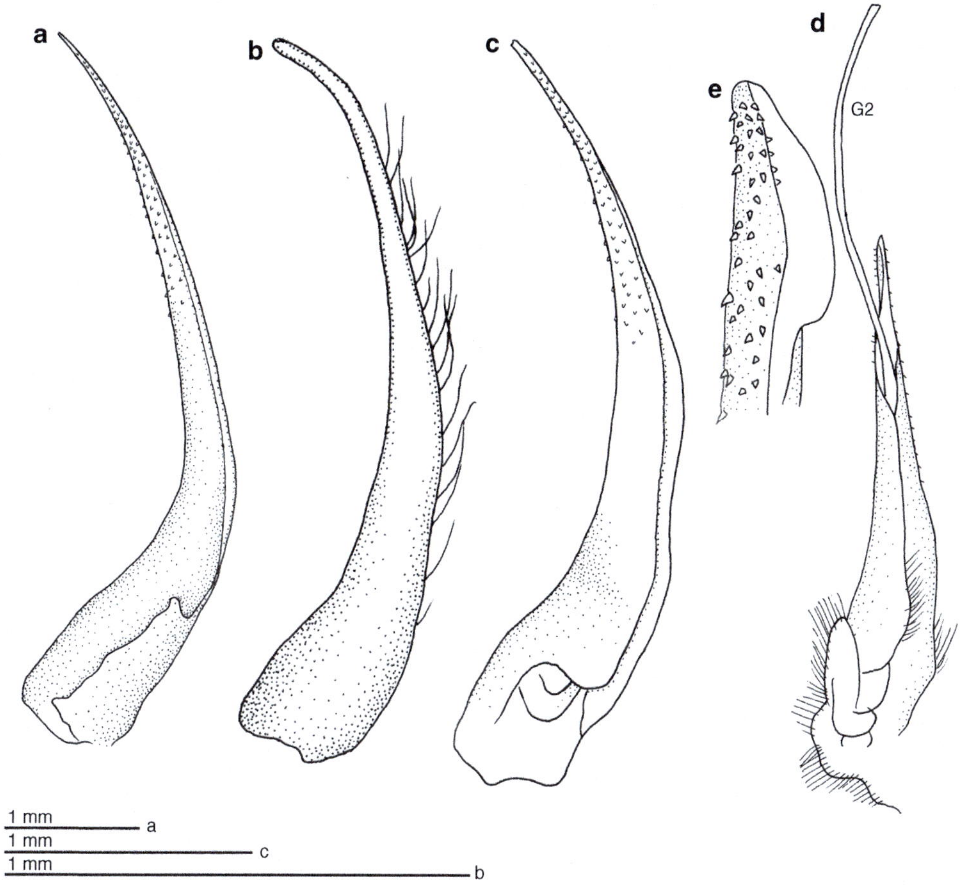

Fig. 11.2 Male G1, ventral view: (**a**) *Eucrate crenata* (De Haan, 1835) (Persian Gulf); (**b**) *Henicoplax eriochir* Castro and Ng, 2010 (Persian Gulf); (**c**) *Trissoplax dentata* (Stimpson, 1858) (ZUTC Brach1086); (**d**) *Carcinoplax sinica* Chen, 1984 (ZUMC); (**e**) apical part of (**d**), dorsal view

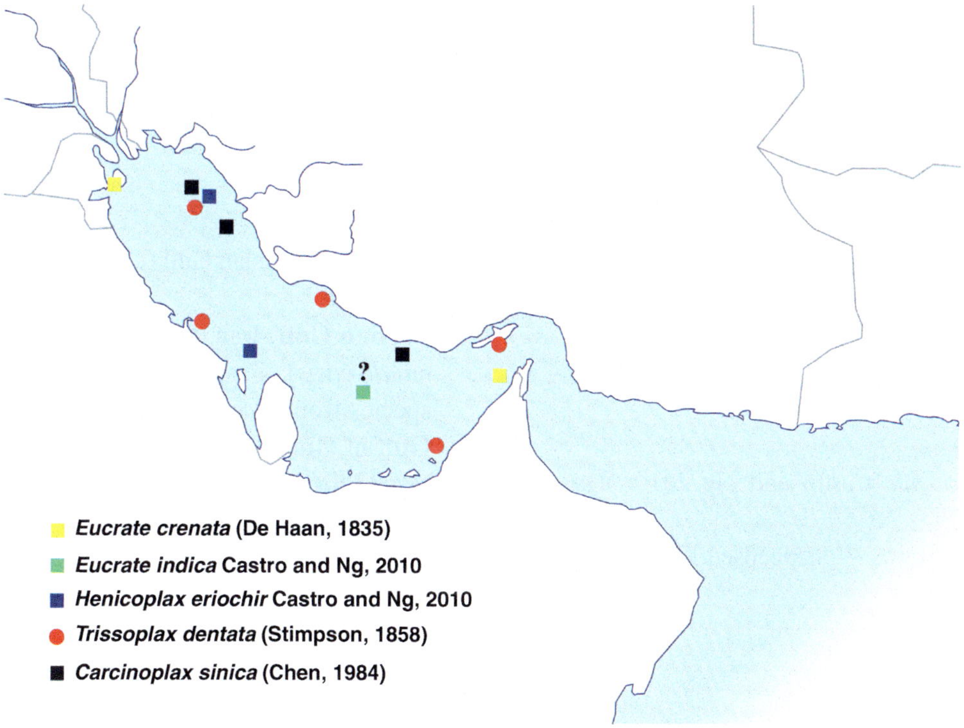

Fig. 11.3 Map showing the recorded localities of *Eucrate crenata* (De Haan, 1835), *Eucrate indica* Castro and Ng, 2010 (the exact locality is not known), *Henicoplax eriochir* Castro and Ng, 2010, *Trissoplax dentata* (Stimpson, 1858), and *Carcinoplax sinica* Chen, 1984

Fig. 11.4 *Eucrate indica* Castro and Ng, 2010. Female, Iran, Persian Gulf, the specimen is not preserved (Reproduced from Castro and Ng (2010) by permission of Magnolia Press)

Genus *Henicoplax* Castro and Ng, 2010

Henicoplax eriochir Castro and Ng, 2010
(Figs. 11.2b, 11.3, and 11.5)

Henicoplax eriochir Castro and Ng 2010: figs 21a–f, 24a–c.
Goneplax maldivensis — Stephensen 1946: 171.

Type locality: Persian Gulf (unknown locality).
Distribution
Persian Gulf: Iran [Stephensen (1946) as *Goneplax maldivensis*]. Castro and Ng (2010) described this species from an unknown locality.
Gulf of Oman: No record.
World distribution: This species is still only known from the Persian Gulf.
Habitat: Subtidal, 13–23 m.

Fig. 11.5 *Henicoplax eriochir* Castro and Ng, 2010. Male (ZMUC CRU965), CL = 9.4, CB = 5.3 mm, Khark Island, Iran, Persian Gulf, 05.03.1937, coll. G. Thorson (Photo: S. Träkner)

Genus *Trissoplax* Castro and Ng, 2010

Trissoplax dentata (Stimpson, 1858)

(Figs. 11.2c, 11.3, and 11.6)

Heteroplax dentata Stimpson 1858b: 94.
Eucrate affinis Haswell 1882: 547.
Eucrate haswelli Campbell 1969: 119 (in key), 130, figs 2, 5.
Eucrate costata Yang and Sun 1979: 4, 10, fig 3, pl, figs 5, 6, 9.
Heteroplax dentata — Ng et al. 2008: 78, 79 (in list).
Eucrate sulcatifrons — Stephensen 1946: 167, fig 45a, b; Titgen 1982: 252 (in list).
Eucrate haswelli — Guinot 1971: 1080 (in list): Tirmizi and Ghani 1988a: 139, figs 2, 3; Tirmizi and Ghani 1996: 77 (in key), 77, figs 29, 30; Apel 2001: 102; Naderloo and Sari 2007a: 344, tab 1.
Eucrate costata — Ng et al. 2008: 78 (in list).
Trissoplax dentata — Castro and Ng 2010: 98–104, figs 39d–f, 40a–f, 41a–g, 43a–c.

Type locality: *Heteroplax dentata*: Hong Kong. — *Eucrate affinis*: Australia. — *Eucrate haswelli*: Australia. — *Eucrate costata*: China.

Distribution

Persian Gulf: Iran [Stephensen (1946) as *Eucrate sulcatifrons*; Naderloo and Sari (2007a) as *Eucrate haswelli*], Saudi Arabia [Apel (2001) as *Eucrate haswelli*].

Gulf of Oman: No record.

World distribution: Indo-West Pacific: South Africa, Gulf of Aden, Persian Gulf, Pakistan, India, Andaman Sea, Gulf of Thailand, China, Hong Kong, Singapore and Australia.

Habitat: Intertidal to 40 m, substrates rocky/sandy.

Fig. 11.6 *Trissoplax dentata* (Stimpson, 1858). Male (ZUTC Brach1086), CL = 13.06, CB = 18.04 mm, Iran, Persian Gulf, 27° 07′N, 56° 25′E, 19.10.2004, coll. R. Naderloo (Photo: R. Abdollahi)

Family Goneplacidae MacLeay, 1838

12

Diagnosis Carapace xanthoid, hexagonal or subquadrate, broader than long, dorsal surface smooth, regions weakly defined; anterolateral margins of carapace usually with broadly distanced spines or without any spine, with epibranchial spine. Front entire, straight. Front-orbital breadth as broad as or broader than carapace breadth. Orbits complete, long, shallow, not concealing eyes; eyestalks moderately long. Antennules folded obliquely or transversely. Basal antennal segment short, never reaching front, flagellum long. Third maxillipeds with carpus articulating in anterointernal margin of merus. Suture between abdominal sternites 6–7 entire, sternite 8 of males visible when abdomen closed. Chelipeds nearly equal, robust. Walking legs narrow, nearly cylindrical, distal segments setose. Abdomen of males and females with seven freely moving somites. G1 usually stout, distally with spicules; G2 narrow, longer than G1.

Genus *Carcinoplax* H. Milne Edwards, 1852

Carcinoplax sinica Chen, 1984
(Figs. 11.2d, e, 11.3, and 12.1)

Carcinoplax sinica Chen 1984: 189 (in key), 190, 200, fig 2, pl 1, figs 6, 10.
Carcinoplax (*purpurea* Rathbun?) — Stephensen 1946: 166, 167, fig 44a, b; Guinot 1967a: 276 (in list; partly); Titgen 1982: 252 (in list).
Carcinoplax sinica — Guinot 1989: 285–287, figs 12a–d, 13a, b, 14a, b, pl 5a–i; Apel 2001: 101; Naderloo and Sari 2007a: 449, tab 1.

Type locality: China Sea (19° 50′N, 109° 00′E).
Distribution
Persian Gulf: Iran [Stephensen (1946) as *Carcinoplax* (*purpurea* Rathbun?); Apel 2001; Naderloo and Sari 2007a].
Gulf of Oman: No record.
World distribution: Indo-West Pacific: Persian Gulf, Thailand, Philippines and China.
Habitat: Subtidal, 13–41 m, substrates sandy and muddy.

R. Naderloo, *Atlas of Crabs of the Persian Gulf*, DOI 10.1007/978-3-319-49374-9_12

Fig. 12.1 *Carcinoplax sinica* Chen, 1984. Female (ZUTC Brach1083), CL = 26.7, CB = 40.7 mm, Iran, Persian Gulf, 29° 28′N, 49° 52′E, depth 13 m, 06.03.2004, coll. R. Naderloo (Photo: R. Abdollahi)

Family Iphiculidae Alcock, 1896

13

Diagnosis Carapace subglobular, or transversely oval, posterior surface covered with dense spongy pubescence, lateral margins with spines; regions weakly defined, intestinal region swollen, marked with circled groove. Front narrow, bilobed, projecting. Eyestalks relatively long; orbital hiatus broad, supraorbital margin with two fissures, inner infraorbital tooth well developed. Antennules folded obliquely; antennules and antennae long. Third maxillipeds with short merus. Chelipeds slender, palm short, proximally swollen, fingers filiform, much longer than palm. Abdomen of males with somites 3–4, 3–6 or 3–5 fused, abdomen of females freely moving. G1 longer than G2.

Genus *Iphiculus* Adams and White, 1849

Iphiculus spongiosus Adams and White, 1849

(Figs. 13.1, 13.2a, and 13.3)

Iphiculus spongiosus Adams and White 1849: 57, pl 13, fig 5.
Iphiculus spongiosus — Nobili 1906b: 170; Stephensen 1946: 70, fig 6d, e; Guinot 1967a: 248 (in list); Apel 2001: 50; Naderloo and Sari 2005a: 33, fig 2; Naderloo and Sari 2007a: 342, tab 1; Ng et al. 2008: 87 (in list), fig. 79.

Type locality: Philippines.
Distribution
Persian Gulf: Iran (Apel 2001; Naderloo and Sari 2005a, 2007a).
Gulf of Oman: Iran (Stephensen 1946).
World distribution: Indo-West Pacific: Red Sea, Gulf of Aden, Persian Gulf, Gulf of Oman, Pakistan, Bay of Bengal, Andamans, Arafura Sea, Gulf of Thailand, China, Philippines and Singapore.
Habitat: Intertidal to shallow subtidal, substrate sandy.

R. Naderloo, *Atlas of Crabs of the Persian Gulf*, DOI 10.1007/978-3-319-49374-9_13

Fig. 13.1 *Iphiculus spongiosus* Adams and White, 1849. Male (ZUTC Brach1002), CL = 10.25, CB = 16.50 mm, Iran, Persian Gulf, 29° 46′N, 49° 55′E, depth 14 m, 31.12.2003, coll. R. Naderloo (Photo: S. Shahbazi)

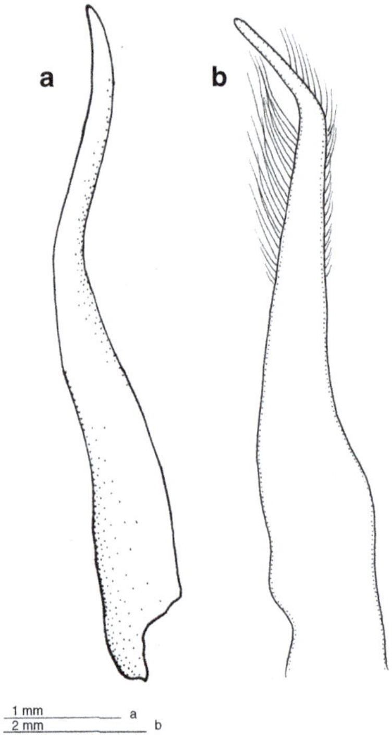

Fig. 13.2 Male G1, ventral view: (**a**) *Iphiculus spongiosus* Adams and White, 1849; (**b**) *Pariphiculus mariannae* (Herklots, 1852)

Genus *Pariphiculus* Alcock, 1896

Pariphiculus mariannae (Herklots, 1852)
(Figs. 13.2b, 13.3 and 13.4)

Ilia mariannae Herklots 1852: 36, 37, pl 1, fig 2.
Pariphiculus rostratus — Alcock 1996: 258, pl 8, fig 2.
Pariphiculus mariannae — Ihle 1918: 249, 250; Tirmizi and Kazmi 1986: 83, fig 24; Chen 1989: 231, fig 23. pl 4, fig 3; Naderloo and Sari 2005a: 37, fig 2, 2007a: 342, tab 1; Ng et al. 2008: 87 (in list), fig. 81.
Type locality: China.
Distribution

Persian Gulf: Iran (Naderloo and Sari 2005a, 2007a).

Gulf of Oman: No record.

World distribution: Indo-West Pacific: Persian Gulf, Pakistan, India, Burma, China, Philippines and Indonesia.

Habitat: Intertidal to shallow subtidal, substrate sandy.

Fig. 13.3 *Pariphiculus mariannae* (Herklots, 1852). Female (ZUTC Brach1014), CL = 23.9, CB = 23.00 mm, Bandar-Abbas, Iran, Persian Gulf, depth 20 m, 11.2003, coll. R. Naderloo (Photo: R. Abdollahi)

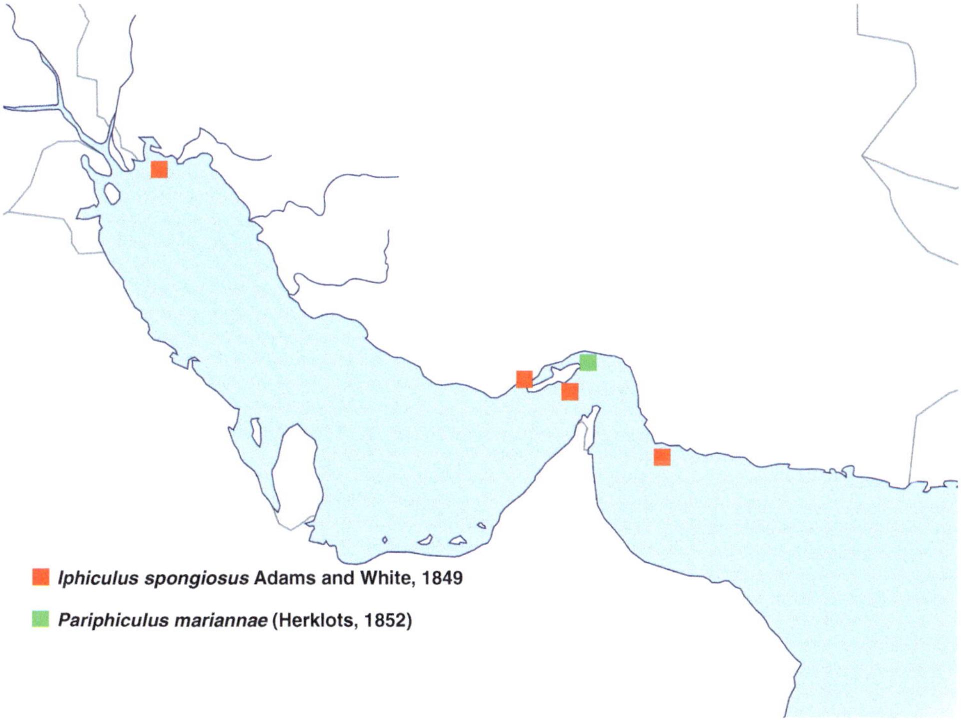

Fig. 13.4 Map showing the recorded localities of *Iphiculus spongiosus* Adams and White, 1849, and *Pariphiculus mariannae* (Herklots, 1852)

Family Leucosiidae Samouelle, 1819 (Pebble Crabs)

14

Diagnosis Carapace with varied forms of pyriform to oval and subcircular, completely flat to strongly convex, usually glabrous, smooth or with teeth and spines. Front narrow, more or less produced. Orbits and eyes markedly small. Third maxillipeds long, with long and triangular merus, completely covering buccal cavern, exopod broad. Afferent branchial canals running along the sides of buccal cavern, opening at bases of third maxillipeds. Chelipeds nearly equal, usually long, robust. Walking legs normally narrow, with simple dactylus. Abdomen of males and females with somites 3–5 or 3–6 tightly fused. G1 longer than G2. Genital opening of males coxal or sternal. Genital opening of females sternal.

Key to genera of the family Leucosiidae in the region

1. Frontal region produced anteriorly, forming a distinct front; thoracic sinuous of carapace forming a shallow part in bases of the chelipeds 2
- Frontal region not markedly produced anteriorly, no distinct front; without thoracic sinuous 6

2. Male abdomen with somites 3–6 tightly fused; G1 straight, twisted once .... *Seulocia*
- Male abdomen with somites 3–5 tightly fused; G1 not twisted or twisted more than once 3

3. Carapace urn-shaped, often laterally pubescent; chelipeds with spongy setal patch proximally on upper and lower surface of merus; G1 not twisted .... *Urnalana*
- Carapace normal, subpentagonal, laterally not pubescent; chelipeds without spongy setal patch on merus; G1 twisted 4

4. G1 screw-like; small apical process concealed with dense setal patch .... *Leucosia*
- G1 twisted three times in its entire length; apical process without dense setal patch 5

5. Carapace with anterior margin of thoracic sinuous flat .... *Coleusia*
- Carapace with anterior margin of thoracic sinuous pursed, loop-shaped .... *Euclosiana*

6. Anterior margin of buccal cavern extending beyond level of anterior border of pterygostomian regions; epistome and infraorbital lobes usually well developed; distinct space between edge of floor of orbit and free edge of buccal cavern 7

R. Naderloo, *Atlas of Crabs of the Persian Gulf*, DOI 10.1007/978-3-319-49374-9_14

- Anterior margin of buccal cavern hardly reaching to level of anterior border of pterygostomian regions; epistome reduced; infraorbital lobe rarely well-developed 8

7. Carapace markedly broader than long, expanded laterally, completely concealing walking legs; upper surface of carapace extremely uneven with irregular hollows on branchial region .... *Oreophorus*

- Carapace spherical, not markedly broader than long; walking legs visible from dorsal veiw; upper surface of carapace moderately uneven without hollows .... *Ebalia*

8. Carapace polygonal, lateral margins of carapace thin and expanded 9

- Carapace rhomboidal, without lateral expansion, with or without spines on lateral margins 10

9. Carapace broadly polygonal; fingers markedly longer than palm, narrow, apically curved, serrated with fine teeth .... *Nursilia*

- Carapace pentagonal; fingers normal .... *Nursia*

10. Carapace with long spine on lateral margins 11

- Carapace without spine on lateral margins 12

11. Carapace extremely broader than long, only with large conical lateral spines, longer than half of carapace width .... *Ixa*

- Carapace not markedly broader than long, with large lateral spine, not longer than half of carapace width, usually with more spines on lateral and posterior margins .... *Arcania*

12. Posterior margin of carapace with median tubercle or spine .... *Myra*

- Posterior margin of carapace without any median tubercle or spine 13

13. Front produced anteriorly, forming a broad front; carapace without hepatic facet; merus of chelipeds with large tubercles on proximal portion .... *Pseudophilyra*

- Frontal region not produced anteriorly, without distinct front; carapace with hepatic facet; merus of chelipeds without large tubercles, or with small on proximal portion 14

14. Male abdomen with first and second somites articulate 15

- Male abdomen with first somite articulate 16

15. Carapace broadly pear-shaped; epistome and lower margins of hepatic facet not projecting; third maxillipeds with exopod subquadrate; first and second somites of male abdomen transversely narrow; G1 with apical process subdistally expanded .... *Philyra*

- Carapace nearly spherical; epistome and lower margins of hepatic facet projecting beyond front; third maxillipeds with kidney-shaped exopod; first and second somites of male abdomen laterally swollen; G1 with apical process slender, tubular .... *Ryphila*

16. Chelipeds relatively slender and long; immovable fingers with dense setae along cutting edge; male abdomen with fused somites 2–6 without denticle .... *Hiplyra*

- Chelipeds stout and relatively short; fingers without dense setae along cutting edge, with large gap between; male abdomen with fused somites 2–6 with denticle .... *Lyphira*

Genus *Arcania* Leach, 1817

Key to species of the genus *Arcania* in the region

1. Carapace rhomboidal; epibranchial spine distinctly longer than other lateral spines 2
- Carapace rounded or pyriform; epibranchial spine not longer than other lateral spines 4
2. Epibranchial spines nearly straight; intestinal tooth tubercle-shape, round, shorter than tubercles of posterior margin .... *Arcania cornuta*
- Epibranchial spines slightly directed backward; intestinal spines longer than spines of posterior margin 3
3. Posterolateral margins of carapace with upward spine on median portion .... *Arcania septemspinosa*
- Posterolateral margins of carapace without upward spine on median portion .... *Arcania gracilis*
4. Carapace round; posterior surface of carapace prominently spinose; epibranchial spines of carapace as long as posterolateral spines .... *Arcania erinacea*
- Carapace pyriform; posterior surface of carapace prominently tuberculate; epibranchial spines of carapace distinctly shorter than posterolateral spines .... *Arcania tuberculata*

Arcania cornuta (MacGilchrist, 1905)

(Figs. 14.1, 14.2a, and 14.3)

Ixoides cornutus MacGilchrist 1905: 255.
Ixoides cornutus — Alcock and MacGilchrist 1905: pl 73, figs 2, 2b; Stephensen 1946: 74; Titgen 1982: 247 (in list); Tan 1996: 1033–1035, figs 3j–l, 4a, b; Apel 2001: 51; Naderloo and Sari 2005a: 34, fig 3, 2007a: 342, tab 1.
Arcania cornuta — Galil 2001: 171 (in key), 173, figs 1b, 4b.

Type locality: Persian Gulf (26° 20′N, 53° 54′E).

Distribution

Persian Gulf: Iran [Apel (2001) and Naderloo and Sari (2005a, 2007a) as *Ixoides cornutus*], UAE [MacGilchrist (1905) as *Ixoides cornutus*].

Gulf of Oman: No record.

World distribution: Indo-West Pacific: South Africa, Mozambique Channel, Madagascar, Persian Gulf, Vietnam, China, Philippines, Japan, New Caledonia and Fiji.

Habitat: Subtidal, 86–300 m, substrate sandy.

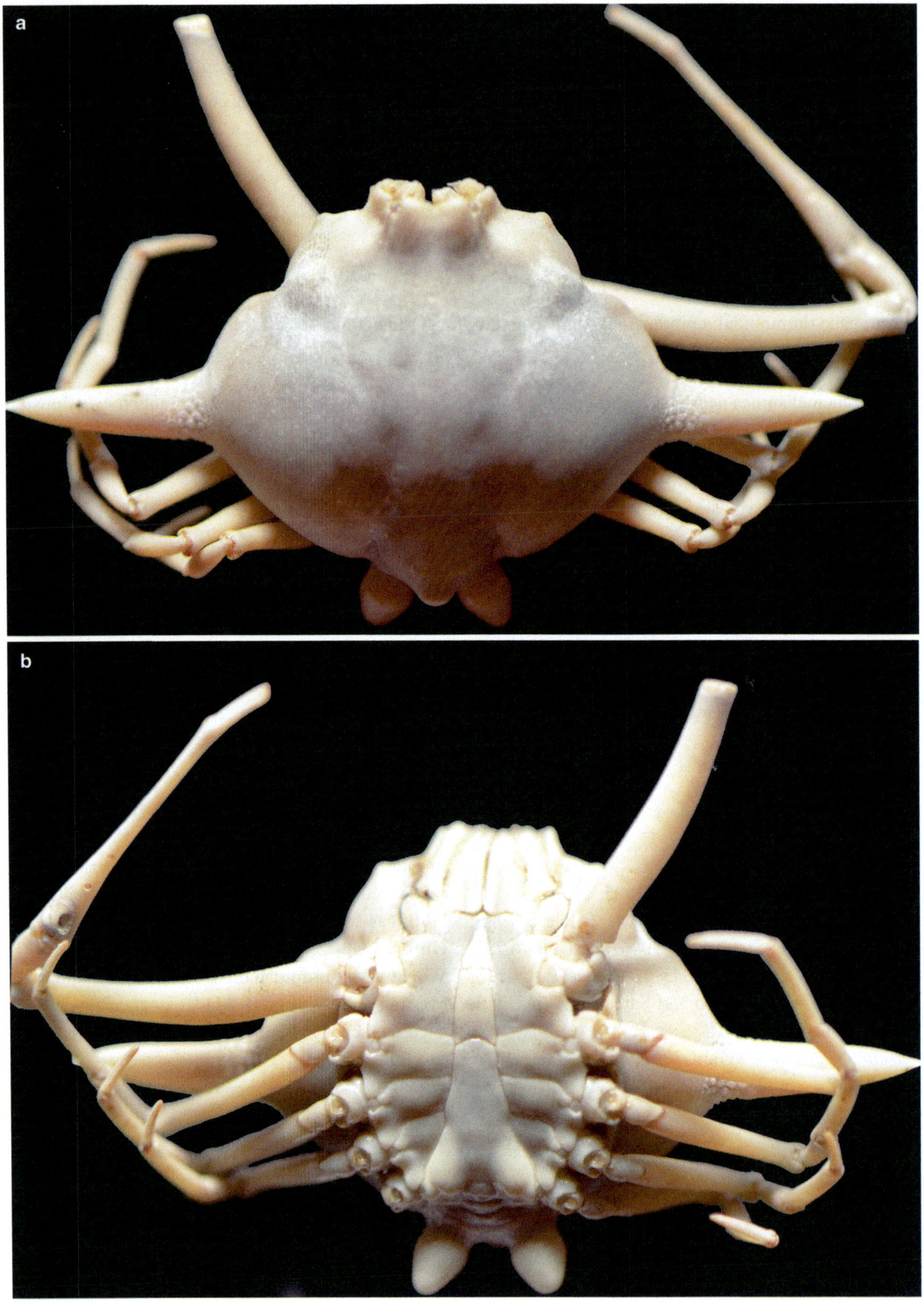

Fig. 14.1 *Arcania cornuta* (MacGilchrist, 1905). Male holotype (Calcutta Museum), Persian Gulf: (**a**) posterior view; (**b**) ventral view

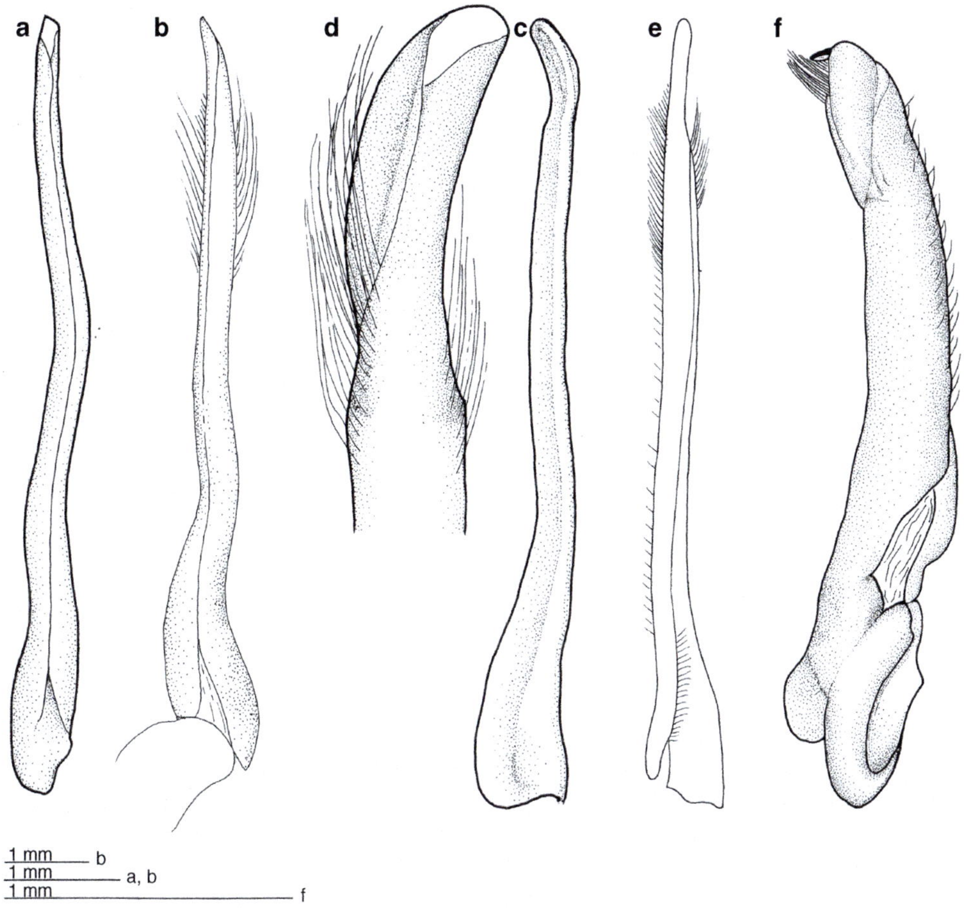

Fig. 14.2 Male G1, ventral view: (**a**) *Arcania cornuta* (MacGilchrist, 1905) (SMF: Iran); (**b**) *Arcania erinacea* (Fabricius, 1787) (ZUTC Brach1001); (**c**) *Arcania gracilis* Henderson, 1893 (ZMUC CRU879); (**d**) apiccal part of (**c**), dorsal view; (**e**) *Arcania septemspinosa* (Fabricius, 1787) [Modified from Stephensen (1946: fig. 7b)]; (**f**) *Ebalia abdominalis* Nobili, 1905 (ZUTC Brach1304)

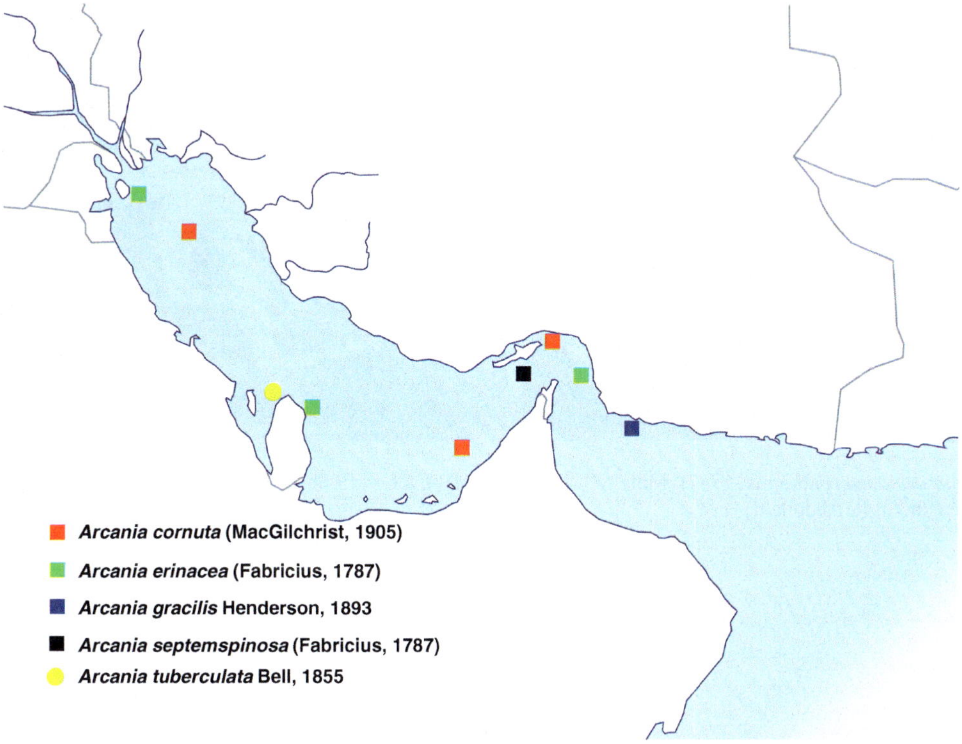

Fig. 14.3 Map showing the recorded localities of *Arcania cornuta* (MacGilchrist, 1905), *Arcania erinacea* (Fabricius, 1787), *Arcania gracilis* Henderson, 1893, *Arcania septemspinosa* (Fabricius, 1787), and *Arcania tuberculata* Bell, 1855

Arcania erinacea (Fabricius, 1787)
(Figs. 14.2b, 14.3, and 14.4)

Cancer erinaceus Fabricius 1787: 325.
Cancer erinaceus — Herbst 1790: 258, pl 20, fig 3.
Arcania erinaceus — Alcock 1896: 268; Tirmizi and Kazmi 1986: 72, fig 20a–d: Apel 2001: 49.
Arcania erinacea — Galil 2001: 171 (in key), 176, figs 1c, 4c; Naderloo and Sari 2005a: 32, fig 3, 2007a: 342, tab 1; Ng et al. 2008: 89 (in list).

Type locality: Indian Ocean.
Distribution
Persian Gulf: Iran (Naderloo and Sari 2005a, 2007a), Kuwait (Apel 2001), Qatar (Apel 2001).
Gulf of Oman: Apel (2001) recorded this species based on a specimen deposited in NHM (BNMH 1971.227) from an unknown locality.
World distribution: Indo-West Pacific: Persian Gulf, Gulf of Oman, Pakistan, India, Sri Lanka, Thailand, China, Japan and Singapore.
Habitat: Subtidal, 7–85 m, substrate sandy.

Arcania gracilis Henderson, 1893
(Figs. 14.2c, d, 14.3, and 14.5)

Arcania septemspinosa var. *gracilis* Henderson 1893: 403.
Arcania quinquespinosa — Alcock and Anderson 1894b: 206, 207; Alcock 1896: 266; Alcock and Anderson 1896: pl 24, fig 6; Ihle 1918: 266, 267, 313 (in list); Stephensen 1946: 72, 73, fig 7d, e; Guinot 1967a: 248 (in list); Titgen 1982: 247 (in list); Apel 2001: 49.

Fig. 14.4 *Arcania erinacea* (Fabricius, 1787). Male (ZUTC Brach1001), CL = 19.00, CB = 18.00 mm, Iran, Persian Gulf, 27.12.2004, coll. R. Naderloo (Photo: R. Abdollahi)

Arcania quinquispinosa — Zarenkov 1969: 23, 1971: 172, 173.
Arcania gracilis — Galil 2001: 171 (in key), 184, figs 2c, 5d; Ng et al. 2008: 94 (in list).
Type locality: Andaman Sea (Martaban).
Distribution
Persian Gulf: Alcock and Anderson (1894b) recorded this species as *Arcania quinquespinosa* from an unknown locality.
Gulf of Oman: Iran [Stephensen (1946) as *Arcania quinquespinosa*].
World distribution: Indo-West Pacific: Madagascar, Red Sea, southern Oman, Persian Gulf, Gulf of Oman, India, Lakshadweep Islands, Sri Lanka, China, Hong Kong, Taiwan, Philippines, Japan, Indonesia, Singapore, Australia, New Caledonia and Vanuatu.
Habitat: Subtidal, 26–366 m, substrate sandy.

Arcania septemspinosa (Fabricius, 1787)
(Figs. 14.2e, 14.3, and 14.6)

Cancer septemspinosus Fabricius 1787: 325.
Arcania septemspinosa — Alcock 1896: 263 (in key), 265; Nobili 1906b: 171; Stephensen 1946: 73,

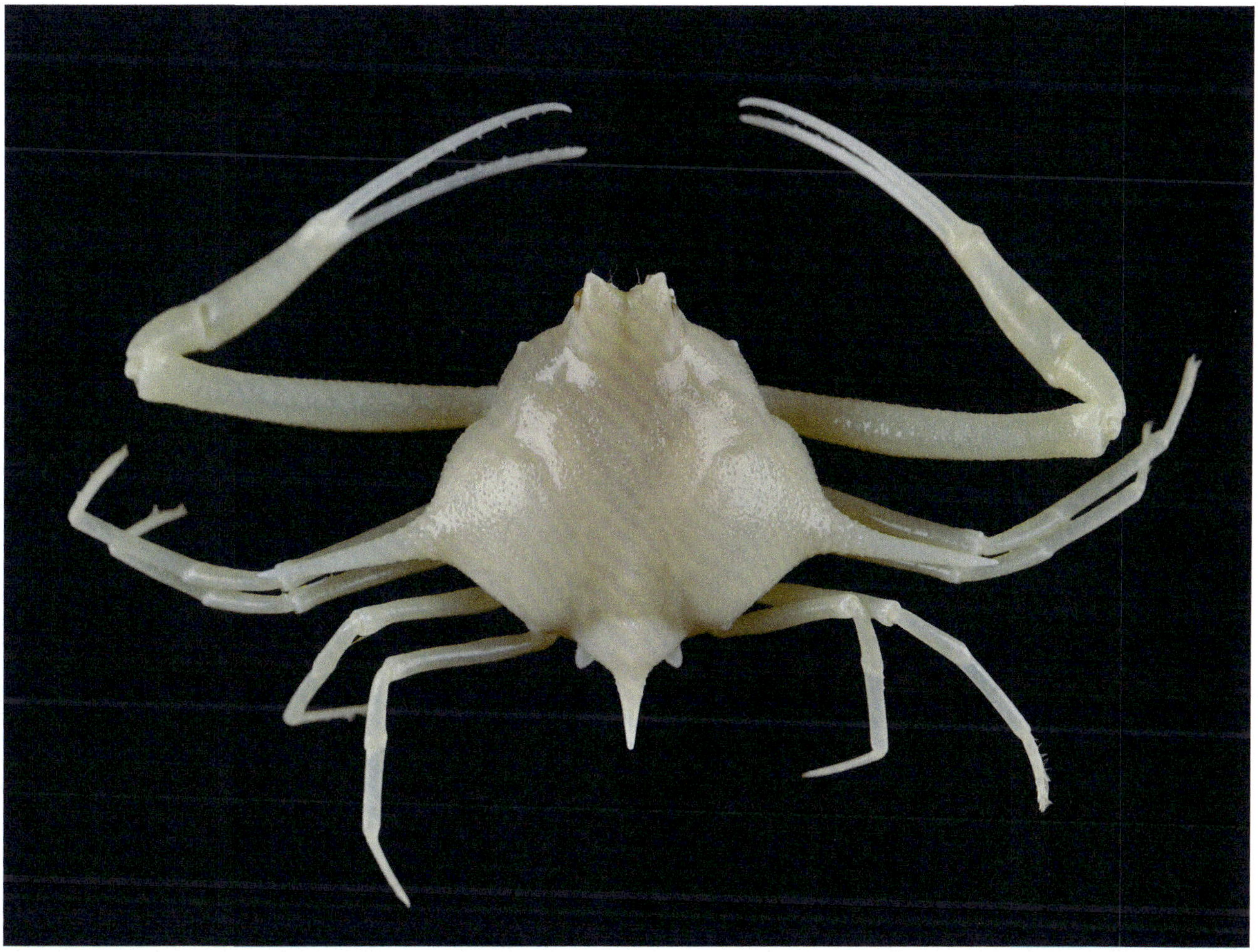

Fig. 14.5 *Arcania gracilis* Henderson, 1893. Male (ZMUC CRU879), CL = 14.8, CB = 23.4 mm, SW of Jask, Iran, Gulf of Oman, depth 30 m, 29.03.1938, coll. B. Loppenthin (Photo: S. Träkner)

fig 7f; Chhapgar 1957a: 408, pl 3a–c; Guinot 1967a: 248 (in list); Titgen 1982: 247 (in list); Tirmizi and Kazmi 1986: 72–76, fig 21a–h; Apel 2001: 50; Galil 2001: 171 (in key), 193, figs 3b, 7b; Ng et al. 2008: 94 (in list).

Type locality: Indian Ocean.

Distribution

Persian Gulf: Iran (Stephensen 1946). Alcock (1896) recorded this species from an unknown locality.

Gulf of Oman: No record.

World distribution: Indo-West Pacific: South Africa, Madagascar, Mozambique Channel, Red Sea, Gulf of Aden, Persian Gulf, Pakistan, India, Gulf of Bengal, Burma, Gulf of Thailand, Philippines, Indonesia, Australia and Fiji.

Habitat: Subtidal, 7–182 m, substrate sandy.

Arcania tuberculata Bell, 1855

(Figs. 14.3 and 14.7)

Arcania tuberculata Bell 1855a: 367, 1855b: 310, pl 34, fig 8.

Arcania tuberculata — Alcock 1896: 268; Ihle 1918: 264; Galil 2001: 172 (in key), 195, fig 3c, 7c; Ng et al. 2008: 94 (in list); Galil et al. 2012: 1, fi. 2a.

Fig. 14.6 *Arcania septemspinosa* (Fabricius, 1787). Female (ZMUC CRU839), CL = 32.2, CB = 52.6 mm, SW of Hengam Island, Iran, Persian Gulf, depth 31 m, 10.04.1937, coll. G. Thorson (Photo: S. Träkner)

Fig. 14.7 *Arcania tuberculata* Bell, 1855. Female (NHM reg.2011.8903), CL = 8.04, CB = 8.49 mm, Al-Aidd Ash Sharqi, Qatar, Persian Gulf, depth 35.09 m, 27.08.2003 (Photo: H. Taylor)

Type locality: *Arcania tuberculata*: Borneo. — *Arcania laevimana*: Philippines.

Distribution

Persian Gulf: Qatar (Galil et al. 2012).

Gulf of Oman: No record.

World distribution: Indo-West Pacific: Madagascar, Seychelles, Persian Gulf, Maldives, Sri Lanka, Andaman Islands, South China Sea, Philippines, Indonesia, Australia and New Caledonia.

Habitat: Subtidal, 10–124 m, substrate sandy.

Genus *Coleusia* Galil, 2006

Coleusia biannulata (Tyndale-Biscoe and George, 1962)

(Figs. 14.8 and 14.9)

Leucosia biannulata Tyndale-Biscoe and George 1962: 92, 93, fig 4.

Leucosia longifrons var. neocaledonica — Alcock 1896: 218.

Leucosia longifrons — Nobili 1906a: 97; Stephensen 1946: 93–94; Chhapgar 1969: 610, pl 1b.

Fig. 14.8 *Coleusia biannulata* (Tyndale-Biscoe and George, 1962). Female (NHM reg.1984.415), CL = 21.20, CB = 18.19 mm, Karachi Fish Market Harbour, Pakistan, 14.04.1982, coll. N.M. Tirmiz and N. Ghani (Photo: H. Taylor)

Leucosia perlata — Stephensen 1946: 95.

Leucosia biannulata — Titgen 1982: 247 (in list); Tirmizi and Kazmi 1986: 94–97, fig 27a–f; Apel 2001: 52; Ng et al. 2008: 94 (in list).

Coleusia biannulata — Galil 2006: 57, figs 1a, 3a, 4a.

Type locality: India (Palk Strait).

Distribution

Persian Gulf: Iran [Stephensen (1946) as *Leucosia perlata*], UAE [Nobili (1906a) as *Leucosia longifrons*; Apel 2001]. Alcock (1986) recorded this species as *Leucosia longifrons var. neocaledonica* A. Milne-Edwards, 1874, from an unknown locality.

Gulf of Oman: No record.

World distribution: Northwestern Indian Ocean: Red Sea, Persian Gulf, Pakistan, and India (Mumbai and Palk Strait).

Habitat: Subtidal.

Genus *Ebalia* Leach, 1817

Ebalia abdominalis Nobili, 1905

(Figs. 14.2f, 14.9, and 14.10)

Ebalia abdominalis Nobili 1905c: 399.

Ebalia abdominalis — Nobili 1906b: 155 (in key), 157, pl 9, fig 2; Ihle 1918: 310 (in list); Apel 2001: 50; Ng et al. 2008: 90 (in list); Naderloo and Türkay 2012: 31; Naderloo et al. 2013: 449, tab 1.

Type locality: Red Sea and Gulf of Aden.

Distribution

Persian Gulf: Iran (Naderloo and Türkay 2012; Naderloo et al. 2013), UAE (Apel 2001).

Gulf of Oman: No record.

World distribution: Northwestern Indian Ocean: Somalia, Red Sea, Gulf of Aden and Persian Gulf.

Habitat: Intertidal to shallow subtidal, substrate sandy.

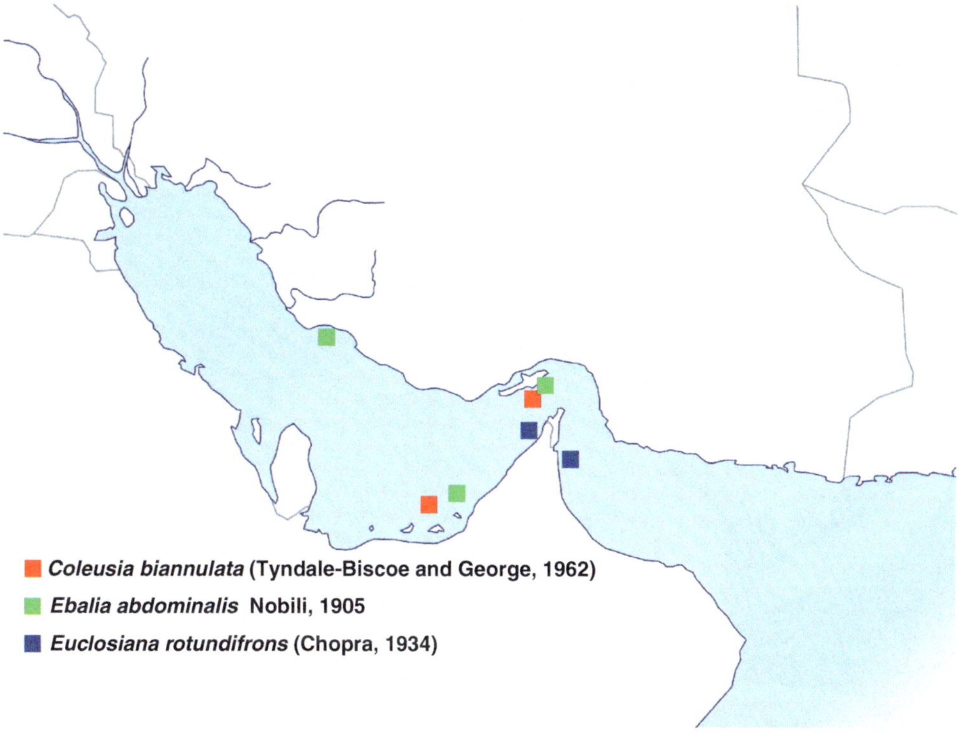

Fig. 14.9 Map showing the recorded localities of *Coleusia biannulata* (Tyndale-Biscoe and George, 1962), *Ebalia abdominalis* Nobili, 1905, and *Euclosiana rotundifrons* (Chopra, 1934)

Fig. 14.10 *Ebalia abdominalis* Nobili, 1905. Male (ZUTC Brach1304), CL = 7.1, CB = 6.5 mm, Bandar-Dayyer, Bushehr, Iran, Persian Gulf, 11.06.2006, coll. R. Naderloo (Photo: S. Träkner)

Fig. 14.11 *Euclosiana rotundifrons* (Chopra, 1934). Male (NHM 1900.4.20.5/6), CL = 31.4, CB = 29.0 mm, Abu Dhabi, UAE, Persian Gulf, 29.10.1900 (Photo: H. Taylor)

Genus *Euclosiana* Galil and Ng, 2010

Euclosiana rotundifrons (Chopra, 1934)

(Figs. 14.9 and 14.11)

Leucosia rotundifrons Chopra 1934: 34, text-fig 1, 2.
Leucosia rotundifrons — Serène 1968: 47.
Euclosiana rotundifrons — Galil 2003: 333 (in key), 338, 339, 1f, 4b, 6f; Ng et al. 2008: 94 (in list).

Type locality: Bay of Bengal.
Distribution
Persian Gulf: UAE (Galil 2003).
Gulf of Oman: Iran (Galil 2003), UAE (Galil 2003).
World distribution: Northern Indian Ocean: Persian Gulf, Gulf of Oman, Arabian Sea and Bay of Bengal.
Habitat: Subtidal, 124–311 m.

Genus *Hiplyra* Galil, 2009

Key to species of the genus *Hiplyra* in the region

1. Male abdomen with elevated arrow-shaped process on somite 6 2
- Male abdomen with somite 6 smooth, without process 3
2. Male abdomen with somite 6 with long arrow-shaped process, creating distinct groove; telson narrow, smooth, with no elevation; G1 distally narrow, with relatively large apical process directed laterally .... *Hiplyra sagitta*

- Male abdomen with somite 6 with triangular arrow-shaped process on distal portion; telson broad-triangular, swollen on basal portion; G1 distally broadned, with small apical process directed dorsally .... *Hiplyra ramli*

3. Male abdomen with telson elevated on lateral portions; G1 broadened distally, with very small subdistal process, directed laterally; first somite of female abdomen not lobate .... *Hiplyra elegans*

- Male abdomen with telson completely smooth; G1 tapering distally, distal process directed ventrally; first somite of female abdomen trilobate .... *Hiplyra variegata*

Hiplyra elegans (Gravier, 1920)

(Figs. 14.12, 14.13a, and 14.14)

Philyra variegata var. *elegans* Gravier 1920: 379, figs 1–7.
Philyra platychira Laurie 1906: 363.
Philyra variegata — Stephensen 1946: 89 [part of the material from st. 73, Jask, Iran].
Hiplyra elegans — Galil 2009: 292, 315 (in key), fig 7; Naderloo and Apel 2012: 249, figs 1, 2.

Type locality: Madagascar.
Distribution
Persian Gulf: No record.
Gulf of Oman: Iran (Naderloo and Apel 2012).
World distribution: Indian Ocean: Madagascar, Gulf of Oman and Sri Lanka.
Habitat: Intertidal to shallow subtidal, substrate sandy.

Fig. 14.12 *Hiplyra elegans* (Gravier, 1920). Male (ZMUC CRU929), CL = 9.90, CB = 9.33 mm, Jask, Iran, Gulf of Oman, 20.04.1937, coll. G. Thorson (Photo: S. Träkner)

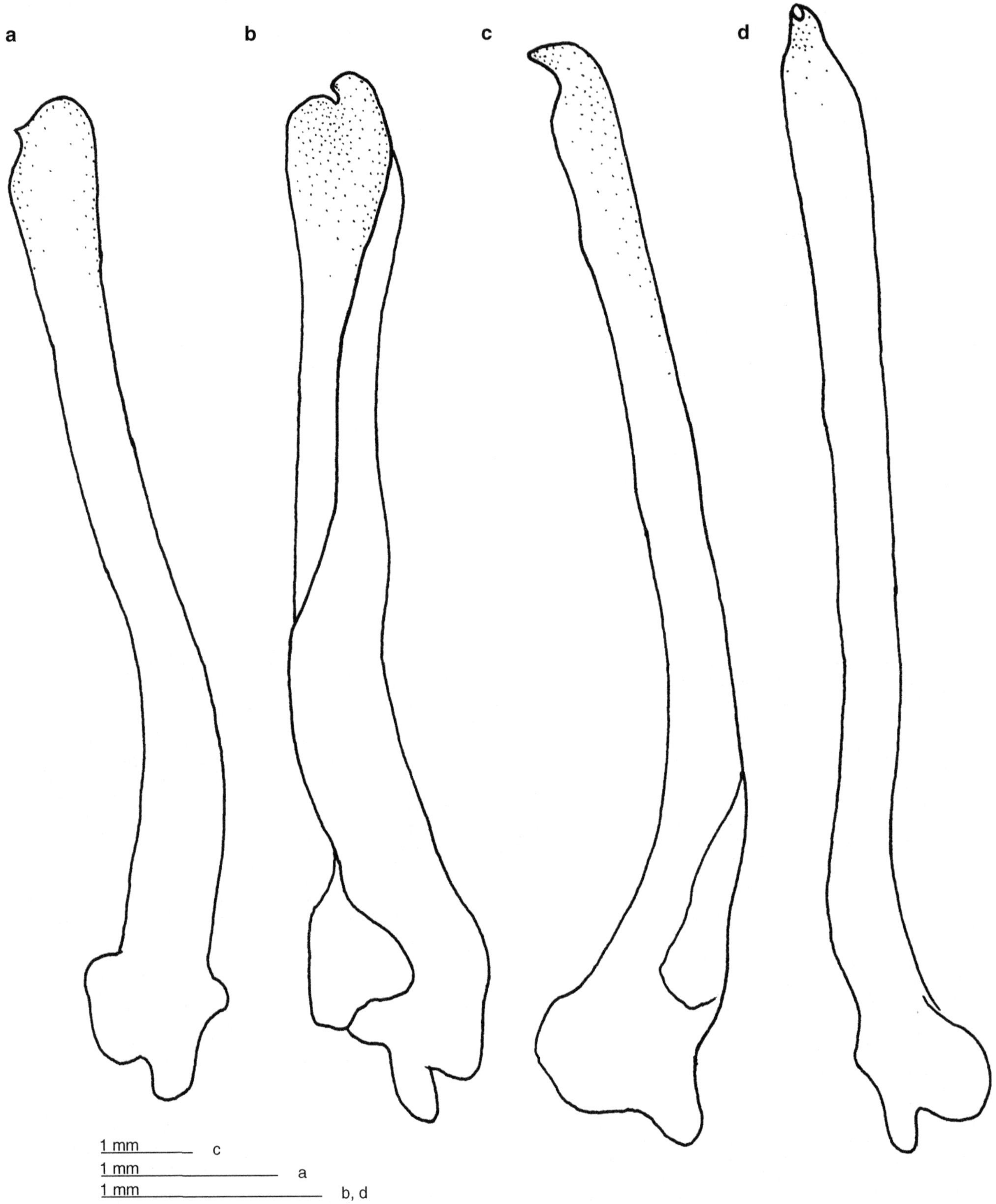

Fig. 14.13 Male G1, ventral view: (**a**) *Hiplyra elegans* (Gravier, 1920) (ZMUC CRU929); (**b**) *Hiplyra ramli* Naderloo and Apel, 2012 (SMF 38466); (**c**) *Hiplyra sagitta* Galil, 2009 (ZMUC CRU880); (**d**) *Hiplyra variegata* (Rüppell, 1830) (SMF 11121)

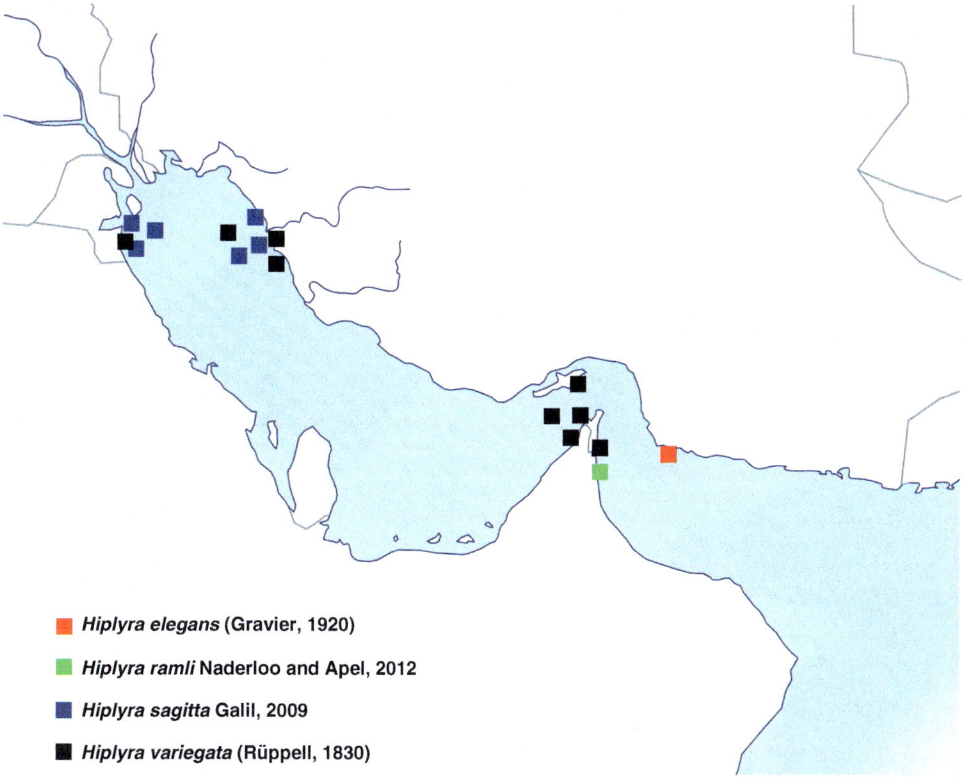

Fig. 14.14 Map showing the recorded localities of *Hiplyra elegans* (Gravier, 1920), *Hiplyra ramli* Naderloo and Apel, 2012, *Hiplyra sagitta* Galil, 2009, and *Hiplyra variegata* (Rüppell, 1830)

Hiplyra ramli Naderloo and Apel, 2012

(Figs. 14.13b, 14.14, and 14.15)

Hiplyra ramli Naderloo and Apel 2012: 251, figs 3, 4.

Type locality: Gulf of Oman (UAE).

Distribution

Persian Gulf: No record.

Gulf of Oman: UAE (Naderloo and Apel 2012).

World distribution: This species is still only recorded from the Gulf of Oman.

Habitat: Subtidal, 3–4 m, substrate sandy.

Hiplyra sagitta Galil, 2009

(Figs. 14.13c, 14.14, and 14.16)

Hiplyra sagitta Galil 2009: 296, 315 (in key), figs 11, 12a.
Philyra platychira — Alcock 1896: 242.
Philyra platycheir — Tirmizi and Kazmi 1986: 100, fig 29.
Philyra variegata — Stephensen 1946: 89 [part of the material from st. 27, near Bushehr, Iran].
Hiplyra sagitta — Naderloo and Apel 2012: 254, figs 5, 6; Naderloo and Türkay 2012: 31; Galil et al. 2012: 1, fig 2b.

Tye locality: Persian Gulf (Bushehr, Iran).

Distribution

Persian Gulf: Iran (Galil 2009; Naderloo and Apel 2012; Naderloo and Türkay 2012), Kuwait (Galil 2009).

Gulf of Oman: No record.

World distribution: Northern Indian Ocean: Persian Gulf, Pakistan, India and Andaman Sea.

Habitat: Subtidal, 7.5–22 m, substrate sandy.

Fig. 14.15 *Hiplyra ramli* Naderloo and Apel, 2012. Male holotype (SMF 34466), CL = 7.4, CB = 6.8 mm, Al-Aqah, near Sandy Beach Hotel, Fujairah, UAE, Gulf of Oman, depth 3–4 m, 04.06.1995, coll. M. Apel (Photo: S. Träkner)

Fig. 14.16 *Hiplyra sagitta* Galil, 2009. Male (SMF 38392), CL = 16.79, CB = 15.13 mm, Kuwait, Persian Gulf, 28° 53′N, 48° 24′E, depth 22 m, 24.04.1995, coll. F. Krupp (Photo: S. Träkner)

Hiplyra variegata (Rüppell, 1830)

(Figs. 14.13d, 14.14, and 14.17)

Myra variegata Rüppell 1830: 17, pl 4–4.
Myra variegata — Nobili 1906b: 169.
Philyra platycheira — Paul'son 1875: 83, pl 10, fig 3; Alcock 1896: 242 (part of the material from the Persian Gulf).
Philyra variegata — Laurie 1915: 410; Stephensen 1946: 89, figs 15f–k, 16; Serène 1968: 46; Guinot 1967a: 249 (in list); Titgen 1982: 248 (in list); Apel 2001: 58; Ng et al. 2008: 93 (in list).
Hiplyra variegata — Galil 2009: 287, 315 (in key), fig 13; Naderloo and Apel 2012: 255, figs 7, 8; Naderloo and Türkay 2012: 31.

Type locality: Red Sea (Egypt).

Distribution

Persian Gulf: Iran (Stephensen 1946; Naderloo and Apel 2012; Naderloo and Türkay 2012), UAE (Apel 2001).

Gulf of Oman: UAE (Apel 2001).

World distribution: Western Indian Ocean: East Africa (Mombasa, Kenya), Red Sea, Gulf of Aden, Persian Gulf and Gulf of Oman.

Habitat: Intertidal to 56 m, substrate sandy.

Fig. 14.17 *Hiplyra variegata* (Rüppell, 1830). Male (ZUTC 5538), CW = 8.32, CL = 8.65 mm, Al-Khairan, Kuwait, Persian Gulf, coll. N. Manickam, 02.2014 (Photo: R. Abdollahi)

Genus *Ixa* Leach, 1816

Ixa holthuisi Tirmizi, 1970

(Figs. 14.18, 14.19 and 14.21a)

Ixa holthuisi Tirmizi 1970: 313, figs 1, 2.

Ixa holthuisi — Tirmizi and Kazmi 1986: 69–71, fig 19a–f; Apel 2001: 50, 51.

Ixa edwardsi — Stephensen 1946: 73, 74.

Ixa sp. 5 — Basson et al. 1977: 243.

Type locality: Pakistan (Karachi).

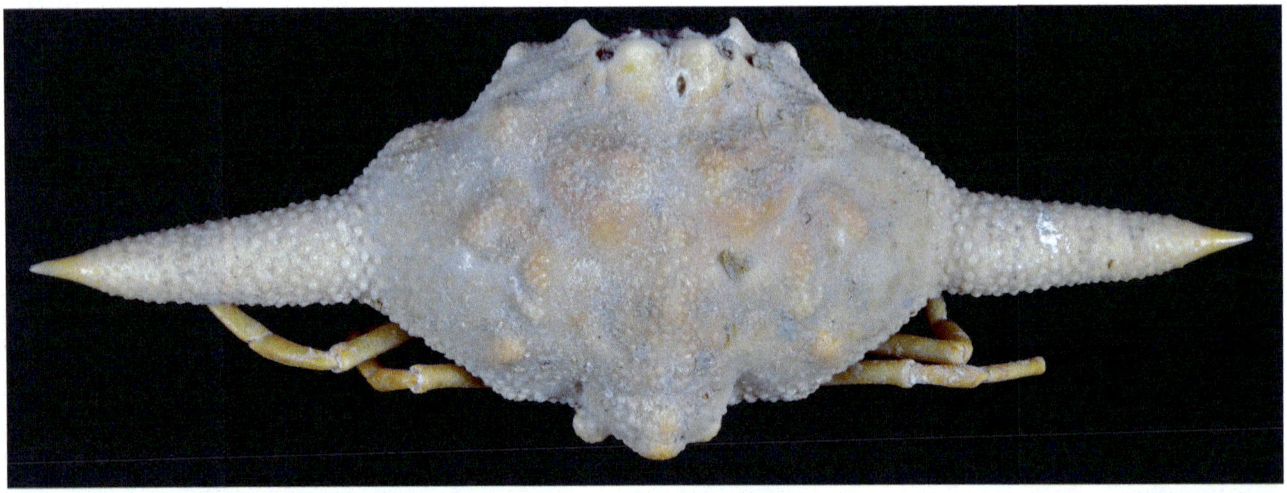

Fig. 14.18 *Ixa holthuisi* Tirmizi, 1970. Male (ZUTC 5318), CL = 21.08, CB = 58.92 mm, Kuwait, Persian Gulf, coll. J. Bishop. Photo: R. Abdollahi

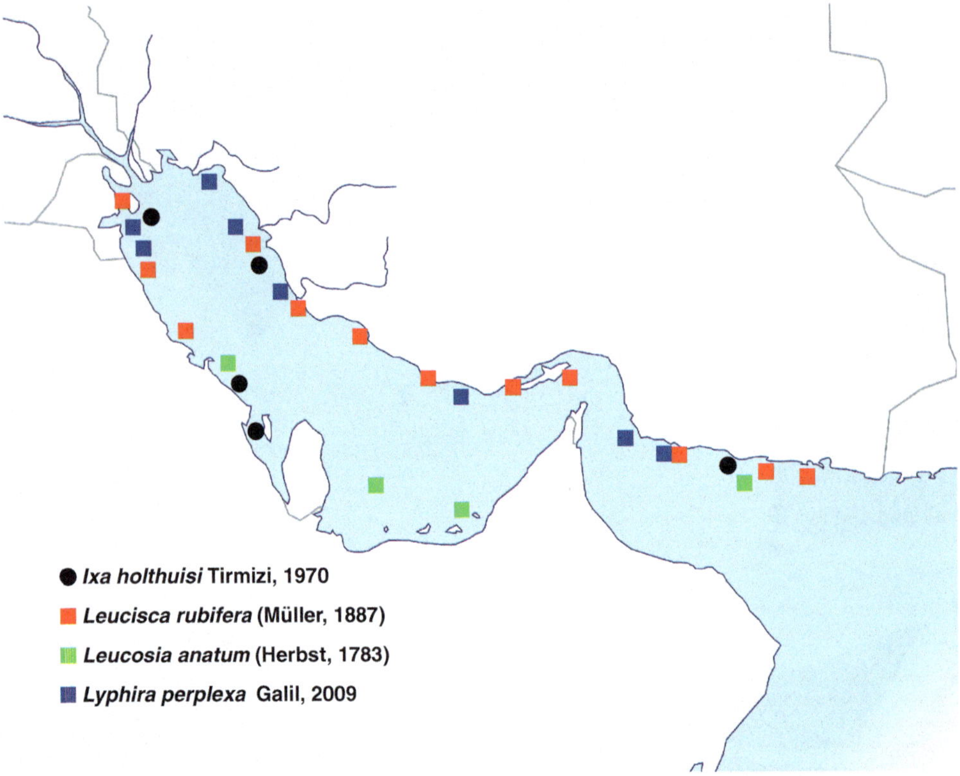

Fig. 14.19 Map showing the recorded localities of *Ixa holthuisi* Tirmizi, 1970, *Leucisca rubifera* (Müller, 1887), *Leucosia anatum* (Herbst, 1783), and *Lyphira perplexa* Galil, 2009

Distribution

Persian Gulf: Iran (present study), Kuwait (present study), Saudi Arabia [Basson et al. (1977) as *Ixa* sp. 5; Apel 2001].

Gulf of Oman: Iran [Stephensen (1946) as *Ixa edwardsi*].

World distribution: Northern Indian Ocean: Persian Gulf, Gulf of Oman and Pakistan.

Habitat: Subtidal to 18 m, substrate sandy.

Genus *Leucisca* MacLeay, 1838

Leucisca rubifera (Müller, 1887)

(Figs. 14.19, 14.20, and 14.21c)

Nursia rubifera Müller 1887: 480, pl 4, figs 4, 4a, b.
Nursia rubifera — Alcock 1896: 180 (in key), 185; Ihle 1918: 236 (in key), 312; Tirmizi and Kazmi 1986: 80, fig 23a–e; Apel 2001: 54; Naderloo and Türkay 2012: 31; Naderloo et al. 2013: 449, tab 1.
Leucisca rubifera — Ng et al. 2008: 89 (in list).
Type locality: Sri Lanka (Trincomalee).

Fig. 14.20 *Leucisca rubifera* (Müller, 1887). Male (the specimen not preserved), CL = 5.6, CB = 6.2 mm, Zeytun Park, Qeshm Island, Iran, Persian Gulf, coll. R. Naderloo (Photo: R. Abdollahi)

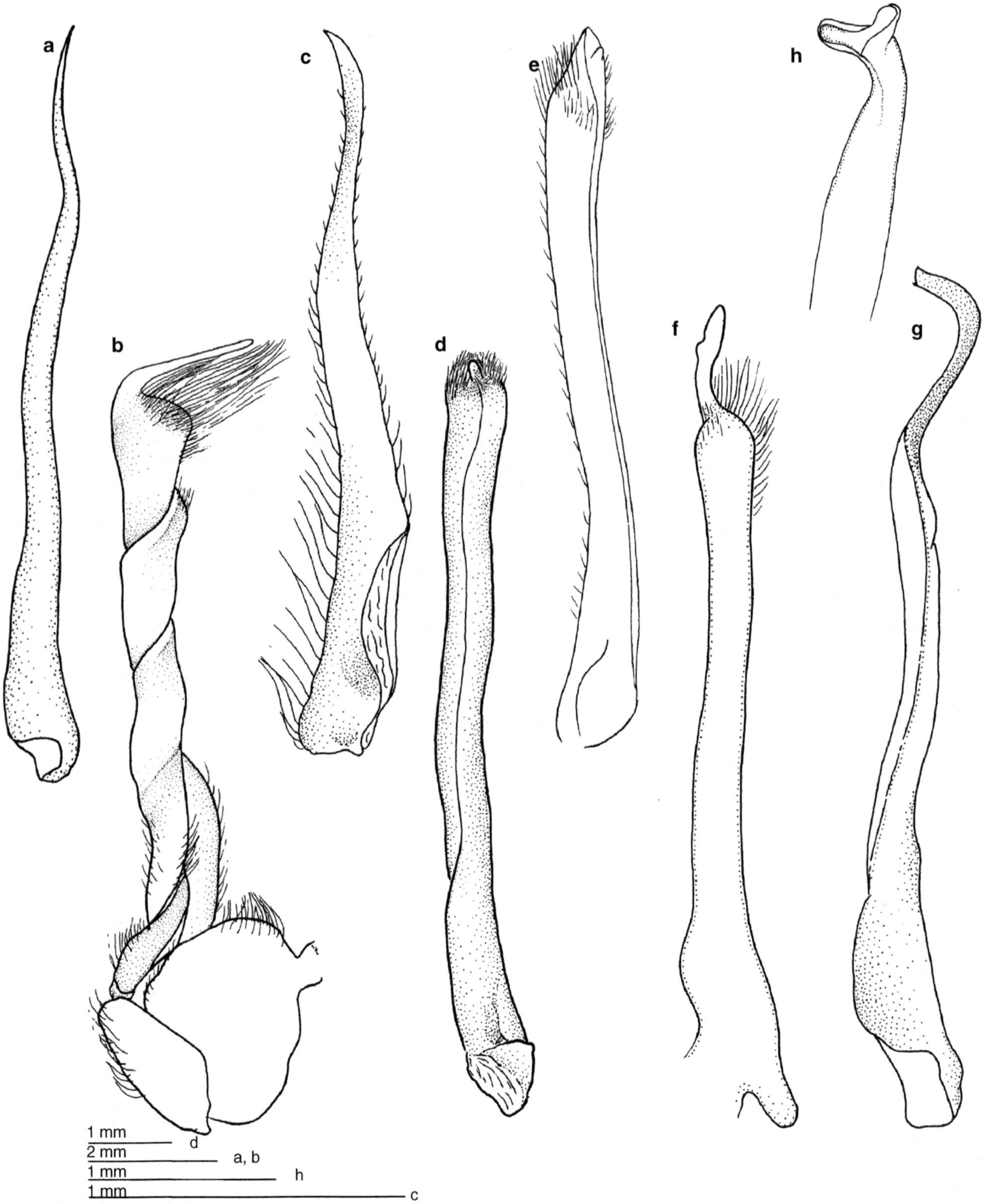

Fig. 14.21 Male G1, ventral view: (**a**) *Ixa holthuisi* Tirmizi, 1970 (ZUTC 5318); (**b**) *Leucosia anatum* (Herbst, 1783) (SMF 25698); (**c**) *Leucisca rubifera* (Müller, 1887) (ZUTC Brach1272); (**d**) *Lyphira perplexa* Galil, 2009 (ZMUC CRU850); (**e**) *Myra affinis* Bell, 1855 [locality: Thailand, modified from Stephensen (1946)]; (**f**) *Myra pernix* Galil, 2001 (ZMUC CRU949); (**g**) *Nursia plicata* (Herbst, 1803) (ZMUC CRU954); (**h**) apical part of (**g**), lateral view

Distribution

Persian Gulf: Iran [Naderloo and Türkay (2012) and Naderloo et al. (2013) as *Nursia rubifera*], Saudi Arabia [Apel (2001) as *Nursia rubifera*].

Gulf of Oman: Iran (Naderloo et al. 2015).

World distribution: Western Indian Ocean: Madagascar, Red Sea, Persian Gulf, Gulf of Oman, Pakistan, India and Sri Lanka.

Habitat: Intertidal, substrate rocky/cobble.

Genus *Leucosia* Weber, 1795

Leucosia anatum (Herbst, 1783)

(Figs. 14.19, 14.22 and 14.21b)

Cancellus Anatum Primus Rumphius 1741: pl 10, fig a.
Cancer anatum Herbst 1783: 93 (partly: based on Rumphius' table).
Cancer urania Herbst 1801: 17, pl 53, fig 3.
Leucosia longifrons De Haan 1841: 132, pl 33, fig 6.
Leucosia neocaledonica A. Milne-Edwards 1874: 40, pl 2, fig 1.
Leucosia pulcherrima Miers 1877: 236, pl 38, figs 4–6.
Leucosia longifrons — Alcock 1896: 217; Ihle 1918: 316.
Leucosia longifrons var. *pulcherrima* — Alcock 1896: 219; Nobili 1906a: 98; Ihle 1918: 316; Stephensen 1946: 94, fig 17c.
Leucosides urania — Rathbun 1910: 309, pl 1, fig 1.
Leucosia anatum — Tyndale-Biscoe and George 1962: 77 (in key), 80, figs 5, 6, pl 3; Titgen 1982: 247 (in list); Tirmizi and Kazmi 1986: 96 (in text); Sakai 1999: 19, figs 1a, c, pl 7f; Apel 2001: 51; Ng et al. 2008: 95 (in list).
Leucosia? *anatum* — Basson et al. 1977: 231, 243.

Type locality: *Cancer anatum*: Indonesia (Ambon, Maluku Islands). — *Cancer urania*: "Ostindien". East India or Indo-Malayan Archipelago. — *Leucosia longifrons*: Japan. — *Leucosia neocaledonica*: New Caledonia.

Fig. 14.22 *Leucosia anatum* (Herbst, 1783). Male (ZUTC), Gulf of Oman (Photo: R. Abdollahi)

Distribution

Persian Gulf: Iran [Stephensen (1946) as *L. longifrons* var. *pulcherrima*], Saudi Arabia [Basson et al. (1977) as *L.? anatum*; Apel 2001], UAE [Nobili (1906a) as *L. longifrons* var. *pulcherrima*]. Alcock (1986) recorded this species as *L. longifrons* and *L. longifrons* var. *pulcherrima* from unknown localities.

Gulf of Oman: Iran (Naderloo et al. 2015).

World distribution: Indo-West Pacific: Persian Gulf, Gulf of Oman, Pakistan, Sri Lanka, Andaman Islands, Mergui Archipelago, Thailand, Indonesia, China, Japan, Australia, New Caledonia and Fiji.

Habitat: Subtidal, 10–40 m, substrate sandy.

Genus *Lyphira* Galil, 2009

Lyphira perplexa Galil, 2009

(Figs. 14.19, 14.21d, and 14.23)

Lyphira perplexa Galil 2009: 303, figs 12c, 17.

Cancer globus Fabricius 1775: 401.

Cancer anatum Herbst 1783: 93, pl 2, fig 19.

Cancer globosus Fabricius 1787: 315, 1793: 441.

Leucosia globosa Fabricius 1798: 349.

Philyra globulosa — Alcock 1896: 245; Chopra 1934: 38; Stephensen 1946: 77, figs 10, 11a–l; Tirmizi and Kazmi 1986: 98, fig 28; Apel 2001: 57; Naderloo and Sari 2005a: 40, fig 9, 2007a: 342, tab 1.

Type locality: India (Malabar).

Distribution

Persian Gulf: Iran [Stephensen (1946) as *Philyra globulosa*; Naderloo and Sari (2005a) as *Philyra globulosa*, (2007a) as *Philyra globulosa*], Kuwait [Apel (2001) as *Philyra globulosa*].

Gulf of Oman: Iran [Stephensen (1946) as *Philyra globulosa*].

World distribution: Northern Indian Ocean: Persian Gulf, Gulf of Oman, Pakistan, India and Anadaman Islands.

Habitat: Subtidal, 7–30 m, substrates sandy and muddy.

Fig. 14.23 *Lyphira perplexa* Galil, 2009. Male (ZUTC Brach1019), CL = 11.45, CB = 11.64 mm, Iran, Persian Gulf, 29° 44′N, 49° 56′E, depth 13 m, 02.03.2004, coll. R. Naderloo (Photo: R. Abdollahi)

Genus *Myra* Leach, 1817

Key to species of the genus *Myra* in the region

1. Male chelipeds with merus shorter than carapace length; median denticle on posterior margin of carapace acuminate, more prominent than laterals; third thoracic sternum of males with horizontal row of granules; female abdomen with granulate margins .... *Myra affinis*
- Male chelipeds with merus longer than carapace length 2
2. Male chelipeds with merus about 1.2 times as long as carapace length; posterior margin of male carapace with median spine one-third as long as carapace, lateral denticles acuminate; G1 with apical process foliate, notched in mesial margin .... *Myra pernix*
- Male chelipeds with merus about 1.4 times as long as carapace length; posterior margin of male carapace with median spine one-fifth as long as carapace, lateral denticles triangular; G1 with apical process curved distally .... *Myra subgranulata*

Myra affinis Bell, 1855
(Figs. 14.21e, 14.24, and 14.25)

Myra affinis Bell 1855b: 296, pl 32, fig 2a–c.
Myra affinis — Alcock 1896: 201 (in key), 205; Nobili 1906a: 95; Ihle 1918: 257; Stephensen 1946: 71, fig 7a–c; Guinot 1976: 248 (in list); Titgen 1982: 248 (in list); Apel 2001: 53; Galil 2001: 413 (in key), 414–416, figs 1a, 4; Ng et al. 2008: 291 (in list).

Type locality: Philippines.
Distribution
Persian Gulf: UAE (Nobili 1906a). Alcock (1986) recorded this species from an unknown locality.
Gulf of Oman: No record.
World distribution: Indo-West Pacific: Persian Gulf, Pakistan, Sri Lanka, Andaman Islands, Mergui Archipelago, Burma, Indonesia, Japan and Australia.
Habitat: Subtidal, 10–36 m, substrate sandy.

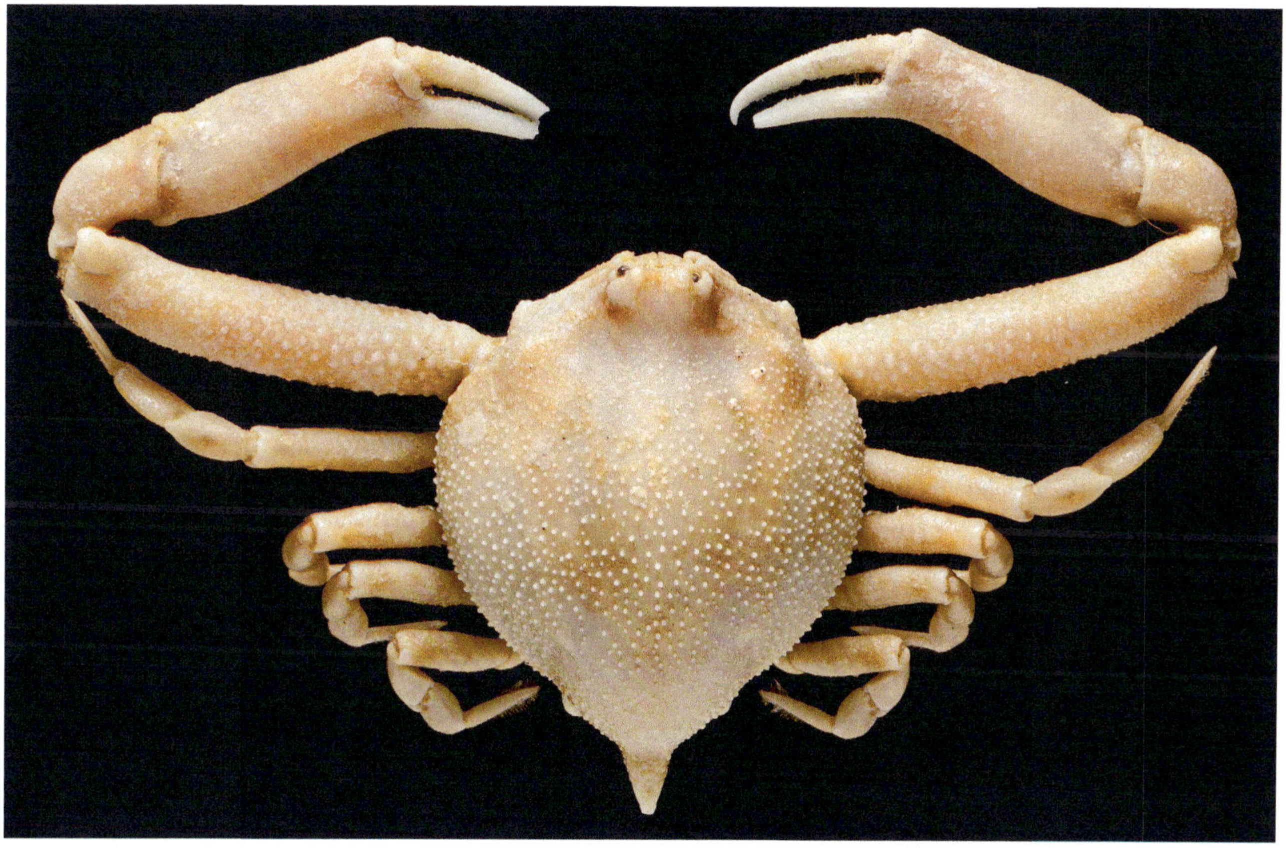

Fig. 14.24 *Myra affinis* Bell, 1855. Male (NHM reg.1907.5.22), CL = 29.55, CB = 22.70 mm, Sri Lanka, Herdman collection (Photo: H. Taylor)

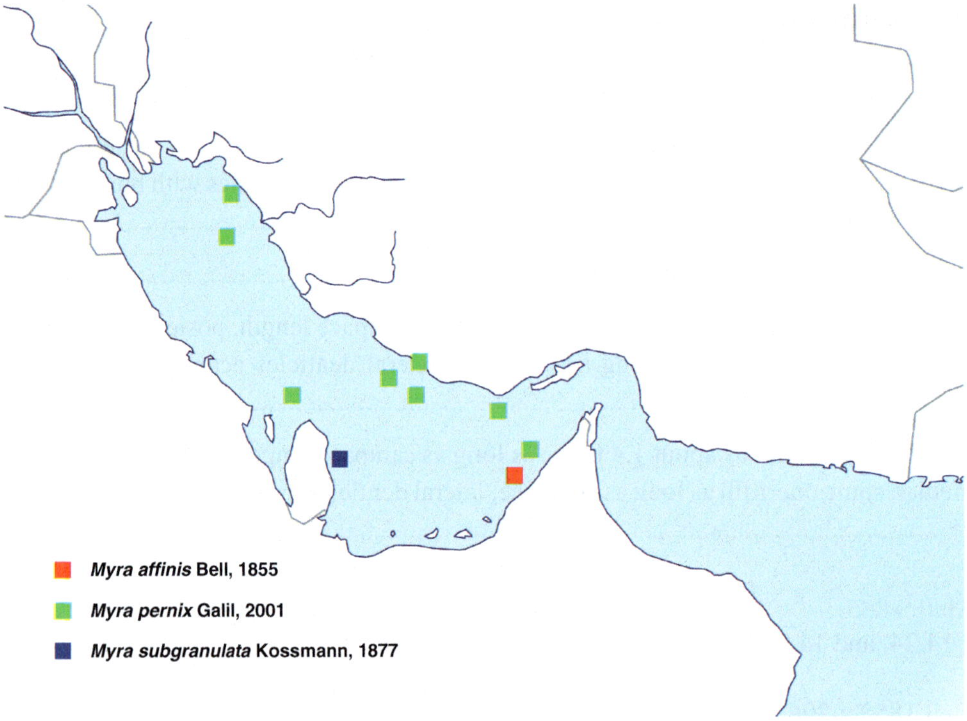

Fig. 14.25 Map showing the recorded localities of *Myra affinis* Bell, 1855, *Myra pernix* Galil, 2001, and *Myra subgranulata* Kossmann, 1877

Myra pernix Galil, 2001
(Figs. 14.21f, 14.25 and 14.26)

Myra pernix Galil 2001: 430, 431, figs 2f, 15.
Myra fugax — Alcock 1896: 202 (partly); Stephensen 1946: 72, fig 7b, c; Tirmizi and Kazmi 1986: 89 (partly), fig 26; Apel 2001: 53.
Myra affinis — Nobili 1906a: 95.
Myra pernix — Naderloo and Sari 2005a: 35, fig 4, 2007a: 342, tab 1; Ng et al. 2008: 92 (in list).
Type locality: Thailand.
Distribution

Persian Gulf: Iran [Stephensen (1946) as *Myra fugas*; Naderloo and Sari 2005a; Naderloo and Sari 2007a], Bahrain [Stephensen (1946) as *Myra fugas*].

Gulf of Oman: No record.

World distribution: Indian Ocean: Persian Gulf, Pakistan, Thailand and Singapore.

Habitat: Subtidal, 5.5–83 m, substrate sandy.

Myra subgranulata Kossmann, 1877
(Figs. 14.25 and 14.27)

Myra subgranulata Kossmann 1877: 65, pl 1, fig 7.
Myra fugax — Nobili 1906a: 164; Tirmizi and Kazmi 1986: 89, fig 26 k.
Myra subgranulata — Galil 2001: 431, fig 3a, 16; Ng et al. 2008: 92 (in list); Galil et al. 2012: 3, fig 2c.
Type locality: Red Sea.
Distribution

Persian Gulf: Qatar (Galil et al. 2012).

Gulf of Oman: No record.

World distribution: Western Indian Ocean and Mediterranean: South Africa, Mozambique, Madagascar, Mauritius, Red Sea, Gulf of Aden, Suez Canal, Eastern Mediterranean (Lessepsian migrant) and Persian Gulf.

Habitat: Subtidal, 5–130 m, substrate sandy.

Fig. 14.26 *Myra pernix* Galil, 2001. Male (ZUTC Brach1006) CL = 23.85, CB = 16.96 mm, Iran, Persian Gulf, 07.03.2004, coll. R. Naderloo (Photo: R. Abdollahi)

Genus *Nursia* Leach, 1817

Key to species of the genus *Nursia* in the region

1. Carapace with median longitudinal and epibranchial ridges, without additional ridge; posterior margin not clearly bilobed; chelipeds with distinct ridge along upper margin of palm .... *Nursia persica*

- Carapace with anterior oblique ridge additional to longitudinal and epibranchial ridges; posterior margin markedly bilobed; chelipeds without ridge along upper margin of palm 2

2. Carapace markedly broader than long, lateral margins granular; front nearly triangular, extending beyond inner orbital angle .... *Nursia blandfordi*

- Carapace nearly as long as broad; lateral margins smooth; frontal margin straight, not extending beyond inner orbital angle .... *Nursia plicata*

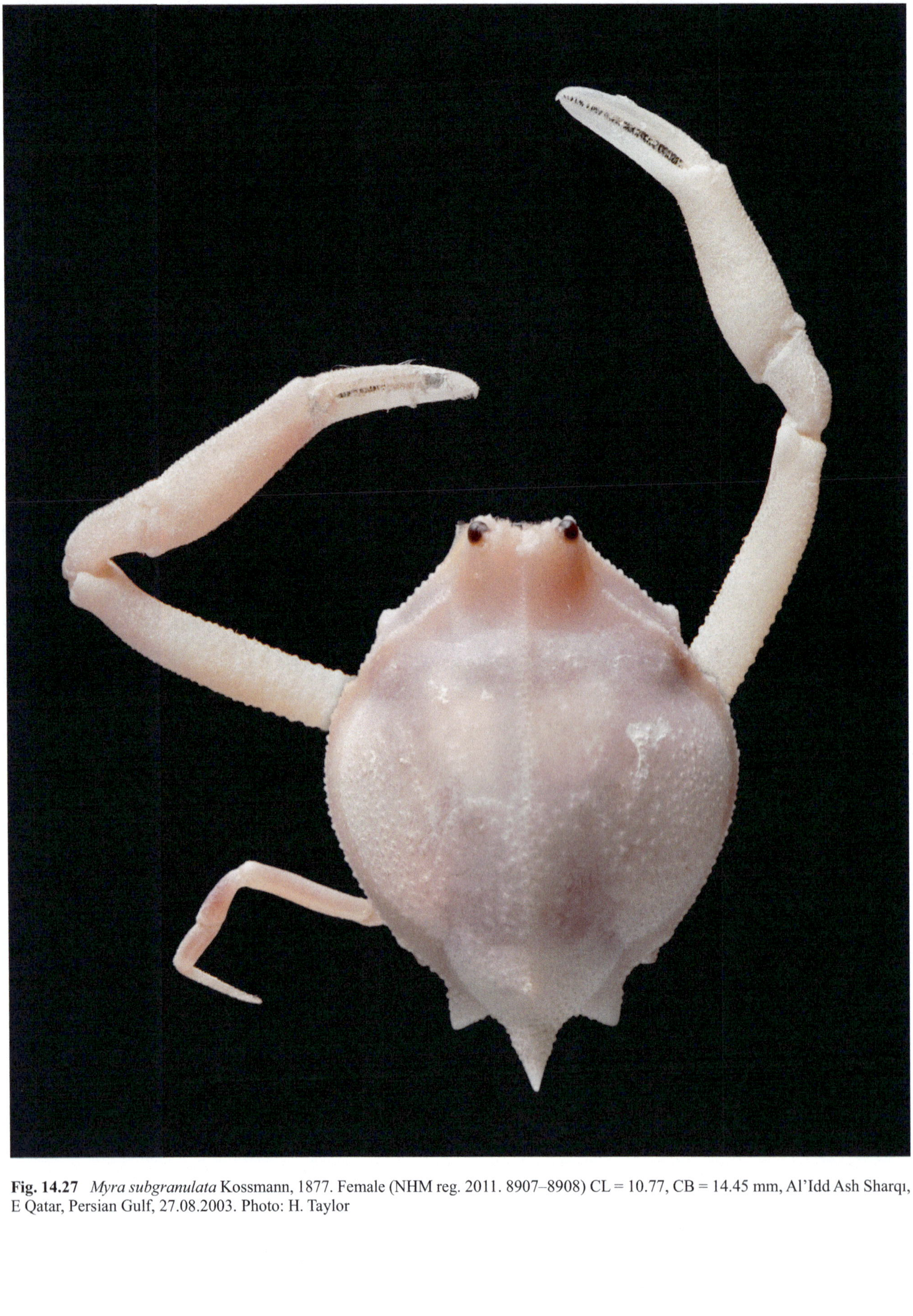

Fig. 14.27 *Myra subgranulata* Kossmann, 1877. Female (NHM reg. 2011. 8907–8908) CL = 10.77, CB = 14.45 mm, Al'Idd Ash Sharqı, E Qatar, Persian Gulf, 27.08.2003. Photo: H. Taylor

Nursia blandfordi Alcock, 1896
(Figs. 14.28 and 14.29)

Nursia blandfordi Alcock 1896: 179 (in key), 182, pl 7, fig 5.
Nursia blandfordi — Alcock and Anderson 1897: pl 29, fig 3; Stephensen 1946: 69; Titgen 1982: 248 (in list); Apel 2001: 54; Ng et al. 2008: 92 (in list).
Nursia sp. — Stephensen 1946: 70.

Typ locality: Persian Gulf (exact locality unknown).
Distribution
Persian Gulf: Iran (Stephensen 1946), Qatar (Apel 2001). Alcock (1896) described this species from an unknown locality.
Gulf of Oman: No record.
World distribution: Northern Indian Ocean: Persian Gulf and Pakistan.
Habitat: Shallow subtidal to 91 m.

Fig. 14.28 *Nursia blandfordi* Alcock, 1896. Female holotype (BMNH 96.9.8.14), CL = 7.9, CB = 8.2 mm, Persian Gulf, exact locality unknown (Photo: Kevin Webb)

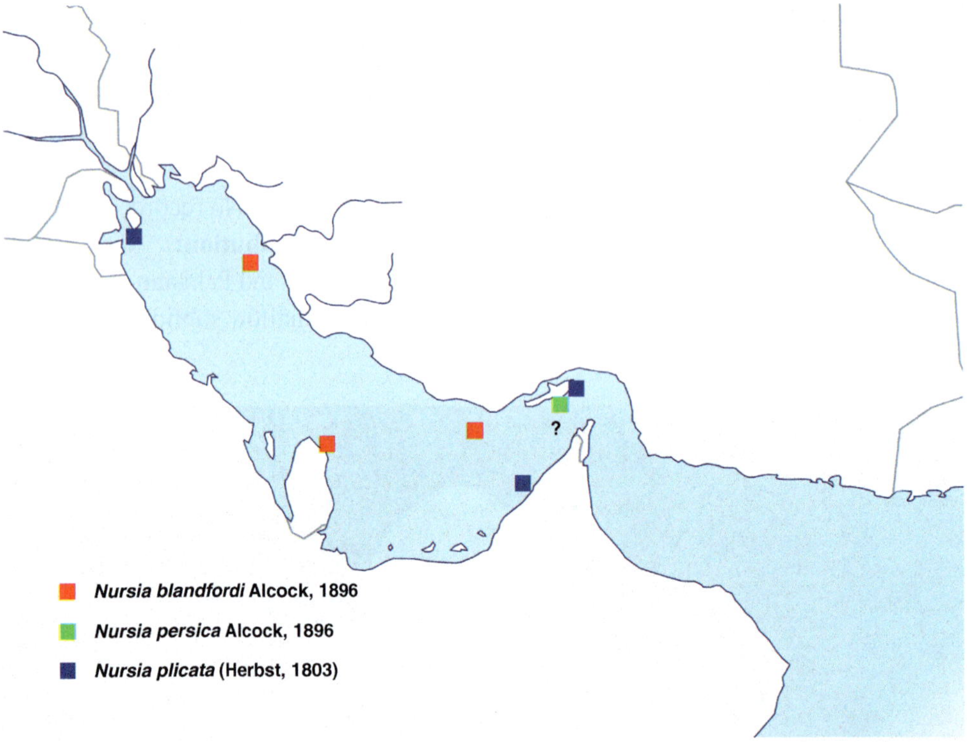

Fig. 14.29 Map showing the recorded localities of *Nursia blandfordi* Alcock, 1896, *Nursia persica* Alcock, 1896, and *Nursia plicata* (Herbst, 1803)

Nursia persica Alcock, 1896

(Figs. 14.29 and 14.30)

Nursia persica Alcock 1896: 183, 184, pl 7, fig 7.
Nursia persica — Alcock and Anderson 1897: pl 29, fig 1; Ihle 1918: 235 (in key), 311 (in list); Stephensen 1946: 69, 70, fig 6b; Titgen 1982: 248 (in list); Apel 2001: 54; Ng et al. 2008: 92 (in list).

Type locality: Persian Gulf (exact locality unknown).

Distribution

Persian Gulf: Alcock (1896) described this species from an unknown locality.

Gulf of Oman: No record.

World distribution: Northern Indian Ocean: Persian Gulf and Gulf of Oman.

Habitat: Subtidal to 11 m, substrates sand, gravel and shells.

Nursia plicata (Herbst, 1803)

(Figs. 14.21g, h, 14.29, and 14.31)

Cancer plicatus Herbst 1803: 2, 3, pl 59, fig 2.
Nursia plicata — Alcock 1896: 179 (in key), 180, 181; Stephensen 1946: 70, fig 6c; Titgen 1982: 248 (in list); Sakai 1999: 16; Apel 2001: 54; Naderloo and Sari 2005a: 35, fig 5, 2007a: 342, tab 1; Ng et al. 2008: 92 (in list).

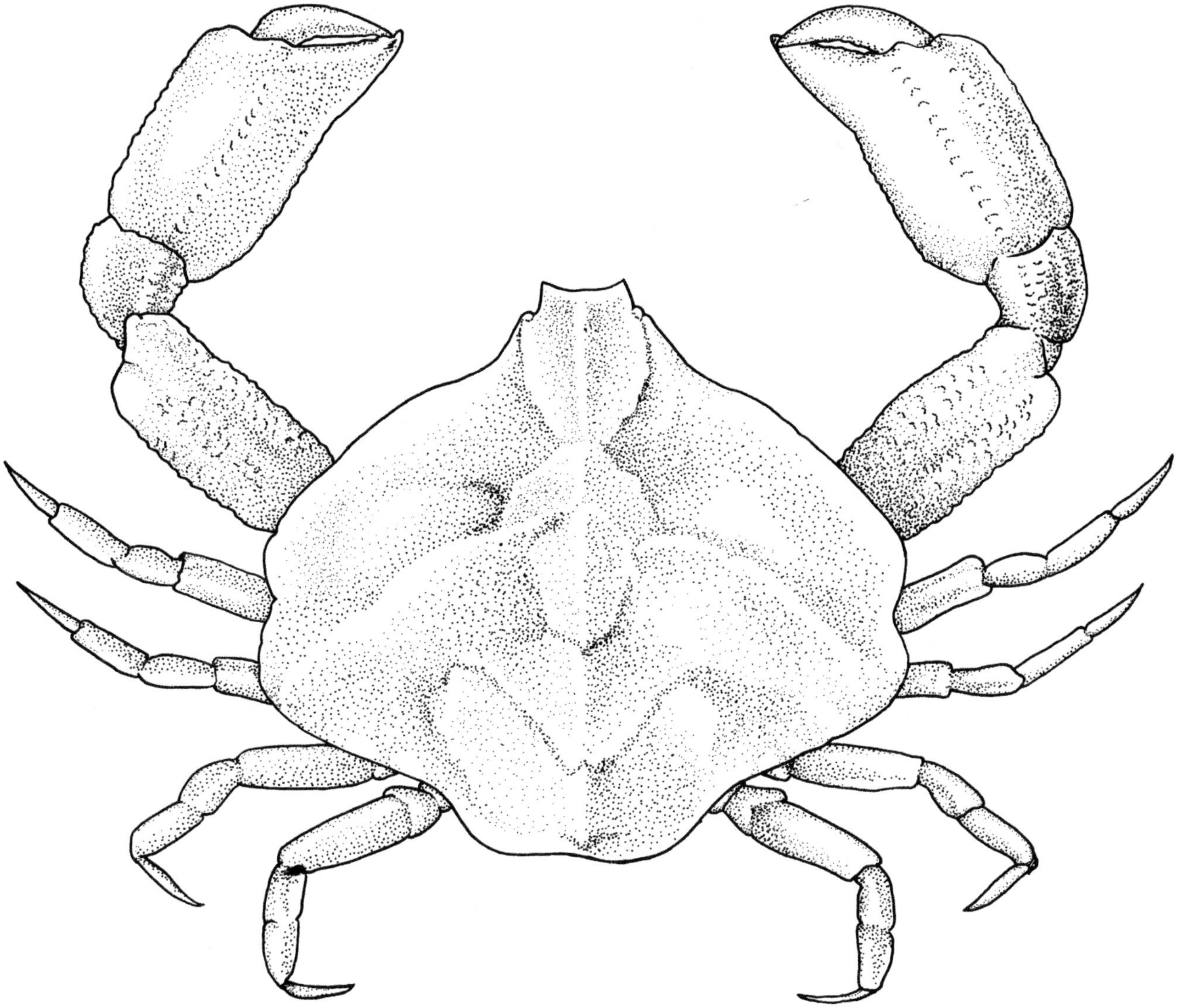

Fig. 14.30 *Nursia persica* Alcock, 1896. Modified after original drawing of holotype from the Persian Gulf. The type material was not found neither in the Calcutta Museum not in NHM

Nursia plicata plicata — Campbell and Stephenson 1970: 249 (in key).

Type locality: "Ostindien". East India or Indo-Malayan Archipelago.

Distribution

Persian Gulf: Iran (Naderloo and Sari 2005a; Naderloo and Sari 2007a), Kuwait (Apel 2001). Alcock (1896) recorded this species from an unknown locality.

Gulf of Oman: No record.

World distribution: Indo-West Pacific: Persian Gulf, India, Palk Straits, Mumbai, China, Hong Kong and Japan.

Habitat: Subtidal, 3–19 m, substrate sandy with clay and stones.

Fig. 14.31 *Nursia plicata* (Herbst, 1803). Male (ZUTC Brach1013), CL = 8.00, CB = 10.50 mm, Bandar-Abbas, Iran, Persian Gulf, depth 16 m, 07.2004, coll. R. Naderloo (Photo: R. Abdollahi)

Genus *Nursilia* Bell, 1855

Nursilia dentata Bell, 1855

(Figs. 14.32 and 14.33)

Nursilia dentata Bell 1855b: 309, pl 34, fig 6.
Nursilia dentata — Alcock 1896: 260; Ihle 1918: 244, 245; Chen 1982: 267, 268, fig 1 (1, 2); Apel 2001: 55; Ng et al. 2008: 92 (in list).
Type locality: Indian Ocean.

Distribution

Persian Gulf: Qatar (Apel 2001).

Gulf of Oman: No record.

World distribution: Indo-West Pacific: Mozambique, Seychelles, Red Sea, Persian Gulf, Maledives, India, Sri Lanka, Thailand, China, Japan, Indonesia, Australia and Fiji.

Habitat: Intertidal to 35 m, substrate dead coral with mud and sea grass.

Fig. 14.32 *Nursilia dentata* Bell, 1855. Male (ZUTC 5930), CL = 7.1, CB = 8.06 mm, SE Al-Hallaiyah Island, Kuria Muria Islands, Oman, Arabian Sea, depth 35 m, 25.03.1995 (Photo: Sohiel Shahbazi)

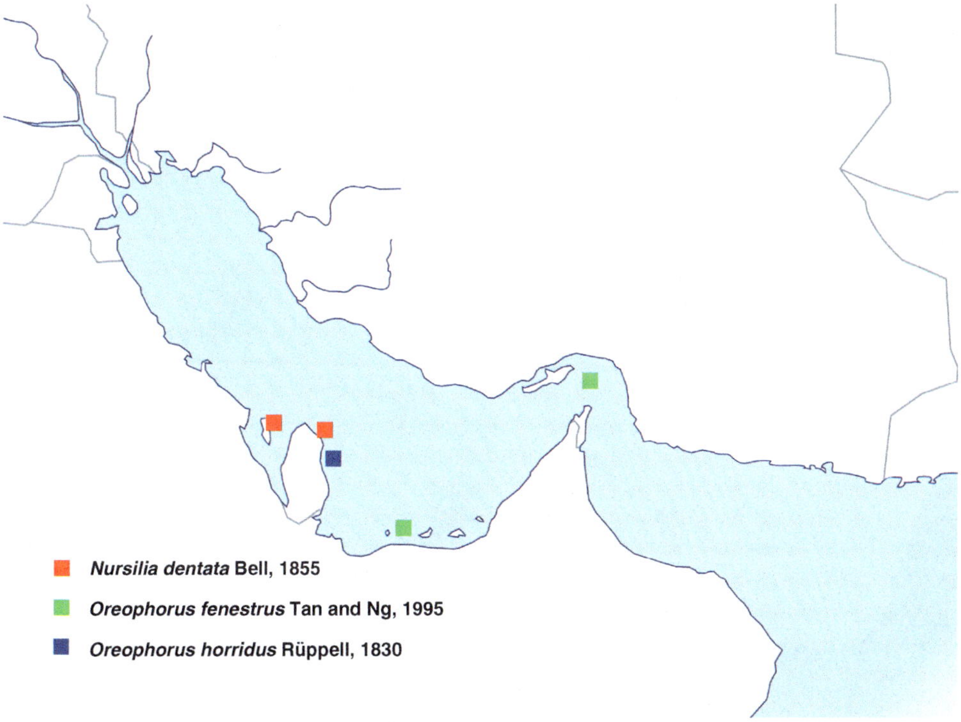

Fig. 14.33 Map showing the recorded localities of *Nursilia dentata* Bell, 1855, *Oreophorus fenestrus* Tan and Ng, 1995, and *Oreophorus horridus* Rüppell, 1830

Genus *Oreophorus* Rüppell, 1830

Oreophorus fenestrus Tan and Ng, 1995

(Figs. 14.33, 14.34, and 14.35a)

Oreophorus fenestrus Tan and Ng 1995: 106 (in key), 111–114, fig 4, pl 2d–f.

?*Oreophorus reticulatus* — Alcock 1896: 174, 175 (material from the Persian Gulf); MacGilchrist 1905: 254; Ihle 1918: 309 (material from the Persian Gulf); Stephensen 1946: 68; Guinot 1967a: 247 (in list); Titgen 1982: 248 (in list).

Oreophorus fenestrus — Apel 2001: 55; Ng et al. 2008: 92 (in list).

Type locality: Gulf of Oman (Muscat, Oman). In original work, it was erroneously cited Red Sea.

Distribution

Persian Gulf: UAE (Apel 2001). Alcock (1896) and MacGilchrist (1905) recorded this species as *Oreophorus reticulatus* from a station ("Investigtor" St. 291) in the Strait of Hormuz.

Gulf of Oman: Oman (Tan and Ng 1995).

World distribution: Northern Indian Ocean: Persian Gulf and Gulf of Oman.

Habitat: Shallow subtidal, substrate sea grass with mud.

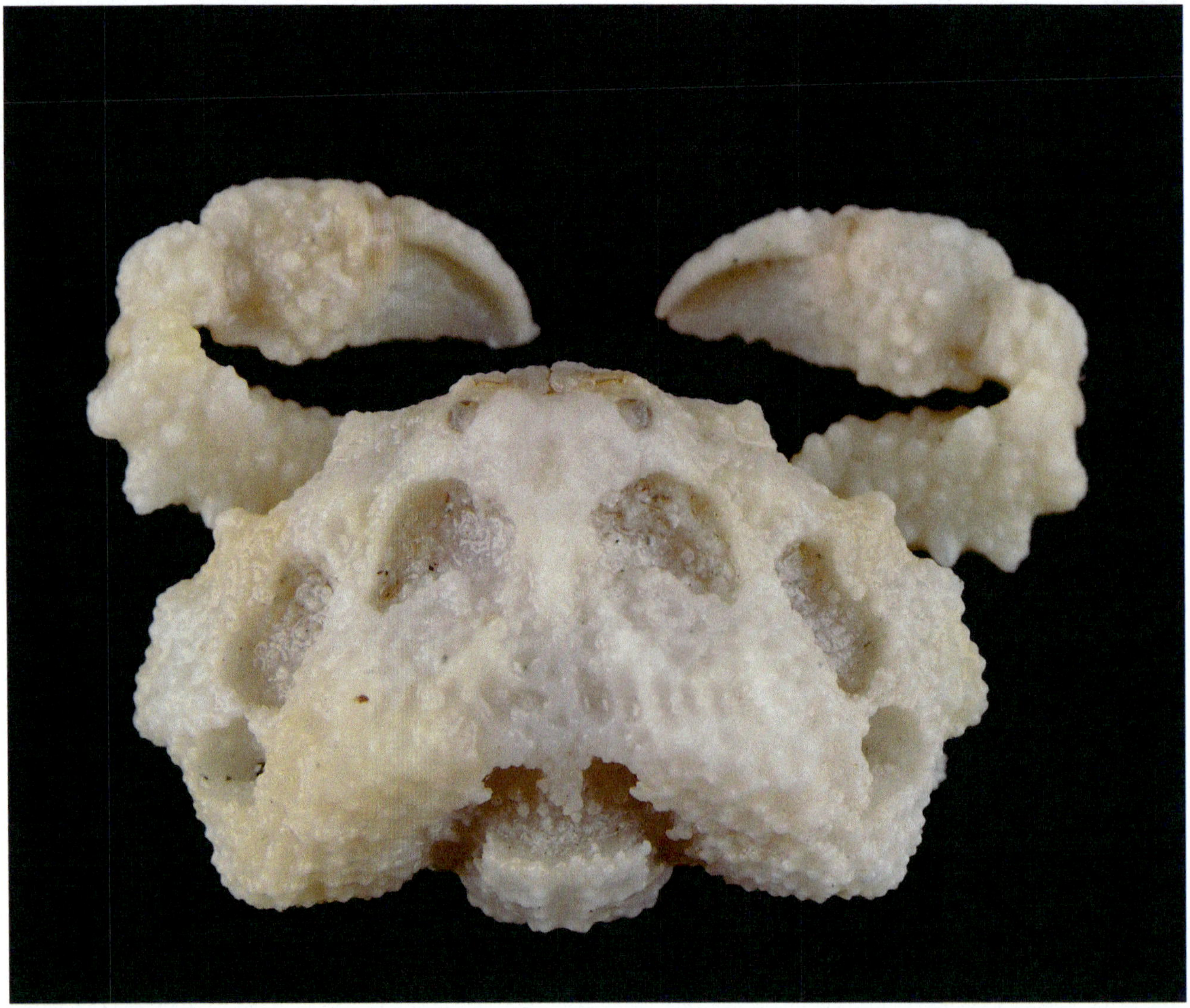

Fig. 14.34 *Oreophorus fenestrus* Tan and Ng, 1995. Male (SMF), CL = 7.8, CB = 11.5 mm, UAE, Persian Gulf, 13.06.1995, coll. M. Apel (Photo: S. Träkner)

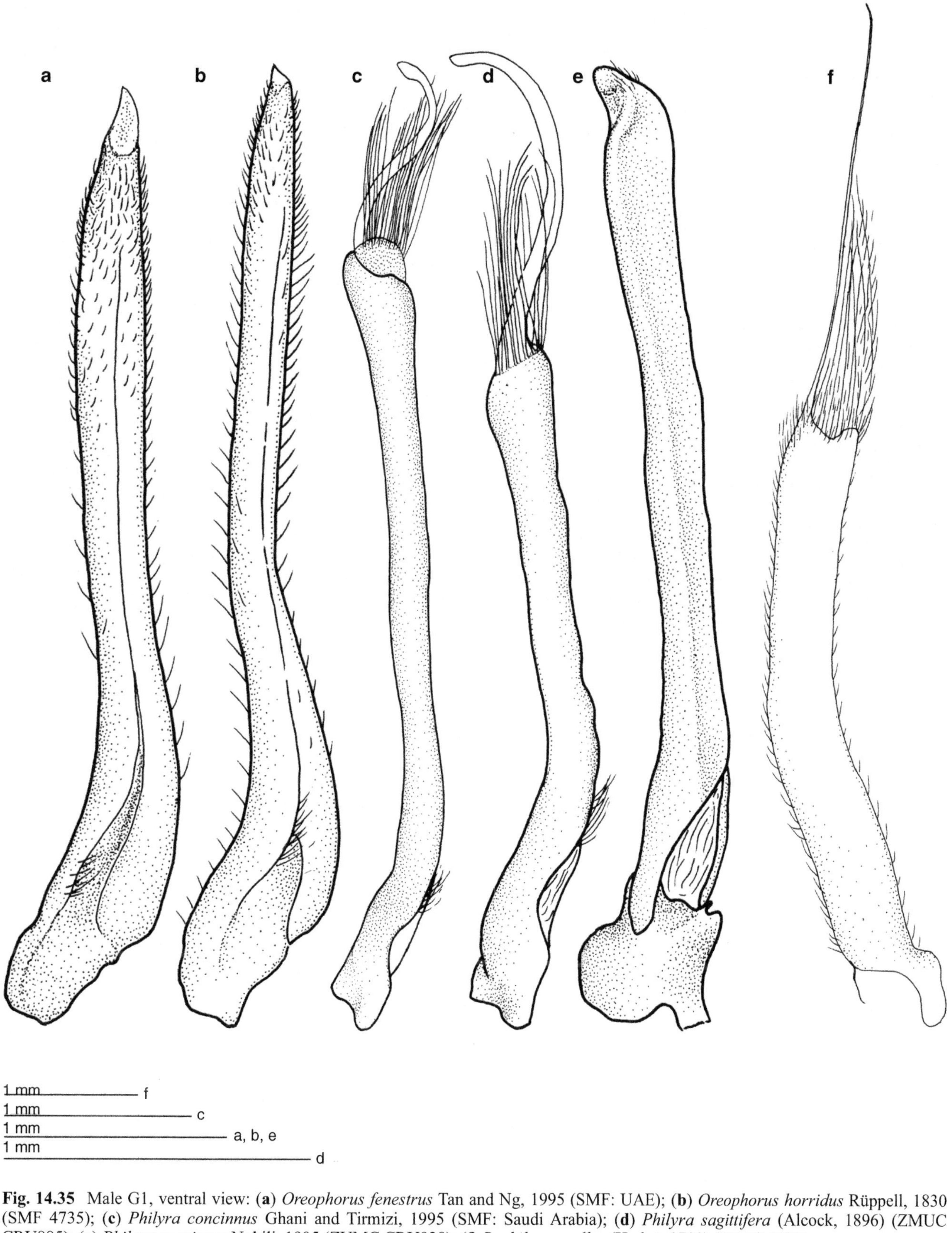

Fig. 14.35 Male G1, ventral view: (**a**) *Oreophorus fenestrus* Tan and Ng, 1995 (SMF: UAE); (**b**) *Oreophorus horridus* Rüppell, 1830 (SMF 4735); (**c**) *Philyra concinnus* Ghani and Tirmizi, 1995 (SMF: Saudi Arabia); (**d**) *Philyra sagittifera* (Alcock, 1896) (ZMUC CRU885); (**e**) *Philyra granigera* Nobili, 1905 (ZUMC CRU928); (**f**) *Ryphila cancellus* (Herbst, 1783) (ZUTC 5529)

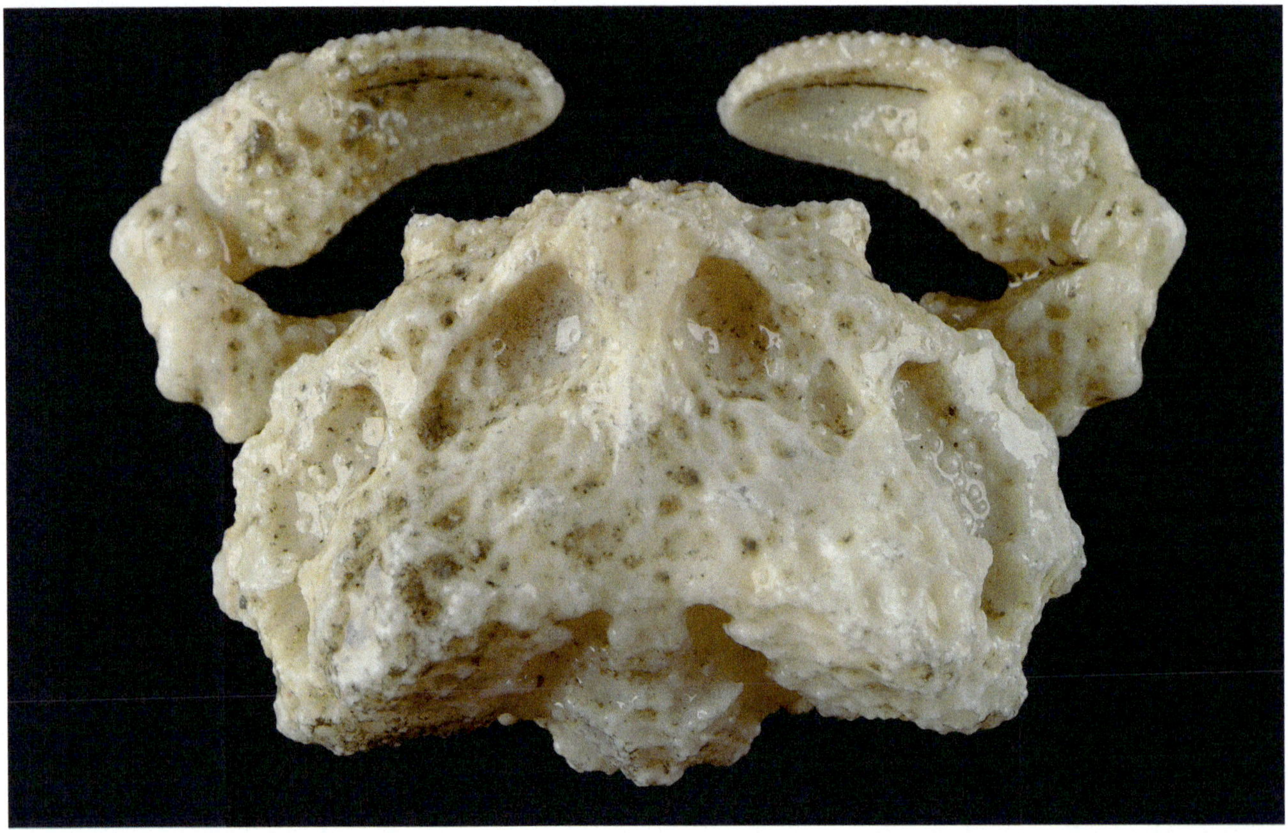

Fig. 14.36 *Oreophorus horridus* Rüppell, 1830. Female (SMF 4535), CL = 16.5, CB = 22.5 mm, Egypt, Red Sea (Photo: S. Träkner)

Oreophorus horridus **Rüppell, 1830**
(Figs. 14.33, 14.35b, and 14.36)

Oreophorus horridus Rüppell 1830: 18.
Oreophorus horridus — Nobili 1906b: 150; Tan and Ng 1995: 106, fig 2, pl 1; Ng et al. 2008: 92 (in list); Galil et al. 2012: 3, fig 2d.
Type locality: Red Sea.

Distribution

Persian Gulf: Qatar (Galil et al. 2012).

Gulf of Oman: No record.

World distribution: Western Indian Ocean: Mayotte Island, Kenya, Red Sea, Gulf of Aden and Persian Gulf.

Habitat: Shallow subtidal to 32 m.

Genus *Philyra* Leach, 1817

Key to species of the genus *Philyra* in the region

1. Carapace nearly hexagonal, junction of anterolateral and posterolateral margins distinct with denticle, posterior margin laterally with denticles or spine; front broad, entire 2

\- Carapace nearly round, lateral margins arched, no distinct junction between anterolateral and posterolateral margins, posterior margin laterally arched; front faintly tridentate 3

2. Carapace with small distinct granules on epibranchial, cardiac and intestinal regions; G1 apical process short, about 0.2 times as long as whole length, more sinuous-shape, with spade-shape ending .... *Philyra concinnus*

- Carapace elevations with cluster of small granules; G1 apical process long, about 0.3 times as long as whole length, slightly sinuous-shape, with tapering ending .... *Philyra sagittifera*

3. Carapace distinctly longer than broad, dorsal surface evenly convex, with small granules; chelipeds medium-size, manus as long as fingers; G1 with particular laterally expanded subterminal cirrate, apical process curved mesially .... *Philyra globus*

- Carapace nearly as long as broad, dorsal surface with epibranchial and cardiac elevations, with larger granules; chelipeds relatively long, manus markedly longer than fingers; G1 apical process short, blunt, curved laterally .... *Philyra granigera*

Philyra concinnus Ghani and Tirmizi, 1995

(Figs. 14.35c, 14.37, and 14.38)

Philyra concinnus Ghani and Tirmizi 1995a: 144–147, fig 1a–h (material from Karachi).
Philyra sexangula — Stephensen 1946: 89.
Ebalia sagittifera — Cooper 1997: 173, 174 (?partly), fig 7.
Philyra concinnus — Apel 2001: 56; Ng et al. 2008: 92 (in list); Galil 2009: 281, tab 1; Naderloo and Türkay 2012: 32.

Type locality: Pakistan (Karachi).

Distribution

Persian Gulf: Iran [Stephensen (1946) as *Philyra sexangula*; Naderloo and Türkay 2012], Saudi Arabia (Apel 2001), UAE [Cooper (1997) as *Ebalia*? *sagittifera*; Apel 2001].

Gulf of Oman: Iran (Naderloo et al. 2015).

World distribution: Northern Indian Ocean: Persian Gulf, Gulf of Oman and Pakistan.

Habitat: Intertidal, substrates sandy and muddy-sand.

Fig. 14.37 *Philyra concinnus* Ghani and Tirmizi, 1995. Male (SMF), CL = 8.1, CB = 7.4 mm, Abu Ali, Saudi Arabia, Persian Gulf, 29.06.1992, coll. M. Apel (Photo: S. Träkner)

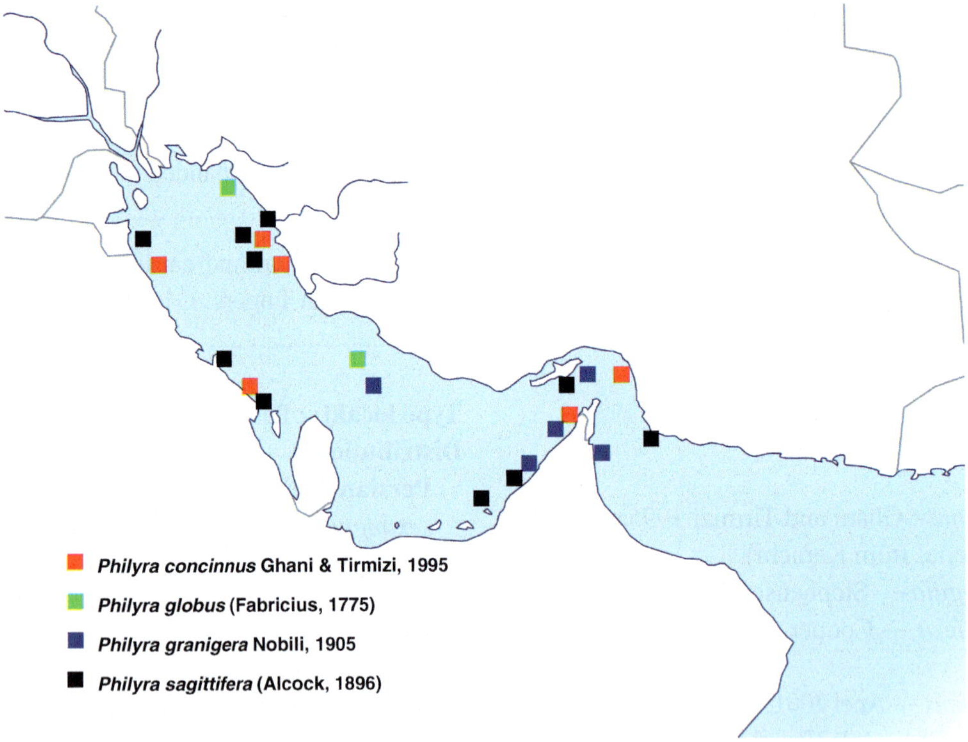

Fig. 14.38 Map showing the recorded localities of *Philyra concinnus* Ghani and Tirmizi, 1995, *Philyra globus* (Fabricius, 1775), *Philyra granigera* Nobili, 1905, and *Philyra sagittifera* (Alcock, 1896)

Remarks *Philyra concinnus* was described by Ghani and Tirmizi (1995a) from Karachi which is very closely related to *Philyra sagittifera* (Alcock, 1986) which was described from the same locality. There are some small differences on the granulations of the carapace between the two species, but a range of variety of the granulation is seen, in particular the form of the lateral teeth on the posterior margin of the carapace is highly variable. Long spine-shaped form and blunt rounded form of teeth on the posterior margin of the carapace are seen in both species. In *P. concinnus*, there are small distinct granules on the epibranchial, cardiac and intestinal regions, whereas in *P. sagittifera* elevation of the regions are with cluster of small granules. More reliable characteristic which is distinguishable between the two species is G1 of males. The apical long process in *P. sagittifera* is longer than that of *P. concinnus* (0.3 times as long as whole length of G1 in *P. sagittifera*, while 0.2 in *P. concinnus*). Further, the apical process in *P. concinnus* is more sinuous-shape with spade-shape ending, whereas in *P. sagittifera* the apical process is slightly sinuous-shape with tapering ending. Furthermore, the generic position of the species within the family Leucosiidae is not satisfactory, as they are morphologically distinctly different from their congers in the genus *Philyra*. In our opinion, a distinct genus must be defined for these two species by a detailed taxonomic revision (Naderloo and Türkay 2012).

Philyra globus (Fabricius, 1775)
(Figs.14.38 and 14.39)

Cancer globus Fabricius 1775: 401.
Cancer globosus — Fabricius 1787: 315.
Leucosia globosa — Fabricius 1798: 349.
Philyra globosa — Alcock 1896: 243; Ihle 1918: 274, 275 (in text), 315 (in list); Stephensen 1946: 83, fig 12a–d; Guinot 1967a: 250 (in list); Titgen 1982: 248 (in list); Tirmizi and Kazmi 1986: 103–106, fig 30a–g.
Philyra globus — Sakai 1999: 19 (in text as *Philyra porcellana*); Apel 2001: 57; Naderloo and Sari 2005a: 41, fig 10; Naderloo and Sari 2007a: 342 (in list); Ng et al. 2008: 92 (in list); Galil 2009: 283, figs 1, 2a.

Type locality: India (Malabar Coast).

Distribution

Persian Gulf: Iran [Stephensen (1946) as *Philyra globosa*; Naderloo and Sari 2005a; Naderloo and Sari 2007a].

Gulf of Oman: No record.

World distribution: Indian Ocean: Persian Gulf, Pakistan, India, Sri Lanka and Mergui Archipelago.

Habitat: Subtidal, 5–85 m, substrate sandy.

Fig. 14.39 *Philyra globus* (Fabricius, 1775). Male (ZUMC CRU887), CL = 14.1, CB = 12.9 mm, Bostano, Parsian, Iran, Persian Gulf, 27° 05′N, 53° 02′E, depth 5–6 m, 06.04.1937, coll. G. Thorson (Photo: S. Träkner)

Philyra granigera Nobili, 1905
(Figs. 14.35e, 14.38, and 14.40)

Philyra granigera Nobili 1905a: 162.
Philyra granigera — Nobili 1906a: 102–104, pl 6, fig 30; Ihle 1918: 315; Stephensen 1946: (?partly) 85–87, figs 13, 14; Guinot 1967a: 249 (in list); Titgen 1982: 248 (in list); Apel 2001: 57, 58; Ng et al. 2008: 92 (in list); Galil 2009: 281 (in list).
Type locality: Persian Gulf (Between Abu Dhabi and Dubai, 25° 10′N, 55° 10′E–24° 55′N, 54° 40′E).
Distribution
Persian Gulf: Iran (Stephensen 1946), UAE (Nobili 1905a, 1906a).
Gulf of Oman: UAE (Apel 2001).
World distribution: Northwestern Indian Ocean: Gulf of Aden, Red Sea, Gulf of Oman and Persian Gulf.
Habitat: Shallow subtidal to 6 m, substrate sandy.

Philyra sagittifera (Alcock, 1896)
(Figs. 14.35d, 14.38, and 14.41)

Ebalia sagittifera Alcock 1896: 186 (in key), 188, 189.
Ebalia sagittifera — Alcock and Anderson 1897: pl 29, fig 9; Apel 2001: 50.
Ebalia? *sagittifera* — Titgen 1982: 113, 247 (in list); Cooper 1997: 173, 174, fig 7.
Philyra concinnus — Ghani and Tirmizi 1995a: 143–147 (partly: material from the Persian Gulf), fig 2a–e.
Philyra rectangularis — Stephensen 1946: 87, 88, fig 15a, b; Guinot 1967a: 249 (in list); Basson et al. 1977: 250.
Philyra sagittifera — Ng et al. 2008: 93; Galil 2009: 281, tab 1; Naderloo and Türkay 2012: 32.
Type locality: Pakistan (Karachi).
Distribution
Persian Gulf: Iran [Stephensen (1946) as *P. rectangularis*; Naderloo and Türkay 2012], Saudi Arabia

Fig. 14.40 *Philyra granigera* Nobili, 1905. Male (ZUMC CRU928), CL = 9.4, CB = 9.4 mm, Bostano, Parsian, Iran, Persian Gulf, 27° 05′N, 53° 02′E, depth 5–6 m, 06.04.1937, coll. G. Thorson (Photo: S. Träkner)

Fig. 14.41 *Philyra sagittifera* (Alcock, 1896). Male (ZUMC CRU885), CL = 4.2, CB = 3.9 mm, Bushehr, Iran, Persian Gulf, depth 6 m, 03.03.1937, coll. G. Thorson (Photo: S. Träkner)

[Basson et al. (1977) as *P. rectangularis*; Apel (2001) as *E. sagittifera*], UAE [Titgen (1982), Cooper (1997) and Apel (2001) all as *E. sagittifera*].

Gulf of Oman: Iran (Naderloo et al. 2015).

World distribution: Northern Indian Ocean: Persian Gulf, Gulf of Oman and Pakistan.

Habitat: Intertidal, 0–6 m, substrates sandy-mud and muddy-sand.

Genus *Pseudophilyra* Miers, 1879

Pseudophilyra blandfordi Alcock, 1896

(Figs. 14.42, 14.43, and 14.44a, b)

Pseudophilyra blandfordi Alcock 1896: 249 (in key), 252, 253, pl 6, fig 7.

Pseudophilyra blandfordi — Ihle 1918: 314 (in list); Stephensen 1946: 74, 75, fig 8a–e; Titgen 1982: 248 (in list); Apel 2001: 59; Ng et al. 2008: 93 (in list).

Type locality: Pakistan.

Fig. 14.42 *Pseudophilyra blandfordi* Alcock, 1896. Male (ZMUC CRU890), CL = 6.00, CB = 6.00 mm, SW Ras-Meydani, Iran, Gulf of Oman, depth 18 m, 01.04.1938, coll. B. Loppenthin (Photo: S. Träkner)

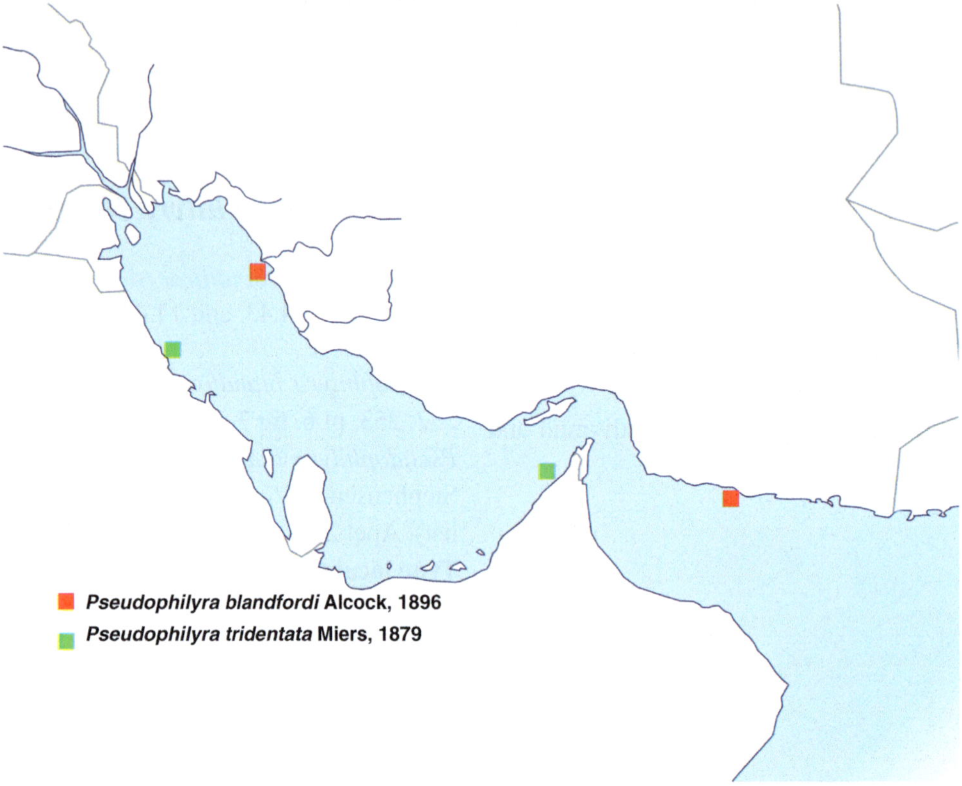

Fig. 14.43 Map showing the recorded localities of *Pseudophilyra blandfordi* Alcock, 1896, and *Pseudophilyra* cf. *tridentata* Miers, 1879

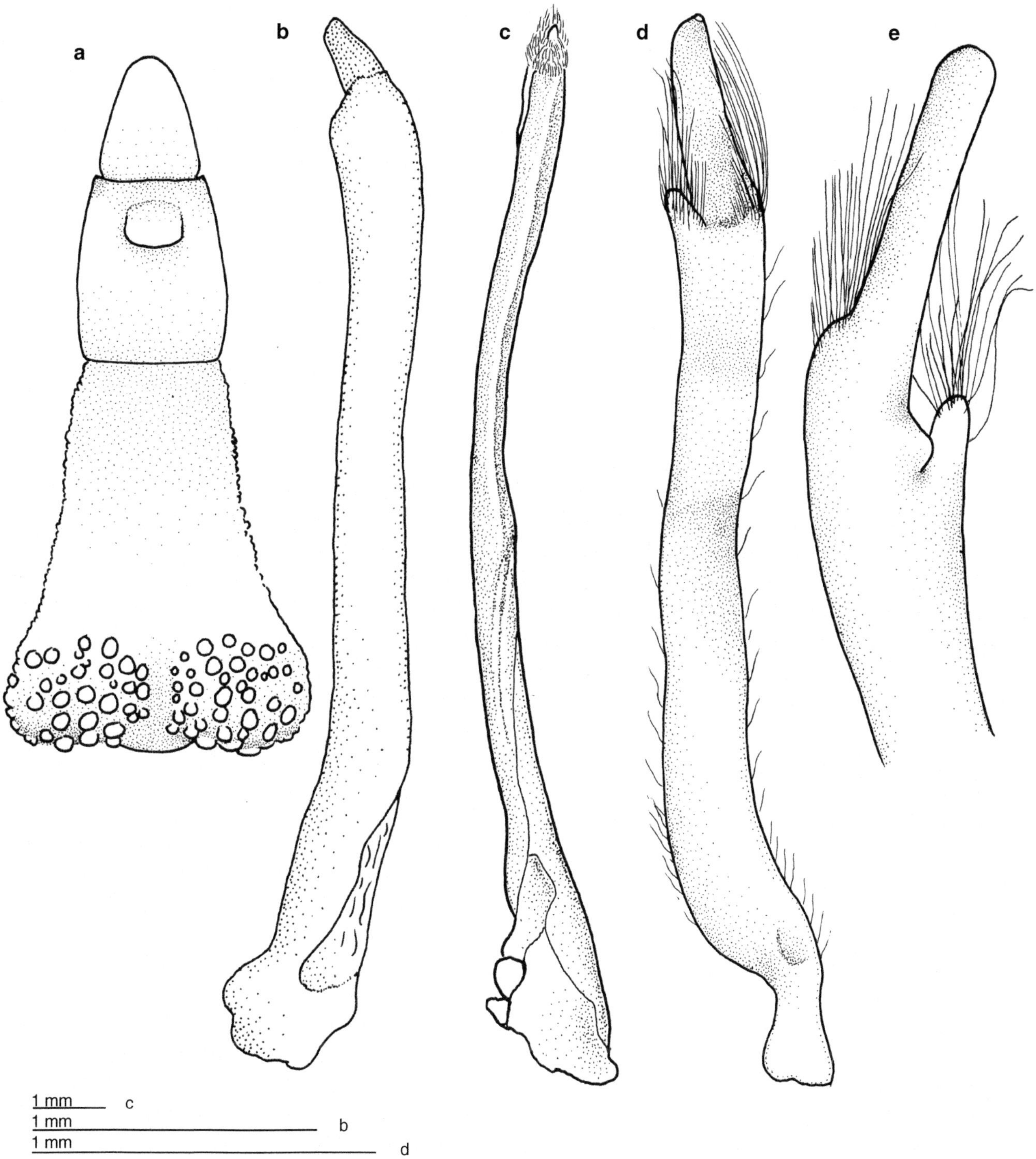

Fig. 14.44 (**a**) male abdomen of *Pseudophilyra blandfordi* Alcock, 1896 (ZMUC CRU890); (**b–e**) male G1, ventral view: (**b**) *Pseudophilyra blandfordi* Alcock, 1896 (ZMUC CRU890); (**c**) *Seulocia anahita* Galil, 2005 (SMF: Bahrain); (**d**) *Urnalana hilaris* (Nobili, 1905) (SMF: UAE); (**e**) apical part of (**d**), dorsal view

Distribution

Persian Gulf: Iran (Stephensen 1946).

Gulf of Oman: Iran (Stephensen 1946).

World distribution: Northern Indian Ocean: Persian Gulf, Gulf of Oman and Pakistan.

Habitat: Subtidal, 18–56 m, substrate sandy.

Pseudophilyra cf. *tridentata* Miers, 1879

(Figs. 14.43 and 14.45)

Pseudophilyra tridentata Miers 1879a: 20, 41, pl 2, figs 4, 4a.

Pseudophilyra tridentata — Alcock 1896: 250; Nobili 1906a: 104, 105; Ihle 1918: 314 (in list); Stephensen 1946: 77, fig 9d; Guinot 1967a: 250 (in list); Titgen 1982: 248 (in list); Apel 2001: 59; Ng et al. 2008: 93 (in list).

Type locality: Korean Channel (33° 04′N, 129° 18′E).

Distribution

Persian Gulf: Saudi Arabia (Apel 2001), UAE (Nobili 1906a). Alcock (1986) recorded this species from an unknown locality.

Gulf of Oman: No record.

World distribution: Indo-West Pacific: Persian Gulf, Sri Lanka, Thailand and Japan.

Habitat: Intertidal to shallow subtidal, substrate sandy.

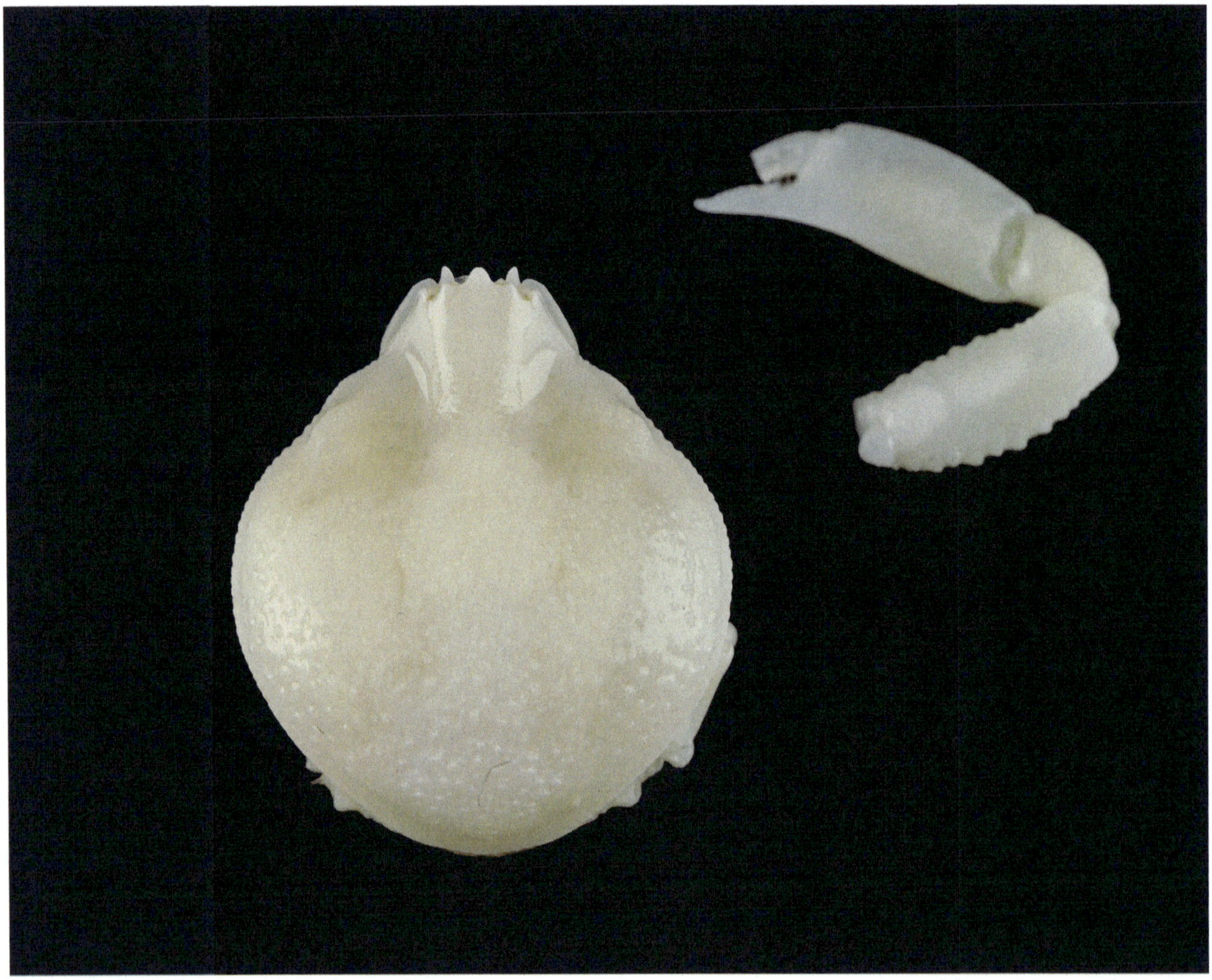

Fig. 14.45 *Pseudophilyra* cf. *tridentata* Miers, 1879. Female (SMF), CL = 12.3, CB = 11.2 mm, Ras Al-Bukhara, Saudi Arabia, Persian Gulf, 11.12.1991, coll. M. Apel (Photo: S. Träkner)

Genus *Ryphila* Galil, 2009

Ryphila cancellus (Herbst, 1783)

(Figs. 14.35f, 14.46, and 14.47)

Cancer cancellus Herbst 1783: 94, 95, pl 2, fig 20.
Leucosia scabriuscula Weber 1795: 92.
Philyra scabriuscula — Alcock 1896: 238 (in key), 239, 240; Ihle 1918: 275, 315, figs 97, 102, 115; Stephensen 1946: 88, 89, fig 15c–e; Guinot 1967a: 249 (in list); Basson et al. 1977: 224; Titgen 1982: 248 (in list); Hogarth 1989: 115; Tirmizi and Kazmi 1986: 98 (in key), 106–109, fig 31a–i; Hornby 1997: 16.
Philyra cancella — Sakai 1999: 17, 18, pl 7b; Apel 2001: 56.
Ryphila cancellus — Galil 2009: 310–312, figs 21a, 22; Naderloo and Türkay 2012: 32; Naderloo et al. 2013: 449, tab 1, 458, tab 3.

Type locality: *Cancer cancellus*: Indian Ocean?. — *Leucosia scabriuscula*: Tharangambadi and Sumatra.

Distribution

Persian Gulf: Iran [Stephensen (1946) as *Philyra scabriuscula*; Naderloo and Türkay 2012; Naderloo et al. 2013], Saudi Arabia [Basson et al. (1977) as *Philyra scabriuscula*; Galil 2009], UAE [Hornby (1997) as *Philyra scabriuscula*; Apel (2001) as *Philyra cancella*].

Gulf of Oman: Iran (Naderloo et al. 2015).

World distribution: Indo-West Pacific: East Africa, Madagascar, southern Oman, Persian Gulf, Gulf of Oman, Pakistan, India, Mergui Archipelago, Malay Archipelago, Burma, Thailand, Borneo, Sumatra and Australia.

Habitat: Intertidal, substrate sandy beach.

Fig. 14.46 *Ryphila cancellus* (Herbst, 1783). Male (ZUTC 5529), CL = 9.79, CB = 9.49 mm, Guwatr, Sistan-Baluchestan, Iran, Gulf of Oman, 10.05.2013, coll. R. Naderloo (Photo: R. Abdollahi)

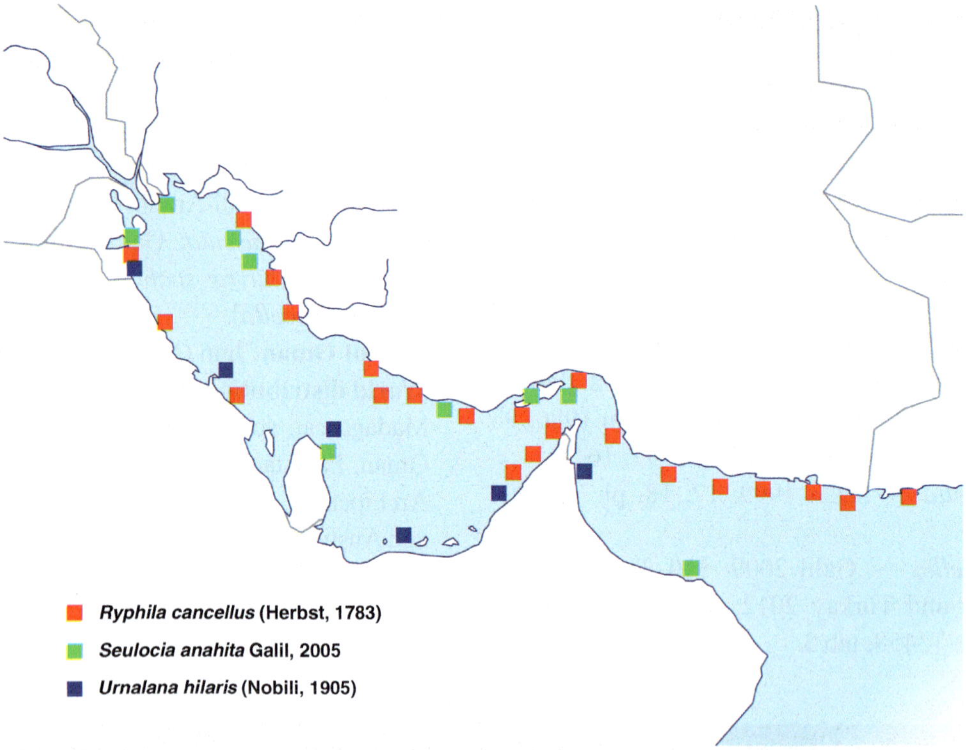

Fig. 14.47 Map showing the recorded localities of *Ryphila cancellus* (Herbst, 1783), *Seulocia anahita* Galil, 2005, and *Urnalana hilaris* (Nobili, 1905)

Genus *Seulocia* Galil, 2005

Seulocia anahita Galil, 2005

(Figs. 14.44c, 14.47, and 14.48)

Seulocia anahita Galil 2005a: 43, fig 1a.
Leucosia pubescens — Alcock 1896: 233; Stephensen 1946: 95, fig 17a, b; Tirmizi and Kazmi 1986: 94.
Leucosia aff. *Pubescens* — Apel 2001: 52.
Seulocia anahita — Naderloo and Sari 2005a: 39, fig 7; Naderloo and Sari 2007a: 342, tab 1; Ng et al. 2008: 95 (in list).

Type locality: Persian Gulf.

Distribution

Persian Gulf: Iran [Stephensen (1946) as *Leucosia pubescens*; Galil 2005a; Naderloo and Sari 2005a; Naderloo and Sari 2007a], Kuwait (Galil 2005a), Saudi Arabia (Galil 2005a).

Gulf of Oman: Oman [Apel (2001) as *Leucosia* aff. *pubescens*].

World distribution: Northern Indian Ocean: Persian Gulf, Gulf of Oman and Pakistan

Habitat: Subtidal, 1–27 m, substrate sandy.

Genus *Urnalana* Galil, 2005

Urnalana hilaris (Nobili, 1905)

(Figs. 14.44d, e, 14.47, and 14.49)

Leucosia hilaris Nobili 1905a: 162.
Leucosia sima Alcock 1896: 227, pl 6, fig 5.
Leucosia hilaris — Nobili 1906a: 100–102, pl 6, fig 29; Guinot 1976: 251 (in list).

Fig. 14.48 *Seulocia anahita* Galil, 2005. Male (SMF), CL = 21.8, CB = 18.9 mm, Persian Gulf (exact locality unknown) (Photo: S. Träkner)

Leucosia sima — Alcock and Anderson 1897: pl 30, fig 1; Kazmi and Tirmizi 1990: 20–22, fig 2a–d; Apel 2001: 53.

Urnalana hilaris Galil 2005b: 23, figs 2c, 6b; Ng et al. 2008: 95 (in list).

Type locality: *Leucosia sima*: Mumbai. — *Leucosia hilaris*: Persian Gulf (Abu Dhabi, 25° 10′N, 55° 10′E–24° 55′N, 54° 40′E) and Djibouti.

Distribution

Persian Gulf: Saudi Arabia [Apel (2001) as *Leucosia sima*], UAE [Nobili 1905a; Apel (2001) as *Leucosia sima*].

Gulf of Oman: UAE [Apel (2001) as *Leucosia sima*].

World distribution: Western Indian Ocean: Mauritius, Gulf of Aden, Persian Gulf, Gulf of Oman, Pakistan and India.

Habitat: Shallow subtidal, 1–3 m, substrate sandy.

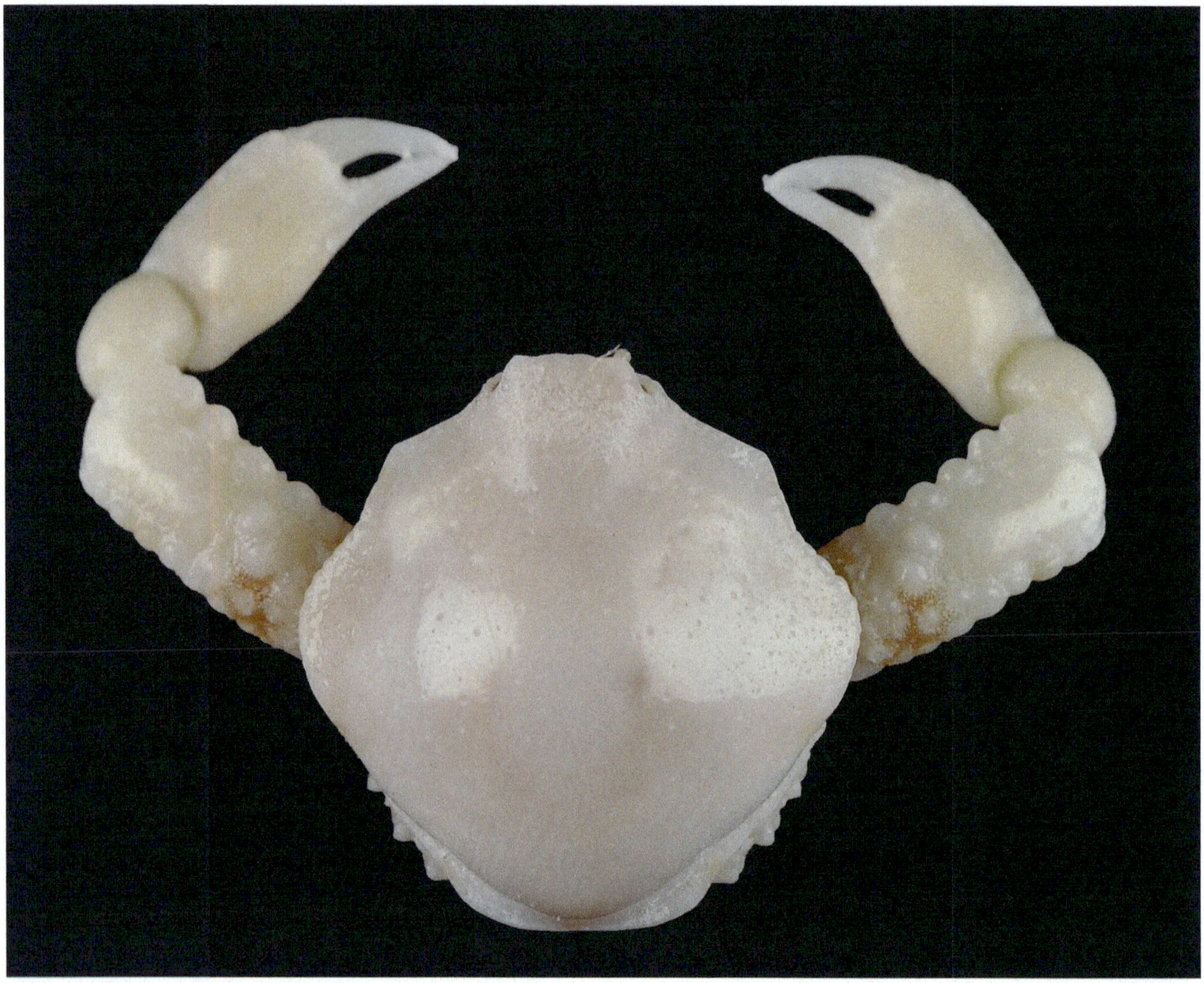

Fig. 14.49 *Urnalana hilaris* (Nobili, 1905). Male (SMF) CL = 8.4, CB = 8.3 mm, UAE, Persian Gulf, 13.06.1995, coll. M. Apel (Photo: S. Träkner)

Family Hymenosomatidae MacLeay, 1838 (Spider Crabs) 15

Diagnosis Carapace circular, pyriform, thin, markedly flat, poorly calcified, without hooked setae, regions weakly or well defined, anterolateral margins entire, without any spine or tooth, junction between anterolateral and posterolateral margins not distinct. Front distinct, produced. Orbits not distinct or incomplete, eyes exposed. Antenullar sockets shallow, weakly distinct, antennules folded longitudinally. Antennae short, first antennal segment short, fused with epistome, second segment slender. Third maxillipeds with well-developed ischium, palp attached to anterolateral angle of merus. Chelipeds sexually dimorphic, those of males large, robust. Walking legs narrow, long. G1 in various sizes; G2 shorter than G1. Genital opening of males coxal. Genital opening of females sternal.

Genus *Elamena* H. Milne Edwards, 1837

Elamena sindensis Alcock, 1900

(Figs. 15.1, 15.2a, c, and 15.3)

Elamena sindensis Alcock 1900: 386.
Elamena sindensis — Alcock and Mcardle 1903: pl 64, fig 4; Chopra and Das 1930: 424, 425; Stephensen 1946: 96, 97; Basson et al. 1977: 256; Titgen 1982: 248 (in list) Tirmizi and Kazmi 1986: 112–114, fig 32; Apel 2001: 60; Ng et al. 2008: 108 (in list); Naderloo and Türkay 2012: 33; Naderloo et al. 2013: 449, tab 1.
Type locality: Pakistan (Karachi).

Distribution
Persian Gulf: Iran (Stephensen 1946; Naderloo and Türkay 2012; Naderloo et al. 2013), Kuwait (Chopra and Das 1930), Saudi Arabia (Basson et al. 1977; Apel 2001).
Gulf of Oman: Iran (Naderloo and Türkay 2012; Naderloo et al. 2015).
World distribution: Northern Indian Ocean: Persian Gulf, Gulf of Oman, Pakistan, India (Mumbai) and Sri Lanka.
Habitat: Intertidal, rocky/cobble substrate, the species is found attached to algae and sea weeds.

R. Naderloo, *Atlas of Crabs of the Persian Gulf*, DOI 10.1007/978-3-319-49374-9_15

Fig. 15.1 *Elamena sindensis* Alcock, 1900. Male (ZUTC Brach1354), CL = 4.9, CB = 4.8 mm, Jod Village, Sistan-Baluchestan, Iran, Gulf of Oman, 12.11.2005, coll. R. Naderloo and A. Kazemi (Photo: S. Träkner)

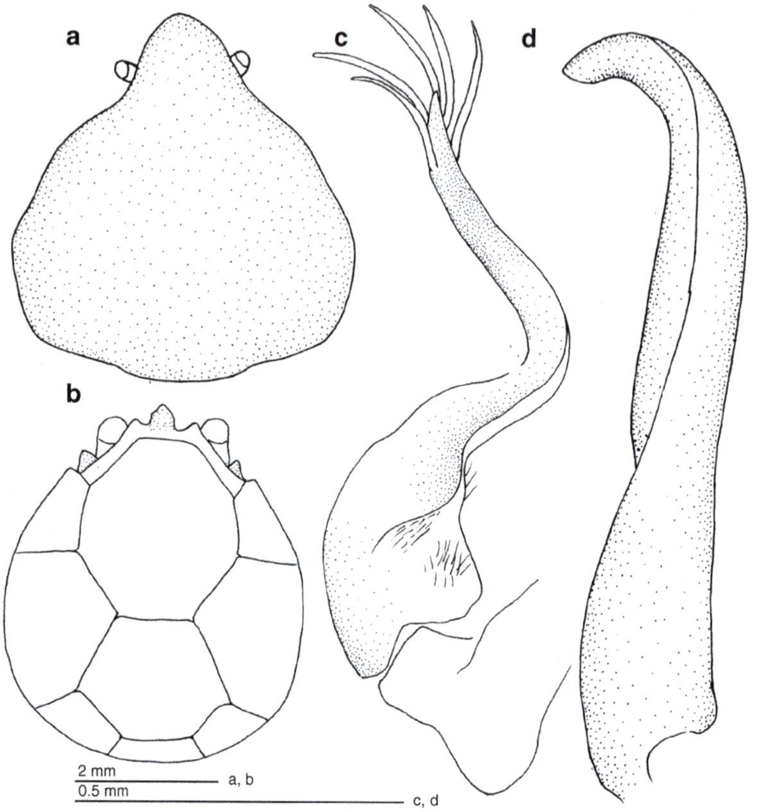

Fig. 15.2 *Elamena sindensis* Alcock, 1900: (**a**) dorsal view of carapace; (**c**) male G1, ventral view; *Neorhynchoplax kempi* (Chopra and Das, 1930): (**b**) dorsal view of carapace; (**d**) male G1, ventral view

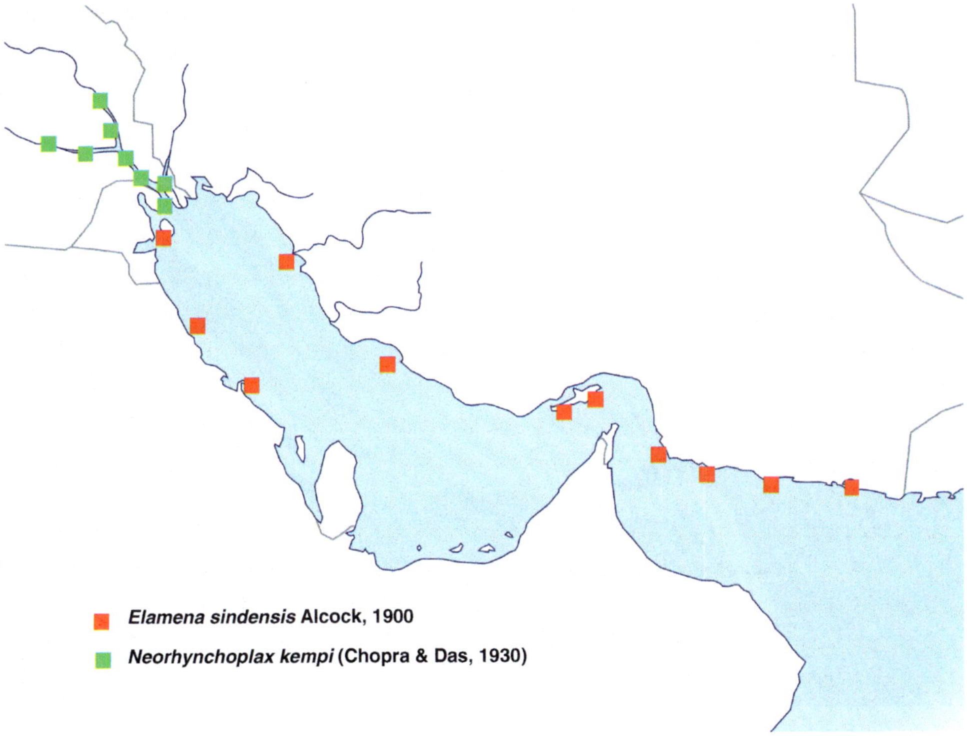

Fig. 15.3 Map showing the distribution localities of *Elamena sindensis* Alcock, 1900, and *Neorhynchoplax kempi* (Chopra and Das, 1930)

Genus *Neorhynchoplax* Sakai, 1938

***Neorhynchoplax kempi* (Chopra and Das, 1930)**
(Figs. 15.2b, d, 15.3 and 15.4)

Rhynchoplax kempi Chopra and Das 1930: 416–419, figs 2–5.
Neorhynchoplax kempi — Sakai 1938: 200 (in list); Ng et al. 2008: 109 (in list); Naderloo 2014: 2 (in text).
Type locality: Persian Gulf (Shat Al-Arab).

Distribution

Persian Gulf: Iran (Naderloo 2014), Iraq (Chopra and Das 1930).

Gulf of Oman: No record.

World distribution: Endemic: This species is endemic to the estuarine system of northern Persian Gulf finding in Shat Al-Arab, Bahmanshir and Karun rivers. *Neorhynchoplax kempi* recorded from Gulf of Mexico in 1960s by Abele (1972) and its population is currently well established in the Panama Canal (Kam et al. 2011).

Habitat: Muddy substrates of the rivers.

Fig. 15.4 *Neorhynchoplax kempi* (Chopra and Das, 1930). Male (ZUTC 5929), CL = 5.16, CB = 5.56 mm, Bahmanshir River, Pole-Tanki Abolhassan, Abadan, Iran, 04.11.2013, coll. R. Naderloo (Photo: S. Shahbazi)

Family Epialtidae MacLeay, 1838 (Spider Crabs, Decorator Crabs)

16

Diagnosis Carapace subcircular to pyriform, moderately to strongly convexed, usually longer than broad. Front with single or double spines. Orbits incomplete, reduced, never concealing cornea, postorbital spines sometimes present, intercalated spine usually absent, eyestalks usually short, but not always, often immovable or slightly movable. Antennules folded longitudinally. Antennae with basal segment relatively long, fused distally. Third maxillipeds with merus as broad as ischium, slightly shorter than latter. Chelipeds robust, nearly equal, palm more or less swollen. First walking legs longer than others; dactylus of walking legs prehensile or subchelate. Abdomen of males and females with seven free-moving somites. G1 robust, slightly curved; G2 shorter than G1.

Key to genera of the family Epialtidae in the region

1.	Rostrum single, sometimes apically slightly divided	2
-	Rostrum with two spines or lobes, spines generally divergent, sometimes contiguous	5
2.	Rostrum not divided apically	*Menaethius*
-	Rostrum divided apically	3
3.	Carapace markedly long pyriform; rostrum stout, long	*Xenocarcinus*
-	Carapace broad pyriform to rounded; rostrum very short	4
4.	Supraorbital hiatus with intercalated spine	*Phalangipus*
-	Supraorbital hiatus without intercalated spine	*Doclea*
5.	Eyestalks long, orbits usually strangely expanded above, usually protected by long hornlike supraorbital spines	*Stilbognathus*
-	Eyestalks short; no hornlike supraorbital spines	6
6.	Supraorbital eave with distinct hiatus; postorbital lobe present	7
-	Supraorbital eave entire, without hiatus; no distinct postorbital lobe	8
7.	Rostrum short, with two contiguous lobes, apically rounded; postorbital lobe distinct; long tubercles on postbranchial region of carapace, small-sized species	*Alcockia*
-	Rostrum usually long with two broadly divided spines	*Hyastenus*
8.	Carapace pyriform; hepatic and branchial margins with tubercles or short spines; postorbital lobe fused to supraorbital eave; walking legs not subchelate, propodus cylindrical	*Menaethiops*
-	Carapace subrectangular or suboval; hepatic and branchial margins with lobes; postorbital lobe indistinct; walking legs subchelate, propodus with ventral setiferous projection	*Acanthonyx*

R. Naderloo, *Atlas of Crabs of the Persian Gulf*, DOI 10.1007/978-3-319-49374-9_16

Genus *Acanthonyx* Latreille, 1828

Key to species of the genus *Acanthonyx* in the region

- Anterolateral margins of carapace with three teeth, becoming smaller posteriorly; rostral spines markedly longer than anterior-orbital teeth; chelipeds with one large median tooth on cutting edge of movable finger .... *Acanthonyx limbatus*
- Anterolateral margins of carapace with two teeth, becoming smaller posteriorly; rostral spines as long as or slighlty longer than anterior-orbital teeth; chelipeds with three teeth on cutting edge of movable finger .... *Acanthonyx elongatous*

Acanthonyx elongatus Miers, 1877

(Figs. 16.1 and 16.2)

Acanthonyx elongatus White 1847b: 11 [nomen nudum].
Acanthonyx elongatus Miers 1877: 673, pl 69, fig 1.
Acanthonyx elongatus — Stephensen 1946: 101 (in key); Griffin and Tranter 1974: 164, 1986: 66 (in key).
Acanthonyx aff. *elongatus* — Tirmizi and Serène 1971: 22, fig 1, pl 2a, b.
Acanthonyx elongatus inglei Tirmizi and Kazmi 1986: 162–166, figs 50, 51.

Type locality: Réunion.

Distribution

Persian Gulf: No record.

Gulf of Oman: Iran (Naderloo et al. 2015).

World distribution: Western Indian Ocean: Réunion, Red Sea, Gulf of Oman and Pakistan.

Habitat: Intertidal to shallow subtidal, substrate rocky/cobble covered with algae.

Fig. 16.1 *Acanthonyx elongatus* Miers, 1877. Female (ZUTC 5508), CL = 54. 3, CB = 55.2 mm, Chabahar, Iran, Gulf of Oman, 19.12.2010, coll. Y. Fatemi (Photo: S. Shahbazi)

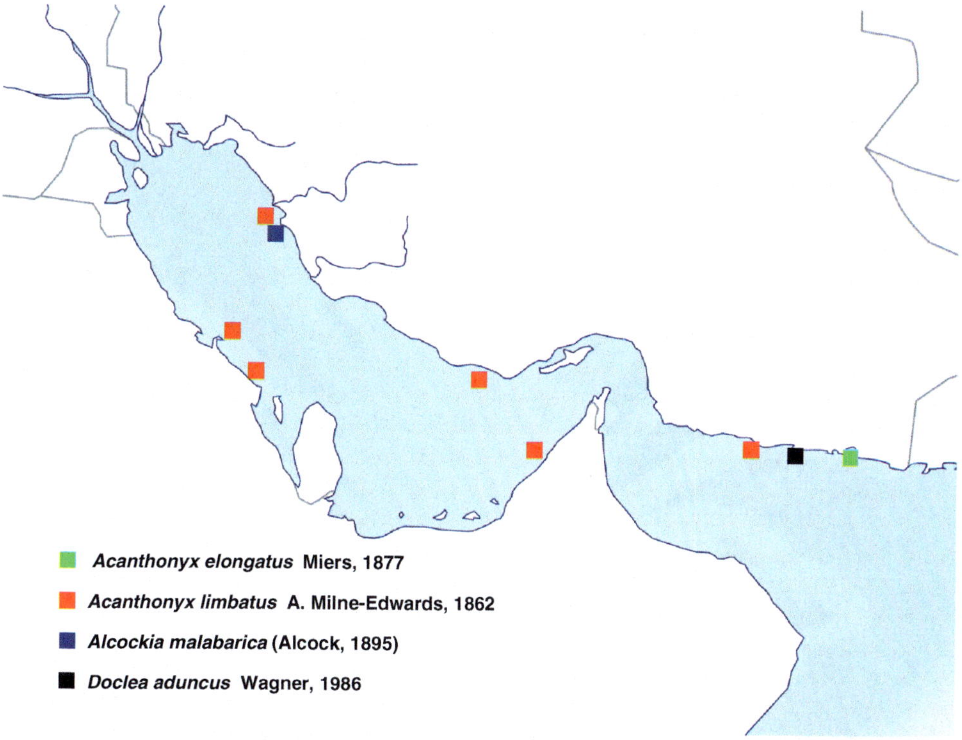

Fig. 16.2 Map showing the distribution localities of *Acanthonyx elongatus* Miers, 1877, *Acanthonyx limbatus* A. Milne-Edwards, 1862, *Alcockia malabarica* (Alcock, 1895), and *Doclea aduncus* Wagner, 1986

Remarks Tirmizi and Kazmi (1986) described a subspecies of this species namely, *Acanthonyx elongatus inglei* from the Pakistani waters by providing a comparison of morphological details with *A. elongatus elongatus* in NHM. There is still no clear taxonomic decision for these two subspecies.

Acanthonyx limbatus A. Milne-Edwards, 1862

(Figs. 16.2, 16.3, and 16.4a)

Dehaanius (*Acanthonyx*) *limbatus* A. Milne-Edwards 1862: 7, 8, pl 17, figs 4, 4a, b.
Acanthonyx (*Dehaanius?*) (*limbatus* A. Milne-Edwards?) — Stephensen 1946: 101 (in key), 102–105, fig 19a–h.
Acanthonyx limbatus — Guinot 1967a: 292 (in list); Basson et al. 1977: 237, 256; Titgen 1982: 114, 115, 248 (in list); Griffin and Tranter 1986: 68, 69; Hogarth 1989: 108; Tirmizi and Kazmi 1986: 159–161, figs 48, 49; Apel 2001: 60; Ng et al. 2008: 100 (in list); Naderloo and Türkay 2012: 33.

Type locality: Réunion.

Distribution

Persian Gulf: Iran (Stephensen 1946; Apel 2001; Naderloo and Türkay 2012), Saudi Arabia (Basson et al. 1977; Apel 2001), UAE (Titgen 1982).

Gulf of Oman: Iran (Naderloo et al. 2015).

World distribution: Western Indian Ocean: Réunion, Mauritius, southern Oman, Persian Gulf, Gulf of Oman, Pakistan and India (Mumbai).

Habitat: Intertidal, substrate rocky/cobble usually covered with algae, usually found on brown algae *Sargassum* spp. in the Persian Gulf.

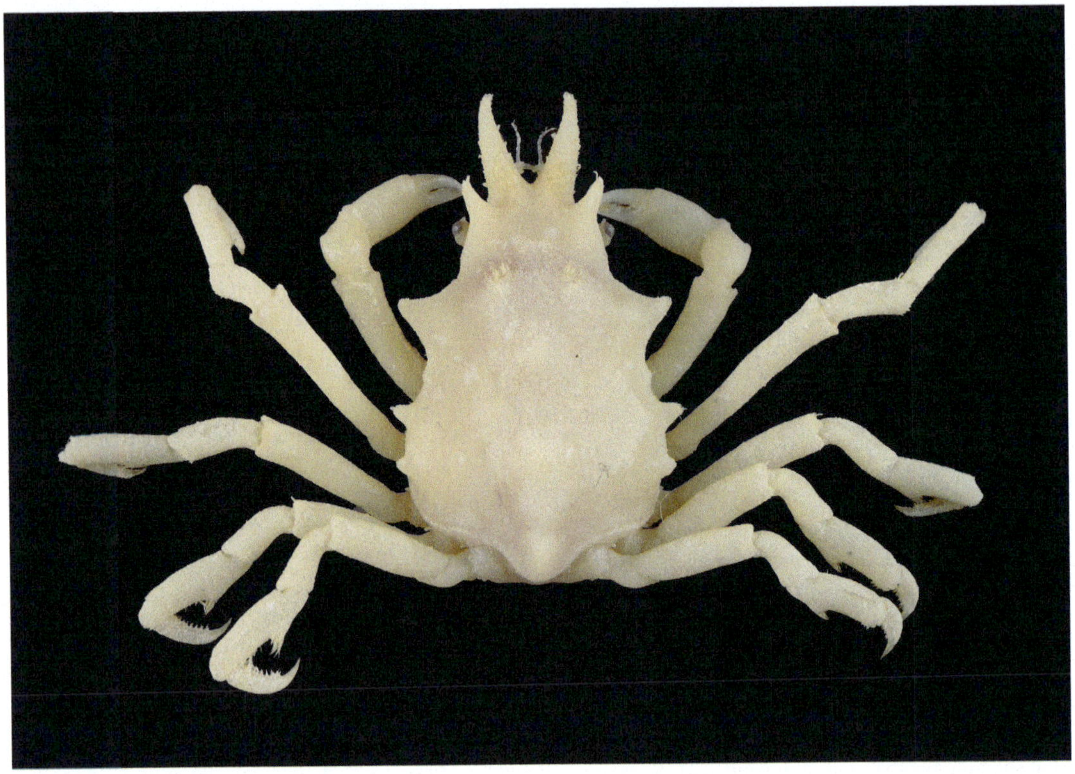

Fig. 16.3 *Acanthonyx limbatus* A. Milne-Edwards, 1862. Male (ZUTC Brach1285), CL = 13.5, CB = 8.2 mm, Jofreh, Bushehr, Iran, Persian Gulf, 19.10.2009, coll. R. Naderloo (Photo: S. Träkner)

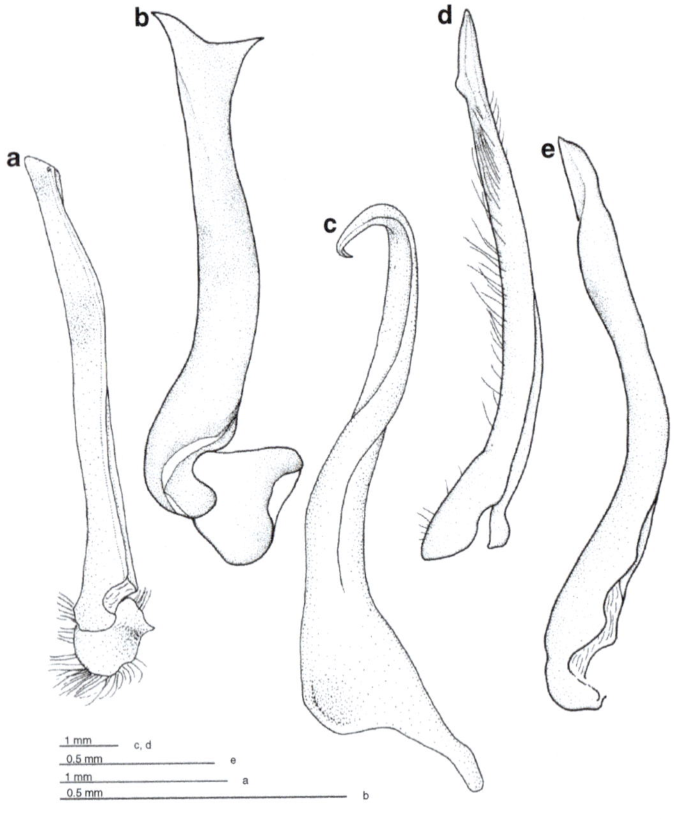

Fig. 16.4 Male G1, ventral view: (**a**) *Acanthonyx limbatus* A. Milne-Edwards, 1862 (ZUTC Brach1285); (**b**) *Alcockia malabarica* (Alcock, 1895) (ZMUC CRU870); (**c**) *Doclea aduncus* Wagner, 1986 (ZUTC 5561); (**d**) *Hyastenus hilgendorfi* De Man, 1887 (SMF: UAE); (**e**) *Hyastenus inermis* (Rathbun, 1911) (SMF: Saudi Arabia)

Genus *Alcockia* Števcic, 2005

Alcockia malabarica (Alcock, 1895)
(Figs. 16.2, 16.4b, and 16.5)

Collodes malabaricus Alcock 1895: 189, pl 5, fig 3.
Collodes malabaricus — Alcock and Anderson 1896: pl 19, fig 1; Stephensen 1946: 97; Titgen, 1982: 248 (in list); Griffin 1974: 9; Griffin and Tranter 1986: 23; Apel 2001: 61; Ng et al. 2008: 1001 (in list).

Type locality: India (Malabar).

Distribution

Persian Gulf: Iran (Stephensen 1946).

Gulf of Oman: No record.

World distribution: Northern Indian Ocean: Persian Gulf, India (Mumbai and Malabar).

Habitat: Shallow subtidal to 49 m, substrate sandy with clay.

Fig. 16.5 *Alcockia malabarica* (Alcock, 1895). Male (ZMUC CRU870), CL = 3.6, CB = 2.4 mm, SW of Bushehr, Iran, Persian Gulf, depth 49 m, 14.03.1937, coll. G. Thorson (Photo: S. Träkner)

Genus *Doclea* Leach, 1815

Doclea aduncus Wagner, 1986

(Figs. 16.2, 16.4c, and 16.6)

Doclea aduncus Wagner 1986: figs 7–29, pl 8.
Doclea aduncus — Tirmizi and Kazmi 1986: 135 (in key), 135–137, figs 40, 41; Ng et al. 2008: 102 (in list).

Type locality: Pakistan (Karachi).
Distribution
Persian Gulf: No record.
Gulf of Oman: Iran (Naderloo et al. 2015).
World distribution: Northern Indian Ocean: Gulf of Oman and Pakistan.
Habitat: Subtidal.

Fig. 16.6 *Doclea aduncus* Wagner, 1986. Male (ZUTC 5561), CL = 54.3, CB = 55.2 mm, Iran, Gulf of Oman, 19.12.2010, coll. N. Ghotbeddin (Photo: S. Träkner)

Genus *Hyastenus* White, 1847

Key to species of the genus *Hyastenus* in the region

1. Rostral spines short, about one-fourth of postrostral carapace length; lateral margins of carapace with small tubercle immediately behind orbit, without any spine; basal antennal segment with spine on anterolateral angle .... *Hyastenus inermis*

- Rostral spines markedly long; lateral margins of carapace with prominent tubercle or spine behind orbit; basal antennal segment without spine on anterolateral angle 2

2. Rostral spines about half as long as postrostral carapace length; branchial margins with tubercles near to base of chelipeds, with two to five submarginal tubercles behind it; small spine on epibranchial margins of carapace .... *Hyastenus hilgendorfi*

- Rostral spines markedly longer than half of postrostral carapace length; branchial margins without any tubercle; large spine on epibranchial margins of carapace .... *Hyastenus spinosus*

Hyastenus hilgendorfi De Man, 1887

(Figs. 16.4d, 16.7, and 16.8)

Hyastenus hilgendorfi De Man 1887b: 14–18, pl i, figs 3, 4.
Hyastenus hilgendorfi — Nobili 1906a: 105, 106; Stephensen 1946: 107, fig. 20d, e; Griffin 1968: 103–105, pl 1, fig 1; Tirmizi and Serène 1971: 25–27 (in text as *H. pleione*), pl 2, fig b; Basson et al. 1977: 243, 250, 256, 265; Titgen 1982: 248 (in list); Griffin and Tranter 1986: 147, 148, fig 42 h, i; Apel 2001: 62; Naderloo and Sari 2007a: 342, tab 1; Ng et al. 2008: 102 (in list); Naderloo and Türkay 2012: 33.
Type locality: Mergui Archipelago.

Fig. 16.7 *Hyastenus hilgendorfi* De Man, 1887. Male (SMF), CL = 30.00, CB = 17.5 mm, Dubai, UAE, Persian Gulf, 1993, coll. S. Lehmann, (Photo: S. Träkner)

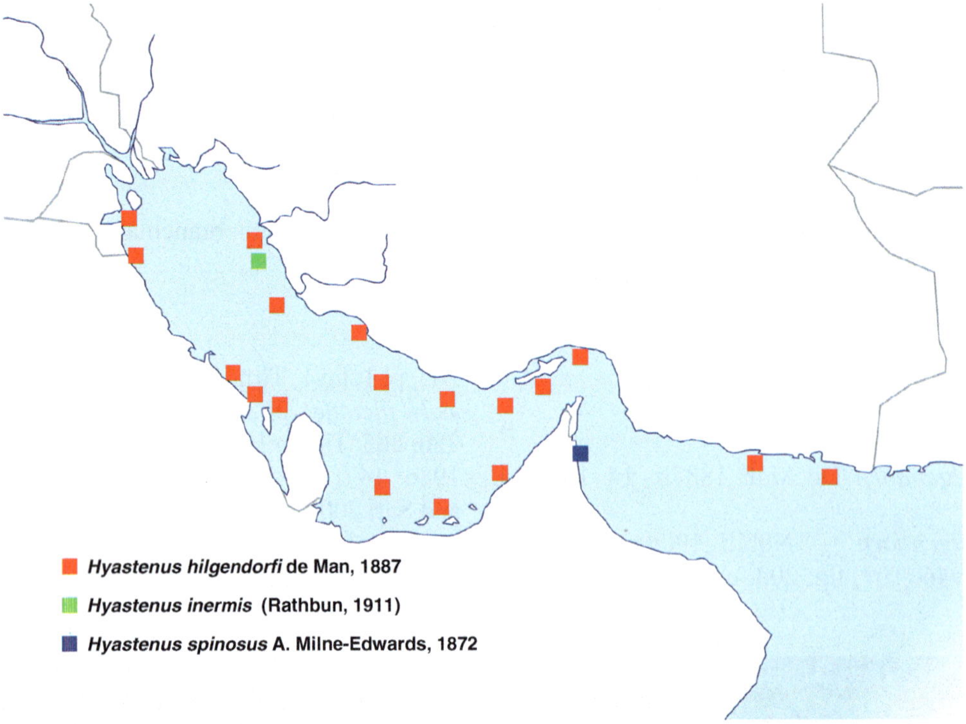

Fig. 16.8 Map showing the distribution localities of *Hyastenus hilgendorfi* De Man, 1887, *Hyastenus inermis* (Rathbun, 1911), and *Hyastenus spinosus* A. Milne-Edwards, 1872

Distribution

Persian Gulf: Iran (Stephensen 1946; Apel 2001; Naderloo and Sari 2007a; Naderloo and Türkay 2012), Kuwait (Apel 2001), Saudi Arabia (Basson et al. 1977; Apel 2001), Bahrain (Stephensen 1946), UAE (Nobili 1906a).

Gulf of Oman: Iran (Naderloo et al. 2015).

World distribution: Indo-West Pacific and Mediterranean: East Africa, Red Sea, Persian Gulf, Andaman Sea, Thailand, Indonesia, Philippines, Australia, Hawaii and Mediterranean (Lessepsian migrant).

Habitat: Intertidal to 41 m, substrates sandy and rocky.

Hyastenus inermis (Rathbun, 1911)

(Figs. 16.4e, 16.8, and 16.9)

Halimus inermis Rathbun 1911: 250, 251, pl 20.

Hyastenus inermis — Griffin 1974: 15; Griffin and Tranter 1974: 171, 1986: 125 (in key), 148; Naderloo and Sari 2007a: 342, tab 1, 2007b: 1261–1264, figs 1, 2; Ng et al. 2008: 102 (in list).

Type locality: Saint Brandon, Saya de Malha Bank, Amirante, Seychelles (type series are from different localities).

Distribution

Persian Gulf: Iran (Naderloo and Sari 2007a, 2007b).

Gulf of Oman: Iran (Naderloo et al. 2015).

World distribution: Western Indian Ocean: Mombasa, Saint Brandon, Saya de Malha Bank, Amirante, Seychelles, Red Sea, Persian Gulf and Gulf of Oman.

Habitat: Shallow subtidal to 18 m.

Fig. 16.9 *Hyastenus inermis* (Rathbun, 1911). Female (ZUTC Brach1025), CL = 9.4, CB = 4.6 mm, Bushehr, Iran, Persian Gulf, depth 18 m, 02.03.2004, coll. R. Naderloo (Photo: S. Träkner)

Hyastenus spinosus A. Milne-Edwards, 1872

(Figs. 16.8 and 16.10)

Hyastenus spinosus A. Milne-Edwards 1872: 250.
Hyastenus spinosus — Griffin and Tranter 1974: 171; Apel 2001: 62; Ng et al. 2008: 102 (in list).

Type locality: Fiji and Mozambique.

Distribution

Persian Gulf: No record.

Gulf of Oman: UAE (Apel 2001).

World distribution: Indo-West Pacific: South Africa, Mozambique, Somalia, Gulf of Aden, Red Sea, Gulf of Oman, Pakistan, India, Sri Lanka, Bay of Bengal, Strait of Malacca, Philippines, Singapore, Australia and Fiji.

Habitat: Subtidal to 20 m, substrates sandy and rocky.

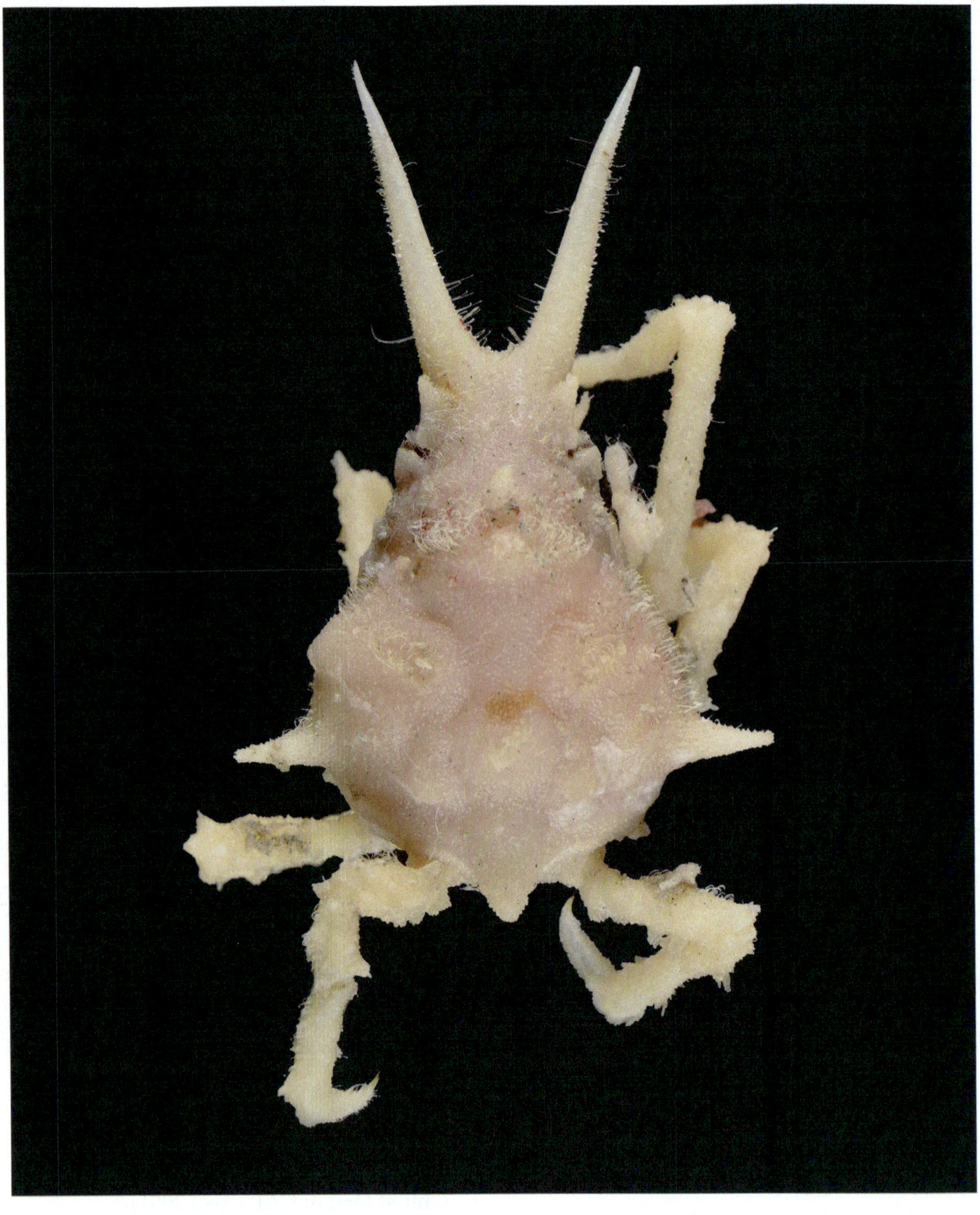

Fig. 16.10 *Hyastenus spinosus* A. Milne-Edwards, 1872. Female (SMF), CL = 33.7, CB = 21.1 mm, Khor Fakkan, UAE, Gulf of Oman, 04.07.1995, coll. M. Apel (Photo: S. Träkner)

Genus *Menaethiops* Alcock, 1895

Key to species of the genus *Menaethiops* in the region

1. Anterior angle of orbital eave not produced, curved .... *Menaethiops abumusa*
- Anterior angle of orbital eave produced anteriorly, angular 2
2. Rostrum with two slender contiguous spines .... *Menaethiops gadaniensis*
- Rostrum with two short broadly divided spines .... *Menaethiops nodulosus*

Menaethiops abumusa Naderloo, 2015
(Figs. 16.11, 16.12, and 16.13a)

Menaethiops abumusa Naderloo 2015a: 431–437, figs 1, 2, 3a–e, 4a–d.
Menaethiops aff. *bicornis* — Naderloo and Türkay 2012: 34.
Type locality: Persian Gulf (Abu Musa Island, Iran).

Distribution
Persian Gulf: Iran [Naderloo and Türkay (2012) as *Menaethiops* aff. *bicornis;* Naderloo 2015a].
Gulf of Oman: Iran (Naderloo 2015a; Naderloo et al. 2015).
World distribution: Northern Indian Ocean: This species is still recorded from the Persian Gulf and Gulf of Oman.
Habitat: Intertidal, substrate rocky/cobble.

Fig. 16.11 *Menaethiops abumusa* Naderloo, 2015. Holotype (ZUTC 5319) CL = 13.60, CB = 9.72 mm, Park-Qadir, Abu Musa Island, Iran, Persian Gulf, 01.05.2014, coll. R. Naderloo and H. Rahimian (Photo: R. Abdollahi. Reproduced from Naderloo (2015a) by permission of Magnolia press)

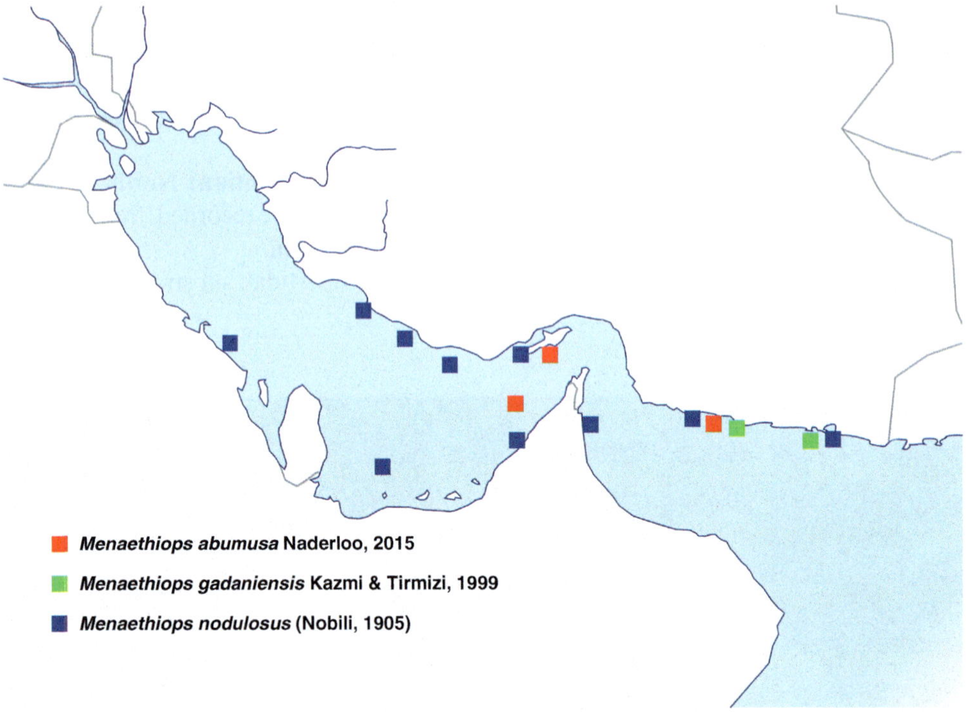

Fig. 16.12 Map showing the distribution localities of *Menaethiops abumusa* Naderloo, 2015, *Menaethiops gadaniensis* Kazmi and Tirmizi, 1999, and *Menaethiops nodulosus* (Nobili, 1905)

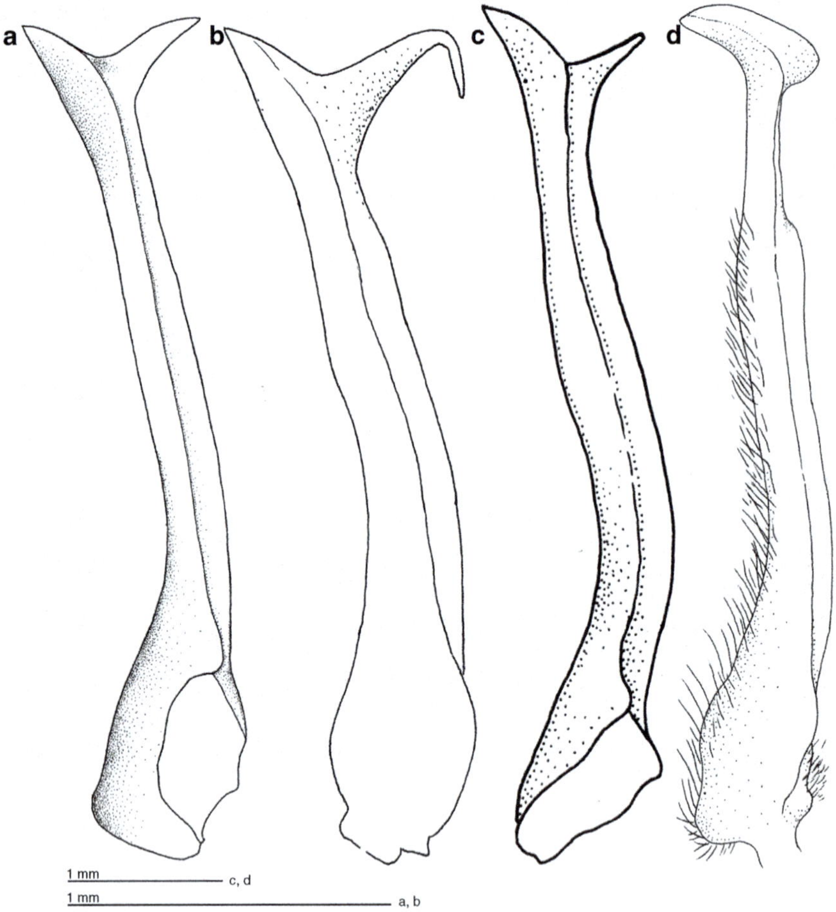

Fig. 16.13 Male G1, ventral view: (**a**) *Menaethiops abumusa* Naderloo, 2015 (ZUTC 5319); (**b**) *Menaethiops gadaniensis* Kazmi and Tirmizi, 1999 (ZUTC 5323); (**c**) *Menaethiops nodulosus* (Nobili, 1905) (ZUTC 5322); (**d**) *Menaethius monoceros* (Latreille, 1825) (ZUTC 5537)

Menaethiops gadaniensis Kazmi and Tirmizi, 1999
(Figs. 16.12, 16.13b, and 16.14)

Menaethiops gadaniensis Kazmi and Tirmizi 1999: 370–372, fig 1.
Menaethiops gadaniensis — Ng et al. 2008: 101 (in list); Naderloo 2015a: 437, fig 5a.
Type locality: Pakistan (Gadani).
Distribution
Persian Gulf: No record.
Gulf of Oman: Iran (Naderloo 2015a; Naderloo et al. 2015).
World distribution: Northern Indian Ocean: Gulf of Oman and Pakistan.
Habitat: Intertidal, substrate rocky/cobble.

Fig. 16.14 *Menaethiops gadaniensis* Kazmi and Tirmizi, 1999. Male (ZUTC 5323), CL = 12.10, CB = 8.40 mm, Jod Village, Sistan-Baluchestan, Iran, Gulf of Oman, 07.02.2009, coll. H. Salehi (Photo: R. Abdollahi. Reproduced from Naderloo (2015a, b) by permission of Magnolia press)

Menaethiops nodulosus (Nobili, 1905)
(Figs. 16.12, 16.13c, and 16.15)

Parahoplophrys nodulosa Nobili 1905b: 239.
Herbstia corniculata Klunzinger 1906: 27, pl 1, fig 4a, b.
Parahoplophrys nodulosa — Nobili 1906a: 106, fig. 2.
Menaethiops nodulosa — Stephensen 1946: 99, fig 18d; Guinot 1967a: 293 (in list); Basson et al. 1977: 261, 265; Titgen 1982: 249 (in list); Griffin and Tranter 1986: 87 (in key); Hogarth 1989: 107, 108; Apel 2001: 63; Naderloo and Türkay 2012: 34; Naderloo et al. 2013: 449, tab 1.
Menaethiops nodulosus — Kazmi and Tirmizi 1999: 372–375, fig 2a–f; Ng et al. 2008: 101 (in list); Naderloo 2015a: 437, fig 5b.
Type locality: *Parahoplophrys nodulosa*: Persian Gulf (Abu Dhabi, UAE). — *Herbstia corniculata*: Red Sea (Al-Qusseir, Egypt).
Distribution.
Persian Gulf: Iran (Stephensen 1946; Naderloo and Türkay 2012; Naderloo et al. 2013), Saudi Arabia (Basson et al. 1977), UAE (Nobili 1905a, 1906a).
Gulf of Oman: UAE (Apel 2001; Naderloo 2015a; Naderloo et al. 2015).
World distribution: Northwestern Indian Ocean: Red Sea, Gulf of Aden, Somalia, southern Oman, Persian Gulf, Gulf of Oman and Pakistan.
Habitat: Intertidal to 6 m, substrate rocky/cobble.

Fig. 16.15 *Menaethiops nodulosus* (Nobili, 1905). Male (ZUTC 5322), CL = 11.02, CB = 7.20 mm, Jod Village, Sistan-Baluchestan, Iran, Gulf of Oman, 07.02.2009, coll. H. Salehi (Photo: R. Abdollahi. Reproduced from Naderloo (2015a) by permission of Magnolia press)

Genus *Menaethius* A. Milne-Edwards, 1834

Menaethius monoceros (Latreille, 1825)

(Figs. 16.13d, 16.16, and 16.17)

Pisa monoceros Latreille 1825: 139.
Menaethius monoceros — Stephensen 1946: 105, fig 20a; Guinot 1967a: 293 (in list); Evans et al. 1973: 276; Basson et al. 1977: 224, 227, 243, 256; Titgen 1982: 249 (in list); Griffin and Tranter 1986: 89, 90; Jones 1986a: 156, pl 45; Hogarth 1989: 107, 108; Tirmizi and Kazmi 1986: 170–173, fig. 53a–g; Cooper 1997: 168, figs 3, 14; Hornby 1997: 16; Apel 2001: 63; Ng et al. 2008: 101 (in list); Naderloo and Türkay 2012: 35.
Type locality: Mauritius and Red Sea.

Distribution

Persian Gulf: Iran (Stephensen 1946; Naderloo and Türkay 2012), Kuwait (Jones 1986a), Saudi Arabia (Basson et al. 1977; Apel 2001), UAE (Evans et al. 1973; Cooper 1997).

Gulf of Oman: Iran (Naderloo et al. 2015).

World distribution: Indo-West Pacific: South Africa, East Africa, Seychelles, Red Sea, southern Oman, Persian Gulf, Gulf of Oman, Pakistan, Maldives, India, Lakshadweep Islands, Nicobar, Andaman Sea, Malaysia, China, Japan, Indonesia, Cocos (Keeling) Islands, Elizabeth and Middleton Reefs, Australia, Fiji, Tahiti, French Polynesia and Hawaii.

Habitat: Intertidal to 33 m, substrates rocky/cobble and oyster bank.

Fig. 16.16 *Menaethius monoceros* (Latreille, 1825). Male (ZUTC 5537), CL = 18.91, CB = 12.32 mm, Park-Qadir, Abu Musa Island, Iran, Persian Gulf, 14.05.2015, coll. R. Naderloo (Photo: R. Abdollahi)

Genus *Phalangipus* Latreille, 1828

Phalangipus persicus Griffin, 1973

(Figs. 16.17, 16.18, and 16.19a)

Phalangipus persicus Griffin 1973: 188, 190, figs 2c, 4e–f, 6 g, 7 g, 9e–f.
Phalangipus arachnoides — Nobili 1906a: 105.
Phalangipus longipes — Stephensen 1946: 105–107, fig 20b, c; Basson et al. 1977: 243, 250; Titgen 1982: 249 (in list).
Phalangipus persicus — Griffin and Tranter 1986: 173 (in key); Apel 2001: 64; Naderloo and Sari 2007a: 342, tab 1; Ng et al. 2008: 105 (in list).

Type locality: Persian Gulf (Bushehr, Iran).

Distribution

Persian Gulf: Iran [Stephensen (1946) as *Phalangipus longipes*; Naderloo and Sari 2007a], Saudi Arabia [Basson et al. (1977) as *Phalangipus longipes*], UAE [Nobili (1906a) as *Phalangipus arachnoides*].

Gulf of Oman: No record.

World distribution: Northwestern Indian Ocean: Red Sea, Gulf of Aden, southern Oman and Persian Gulf.

Habitat: Subtidal, 15–45 m, substrates sandy and muddy.

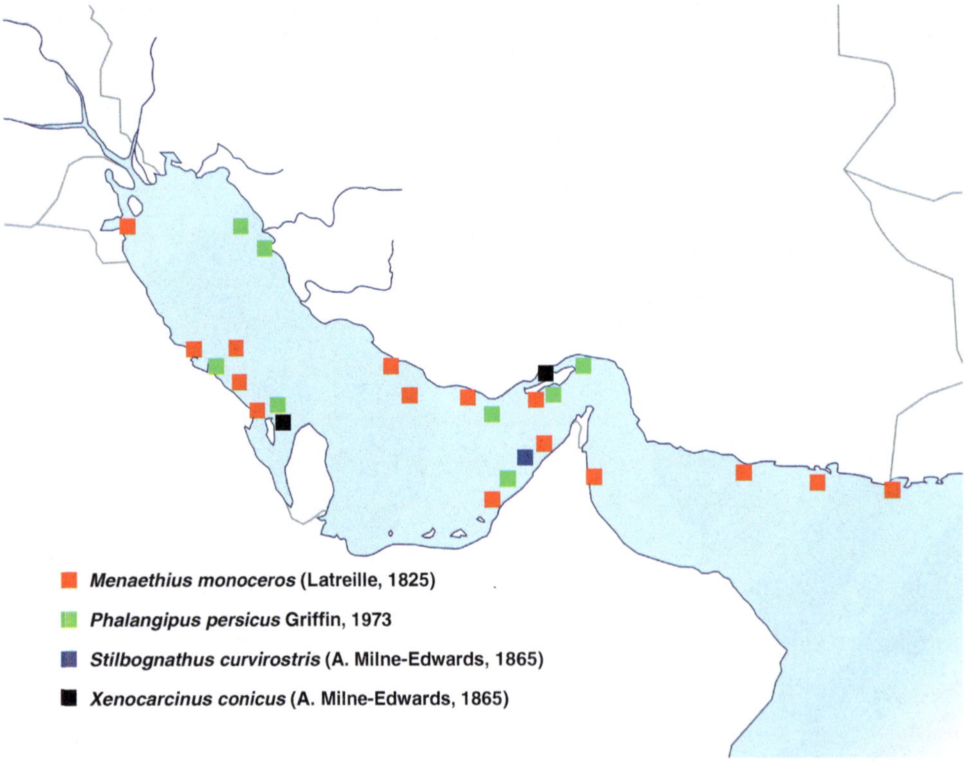

Fig. 16.17 Map showing the distribution localities of *Menaethius monoceros* (Latreille, 1825), *Phalangipus persicus* Griffin, 1973, *Stilbognathus curvirostris* (A. Milne-Edwards, 1865), and *Xenocarcinus conicus* (A. Milne-Edwards, 1865)

Fig. 16.18 *Phalangipus persicus* Griffin, 1973. Male (ZUTC 5540), CL = 24.18, CB = 20.80 mm, Bushehr, Iran, Persian Gulf. coll. V. Sepahvand (Photo: R. Abdollahi)

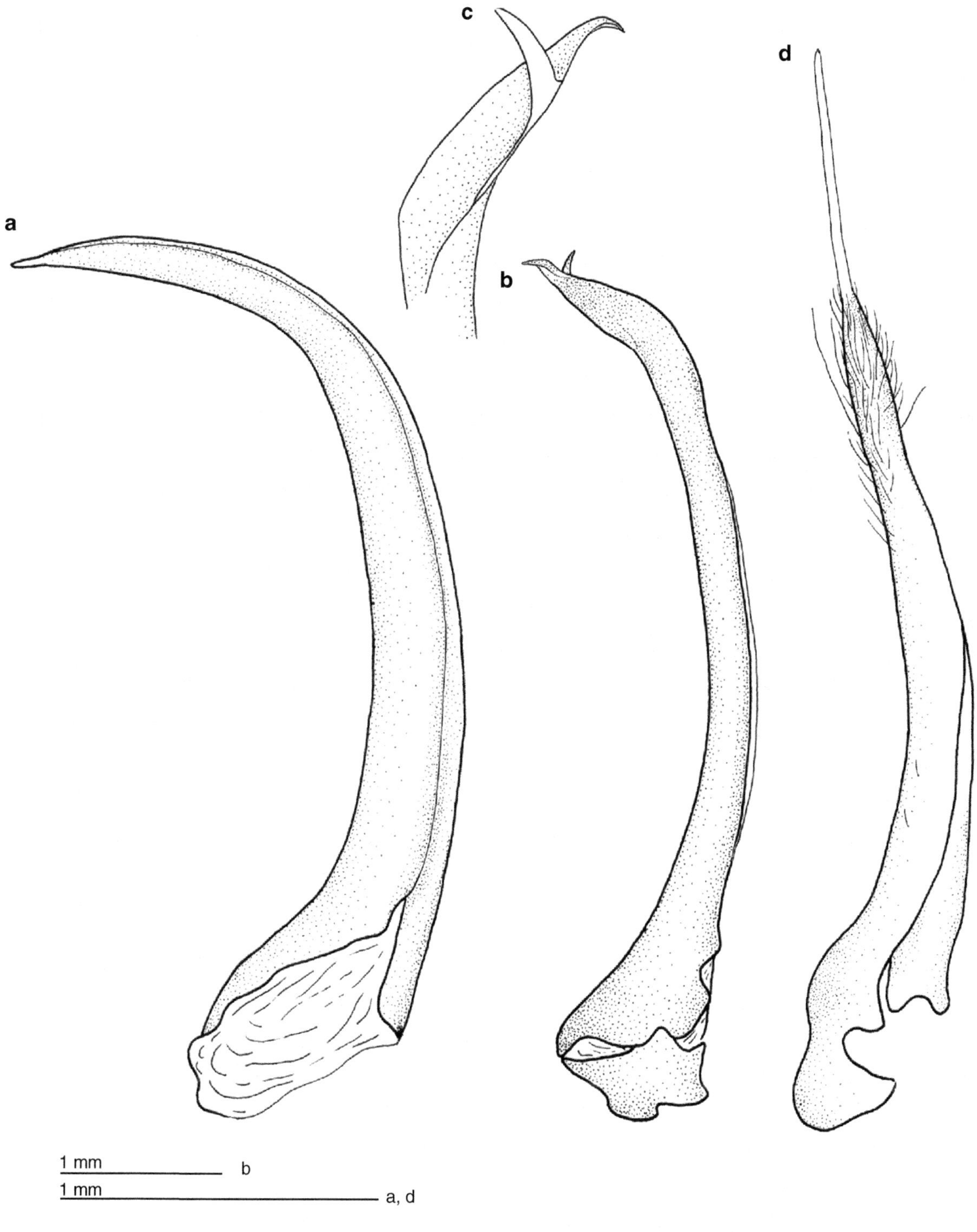

Fig. 16.19 Male G1, ventral view: (**a**) *Phalangipus persicus* Griffin, 1973 (ZUTC 5540); (**b**) *Stilbognathus curvirostris* (A. Milne-Edwards, 1865) (SMF: Qatar); (**c**) apical part of (**b**); (**d**) *Xenocarcinus conicus* (A. Milne-Edwards, 1865) (ZMUC CRU876)

Genus *Stilbognathus* von Martens, 1866

Stilbognathus curvirostris (A. Milne-Edwards, 1865)

(Figs. 16.17, 16.19b, c, and 16.20)

Stenocinops curvirostris A. Milne-Edwards 1865a: 135, pl 5, figs 1, 1a–e.
Stenocinops scurvirostris [sic!] — Nobili 1906a: 108.
Ophthalmias curvirostris — Laurie 1915: 431, pl 45, fig 4; Stephensen 1946: 98; Guinot 1967a: 297 (in list); Griffin and Tranter 1974: 177, 178; Titgen 1982: 249 (in list); Hogarth 1989: 108.
Stilbognathus curvirostris — Griffin and Tranter 1986: 64 (in key); Apel 2001: 65; Ng et al. 2008: 106 (in list).

Type locality: Red Sea.

Distribution

Persian Gulf: Iran (present study), UAE (Nobili 1906a).

Gulf of Oman: No record.

World distribution: Northwestern Indian Ocean: Red Sea, Gulf of Aden, southern Oman and Persian Gulf.

Habitat: Subtidal, 10–90 m.

Fig. 16.20 *Stilbognathus curvirostris* (A. Milne-Edwards, 1865). Female (ZUTC 5496), CL = 54.19, CB = 24.92 mm, Iran, Persian Gulf (Photo: R. Abdollahi)

Genus *Xenocarcinus* White, 1847

Xenocarcinus conicus (A. Milne-Edwards, 1865)
(Figs. 16.17, 16.19d, and 16.21)

Huenioides conica A. Milne-Edwards 1865a: 145, 146, pl 4, fig 3.
Xenocarcinus tuberculatus — Alcock 1895: 192; Alcock and Anderson 1898: pl 33, figs 3, 3a; Stephensen 1946: 109, 110, fig 21c, d; Guinot 1967a: 293 (in list); Titgen 1982: 29 (in list).
Xenocarcinus conicus — Griffin and Tranter 1986: 100–103, fig. 30c, d; Apel 2001: 65; Naderloo and Sari 2007a: 342, tab 1; Ng et al. 2008: 101 (in list).
Type locality: Indian Ocean.
Distribution
Persian Gulf: Iran (Naderloo and Sari 2007a), Bahrain [Stephensen (1946) as *Xenocarcinus tuberculatus*].
Gulf of Oman: No record.
World distribution: Indo-West Pacific: Réunion, Aldabra, Red Sea, Persian Gulf, Sri Lanka, Indo-Malayan Archipelago and Japan.
Habitat: Subtidal to 70 m, living on soft corals.

Fig. 16.21 *Xenocarcinus conicus* (A. Milne-Edwards, 1865). Male (ZUTC Brach1031), CL = 17.30, CB = 6.56 mm, Daragahan, Qeshm Island, Iran, Persian Gulf, depth 70 m, 05.11.2003, coll. R. Naderloo (Photo: S. Shahbazi)

Family Inachidae MacLeay, 1838 (Spider Crabs, Decorator Crabs)

17

Diagnosis Carapace subcircular to subpyriform. Rostrum with short truncated lobes or one long spine. Orbits absent, eyestalks long, fully exposed, more or less movable. Basal antennal segment slender, long, distally free. Third maxillipeds with merus narrower than ischium. Walking legs long, slender. Abdomen of males and females with seven free-moving somites. G1 in different forms from short stout apically curved to long slender with simple or bifurcated apical part; G2 shorter than G1. Genital opening of males coxal. Genital opening of females sternal.

Key to genera of the family Inachidae in the region

1. Eyestalks short; rostrum strong, long triangular; carapace round pyriform .... *Encephalloides*

- Eyestalks long; rostrum not markedly strong; carapace long pyriform 2

2. Walking legs relatively short in comparison to carapace; antennal gland on proximal portion of basal antennal segment; G1 short, stout, bent apically .... *Paratymolus*

- Walking legs markedly long in comparison to carapace; antennal gland distinct, on subproximal portion of basal antennal segment 3

3. Rostrum with two contiguous spines, apically acute, extending beyond antennal fossae; basal antennal segment proximally fused with epistome, distally free .... *Macropodia*

- Rostrum with two deeply divided lobes, short, apically rounded; basal antennal segment proximally and distally fused to epistome .... *Achaeus*

Genus *Achaeus* Leach, 1817

Achaeus lacertosus Stimpson, 1858
(Figs. 17.1, 17.2a, b, and 17.3)

Achaeus lacertosus Stimpson 1858: 218.
Achaeus lacertosus — Stimpson 1907: 20, pl 3, fig 7; Stephensen 1946: 98, fig 18c; Guinot 1967a: 290 (in list); Titgen 1982: 248 (in list); Griffin and Tranter 1986: 4, 5 (in key), 10, 11; Tirmizi and Kazmi 1986: 128–133, figs 38a, b, 39a–g; Ng et al. 2008: 110 (in list); Naderloo and Türkay 2012: 35; Naderloo et al. 2013: 449, tab 1.
Type locality: Australia (Port Jackson).
Distribution.
Persian Gulf: Iran (Stephensen 1946; Apel 2001; Naderloo and Türkay 2012: Naderloo et al. 2013), Saudi Arabia (Apel 2001).
Gulf of Oman: No record.

R. Naderloo, *Atlas of Crabs of the Persian Gulf*, DOI 10.1007/978-3-319-49374-9_17

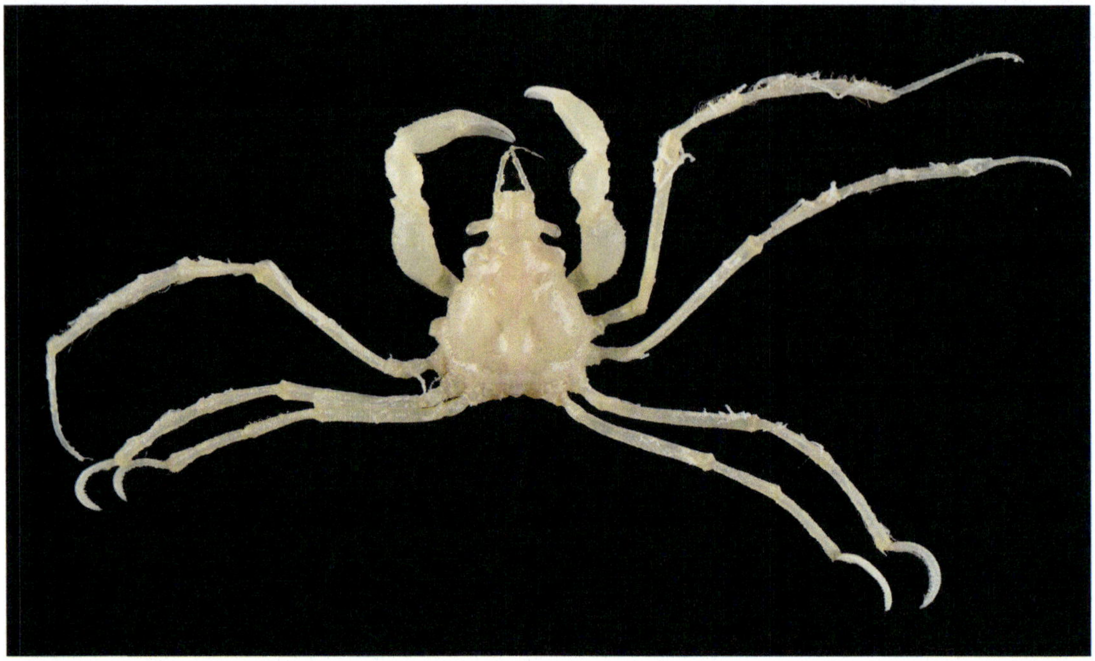

Fig. 17.1 *Achaeus lacertosus* Stimpson, 1858. Male (ZMUC CRU851), CL = 8.0, CB = 6.2 mm, Bushehr, Iran, Persian Gulf, depth 3–5 m, 08.03.1937, coll. G. Thorson (Photo: S. Träkner)

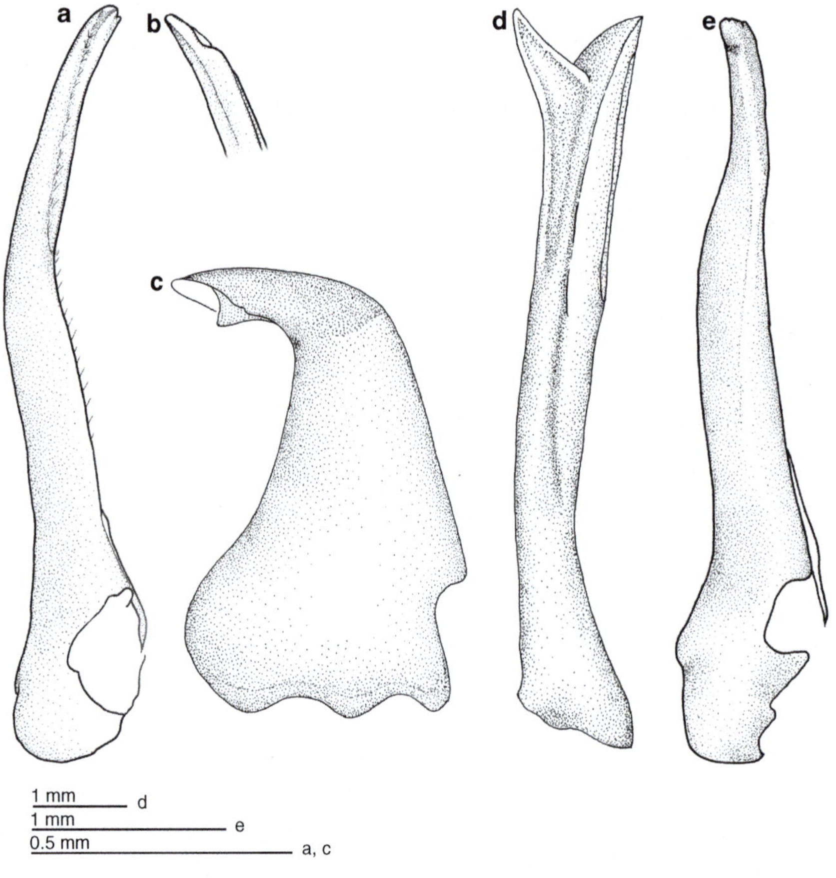

Fig. 17.2 Male G1, ventral view: (**a**) *Achaeus lacertosus* Stimpson, 1858 (ZMUC CRU851); (**b**) apical part of (**a**), dorsal view; (**c**) *Paratymolus apeli* Naderloo and Türkay, 2015 (SMF 48751); (**d**) *Encephalloides armstrongi* Wood-Mason, 1890 (ZUTC 5912); (**e**) *Macropodia formosa* Rathbun, 1911 (ZUMC CRU853)

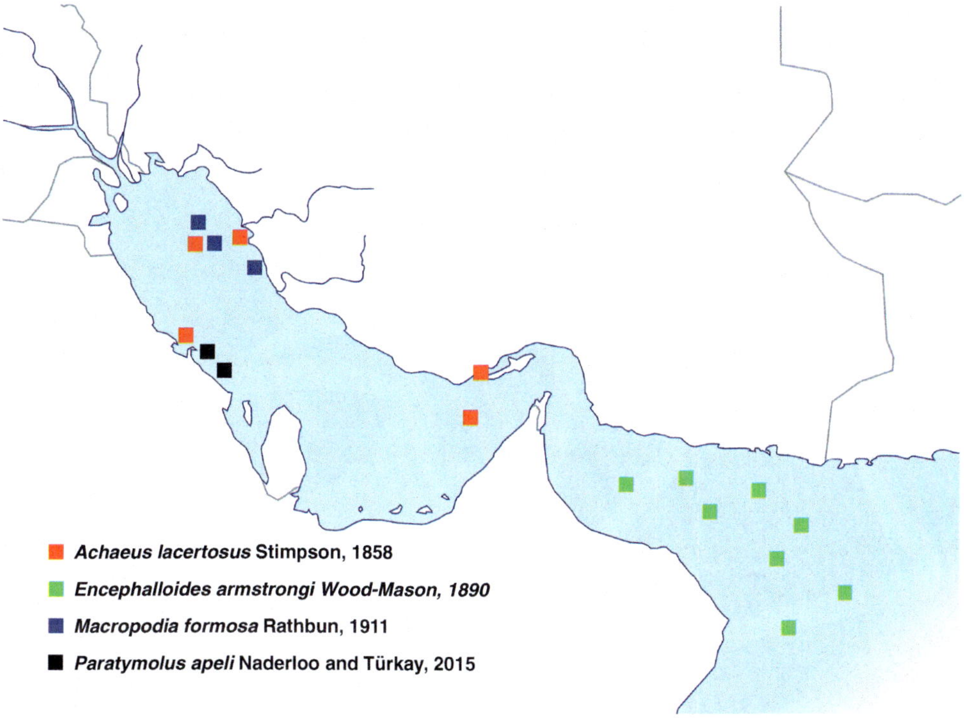

Fig. 17.3 Map showing the distribution localities of *Achaeus lacertosus* Stimpson, 1858, *Encephalloides armstrongi* Wood-Mason, 1890, *Macropodia formosa* Rathbun, 1911, and *Paratymolus apeli* Naderloo and Türkay, 2015

World distribution: Indo-West Pacific: South Africa, Persian Gulf, Pakistan, India, Japan, Indonesia and Australia.

Habitat: Intertidal to 5 m, substrate rocky/cobble.

Genus *Encephalloides* Wood-Mason, 1890

Encephalloides armstrongi Wood-Mason 1890

(Figs. 17.2d, 17.3, and 17.4)

Encephalloides armstrongi Wood-Mason and Alcock 1891: 259, 260, figs 2, 2a.

Encephalloides armstrongi — Alcock 1895: 187, 1899a: 47–49; Creasey et al. 1997: 1–15; Ng et al. 2008: 111 (in list); Khvorov et al. 2012: 1739.

Type locality: India (Bay of Bengal).

Distribution

Persian Gulf: No record.

Gulf of Oman: Iran (Naderloo et al. 2015), Oman (Creasey et al. 1997).

World distribution: Northern Indian Ocean: southern Oman, Gulf of Oman, Arabian Sea and India.

Habitat: Subtidal, 110–1230 m.

Fig. 17.4 *Encephalloides armstrongi* Wood-Mason, 1890. Male (ZUTC 5912), CL = 33.28, CB = 31.47 mm, Iran, Gulf of Oman, coll. Ghotbeddin (Photo: S. Träkner)

Genus *Macropodia* Leach, 1817

***Macropodia formosa* Rathbun, 1911**
(Figs. 17.2e, 17.3, and 17.5)

Macropodia formosa Rathbun 1911: 242–244, fig 1.
Macropodia formosa — Griffin 1974: 19, fig 4c, d; Griffin and Tranter 1986: 33; Kazmi and Tirmizi 1995: 62–64, fig 2a–i; Apel 2001: 62, 63; Ng et al. 2008: 112 (in list).
Achaeus tenuicollis — Stephensen 1946: 97, fig 18a, b; Titgen 1982: 248 (in list).

Type locality: Saint Brandon.

Distribution

Persian Gulf: Iran [Stephensen (1946) as *Achaeus tenuicollis*; Apel 2001].

Gulf of Oman: No record.

World distribution: Western Indian Ocean: Saint Brandon, Gulf of Aden, Persian Gulf and Pakistan.

Habitat: Subtidal to 50 m.

Genus *Paratymolus* Miers, 1879

***Paratymolus apeli* Naderloo and Türkay, 2015**
(Figs. 17.2c, 17.3, and 17.6)

Paratymolus apeli Naderloo and Türkay 2015: 1–5, figs, 1, 2.
Paratymolus sp. nov. [aff. *pubescens* Miers, 1879, and *hastatus* Alcock, 1895] — Apel 2001: 64.

Type locality: Persian Gulf (Saudi Arabia).

Distribution

Persian Gulf: Saudi Arabia [Apel (2001) as *Paratymolus* sp. nov.; Naderloo and Türkay 2015].

Gulf of Oman: No record.

World distribution: This species is still only known from the Persian Gulf.

Habitat: Subtidal, 5–10 m, substrate sandy with patchy corals.

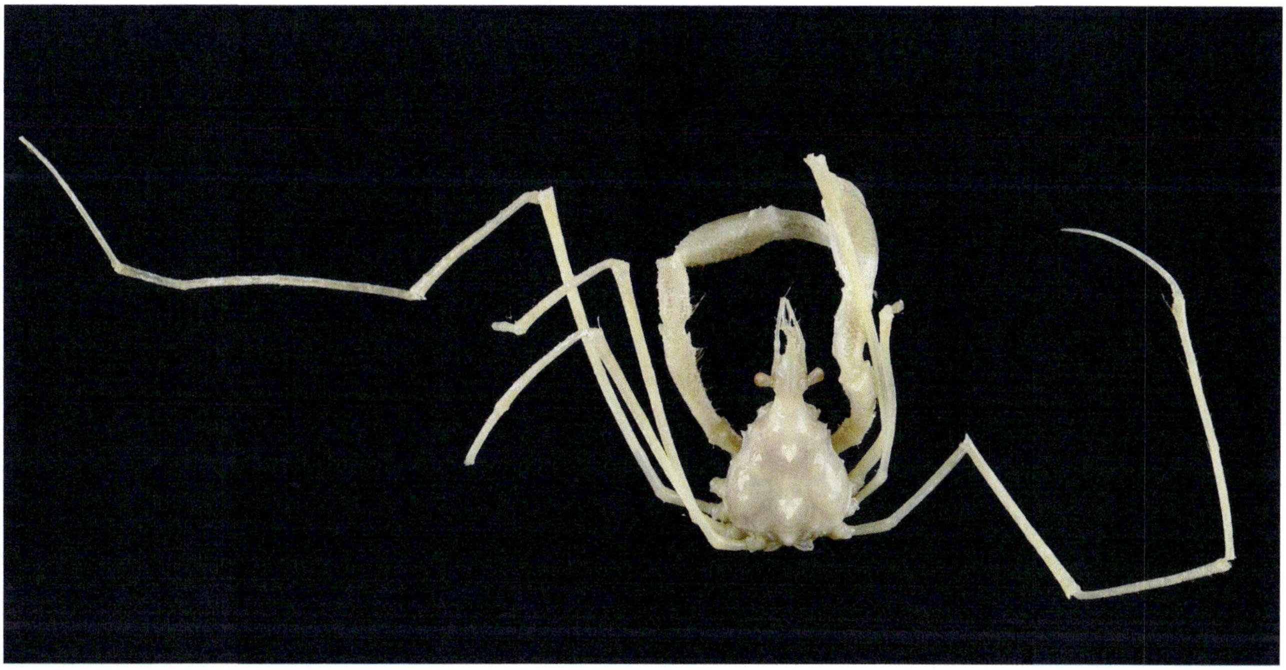

Fig. 17.5 *Macropodia formosa* Rathbun, 1911. Male (ZUMC CRU853), CL = 10.55, CB = 6.2 mm, Iran, Persian Gulf, 29° 07′N, 49° 56′E, depth 40 m, 13.03.1937, coll. G. Thorson (Photo: S. Träkner)

Fig. 17.6 *Paratymolus apeli* Naderloo and Türkay, 2015. Male holotype (SMF 48751), CL = 4.6 mm, CB = 3.4 mm, E Abu Ali, Saudi Arabia, Persian Gulf, depth 5–10 m, 01.06.1995, coll. M. Apel (Photo: S. Träkner. Reproduced from Naderloo and Türkay (2015) by permission of Taylor & Framcis)

Family Majidae Samouelle, 1819 (Spider Crabs, Decorator Crabs)

18

Diagnosis Carapace subquadrate, subcircular to pyriform. Rostrum usually with two spines. Orbits complete with supraorbital eave and postorbital spine, but not closed; eyestalks long, movable in orbits. Antennules folded longitudinally. Antennae with elongated basal antennal segment, usually produced on anterolateral angle. Third maxillipeds with merus as broad as ischium. Abdomen of males and females with seven free-moving somites. G1 usually slender, straight, apically simple, without special apparatus; G2 shorter than G1. Genital opening of males coxal. Genital opening of females sternal.

Key to genera of the family Majidae in the region

1. Carapace narrow, postrostral carapace about two times as long as broad; preorbital spine small .... *Cyphocarcinus*
- Carapace broad, postrostral carapace about as long as broad 2
2. Rostral spines broad, spines fused proximally; upper orbital margin with intercalated spine .... *Micippa*
- Rostral spines not fused proximally; upper orbital margin with intercalated spines .... *Schizophrys*

Genus *Cyphocarcinus* A. Milne-Edwards, 1868

***Cyphocarcinus capreolus* (Paul'son, 1875)**
(Figs. 18.1 and 18.2)

Ixion capreolus Paul'son 1875: 3, pl 2, fig 1a–f.
Ixion capreolus — Klunzinger 1906: 44.
Cyphocarcinus minutus — Alcock 1895: 254 (partly); Nobili 1906a: 109, 1906b: 177; Stephensen 1946: 108, fig 21a, b; Guinot 1967a: 296 (in list; partly); Evans et al. 1973: 276; Griffin and Tranter 1974: 166; Jones 1986a: 156, pl 45; Hogarth 1989: 108, 113.
Cyphocarcinus capreolus — Titgen 1982: 116, 117, 248 (in list); Griffin and Tranter 1986: 266–269, figs 98a, f, 99a; Apel 2001: 61; Ng et al. 2008: 119 (in list); Naderloo and Türkay 2012: 35.
Type locality: Red Sea.
Distribution

R. Naderloo, *Atlas of Crabs of the Persian Gulf*, DOI 10.1007/978-3-319-49374-9_18

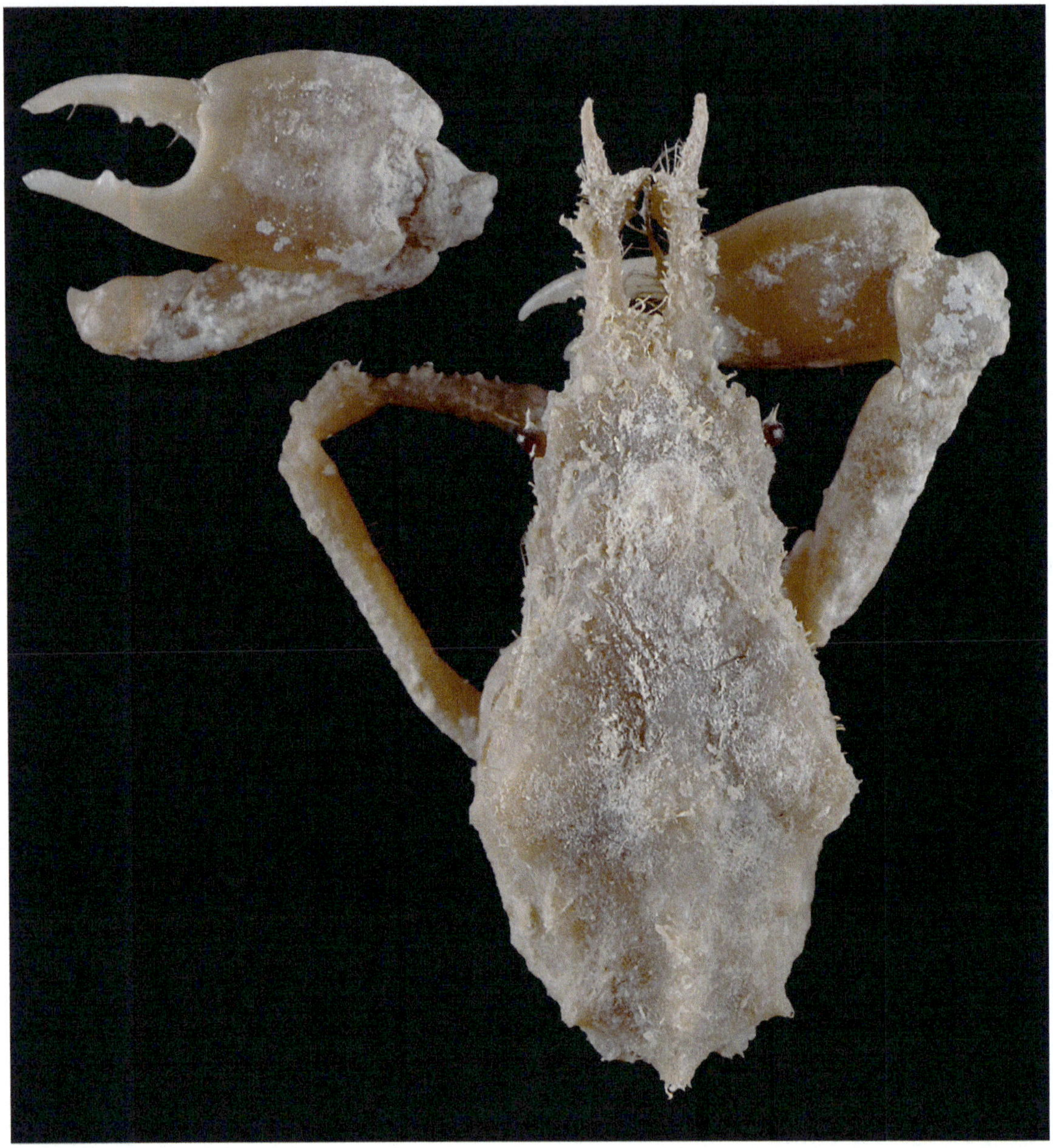

Fig. 18.1 *Cyphocarcinus capreolus* (Paul'son, 1875). Male (ZUTC Brach1276), CL = 7.58, CB = 21.28 mm, Haleh, Khalij-Nayband, Iran, Persian Gulf, 07.05.2008, coll. Kamyab (Photo: S. Träkner)

Persian Gulf: Iran [Stephensen (1946) as *Cyphocarcinus minutus*; Naderloo and Türkay 2012], Kuwait [Jones (1986a) as *Cyphocarcinus minutus*], Saudi Arabia (Griffin and Tranter 1986; Apel 2001), Qatar (Apel 2001), UAE [Nobili (1906a) and Evans et al. (1973) as *Cyphocarcinus minutus*; Titgen 1982].

Gulf of Oman: No record.

World distribution: Northwestern Indian Ocean: Red Sea and Persian Gulf.

Habitat: Intertidal to 6 m, substrate rocky/cobble.

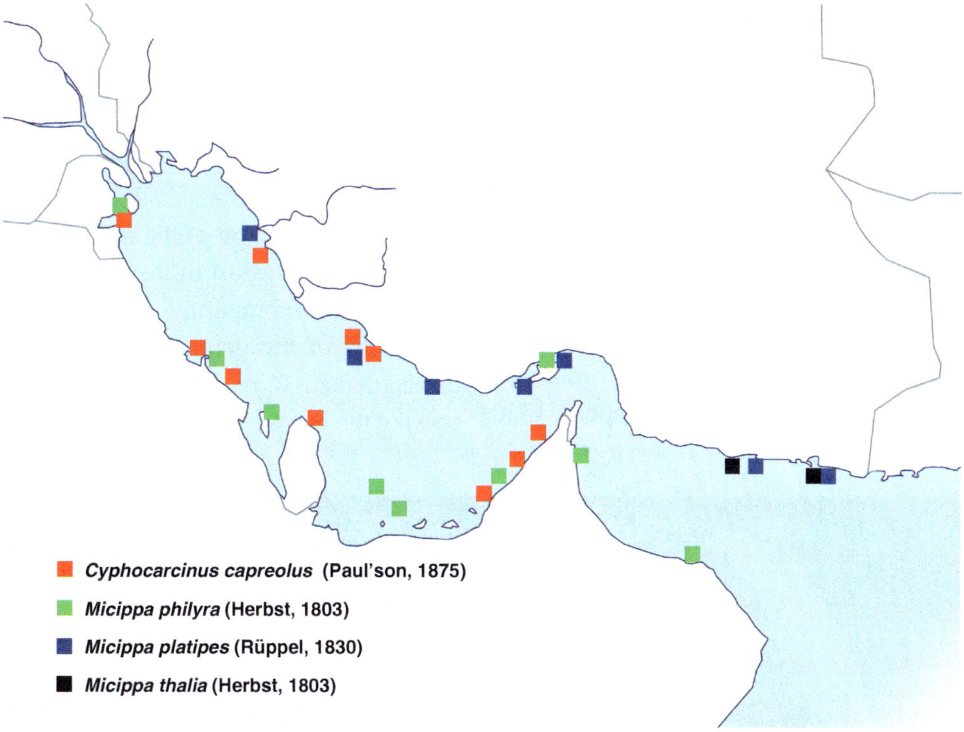

Fig. 18.2 Map showing the distribution localities of *Cyphocarcinus capreolus* (Paul'son, 1875), *Micippa philyra* (Herbst, 1803), *Micippa platipes* Rüppell, 1830, and *Micippa thalia* (Herbst, 1803)

Genus *Micippa* Leach, 1817

Key to species of the genus *Micippa* in the region

1. Rostrum moderately broad, with two sharp long lobes, slightly downflexed; carapace with sharp spines on lateral margins .... *Micippa thalia*

- Rostrum very broad, with four sharp lobes, markedly downflexed; carapace with denticles on lateral margins 2

2. Carapace with eight to ten denticles on anterolateral margin; basal antennal segment smooth; distinct gap between basal antennal segment and postorbital lobe .... *Micippa platipes*

- Carapace with three to six denticles on anterolateral margin; basal antennal segment tuberculate; no gap between basal antennal segment and postorbital lobe .... *Micippa philyra*

Micippa philyra (Herbst 1803)
(Figs. 18.2, 18.3 and 18.4a)

Cancer philyra Herbst 1803: 51, 52, pl 58, fig 4.
Micippa philyra — Buitendijk 1939: 253–256, fig 21, pl 10, figs 1–3; Guinot 1967a: 296 (in list); Evans et al. 1973: 275; Basson et al. 1977: 243, 250, 256; Titgen 1982: 249 (in list); Griffin and Tranter 1986: 277–279; Apel 2001: 63, 64; Ng et al. 2008: 119 (in list).
Micippa philyra var. *mascarenica* — Nobili 1906a: 112; Stephensen 1946: 108, fig 21e, f.

Type locality: "Ostindien". East India or Indo-Malayan Archipelago.

Distribution

Persian Gulf: Iran [Naderloo and Türkay (2012) as *M. platipes*], Bahrain [Stephensen (1946) as *M. philyra* var. *mascarenica*], Saudi Arabia (Basson et al.

1997; Apel 2001), UAE [Nobili (1906a) as *M. philyra* var. *mascarenica*; Evans et al. 1973; Titgen 1982; Apel 2001].

Gulf of Oman: UAE (Apel 2001).

World distribution: Indo-West Pacific: Mozambique, Persian Gulf, Gulf of Oman, Thailand, Indonesia, Philippines, Australia and Japan.

Habitat: Intertidal to 23 m, substrate rocky/cobble.

Remarks *Micippa philyra* (Herbst, 1803) is morphologically very similar to *M. platipes* Rüppell, 1830 and has been frequently confused with the latter. Therefore, general distribution of these two species is not clearly known. Apel (2001) mentioned that all records of the genus from the Persian Gulf is likely assigned to *M. philyra*, whereas Naderloo and Türkay (2012) heistately assigned all the records to *M. platipes*. Detailed examination of the recently collected material from the Iranian coast of the Persian Gulf and the Gulf of Oman and comparing them with already sampled specimens of the genus from the Arabian coast confirmed that all *M. platipes* is restricted along the Iranian coast of the Persian Gulf and the Gulf of Oman extending

Fig. 18.3 *Micippa philyra* (Herbst, 1803). Male (ZUTC Brach1355), CL = 28.87, CB = 26.30 mm, Baghestan, Bandar-Abbas, Iran, Persian Gulf, 21.10.2010, coll. R. Naderloo (Photo: R. Abdollahi)

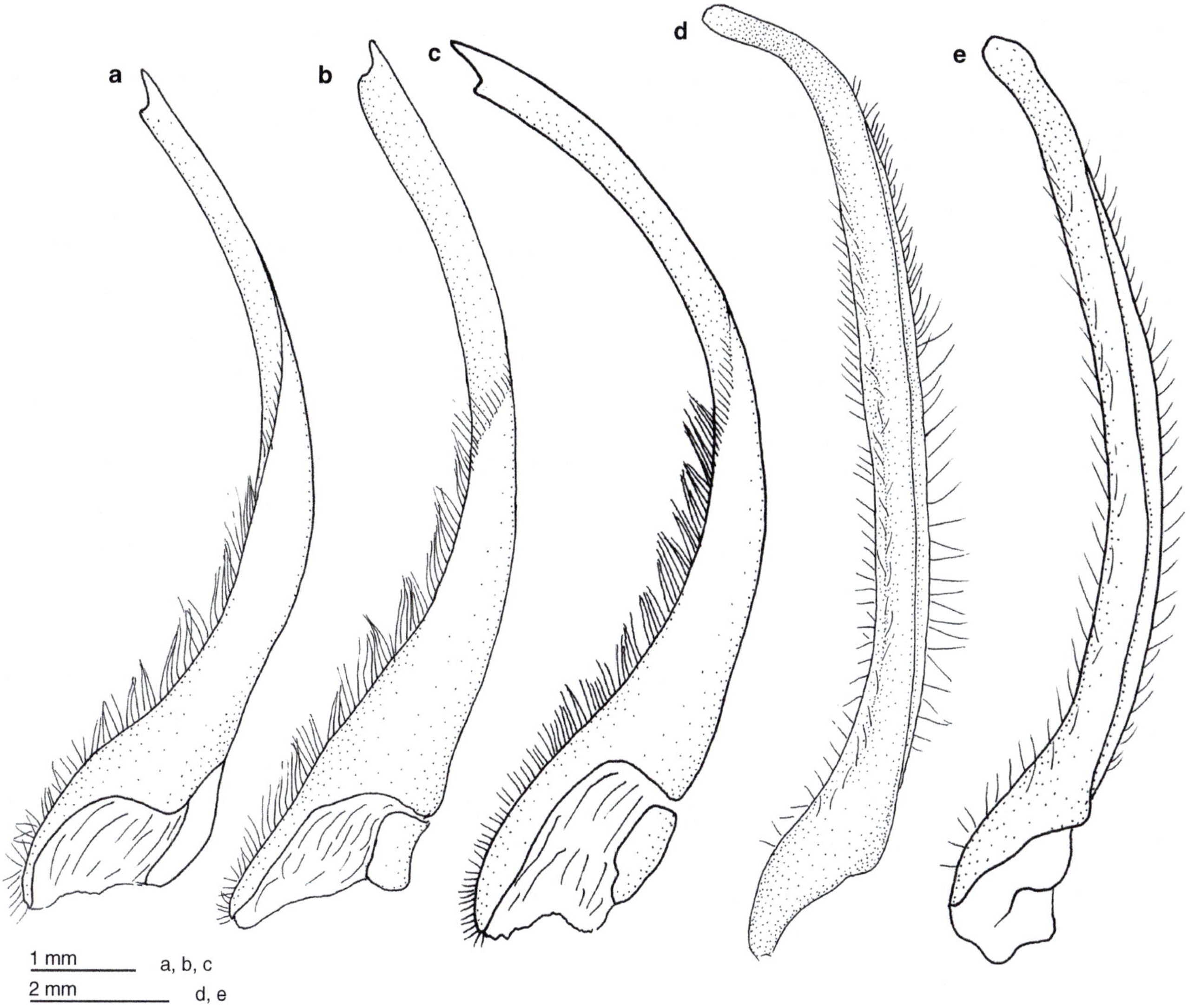

Fig. 18.4 Male G1, ventral view: (**a**) *Micippa philyra* (Herbst, 1803) (ZUTC Brach1355); (**b**) *Micippa platipes* Rüppell, 1830 (ZUTC 5067); (**c**) *Micippa thalia* (Herbst, 1803) (ZUTC Brach1819); (**d**) *Schizophrys aspera* (H. Milne Edwards, 1834) (SMF: Saudi Arabia); (**e**) *Schizophrys pakistanensis* Tirmizi and Kazmi, 1995 (ZUTC 5541)

eastwards to Pakistani coast (Tirmizi and Kazmi 1986), while *M. philyra* is mainly distributed along the Arabian coast of the Persian Gulf and the Gulf of Oman, but was also collected from one locality (Baghestan, Bandar-Abbas) on the Iranian coast of the Persian Gulf.

Micippa platipes Rüppell, 1830

(Figs. 18.2, 18.4b, and 18.5)

Micippa platipes Rüppell 1830: 8, pl 1, fig 4, pl 6, fig a.
Micippa platipes — Tirmizi and Kazmi 1986: 185–187, fig 58; Ng et al. 2008: 119 (in list); Simoes et al. 2001: 84, tab 1; Naderloo and Türkay 2012: 35, 36; Naderloo et al. 2013: 449, tab 1.

Type locality: Red Sea.

Distribution

Persian Gulf: Iran (Naderloo and Türkay 2012: Naderloo et al. 2013).

Gulf of Oman: Iran (Ghotbeddin et al. 2012a; Naderloo et al. 2015).

World distribution: Indo-West Pacific: South Africa, East Africa, Seychelles, Red Sea, Persian Gulf, Pakistan, China, Japan, Malaysia, Indonesia, Australia and Hawaii.

Habitat: Intertidal to 23 m, substrate rocky/cobble.

Fig. 18.5 *Micippa platipes* Rüppell, 1830. Male (ZUTC 5067), CL = 33.40, CB = 26.49 mm, Shib-Deraz, Qeshm Island, Iran, Persian Gulf, 16.02.2012, coll. H. Rahimian and H. Salehi (Photo: R. Abdollahi)

Micippa thalia (Herbst, 1803)

(Figs. 18.2, 18.4c, 18.6 and 18.7)

Cancer thalia Herbst 1803: 50, pl 58, fig 3.
Micippa aculeate Bianconi 1851.
Micippa pusilla Bianconi 1869.
Micippa miliaris Gerstaecker 1856.
Micippa haani Stimpson 1857.
Micippa thalia var. *caledonica* Kossmann 1877.
Micippa thalia var. *indica* Kossmann 1877.
Micippa inermis Haswell 1880.
Pisa (*Micippa*) *thalia* De Haan 1839: 98, pl 23
Micippa thalia — Alcock 1895: 251: Klunzinger 1906: 29; Griffin and Tranter 1974: 177, 1986: 279; Tirmizi and Kazmi 1986: 182–184, fig 57; Simoes et al. 2001: 84, tab 1.

Type locality: “Ostindien”. East India or Indo-Malayan Archipelago.

Distribution

Persian Gulf: No record.

Gulf of Oman: Iran (Naderloo et al. 2015).

World distribution: Indo-West Pacific: East Africa, Red Sea, Gulf of Oman, Pakistan and eastern India.

Habitat: Intertidal, substrate rocky/cobble.

Fig. 18.6 *Micippa thalia* (Herbst, 1803). Male (ZUTC Brach1819), CL = 37.61.40, CB = 36.11 mm, Beris, Sistan-Baluchestan, Iran, Gulf of Oman, depth 29–30 m, 16.10.2010, coll. N. Ghotbeddin, dorsal view of carapace (Photo: R. Abdollahi)

Fig. 18.7 *Micippa thalia* (Herbst, 1803). The same specimen, frontal view (Photo: R. Abdollahi)

Genus *Schizophrys* White, 1848

Key to species of the genus *Schizophrys* in the region

- Rostral spines distinctly curved inwards, lateral accessory spines large, as half as main rostral spines; merus and carpus of chelipeds with large denticle-shaped tubercles; suborbital region nearly smooth; G1 with long apical part, markedly bending laterally .... *Schizophrys aspera*
- Rostral spines nearly straight, lateral accessory spines distinctly small; merus and carpus of chelipeds with mostly granule-shaped tubercles; suborbital region distinctly tuberculate; G1 with short apical part, slightly bending laterally .... *Schizophrys pakistanensis*

Schizophrys aspera (H. Milne Edwards, 1834)
(Figs. 18.4d, 18.8 and 18.9)

Mithrax asper H. Milne Edwards 1834: 320.
Schizophrys aspera — A. Milne-Edwards 1872: 231–234, pl 10, figs 1, 1a–f; Nobili 1906a: 108; Guinot 1967a: 297 (in list); Griffin and Tranter 1974: 181, 182; Titgen 1982: 249 (partly); Griffin and Tranter 1986: 245, 246 (partly), figs 88a, 91 g, h; Hogarth 1989: 108; Apel 2001: 64, 65; Naderloo and Türkay 2012: 36; Naderloo et al. 2013: 449, tab 1.

Fig. 18.8 *Schizophrys aspera* (H. Milne Edwards, 1834). Male (SMF), CL = 56.9, CB = 55.6 mm, Ras Az-Zawr, Saudi Arabia, Persian Gulf, 17.05.1995, coll. M. Apel (Photo: S. Träkner)

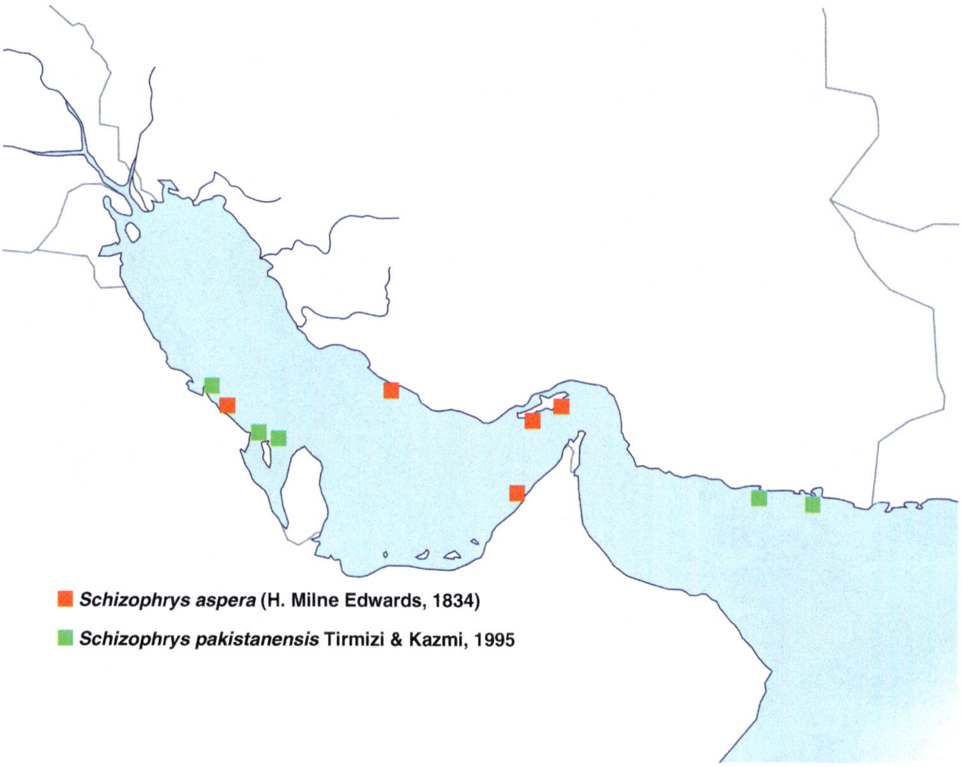

Fig. 18.9 Map showing the recorded localities of *Schizophrys aspera* (H. Milne Edwards, 1834), and *Schizophrys pakistanensis* Tirmizi and Kazmi, 1995

Schizophrys dama — Stephensen 1946: 108 (partly: Nobili's record from the Persian Gulf).

not *Schizophrys aspera* — Basson et al. 1977: 227; Titgen 1982: 249 (partly); Tirmizi and Kazmi 1986: 175–180, figs 54–56. [= *Schizophrys pakistanensis* Tirmizi and Kazmi 1995].

Type locality: Unknown.

Distribution

Persian Gulf: Iran (Naderloo and Türkay 2012: Naderloo et al. 2013), Saudi Arabia (Basson et al. 1977), UAE (Nobili 1906a).

Gulf of Oman: Oman (Apel 2001), UAE (Apel 2001).

World distribution: Indo-West Pacific: East Africa, Red Sea, southern Oman, Persian Gulf, Gulf of Oman, Pakistan, Andaman Sea, Malaysia, Thailand, China, Japan, Indonesia, Christmas Island, Australia, Fiji and Hawaii.

Habitat: Intertidal, substrate rocky/cobble.

Schizophrys pakistanensis Tirmizi and Kazmi, 1995

(Figs. 18.4e, 18.9, and 18.10)

Schizophrys pakistanensis Tirmizi and Kazmi 1995: 136–141, figs 1, 2.

Schizophrys pakistanensis — Apel 2001: 65; Ng et al. 2008: 118 (in list).

Schizophrys aspera — Basson et al. 1977: 227; Titgen 1982: 249 (partly); Tirmizi and Kazmi 1986: 175–180, figs 54–56.

Type locality: Pakistan (Karachi).

Distribution:

Persian Gulf: Saudi Arabia (Apel 2001).

Gulf of Oman: Iran (Naderloo et al. 2015).

World distribution: Northern Indian Ocean: Persian Gulf, Gulf of Oman and Pakistan.

Habitat: Intertidal, substrate rocky/cobble.

Fig. 18.10 *Schizophrys pakistanensis* Tirmizi and Kazmi, 1995. Male (ZUTC 5541), CL = 43.37, CB = 33.77 mm, Tis, Chabahar, Iran, Gulf of Oman, 01.01.2015, coll. Z. Rajaee and R. Abdollahi (Photo: R. Abdollahi)

Family Parthenopidae MacLeay, 1938 (Elbow Crabs)

19

Diagnosis Carapace subtriangular, subpentagonal to subcircular, mostly broader than long, lessly slighlty longer than broad; dorsal surface uneven, regions well defined, branchial regions swollen. Front simple, short, sometimes minutely tridentate. Orbits complete, usually round, short; eyestalks small. Antennules folded obliquely. Third maxillipeds with subquadrate merus. Chelipeds robust, markedly long, longer than walking legs, nearly triangular in cross section, fingers shorter than manus. Abdomen of males and females with somites 3–5 tightly fused, sometimes with distinct sutures. G1 markedly stout, slightly curved, apically with strong spines; G2 usually shorter than G1, sometimes subequal to latter. Genital opening of males coxal. Genital opening of females sternal.

Key to species of the family Parthenopidae in the region

1. Carapace laterally expanded, distinctly broader than long; walking legs not visible from dorsal view 2
- Carapace not laterally expanded, not distinctly broader than long; walking legs visible from dorsal view 3
2. Carapace regions well defined with granular ridges and elevated regions; carapace margins strongly denticulated; G1 S-shape ...... *Cryptopodia echinosa*
- Carapace regions weekly defined, regions not distinctly elevated; carapace margins not strongly denticulated; G1 not S-shape ...... *Cryptopodia patula*
3. Carapace distinctly longer than broad, with neck-like structure; epistome longitudinally rectangular ...... *Rhinolambrus contrarius*
- Carapace slightly broader than long, without neck-like structure; epistome transversely rectangular 4
4. Carapace roughly triangular, without regular teeth or spines on lateral margins; chelipeds not distinctly longer than carapace ...... *Pseudolambrus calappoides*
- Carapace nearly round or ovate, with distinct teeth and spines on lateral margins; chelipeds distinctly longer than carapace 5
5. Front unidentate; posterolateral margins of carapace and manus of chelipeds with large spines; merus of chelipeds not longer than carapace breadth ...... *Enoplolambrus carenatus*
- Front tridentate; posterolateral margins of carapace without spines; merus of chelipeds distinctly longer than carapace breadth ...... *Parthenope longimanus*

R. Naderloo, *Atlas of Crabs of the Persian Gulf*, DOI 10.1007/978-3-319-49374-9_19

Genus *Cryptopodia* H. Milne Edwards, 1834

Cryptopodia echinosa Chiong and Ng, 1998
(Figs. 19.1 and 19.2)

Cryptopodia echinosa Chiong and Ng 1998: 184–189, figs 18–20, 21b, c.
Cryptopodia echinosa — Apel 2001: 66; Ng et al. 2008: 130 (in list); Fatemi et al. 2012: 1–3, figs 2, 3.
Cryptopodia angulata — Alcock 1895: 282 (partly: Malabar Coast, southwestern India); Flipse 1930: 62 (in key).
Cryptopodia angulata var. *cippifer* — Stephensen 1946: 110; Titgen 1982: 249 (in list).
Type locality: India (Madras).
Distribution
Persian Gulf: Iran [Stephensen (1946) as *Cryptopodia angulata* var. *cippifer*].
Gulf of Oman: Iran (Fatemi et al. 2012).
World distribution: Northern Indian Ocean: Persian Gulf, Gulf of Oman, India (Mumbai), Malabar Coast and Madras.
Habitat: Subtidal, 46–62 m.

Cryptopodia patula Chiong and Ng, 1998
(Figs. 19.2, 19.3, and 19.4a)

Cryptopodia patula Chiong and Ng 1998: 174–177, figs 11–13.
Cryptopodia patula — Apel 2001: 66; Ng et al. 2008: 130 (in list).
Cryptopodia fornicata — Alcock 1895: 282 (partly: material from the Persian Gulf); Nobili 1906a: 114; Stephensen 1946: 110, 111; Evans et al. 1973: 275; Titgen 1982: 249 (in list).
Type locality: Persian Gulf.
Distribution
Persian Gulf: Iran [Stephensen (1946) as *C. fornicata*; Apel 2001], UAE [Nobili (1906a) as *C. fornicata*; Evans et al. (1973) as *C. fornicate*; Chiong and Ng 1998].
Gulf of Oman: Iran (Fatemi et al. 2012).
World distribution: Northern Indian Ocean: southern Oman, Persin Gulf, Gulf of Oman and Sri Lanka.
Habitat: Subtidal to 45 m.

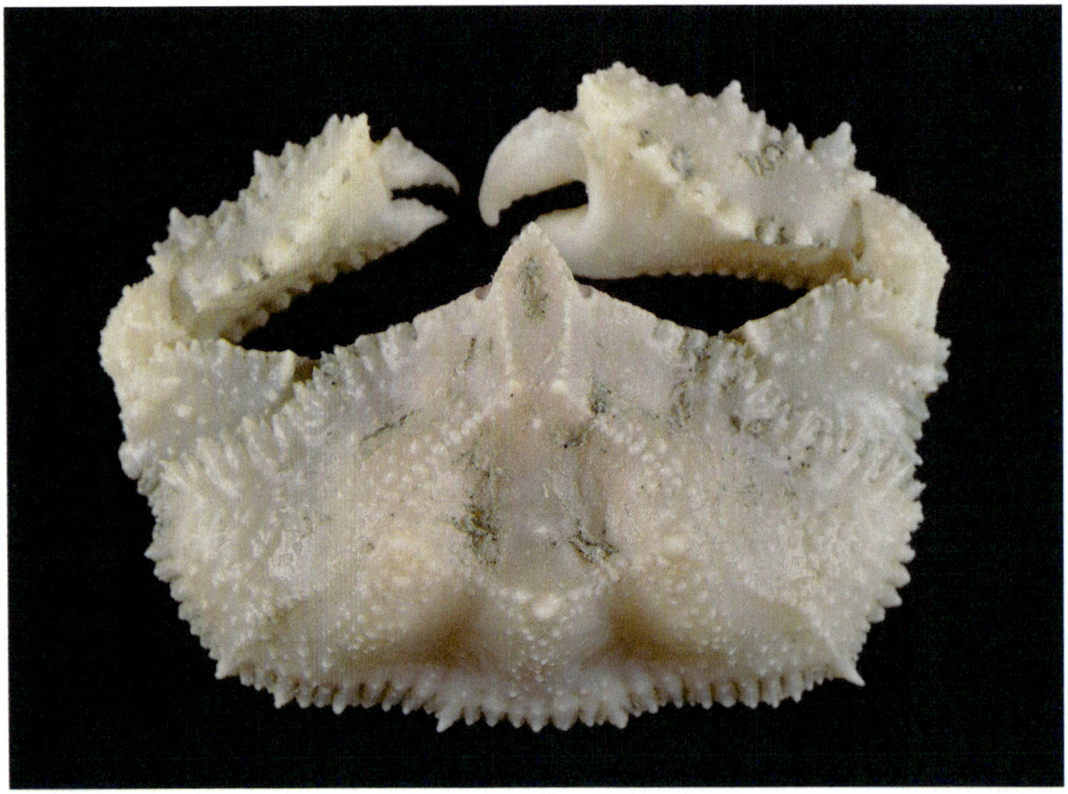

Fig. 19.1 *Cryptopodia echinosa* Chiong and Ng, 1998. Female (ZUMC CRU846), CL = 3.4, CB = 5.25 mm, SW of Bushehr, Iran, Persian Gulf, 14.03.1937, coll. G. Thorson (Photo: S. Träkner)

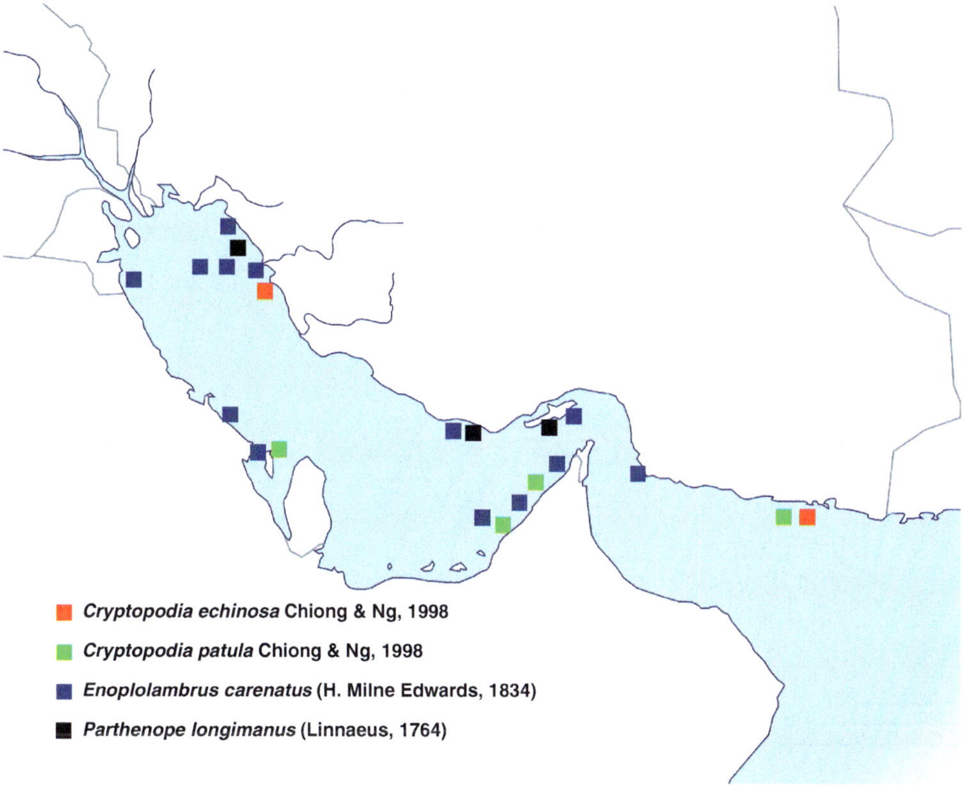

Fig. 19.2 Map showing the recorded localities of *Cryptopodia echinosa* Chiong and Ng, 1998, *Cryptopodia patula* Chiong and Ng, 1998, *Enoplolambrus carenatus* (H. Milne Edwards, 1834), and *Parthenope longimanus* (Linnaeus, 1758)

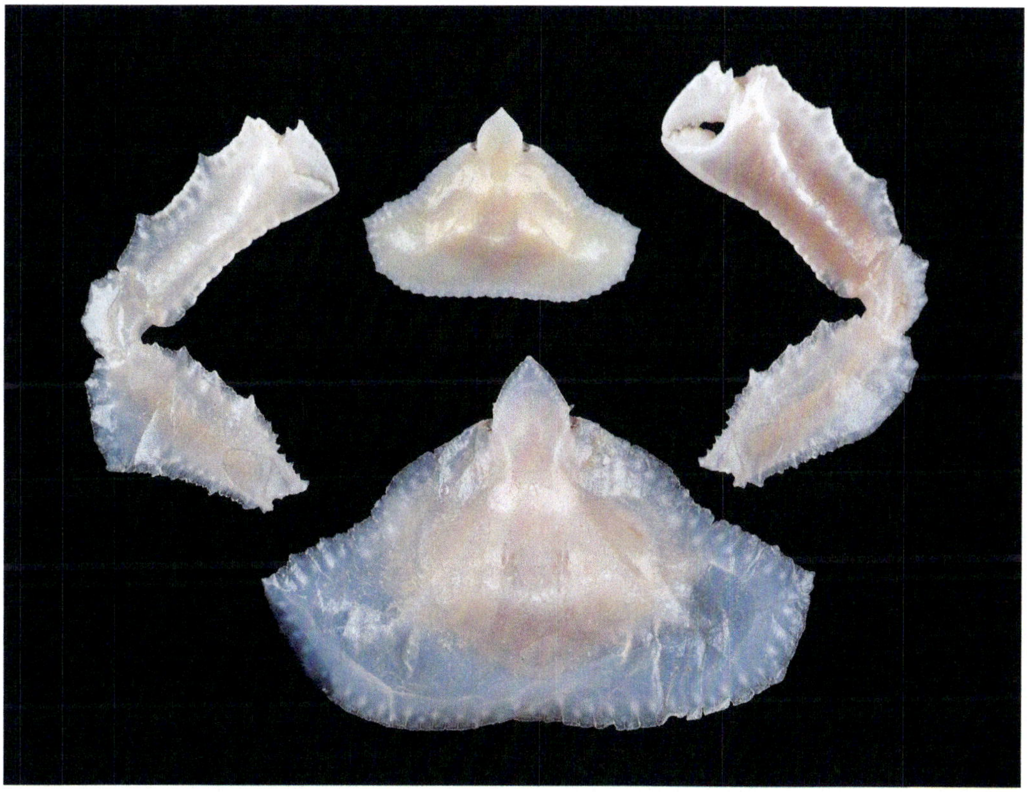

Fig. 19.3 *Cryptopodia patula* Chiong and Ng, 1998. Male (SMF), CL = 2.2, CB = 3.4 mm, Iran, Persian Gulf, collected by "Akademik Expedition", 30.12.1999 (Photo: S. Träkner)

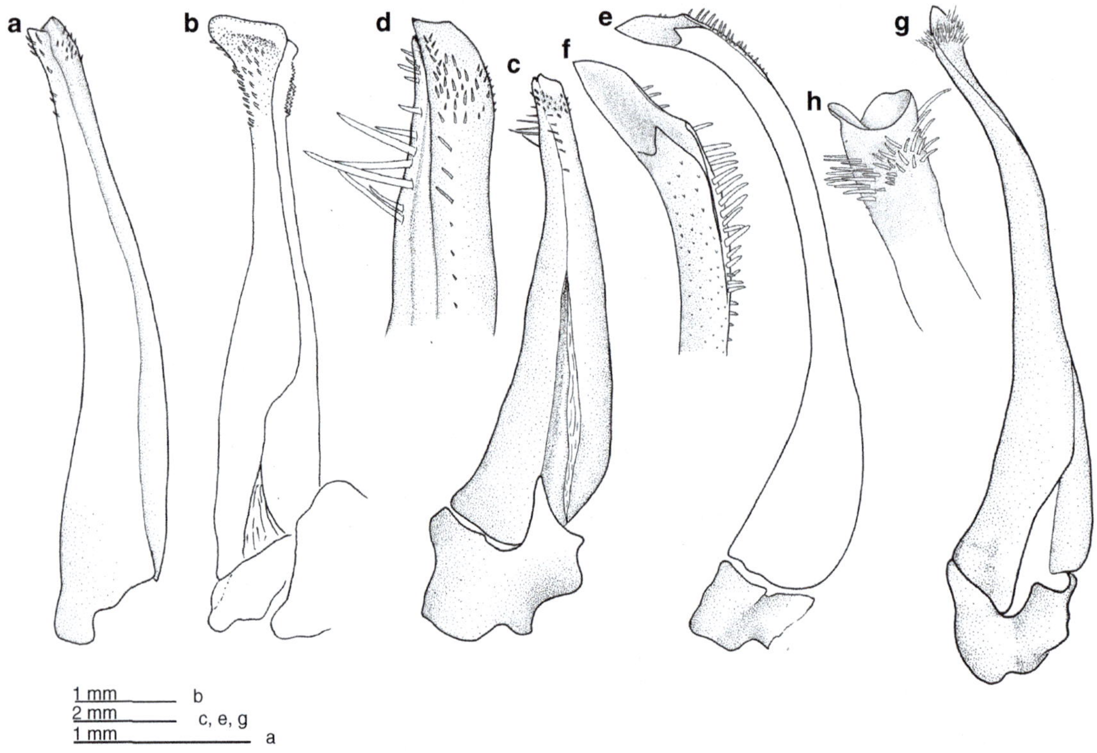

Fig. 19.4 Male G1, ventral view: (**a**) *Cryptopodia patula* Chiong and Ng, 1998 (SMF: Bahrain); (**b**) *Pseudolambrus calappoides* (Adams and White, 1849) (ZMG 939: Philippines); (**c**) *Enoplolambrus carenatus* (H. Milne Edwards, 1834) (ZMUC CRU859); (**d**) apical part of (**c**); (**e**) *Parthenope longimanus* (Linnaeus, 1758) (ZMUC CRU926); (**f**) apical part of (**e**); (**g**) *Rhinolambrus contrarius* (Herbst, 1804) (SMF: Saudi Arabia); (**h**) apical part of (**g**)

Genus *Enoplolambrus* A. Milne-Edwards, 1878

Enoplolambrus carenatus (H. Milne Edwards, 1834)

(Figs. 19.2, 19.4c, d, and 19.5)

Lambrus carenatus H. Milne Edwards 1834: 358.
Lambrus holdsworthi Miers 1879b: 19, pl 5, fig 3.
Lambrus carinatus — Adams and White 1849: 27, pl 5, fig 3; Basson et al. 1977: 243.
Lambrus holdsworthi — Nobili 1906a: 112, 113.
Platylambrus carinatus — Flipse 1930: 85.
Lambrus (*Platylambrus*) *carinatus* — Stephensen 1946: 111–113, figs 22, 23c, d; Guinot 1967a: 297 (in list); Titgen 1982: 249 (in list).
Lambrus carinatus — Basson et al. 1977: 250.
Parthenope cf. *carenatus* — Apel 2001: 67.
Parthenope carenatus — Naderloo and Sari 2007a: 342, tab 1.
Enoplolambrus carenatus —Tan and Ng 2007: 104, fig 6; Ng et al. 2008: 130 (in list).

Type locality: *Lambrus carenatus*: east coast of India (Pondicherry). — *Lambrus holdsworthi*: Sri Lanka.

Distribution

Persian Gulf: Iran [Stephensen (1946) as *Lambrus* (*Platylambrus*) *carinatus*; Apel (2001) as *Parthenope* cf. *carenatus*; Naderloo and Sari (2007a) as *Parthenope carenatus*], Kuwait [Apel (2001) as *Parthenope* cf. *carenatus*], Bahrain [Stephensen (1946) as *Lambrus* (*Platylambrus*) *carinatus*], UAE [Nobili (1906a) as *Lambrus holdsworthi*].

Gulf of Oman: Iran [Stephensen (1946) as *Lambrus* (*Platylambrus*) *carinatus*].

World distribution: Indian Ocean: Persian Gulf, Gulf of Oman, Pakistan, India, Sri Lanka, Andamans, Malaysia and Singapore.

Habitat: Subtidal, 29–45 m.

Fig. 19.5 *Enoplolambrus carenatus* (H. Milne Edwards, 1834). Female (ZUMC CRU842), CL = 25.00, CB = 31.5 mm, N Khark Island, Iran, Persian Gulf, 23.03.1937, coll. G. Thorson (Photo: S. Träkner)

Genus *Parthenope* Weber, 1795

Parthenope longimanus (Linnaeus, 1758)

(Figs. 19.2, 19.4e, f, and 19.6)

Cancer longimanus Linnaeus 1758: 441.
Cancer longimanus — Herbst 1790: 252, pl 19, figs 105, 106.
Lambrus (*Lambrus*) *longimanus* — Flipse 1930: 29, 84 (in list), figs 3, 11, 12, 16, 20, 21, 23a, b; Stephensen 1946: 113, fig 23a, b; Guinot 1967a: 297 (in list); Titgen 1982: 249 (in list).
Parthenope longimana — Tirmizi and Kazmi 1986: 198–201, fig 61a–d.
Parthenope longimanus — Apel 2001: 67; Naderloo and Sari 2007a: 342, tab 1; Ng et al. 2008: 131 (in list).
Type locality: Indian Ocean.
Distribution

Persian Gulf: Iran [Stephensen (1946) as *Lambrus* (*Lambrus*) *longimanus*; Naderloo and Sari 2007a].

Gulf of Oman: No record.

World distribution: Indian Ocean: Persian Gulf, Pakistan, India, Sri Lanka, Andamans, Malaysia and Singapore.

Habitat: Subtidal, 27–80 m, substrates sandy and muddy.

Genus *Pseudolambrus* Paul'son, 1875

Pseudolambrus calappoides (Adams and White, 1849)

(Figs. 19.4b, 19.7, and 19.8)

Parthenope calappoides Adams and White 1849: 34, pl 5, fig 5.
Parthenope calappoides —Titgen 1982: 249 (in list); Apel 2001: 66.
Lambrus (*Pseudolambrus*) *calappoides* — Nobili 1906a: 113–114; Guinot 1967a: 298 (in list).
Pseudolambrus calappoides — Flipse 1930: 87 (in list); Stephensen 1946: 113, 114; Ng et al. 2008: 131 (in list).
Type locality: Eastern Seas.
Distribution

Persian Gulf: Bahrain (Stephensen 1946), UAE [Nobili (1906a) as *Lambrus* (*Pseudolambrus*) *calappoides*].

Gulf of Oman: No record.

World distribution: Indo-West Pacific: Madagascar, Red Sea, Persian Gulf, India, Sri Lanka, China, Japan and Australia.

Habitat: Subtidal to 23 m.

Fig. 19.6 *Parthenope longimanus* (Linnaeus, 1758). Male (ZMUC CRU926), CL = 22.6, CB = 24.8 mm, NE Kish Island, Iran, Persian Gulf, depth 27 m, 08.04.1938, coll. G. Thorson (Photo: S. Träkner)

Fig. 19.7 *Pseudolambrus calappoides* (Adams and White, 1849). Female (ZMUC CRU927), CL = 23.1, CB = 27.4 mm, Bahrain, Persian Gulf, depth 23 m, 27.03.1937, coll. G. Thorson (Photo: S. Träkner)

Genus *Rhinolambrus* A. Milne-Edwards, 1878

Rhinolambrus contrarius (Herbst, 1804)

(Figs. 19.4g, h, 19.8, and 19.9)

Cancer contrarius Herbst 1804: 18, pl 60, fig 3.
Rhinolambrus contrarius — Tan et al. 1999: 200, figs 6, 7; Ng et al. 2008: 132 (in list).
Parthenope sp.? nov. — Apel 2001: 68.

Type locality: "Ostindien". East India or Indo-Malayan Archipelago.

Distribution

Persian Gulf: Saudi Arabia [Apel (2001) as *Parthenope* sp.? nov].

Gulf of Oman: No record.

World distribution: Indo-West Pacific: Seychelles, Persian Gulf, Pakistan, India, Sri Lanka, Japan, Australia and Hawaii.

Habitat: Subtidal, 3–60 m.

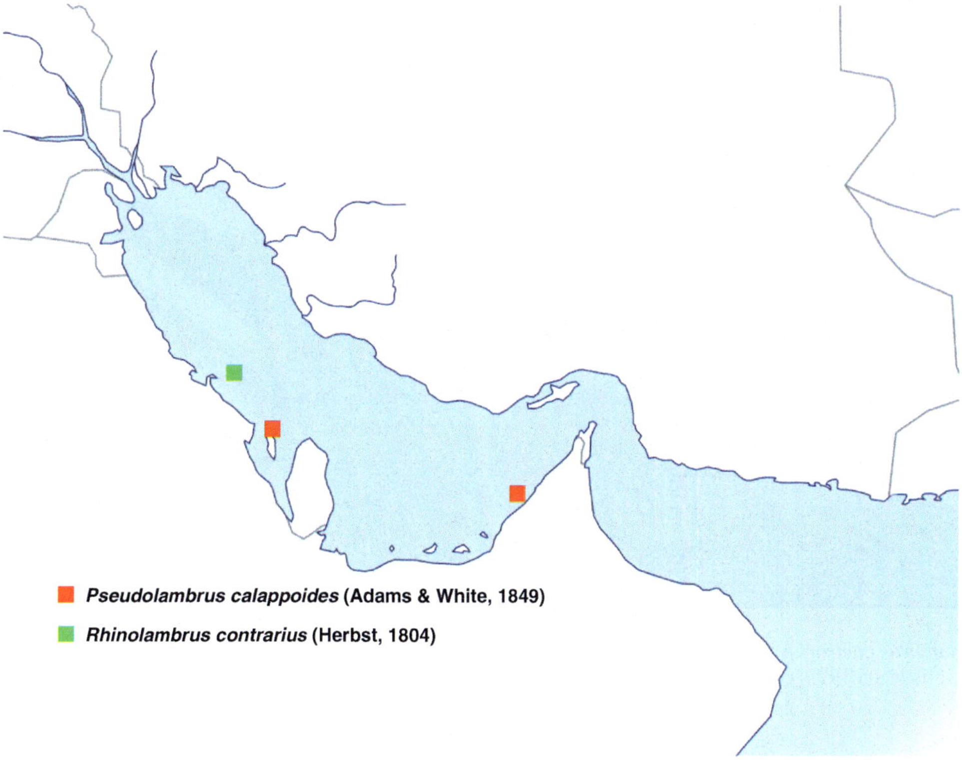

Fig. 19.8 Map showing the recorded localities of *Pseudolambrus calappoides* (Adams and White, 1849), and *Rhinolambrus contrarius* (Herbst, 1804)

Fig. 19.9 *Rhinolambrus contrarius* (Herbst, 1804). Male (SMF), CL = 3.00, CB = 2.60 mm, Abd-al-Kuri, Anjara Bay, Jana Island, Saudi Arabia, Persian Gulf, 31.05.1995, coll. M. Apel (Photo: S. Träkner)

Family Portunidae Rafinesque, 1815 (Swimming Crabs)

20

Diagnosis Carapace subhexagonal to transversely oval, usually broader than long, flat, slighlty convex, regions not well-defined, posterior surface with transverse ridges, epibranchial ridge present; anterolateral margins with spines, with distint junction with posterolateral margin, usually marked with epibranchial spine. Front broad, straight, dentate or lobate. First maxillipeds with lobe on endopodite known as "portunid lobe". Third maxillipeds completely covering buccal cavern, ischium distinctly longer than merus. Antennules folded transversely. Antennae with long basal antennal segment. Chelipeds distinctly robust, long, segments normally spinose, movable finger with longitudinal ridges. Walking legs 3 and 4 with propodus and dactylus paddle-shaped. Abdomen of males with somite 3–5 tightly fused; abdomen of females with free-moving somites. G1 long, rubost; G2 shorter than G1. Genital opening of males coxal. Genital opening of females sternal.

Key to genera of the family Portunidae in the region

1. Eyestalks extremely long .... *Podophthalmus*
- Eyestalks not markedly long 2
2. Supraorbital margins without fissure; chelipeds without spine .... *Libystes*
- Supraorbital margins with two fissures; chelipeds with spine on inner margin of carpus 3
3. Carapace with seven teeth on anterolateral margins; infraorbital margins with four round teeth, nearly equal .... *Carupa*
- Carapace with different number of teeth on anterolateral margins; infraorbital margins with fewer teeth, irregularly sized 4
4. Carapace nearly as long as broad; only epibranchial ridge on carapace; fingers of chelipeds with week grooves on upper surface .... *Lissocarcinus*
- Carapace markedly broader than long; more transverse ridges on carapace; fingers with deep grooves on upper surface 5
5. Anterolateral margins of carapace with nine teeth; basal antennal segment narrow; carpus of chelipeds with one or two spines on outer surface 6
- Anterolateral margins of carapace with four to six teeth; basal antennal segment relatively broad; carpus of chelipeds with three spines on outer surface 7
6. Carapace with weekly defined regions; chelipeds with smooth and flatted manus .... *Scylla*

R. Naderloo, *Atlas of Crabs of the Persian Gulf*, DOI 10.1007/978-3-319-49374-9_20

- Carapace with well-defined regions; manus of chelipeds with ridges and spines .... *Portunus*

7. Anterolateral margins of carapace with six teeth; front with six lobes .... *Charybdis*
   Anterolateral margins of carapace with three to five teeth; front with two to six lobes 8

8. Anterolateral margins of carapace with four teeth, first teeth with one small supplementary tooth on posterior base; front with six lobes, laterals separated from sublateral by deep notch; basal antennal segment with spine .... *Gonioinfradens*

- Anterolateral margins of carapace with five teeth; if less than five anterolateral teeth, no supplementary tooth on base of first tooth and basal antennal segment without spine; front with two to six lobes .... *Thalamita*

Genus *Carupa* Dana, 1851

Carupa tenuipes Dana, 1852

(Figs. 20.1, 20.2, and 20.3a)

Carupa tenuipes Dana 1852a: 85.
Carupa tenuipes — Dana 1852b: 279, 280, pl 17, fig 4a–e; Apel and Spiridonov 1998: 172, 173, fig 4, pl 1; Apel 2001: 69; Ng et al. 2008: 148 (in list).
Type locality: Tuamotu.

Distribution

Persian Gulf: Saudi Arabia (Apel and Spiridonov 1998; Apel 2001).
Gulf of Oman: UAE (Apel and Spiridonov 1998; Apel 2001).
World distribution: Indo-West Pacific: Madagascar, East Africa, Red Sea, Persian Gulf, Gulf of Oman, India, Maldives, Andaman Islands, Indo-Malayan Archipelago, Japan and Australia.

Habitat: Subtidal, 3–60 m.

Fig. 20.1 *Carupa tenuipes* Dana, 1852. Female (SMF 23223), CL = 7.9, CB = 11. 9 mm, Abu Ahmad, Fujairah, UAE, Gulf of Oman, 02.07.1995, coll. M. Apel, 1995 (Photo: S. Träkner)

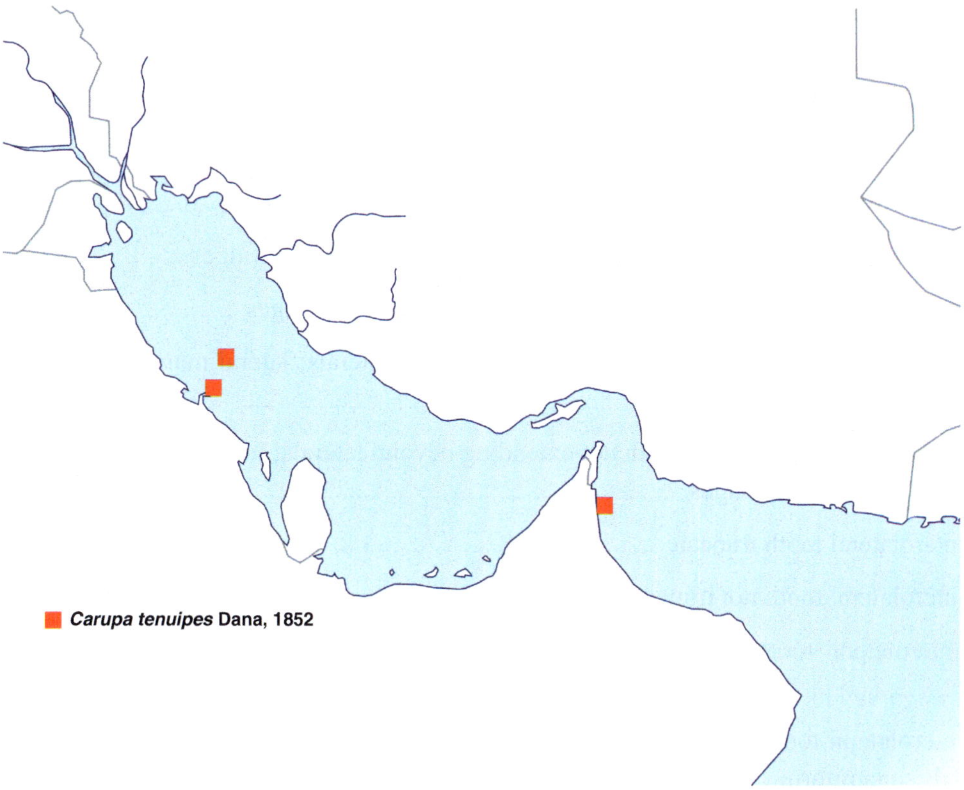

Fig. 20.2 Map showing the recorded localities of *Carupa tenuipes* Dana, 1852

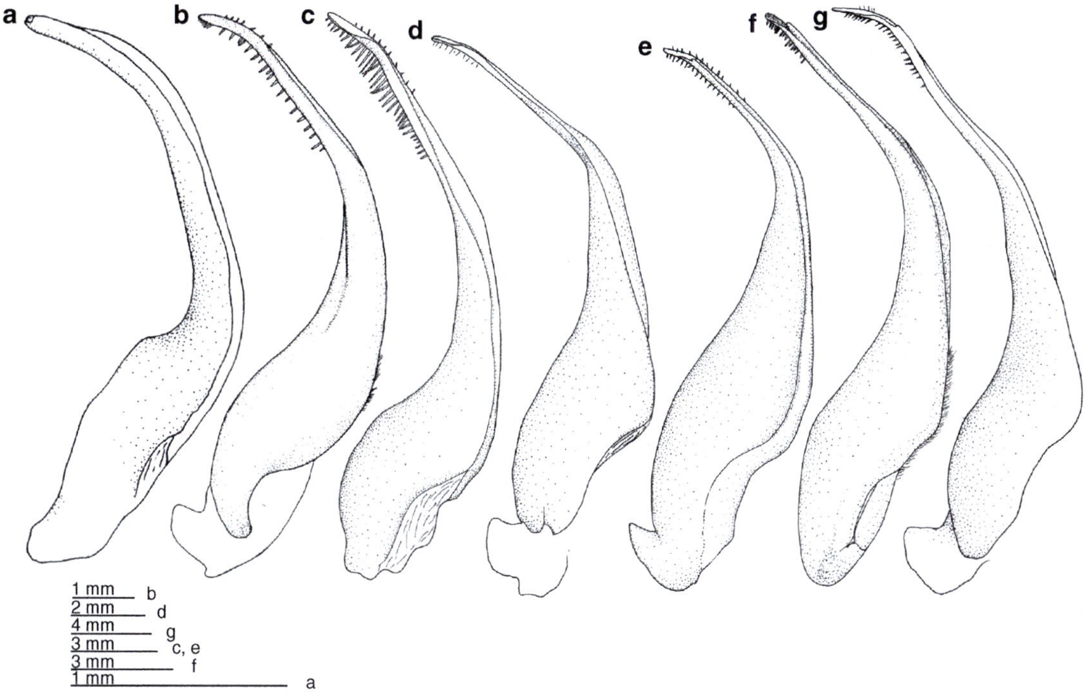

Fig. 20.3 Male G1, ventral view: (**a**) *Carupa tenuipes* Dana, 1852 (SMF 23223); (**b**) *Charybdis* (*Charybdis*) *annulata* (Fabricius, 1798) (SMF: Arabian Sea); (**c**) *Charybdis* (*Charybdis*) *feriata* (Linnaeus, 1758) (ZUTC 5059); (**d**) *Charybdis* (*Charybdis*) *hellerii* (A. Milne-Edwards, 1867) (SMF 40971); (**e**) *Charybdis* (*Charybdis*) *lucifera* (Fabricius, 1798) (ZUTC 5046); (**f**) *Charybdis* (*Charybdis*) *miles* (De Haan, 1835) (ZUTC Brach1047); (**g**) *Charybdis* (*Charybdis*) *natator* (Herbst, 1794) (SMF 24279)

Genus *Charybdis* De Haan, 1833

Key to species of the genus *Charybdis* in the region

1. Junction between posterior margin and posterolateral margins arched 2
- Junction between posterior margin and posterolateral margins angular 7
2. Dorsal surface of carapace with distinct cardiac and mesobranchial ridges 3
- Dorsal surface of carapace without cardiac and mesobranchial ridges 4
3. Median and submedian frontal teeth extending beyond laterals; lateral margins of sixth somite of male abdomen convex ..... *Charybdis* (*Charybdis*) *variegata*
- Median and submedian frontal teeth not extending beyond laterals; two series of granular carapace ridges anterior to epibranchial ridges ..... *Charybdis* (*Charybdis*) *natator*
4. First anterolateral tooth truncate 5
- First anterolateral tooth not truncate 6
5. First anterolateral tooth nearly bifid, following teeth broad; carapace with conspicuous cross-shaped mark ..... *Charybdis* (*Charybdis*) *feriata*
- First anterolateral tooth truncated but not bifid; carapace pilose, without special mark; lower surface of chelipeds squamiform ..... *Charybdis* (*Charybdis*) *miles*
6. Carpus of fifth legs with posterior spine ..... *Charybdis* (*Charybdis*) *hellerii*
- Carpus of fifth legs without posterior spine ..... *Charybdis* (*Charybdis*) *annulata*
7. Last anterolateral tooth markedly longer than other teeth 8
- Last anterolateral tooth not markedly longer than other teeth 10
8. Last anterolateral teeth of moderate length; cardiac and mesobranchial portions of carapace with ridge-like elevations ..... *Charybdis* (*Goniohellenus*) *longicollis*
- Last anterolateral teeth very much longer than other teeth; cardiac and mesobranchial portions of carapace with granular patches 9
9. Merus of last legs about 1.6 times as long as broad; medium-sized species .. *Charybdis* (*Goniohellenus*) *hoplites*
- Merus of last legs about twice as long as broad; small-sized species ... *Charybdis* (*Goniohellenus*) *pusilla*
10. Carapace with granular patches, densely pilose ..... *Charybdis* (*Goniohellenus*) *omanensis*
- Carapace without granular patches, smooth, reddish ..... *Charybdis* (*Goniohellenus*) *smithii*

Charybdis (*Charybdis*) *annulata* (Fabricius, 1798)
(Figs. 20.3b, 20.4, and 20.5)

Portunus annulatus Fabricius 1798: 364.
Charybdis annulata — Gordon 1931: 537, fig 13d; Kazmi and Kazmi 1987: 56; Tirmizi and Kazmi 1996: 29 (in key), 50–52, fig 25a–e, pl 2b.
Charybdis (*Charybdis*) *annulata* — Hogarth 1989: 104 (in list); Apel and Spiridonov 1998: 185 (in key), 187–190, figs 10, 11, 16.

Type locality: "Oceano Indico" (the type specimens probably originate from Tharangambadi, south-east India).

Distribution

Persian Gulf: Iran (present study).

Gulf of Oman: Iran (Naderloo et al. 2015).

World distribution: Indo-West Pacific: South Africa, East Africa, Persian Gulf, Gulf of Oman, Pakistan, India, Sri Lanka, Vietnam, China, Japan, Indonesia, Singapore and Australia.

Habitat: Subtidal, 3–60 m.

Fig. 20.4 *Charybdis* (*Charybdis*) *annulata* (Fabricius, 1798). Female (ZUTC 5486), CL = 49.15, CB = 71.47 mm, Abu Musa Island, Iran, Persian Gulf, 25.12.2014, coll. S. Ebrahimnejad and R. Abdollahi (Photo: R. Abdollahi)

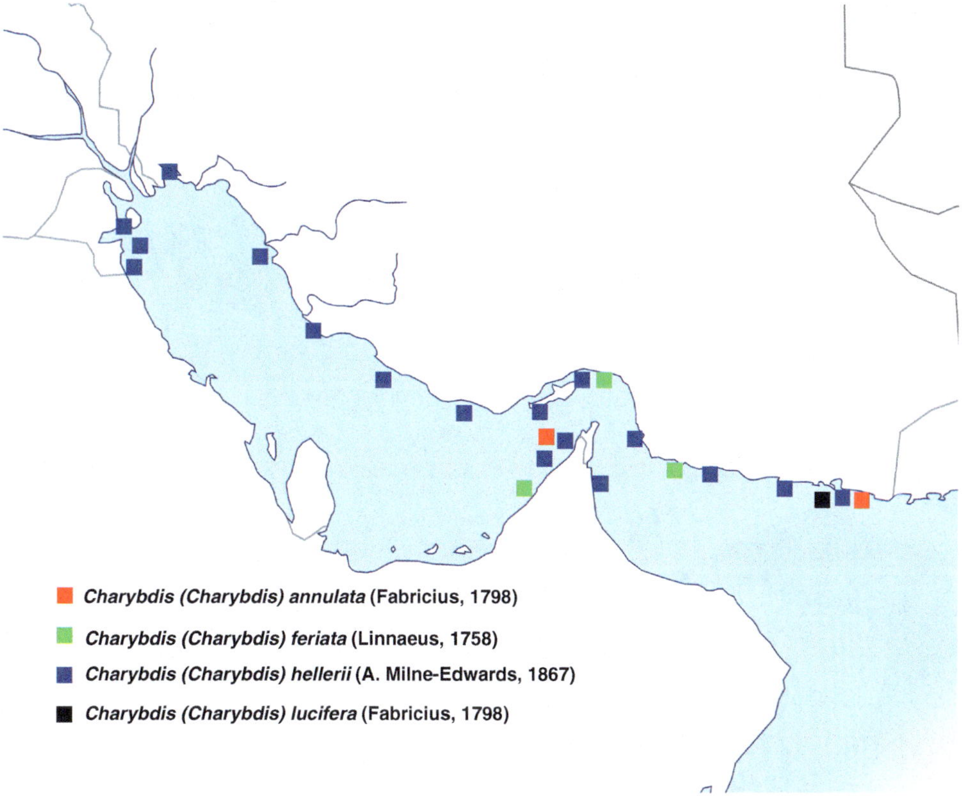

Fig. 20.5 Map showing the recorded localities of *Charybdis* (*Charybdis*) *annulata* (Fabricius, 1798), *Charybdis* (*Charybdis*) *feriata* (Linnaeus, 1758), *Charybdis* (*Charybdis*) *hellerii* (A. Milne-Edwards, 1867), and *Charybdis* (*Charybdis*) *lucifera* (Fabricius, 1798)

Charybdis (*Charybdis*) *feriata* (Linnaeus, 1758)
(Figs. 20.3c, 20.5, and 20.6)

Cancer feriatus Linnaeus 1758: 627.
Cancer cruciata Herbst 1794: pl 38, fig 1; Herbst 1796: 155, 156.
Charybdis cruciata — Guinot 1967a: 255 (in list).
Charybdis (*Charybdis*) *cruciata* — Stephensen 1946: 114, 115.
Charybdis (*Charybdis*) *feriatus* — Titgen 1982: 249 (in list).
Charybdis (*Charybdis*) *feriata* — Apel and Spiridonov 1998: 185 (in key), 192–194, pls 2, 3; Apel 2001: 69; Naderloo and Sari 2007a: 342, tab 1; Ng et al. 2008: 153 (in list).

Type locality: *Cancer feriatus*: Indonesia (Ambon, Maluku Islands). — *Cancer cruciata*: "Ostindien", East India or possibly Indo-Malayan Archipelago.

Distribution

Persian Gulf: Iran [Stephensen (1946) as *Charybdis* (*Charybdis*) *cruciata*; Naderloo and Sari 2007a], UAE (Apel and Spiridonov 1998; Apel 2001).

Gulf of Oman: Iran [Stephensen (1946) as *Charybdis* (*Charybdis*) *cruciata*; Naderloo et al. 2015].

World distribution: Indo-West Pacific: South Africa, East Africa, Persian Gulf, Gulf of Oman, Pakistan, India, Sri Lanka, Vietnam, China, Japan, Indonesia, Singapore and Australia.

Habitat: Subtidal, 3–60 m.

Charybdis (*Charybdis*) *hellerii* (A. Milne-Edwards, 1867)
(Figs. 20.3d, 20.5, and 20.7)

Goniosoma hellerii A. Milne-Edwards 1867: 282, 283.
Goniosoma merguiense De Man 1888b: 82, pl 5, figs 3, 4.
Charybdis (*Goniosoma*) *merguiense* — Alcock 1899a: 55.
Charybdis (*Charybdis*) *hellerii* — Stephensen 1946: 117; Titgen 1982: 249 (in list); Apel and Spiridonov 1998: 185 (in key), 194–197, figs 13–15, 17; Apel 2001: 70; Naderloo and Sari 2007a: 342, tab 1; Ng et al. 2008: 153 (in list); Naderloo and Türkay 2012: 39; Naderloo et al. 2013: 449, tab 1.

Fig. 20.6 *Charybdis* (*Charybdis*) *feriata* (Linnaeus, 1758). Male (ZUTC 5059), *CL* = 65.35, *CB* = 98.90 mm, Tang, Sistan-Baluchestan, Iran, Gulf of Oman, 07.07.2009, coll. N. Ghotbeddin (Photo: R. Abdollahi)

Charybdis (*Charybdis*) *lucifera* — Stephensen 1946: 115, 116; Guinot 1967a: 255 (partly); Titgen 1982: 249 (in list).

Charybdis helleri — Guinot 1967a: 255 (in list); Jones 1986a: 161, pl 47.

Type locality: *Goniosoma hellerii*: New Caledonia. — *Goniosoma merguiense*: Mergui Archipelago.

Distribution

Persian Gulf: Iran (Stephensen 1946; Naderloo and Sari 2007a; Naderloo and Türkay 2012; Naderloo et al. 2013), Kuwait (Jones 1986a; Apel and Spiridonov 1998; Apel 2001), UAE (Apel and Spiridonov 1998; Apel 2001).

Gulf of Oman: Iran (Naderloo et al. 2015), UAE (Apel and Spiridonov 1998; Apel 2001).

World distribution: Pantropical: South Africa, East Africa, Madagascar, Red Sea, Gulf of Aden, Socotra, southern Oman, Persian Gulf, Gulf of Oman, Pakistan, India, Sri Lanka, Mergui-Archipelago, Andaman Sea, Thailand, Malaysia, Philippines, Indonesia, Singapore, Australia, New Caledonia, Mediterranean, Caribbean coast of Colombia, Atlantic coasts of Florida and Brazil.

Habitat: Intertidal to 30 m, substrates rocky/cobble, sandy, muddy and coral reefs.

Remarks *Charybdis hellerii* is one of the common portunid species occurring in the intertidal zone of the Persian Gulf. Naderloo (2005) recorded three different forms of the species from the Iranian waters of the Persian Gulf including 1- large specimens with dorsal surface of carapace glabrous, 2- small specimens with glabrous carapace, 3- large specimens with dorsal surface of the carapace covered with dense pubescence. These forms of the carapace have also been recorded from Pakistan by Tirmizi and Kazmi (1996) and from Red Sea. There is no study providing information on an age- or size-related reason for pubescent/non-pubescent forms in this species.

Fig. 20.7 *Charybdis* (*Charybdis*) *hellerii* (A. Milne-Edwards, 1867). Male (SMF 40791), CL = 48.2, CB = 72.3 mm, Basaeedu, Qeshm Island, Iran, Persian Gulf, 07.05.2008, coll. R. Naderloo and A. Kazmi (Photo: S. Träkner)

Charybdis (Charybdis) lucifera (Fabricius, 1798)
(Figs. 20.3e, 20.5, and 20.8)

Portunus lucifer Fabricius 1798: 364.
Goniosoma quadrimaculatum A. Milne-Edwards, 1861: 375, pl 34, fig 3.
Charybdis lucifera — Kazmi and Kazmi 1987: 56; Tirmizi and Kazmi 1996: 29 (in key), 49, 50, fig 24a–e, pl 1d.
not *Charybdis* (*Charybdis*) *lucifera* — Stephensen 1946: 115, 116 [= *Charybdis* (*Charybdis*) *helleri* (A. Milne-Edwards, 1867)].
Charybdis (*Charybdis*) *lucifera* — Apel and Spiridonov 1998: 185 (in key), 197–199, figs 18, 20, 21; Ng et al. 2008: 153 (in list).

Type locality: *Portunus lucifer* Fabricius, 1798: Indian Ocean (probably Tharangambadi). — *Goniosoma quadrimaculatum* A. Milne-Edwards, 1861: Malabar and Java.

Distribution

Persian Gulf: No record.

Gulf of Oman: Iran (Naderloo et al. 2015).

World distribution: Indo-West Pacific: South Africa, East Africa, Gulf of Oman, Pakistan, India, Sri Lanka, Vietnam, China, Japan, Indonesia, Singapore, Australia.

Habitat: Subtidal, 3–60 m.

Fig. 20.8 *Charybdis* (*Charybdis*) *lucifera* (Fabricius, 1798). Male (ZUTC 5046), CL = 60.69, CB = 90.30 mm, Konarak, Iran, Gulf of Oman, 26.07.2009, coll. N. Ghotbeddin (Photo: R. Abdollahi)

Charybdis (Charybdis) miles (De Haan, 1835)

(Figs. 20.3f, 20.9, and 20.10)

Portunus (*Charybdis*) *miles* De Haan 1835: 41, pl 11, fig 1.

Charybdis (*Charybdis*) *miles* — Stephensen 1946: 116; Titgen 1982: 249 (in list); Apel and Spiridonov 1998: 185 (in key), 199–201, figs 19, 22, 23, pl 4; Apel 2001: 70; Naderloo and Sari 2007a: 343, tab 1; Ng et al. 2008: 153 (in list).

Type locality: Japan.

Distribution

Persian Gulf: Iran (Stephensen 1946; Naderloo and Sari 2007a), UAE (Apel and Spiridonov 1998; Apel 2001).

Gulf of Oman: UAE (Apel and Spiridonov 1998; Apel 2001).

World distribution: Indo-West Pacific: Persian Gulf, Gulf of Oman, India, Bay of Bengal, Andamans, South Asia, China, Japan and Australia.

Habitat: Subtidal, 20–200 m.

Fig. 20.9 *Charybdis* (*Charybdis*) *miles* (De Haan, 1835). Male (ZUTC Brach1047), CL = 58.5, CB = 97.5 mm, Iran, Persian Gulf, 09.10.2003, coll. R. Naderloo (Photo: R. Abdollahi)

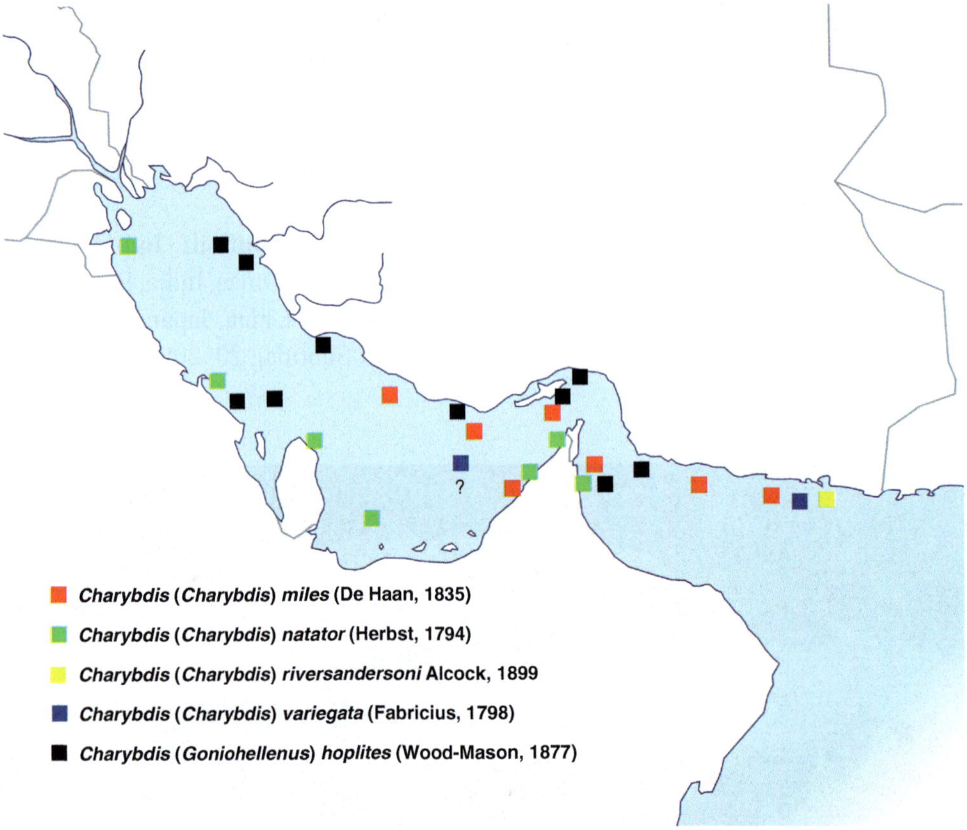

Fig. 20.10 Map showing the recorded localities of *Charybdis* (*Charybdis*) *miles* (De Haan, 1835), *Charybdis* (*Charybdis*) *natator* (Herbst, 1794), *Charybdi* (*Charybdis*) *riversandersoni* Alcock, 1899, *Charybdis* (*Charybdis*) *variegata* (Fabricius, 1798), and *Charybdis* (*Goniohellenus*) *hoplites* (Wood-Mason, 1877)

Charybdis (*Charybdis*) *natator* (Herbst, 1794)

(Figs. 20.3g, 20.10, and 20.11)

Cancer natator Herbst 1794: pl 40, fig 1.
Charybdis (*Goniosoma*) *natator* — Alcock 1899a: 61; Nobili 1906a: 115.
Charybdis natator — Guinot 1967a: 254 (in list); Basson et al. 1977: 250; Jones 1986a: 161, pl 48; Carpenter et al. 1997: 38.
Charybdis (*Charybdis*) *natator* — Stephensen 1946: 116; Guinot 1967a: 254 (in list); Titgen 1982: 249 (in list); Ng et al. 2008: 153 (in list).
Charybdis (*Charybdis*) *natator natator* — Apel and Spiridonov 1998: 185 (in key), 201–206, fig 28, pls 5, 6; Apel 2001: 70.
Charybdis natatur [sic!] — Al-Khayat and Jones 1999: 61.

Type locality: "Ostindien". East India or Indo-Malayan Archipelago.

Distribution

Persian Gulf: Kuwait (Jones 1986a), Saudi Arabia (Basson et al. 1977), Qatar (Al-Khayat and Jones 1999), UAE (Nobili 1906a; Apel and Spiridonov 1998; Apel 2001).

Gulf of Oman: UAE (Apel and Spiridonov 1998; Apel 2001).

World distribution: Indo-West Pacific: South Africa, East Africa, Madagascar, Red Sea, Socotra, southern Oman, Persian Gulf, Gulf of Oman, Pakistan, India, Thailand, Malaysia, Japan and Australia.

Habitat: Intertidal to 100 m, rocky/cobble substrate.

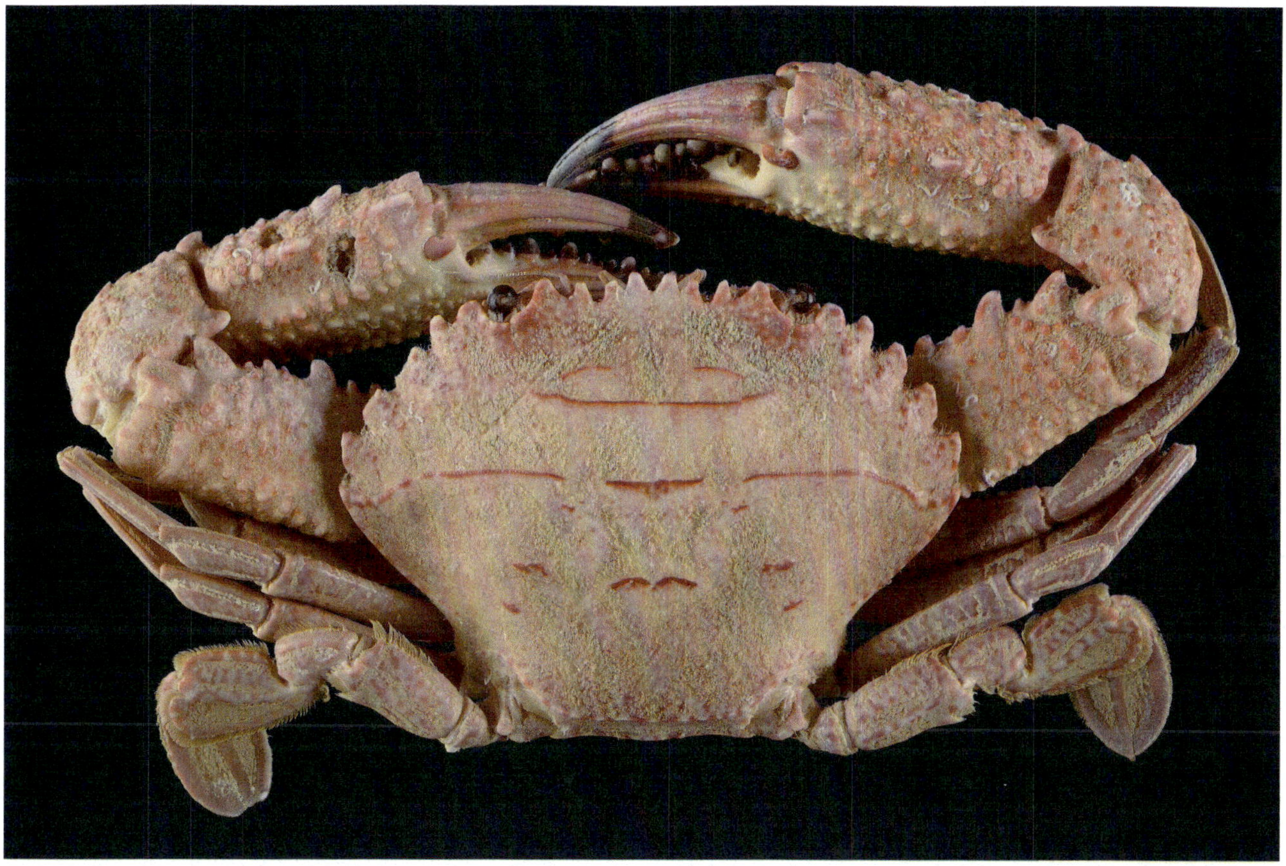

Fig. 20.11 *Charybdis* (*Charybdis*) *natator* (Herbst, 1794). Male (SMF 24279), CL = 77.5, CB = 103.1 mm, Dubai, UAE, Persian Gulf, 22.03.1998, coll. M. Apel (Photo: S. Träkner)

Charybdis (*Charybdis*) *riversandersoni* Alcock, 1899

(Figs. 20.10, 20.12, and 20.13a)

Charybdis (*Goniosoma*) *rivers-Andersoni* Alcock 1899c: 53.

Charybdis (*Goniosoma*) *rivers-Andersoni* — Alcock and McArdle 1900: pl 46, fig 3.

Charybdis (*Charybdis*) *rivers-andersoni* — Leene 1938: 28–30, figs 3, 4a, b.

Charybdis riversandersoni — Stephenson and Rees 1968: 95–100, figs 1b, 1f, 2b, 2d, pl 12b.

Charybdis (*Charybdis*) *riversandersoni* — Apel and Spiridonov 1998: 185 (in key), 208, 209, figs 27, 28; Spiridonov and Türkay 2001: 452–455, figs 1c, 2c, 3c, 4c, 8a–d; Ng et al. 2008: 153 (in list).

Type locality: India (17° 27′N, 71° 41′E, Konkan, NW of India, Arabian Sea).

Distribution

Persian Gulf: No record.

Gulf of Oman: Iran (Naderloo et al. 2015).

World distribution: Northern Indian Ocean: Gulf of Oman, northern Arabian Sea and western India.

Habitat: Subtidal, 94–291 m, substrates muddy and sandy.

Fig. 20.12 *Charybdis* (*Charybdis*) *riversandersoni* Alcock, 1899. Male (ZUTC 5349), CL = 57.31, CB = 85.75 mm, Iran, Gulf of Oman, 13.09.2009, coll. N. Ghotbeddin (Photo: R. Abdollahi)

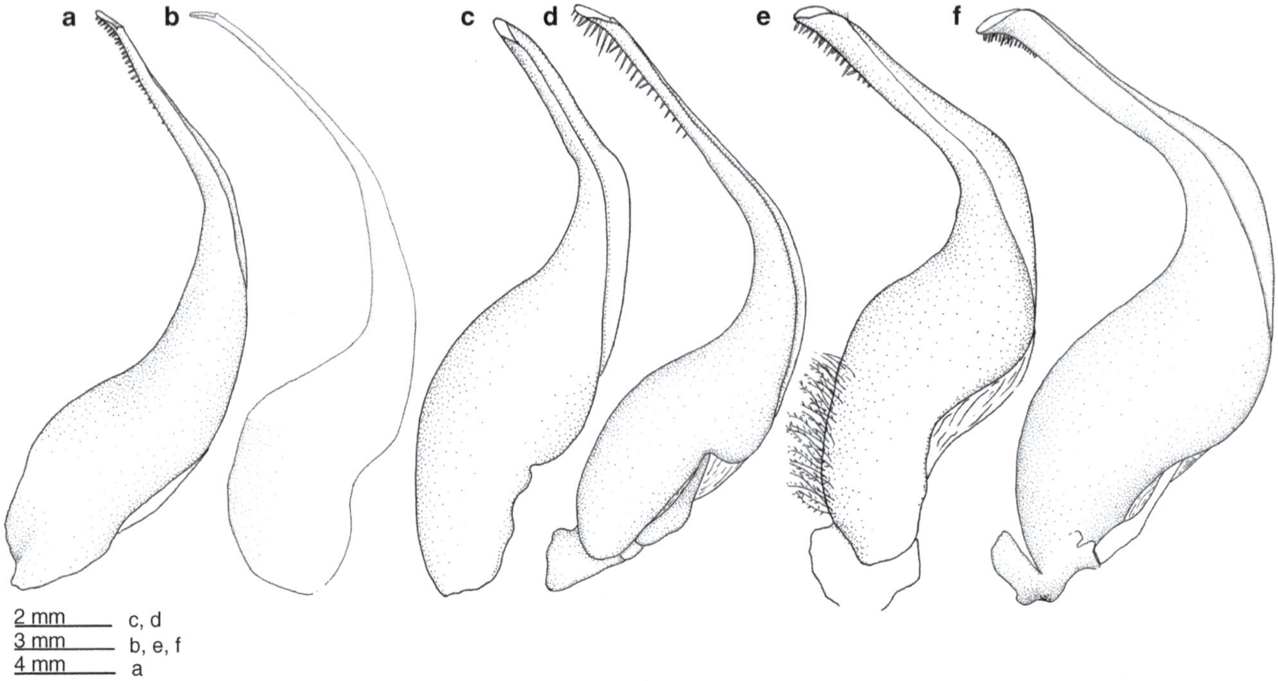

Fig. 20.13 Male G1, ventral view: (**a**) *Charybdis* (*Charybdis*) *riversandersoni* Alcock, 1899 (ZUTC 5349); (**b**) *Charybdis* (*Charybdis*) *variegata* (Fabricius, 1798) (ZUTC Brach1478); (**c**) *Charybdis* (*Goniohellenus*) *hoplites* (Wood-Mason, 1877) (ZUTC Brach1795); (**d**) *Charybdis* (*Goniohellenus*) *longicollis* Leene, 1938 (ZUTC Brach1108); (**e**) *Charybdis* (*Goniohellenus*) *omanensis* Leene, 1938 (SMF 22980); (**f**) *Charybdis* (*Goniohellenus*) *smithii* MacLeay, 1838 (ZUTC 5049)

Charybdis (Charybdis) variegata (Fabricius, 1798)
(Figs. 20.10, 20.13b, and 20.14)

Portunus variegatus Fabricius 1798: 364.
Charybdis (*Goniosoma*) *variegata* — Alcock 1899a: 60; Nobili 1906b: 196; Chopra 1935: 488, fig 10a, b.
Charybdis (*Charybdis*) *variegata* — Stephensen 1946: 116; Titgen 1982: 249 (in list); Apel and Spiridonov 1998: 185 (in key), 209–211, figs 25, 30; Apel 2001: 71; Ng et al. 2008: 153 (in list).
Type locality: Indian Ocean, possibly Tharangambadi in the southeast coast of India.

Distribution

Persian Gulf: Alcock (1899a) and Chopra (1935) recorded this species from an unknown locality.

Gulf of Oman: Iran (Naderloo et al. 2015). Alcock (1899a) and Chopra (1935) recorded this species from an unknown locality.

World distribution: Indo-West Pacific: South Africa, Madagascar, Red Sea, Persian Gulf, Gulf of Oman, Pakistan, India, Malaysia, China and Japan.

Habitat: Subtidal, 9–50 m, substrates sandy, muddy and rocky.

Fig. 20.14 *Charybdis* (*Charybdis*) *variegata* (Fabricius, 1798). Male (ZUTC Brach1478), CL = 39.41, CB = 56.68 mm, Iran, Gulf of Oman, coll. N. Ghotbeddin (Photo: R. Abdollahi)

Charybdis (*Goniohellenus*) *hoplites* (Wood-Mason, 1877)

(Figs. 20.10, 20.13c, and 20.15)

Goniosoma hoplites Wood-Mason 1877: 422.

Goniosoma hoplites — Alcock and Anderson 1894a: 184; Alcock and Anderson 1896: pl 23, fig 6.

Charybdis (*Goniohellenus*) *hoplites* — Alcock 1899a: 66, 67; Nobili 1906a: 119; Titgen 1982: 249 (in list); Apel and Spiridonov 1998: 211–213, figs 31, 33; Apel 2001: 71; Ng et al. 2008: 153 (in list).

Charybdis hoplites forma *typica* — Leene 1938: 99–102, figs 53, 54; Stephensen 1946: 117, 118.

Charybdis hoplites — Guinot 1967a: 254 (in list); Basson et al. 1977: 243, 250; Kazmi and Kazmi 1987: 56; Hogarth 1989: 104; Tirmizi and Kazmi 1996: 28 (in key), 29–32, figs 13a–d, 14a–d.

Type locality: South India (Madras).

Distribution

Persian Gulf: Iran (Nobili 1906a; present study), Saudi Arabia (Basson et al. 1977), Bahrain [Stephensen (1946) as *Charybdis hoplites* forma *typica*].

Gulf of Oman: Iran (Apel and Spiridonov 1998).

World distribution: Northern Indian Ocean: southern Oman, Persian Gulf, Gulf of Oman, Pakistan, India, Lakshadweep Islands and Bay of Bengal.

Habitat: Subtidal, 24–366 m.

Fig. 20.15 *Charybdis* (*Goniohellenus*) *hoplites* (Wood-Mason, 1877). Male (ZUTC Brach1795), CL = 32.28, CB = 63.4 mm, Iran, Gulf of Oman, coll. N. Ghotbeddin (Photo: R. Abdollahi)

Charybdis (Goniohellenus) longicollis Leene, 1938
(Figs. 20.13d, 20.16, and 20.17)

Charybdis (*Goniohellenus*) *hoplites* var. *longicollis* Leene 1938: 107–110, figs 59, 60a–c.
Charybdis (*Goniohellenus*) *hoplites* var. *longicollis* — Stephensen 1946: 118, 119, figs 24f, g, 25.
Charybdis hoplites var. *longicollis* — Guinot 1967a: 255 (in list).
Charybdis (*Goniohellenus*) *longicollis* — Titgen 1982: 250 (in list); Apel and Spiridonov 1998: 214–216, figs 32, 34; Apel 2001: 72; Naderloo and Sari 2007a: 343, tab 1; Ng et al. 2008: 154 (in list).
Type locality: Red Sea (Ras Abu-Soma).

Distribution

Persian Gulf: Iran (Stephensen 1946; Apel and Spiridonov 1998; Apel 2001; Naderloo and Sari 2007a), Kuwait (Apel and Spiridonov 1998; Apel 2001), Saudi Arabia (Apel and Spiridonov 1998; Apel 2001).

Gulf of Oman: Iran (Stephensen 1946), UAE (Apel and Spiridonov 1998; Apel 2001).

World distribution: Western Indian Ocean: East Africa, Red Sea, Gulf of Aden, Persian Gulf, Gulf of Oman and Mediterrenean (Lesspsian migrant).

Habitat: Subtidal, 8–56 m, substrates muddy and sandy.

Fig. 20.16 *Charybdis* (*Goniohellenus*) *longicollis* Leene, 1938. Male (ZUMC CRU3268), CL = 22.4, CB = 38.3 mm, Bandar-Lengeh, Iran, Persian Gulf, depth 10 m, coll. B. Loppenthin (Photo: S. Träkner)

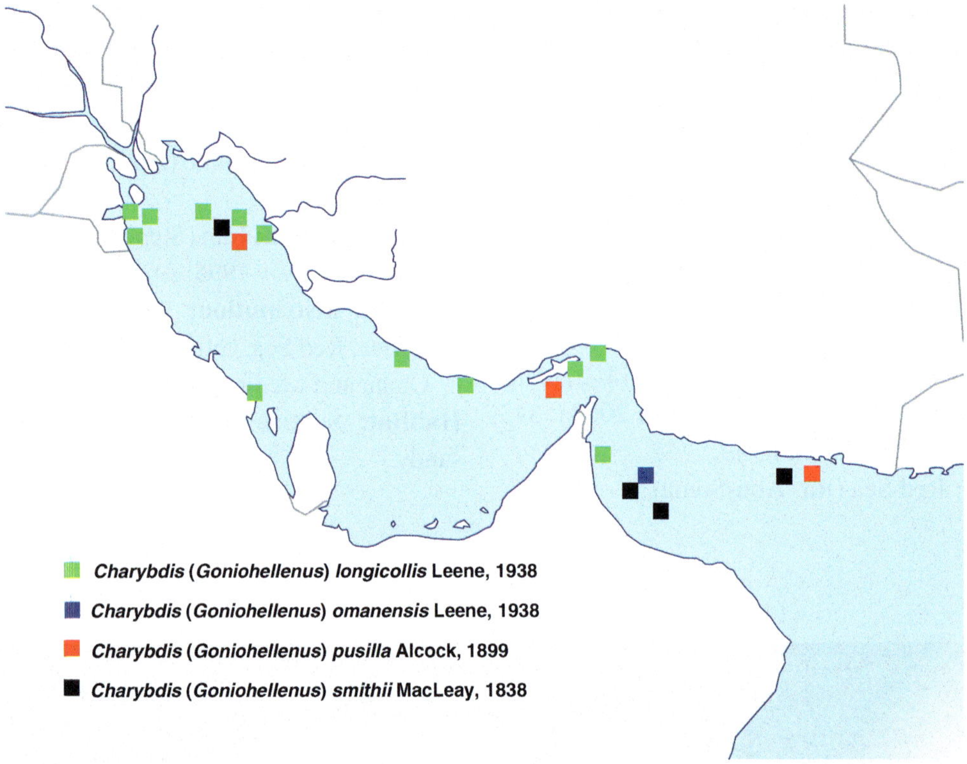

Fig. 20.17 Map showing the recorded localities of *Charybdis* (*Goniohellenus*) *longicollis* Leene, 1938, *Charybdis* (*Goniohellenus*) *omanensis* Leene, 1938, *Charybdis* (*Goniohellenus*) *pusilla* Alcock, 1899, and *Charybdis* (*Goniohellenus*) *smithii* MacLeay, 1838

Charybdis (Goniohellenus) omanensis Leene, 1938

(Figs. 20.13e, 20.17, and 20.18)

Charybdis (*Goniohellenus*) *hoplites* var. *omanensis* Leene 1938: 104–107, figs 57, 58.
Charybdis (*Goniohellenus*) *hoplites* var. *omanensis* — Stephenson and Rees 1967b: 285.
Charybdis hoplites var. *omanensis* — Zarenkov 1971: 181.
Charybdis (*Goniohellenus*) *omanensis* — Stephenson 1972b: 13 (in key), 35; Türkay 1986: 156–158, figs 43–45; Apel and Spiridonov 1998: 216, 217, fig 35; Neumann and Spiridonov 1999: 15.
Charybdis (*Goniohellenus*) *omanensis omanensis* — Ng et al. 2008: 154 (in list).

Type locality: Gulf of Oman (23° 47′N, 58° 30′E).

Distribution

Persian Gulf: No record.

Gulf of Oman: Iran (Naderloo et al. 2015), Oman (Leene 1938).

World distribution: Northwestern Indian Ocean: Red Sea, Gulf of Aden, southern Oman and Gulf of Oman.

Habitat: Subtidal to 400 m, substrates muddy and sandy.

Fig. 20.18 *Charybdis* (*Goniohellenus*) *omanensis* Leene, 1938. Male (SMF 22980), CL = 35.1, CB = 53.7 mm, Gulf of Aden, depth 359–362 m, 13.03.1987, collected by FS "Meteor" (Photo: S. Träkner)

Charybdis (Goniohellenus) pusilla Alcock, 1899
(Fig. 20.17)

Charybdis (*Goniohellenus*) *hoplites* var. *pusilla* Alcock 1899a: 67.

Charybdis (*Goniohellenus*) *hoplites* var. *pusilla* — Leene 1938: 102–104, figs 55, 56.

Charybdis (*Goniohellenus*) *hoplites* var. *vadorum* Alcock 1899a: 67 (partly: material from the Persian Gulf).

Charybdis (*Goniohellenus*) *vadorum* — Stephensen 1946: 119; Stephenson 1972a: 133 (partly: material from "Iranian Strait").

Charybdis vadorum — Guinot 1967a: 254 (in list) (partly: record from the Persian Gulf).

Portunidae spp. indet — Stephensen 1946: 138 (partly: material from station 22A).

Charybdis (*Goniohellenus*) *hoplites pusilla* — Apel and Spiridonov 1998: 213, 214; Naderloo and Sari 2007a: 343, tab 1.

Charybdis (*Goniohellenus*) *pusilla* — Ng et al. 2008: 154 (in list).

Type locality: India (Konkan Coast, 17° 27′N, 71° 41′E).

Distribution

Persian Gulf: Iran [Stephensen (1946) as *Portunidae* spp. indet.; Apel 2001; Naderloo and Sari 2007a]. Alcock (1899a) recorded this species as *Charybdis* (*Goniohellenus*) *hoplites* var. *vadorum* from an unknown locality.

Gulf of Oman: Iran (Naderloo et al. 2015).

World distribution: Northern Indian Ocean: Persian Gulf, Gulf of Oman and Arabian Sea.

Habitat: Subtidal to 56 m.

Charybdis (*Goniohellenus*) *smithii* MacLeay, 1838

(Figs. 20.13f, 20.17, and 20.19)

Charybdis smithii MacLeay 1838: 61.

Charybdis (*Goniohellenus*) *edwardsi* Leene and Buitendijk 1949: 296–298, figs 3, 4c.

Charybdis smithii — Couwelaar et al. 1997: 1251–1280.

Charybdis (*Goniohellenus*) *smithii* — Apel and Spiridonov 1998: 217–220, fig 36; Apel 2001: 72; Ng et al. 2008: 154.

Type locality: *Charybdis smithii*: South Africa (Cape of Good Hope). — *Charybdis* (*Goniohellenus*) *edwardsi*: India (Coast of Malabar).

Distribution

Persian Gulf: Iran (Apel and Spiridonov 1998; Apel 2001).

Gulf of Oman: Iran (Naderloo et al. 2015), Oman (Daniel and Chakrapany 1983; Couwelaar et al. 1997).

World distribution: Indian Ocean: South Africa, East Africa, Madagascar, Seychelles, Red Sea, Gulf of Aden, southern Oman, Persian Gulf, Gulf of Oman, Pakistan, India and Bay of Bengal.

Habitat: Subtidal, 360–400 m, substrates sandy and muddy.

Fig. 20.19 *Charybdis* (*Goniohellenus*) *smithii* MacLeay, 1838. Male (ZUTC Brach1818), CL = 53.26, CB = 78.12 mm, Gordin, Sistan-Baluchestan, Iran, Gulf of Oman, depth 62 m, coll. N. Ghotbeddin (Photo: R. Abdollahi)

Genus *Gonioinfradens* Leene, 1938

Gonioinfradens paucidentata (A. Milne-Edwards, 1861)
(Figs. 20.20, 20.21, and 20.22a)

Goniosoma paucidentatum A. Milne-Edwards 1861: 381, pl 35, fig 3.
Thalamita Giardi Nobili 1905a: 164.
Charybdis (*Goniosoma*) *Giardi* — Nobili 1906a: 115–119, pl 4, fig 23, pl 5, fig. 34.
Charybdis (*Gonioinfradens*) *paucidentata* — Stephensen 1946: 119, 120; Basson et al. 1977: 261; Titgen 1982: 117, 118, 250 (in list).
Charybdis paucidentata — Guinot 1967a: 255 (in list).
Thalamita cf. *spinifera* — Apel 1994b: 433.
Gonioinfradens paucidentata — Apel and Spiridonov 1998: 223–226, figs 40, 41, pl 7; Apel 2001: 72; Naderloo and Sari 2007a: 343, tab 1; Ng et al. 2008: 154 (in list); Naderloo and Türkay 2012: 39; Naderloo et al. 2013: 449, tab 1.

Type locality: *Goniosoma paucidentatum*: Mauritius. — *Thalamita giardi*: Persian Gulf (Arzanah Island, Abu-Dhabi, UAE).

Distribution

Persian Gulf: Iran (Naderloo and Sari 2007a; Naderloo and Türkay 2012; Naderloo et al. 2013), Saudi Arabia (Basson et al. 1977; Apel and Spiridonov 1998; Apel 2001), Bahrain (Apel 2001), UAE [Nobili (1906a) as *Thalamita giardi*; Titgen 1982; Apel and Spiridonov 1998; Apel 2001].

Gulf of Oman: UAE (Apel and Spiridonov 1998; Apel 2001).

World distribution: Indo-West Pacific: East Africa, Mauritius, Seychelles, Réunion, Red Sea, Socotra, Persian Gulf, Gulf of Oman, Andaman Sea, Thailand, Japan, Australia, New Caledonia and French Polynesia.

Habitat: Intertidal to 100 m, substrate rocky/cobble.

Fig. 20.20 *Gonioinfradens paucidentata* (A. Milne-Edwards 1861). Female (ZUTC 5557), CL = 27.8, CB = 37.41 mm, Abu Musa Island, Iran, Persian Gulf, 25.12.2014, coll. R. Abdollahi and S. Ebrahimnejad (Photo: R. Abdollahi)

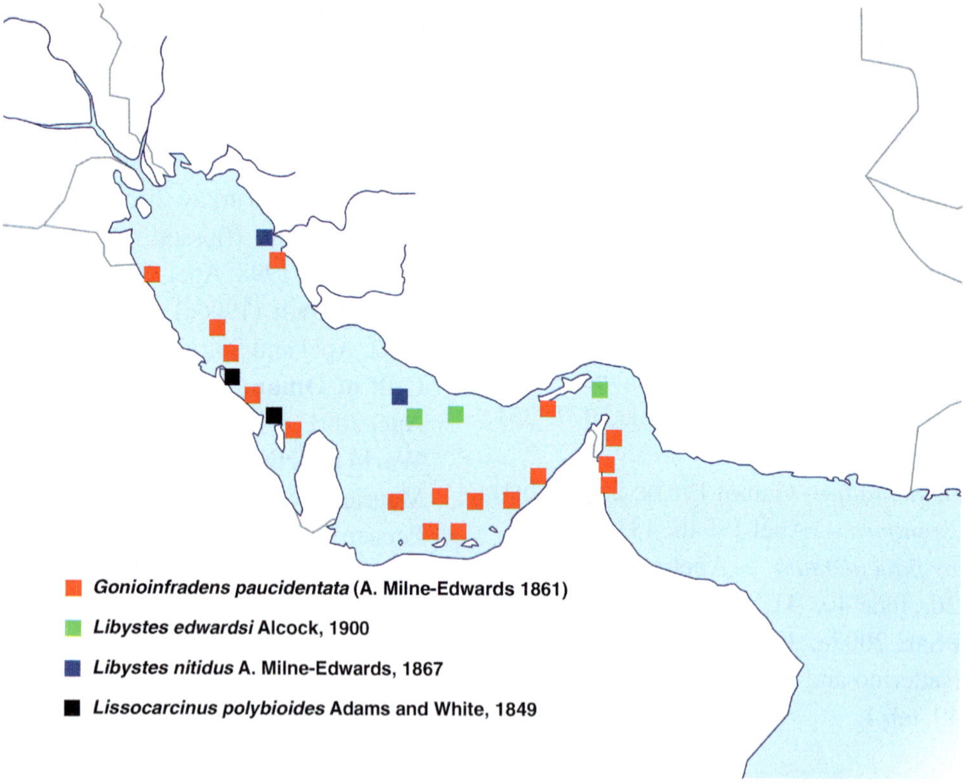

Fig. 20.21 Map showing the recorded localities of *Gonioinfradens paucidentata* (A. Milne-Edwards 1861), *Libystes edwardsi* Alcock, 1900, *Libystes nitidus* A. Milne-Edwards, 1867, and *Lissocarcinus polybioides* Adams and White, 1849

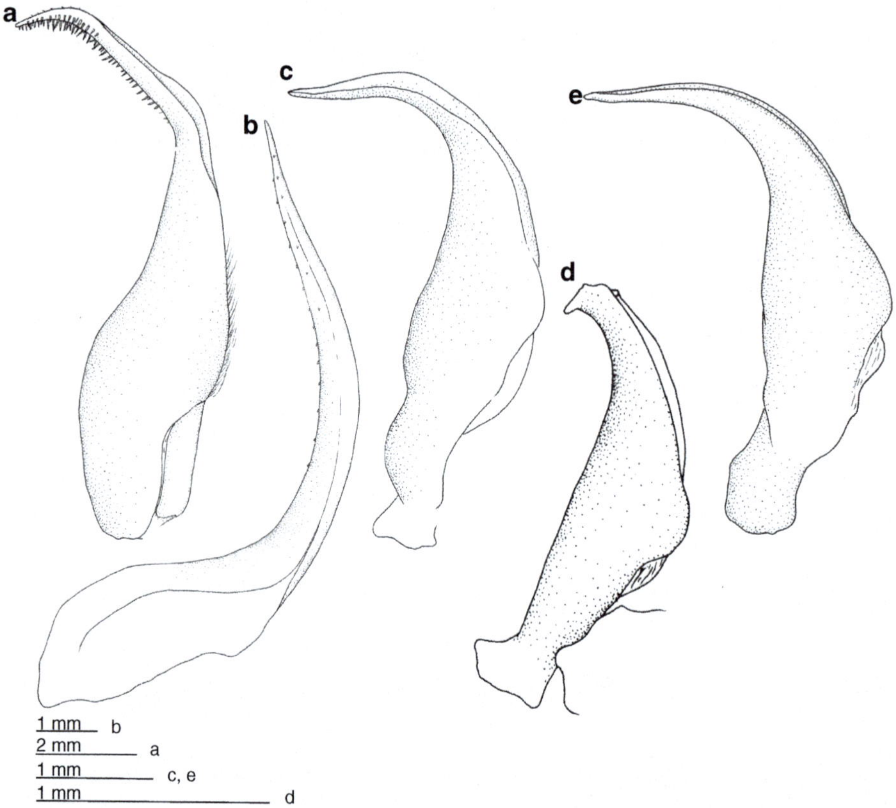

Fig. 20.22 Male G1, ventral view: (**a**) *Gonioinfradens paucidentata* (A. Milne-Edwards, 1861) (ZUTC Brach1062); (**b**) *Podophthalmus vigil* (Fabricius, 1798) (ZUTC: Iran, Gulf of Oman); (**c**) *Portunus* (*Xiphonectes*) *arabicus* (Nobili, 1905) (SMF 24304); (**d**) *Portunus* (*Achelous*) *granulatus* (H. Milne Edwards, 1834) (SMF 24283); (**e**) *Portunus* (*Xiphonectes*) *hastatoides* Fabricius, 1798 (ZUTC Brach1066)

Genus *Libystes* A. Milne-Edwards, 1867

Libystes edwardsi Alcock, 1900

(Figs. 20.21 and 20.23)

Libystes edwardsi Alcock 1900: 306.

Libystes edwardsi — Alcock and Mcardle 1903: pl 61, fig 1; Stephensen 1946: 168; Titgen 1982: 250 (in list); Apel and Spiridonov 1998: 174–176, fig 5; Apel 2001: 73; Naderloo and Sari 2007a: 343, tab 1; Ng et al. 2008: 149 (in list).

Type locality: Persian Gulf and Andaman Sea.

Distribution

Persian Gulf: Iran (Stephensen 1946; Naderloo and Sari 2007a). One specimen of syntype described by Alcock (1900) is from an unknown locality in the Persian Gulf.

Gulf of Oman: No record.

World distribution: Indo-West Pacific: Persian Gulf, India, Andaman Sea, Thailand, Indo-Malayan Archipelago, Vietnam, South China Sea and Japan.

Habitat: Subtidal, 26–50 m, substrates muddy, sandy and gravel.

Fig. 20.23 *Libystes edwardsi* Alcock, 1900. Male (ZUTC Brach1039), CL = 11.3, CB = 6.7 mm, Iran, Persian Gulf, depth 39–41 m, 20.11.2003, coll. R. Naderloo (Photo: S. Shahbazi)

Libystes nitidus A. Milne-Edwards, 1867
(Figs. 20.21 and 20.24)

Libystes nitidus A. Milne-Edwards 1867: 285.
Libystes nitidus — Alcock 1900: 306; Stephensen 1946: 168–169, fig 45c–f; Guinot 1967a: 253 (in list); Titgen 1982: 250 (in list).
Libystes aff. *nitidus* — Apel and Spiridonov 1998: 176–179, fig 6; Naderloo and Sari 2007a: 343, tab 1.
Type locality: Zanzibar.

Distribution

Persian Gulf: Iran (Stephensen 1946; Naderloo and Sari 2007a).

Gulf of Oman: No record.

World distribution: Northwestern Indian Ocean: Red Sea and Persian Gulf. The exact distribution range of the species is not clear, due to the taxonomic complexity of the species.

Habitat: Subtidal to 49 m, substrates sandy and muddy.

Fig. 20.24 *Libystes nitidus* A. Milne-Edwards, 1867. Female (ZUTC Brach1041), CL = 5.00, CB = 9.50 mm, Iran, Persian Gulf, depth 60 m, 18.08.2004, coll. R. Naderloo (Photo: S. Träkner)

Genus *Lissocarcinus* Adams and White, 1849

Lissocarcinus polybioides Adams and White, 1849
(Figs. 20.21 and 20.25)

Lissocarcinus polybioides Adams and White 1849: 46, pl 11, fig 5.

Lissocarcinus polybioides — Apel and Spiridonov 1998: 180, 181, figs 7, 9; Apel 2001: 73; Ng et al. 2008: 148 (in list).

? *Lissocarcinus* sp. — Mohammed and Al-Ssadh 1996: 37.

Type locality: "Eastern Seas".

Distribution

Persian Gulf: Saudi Arabia (Apel and Spiridonov 1998), Bahrain [Mohammed and Al-Ssadh (1996) as *Lissocarcinus* sp.].

Gulf of Oman: No record.

World distribution: Indo-West Pacific: Madagascar, Seychelles, Red Sea, Persian Gulf, India, Sri Lanka, Andaman Sea, Indonesia, Philippines, Japan, Australia, New Caledonia and Vanuatu.

Habitat: Subtidal, 2–100 m, substrates sandy and muddy with shells.

Fig. 20.25 *Lissocarcinus polybioides* Adams and White, 1849. Female (SMF 40380), CL = 14.3, CB = 15 mm, Japan, 06.11.1985, coll. Toyohata-Maru (Photo: S. Träkner)

Genus *Podophthalmus* Lamarck, 1801

Podophthalmus vigil (Fabricius, 1798)
(Figs. 20.22b, 20.26, and 20.27)

Portunus vigil Fabricius 1798: 363, 364.
Podophthalmus vigil — Leach 1815: 149, pl 108; Stephensen 1946: 137, 138, fig 32a, b; Guinot 1967a: 259 (in list); Titgen 1982: 250 (in list); Apel and Spiridonov 1998: 169, 170, fig 8; Apel 2001: 74; Naderloo and Sari 2007a: 343, tab 1; Ng et al. 2008: 149 (in list).
Type locality: Indian Ocean (possibly Tharangambadi, in south coast of India).

Distribution

Persian Gulf: Iran (Stephensen 1946; Naderloo and Sari 2007a).

Gulf of Oman: Iran (Stephensen 1946; Naderloo et al. 2015), UAE (Apel and Spiridonov 1998).

World distribution: Indo-West Pacific: South Africa, Mauritius, Red Sea, Persian Gulf, Gulf of Oman, India, Nicobars, Thailand, Indonesia, Philippines, Japan, Australia, New Caledonia and French Polynesia.

Habitat: Subtidal, 10–70 m, substrates sandy and muddy.

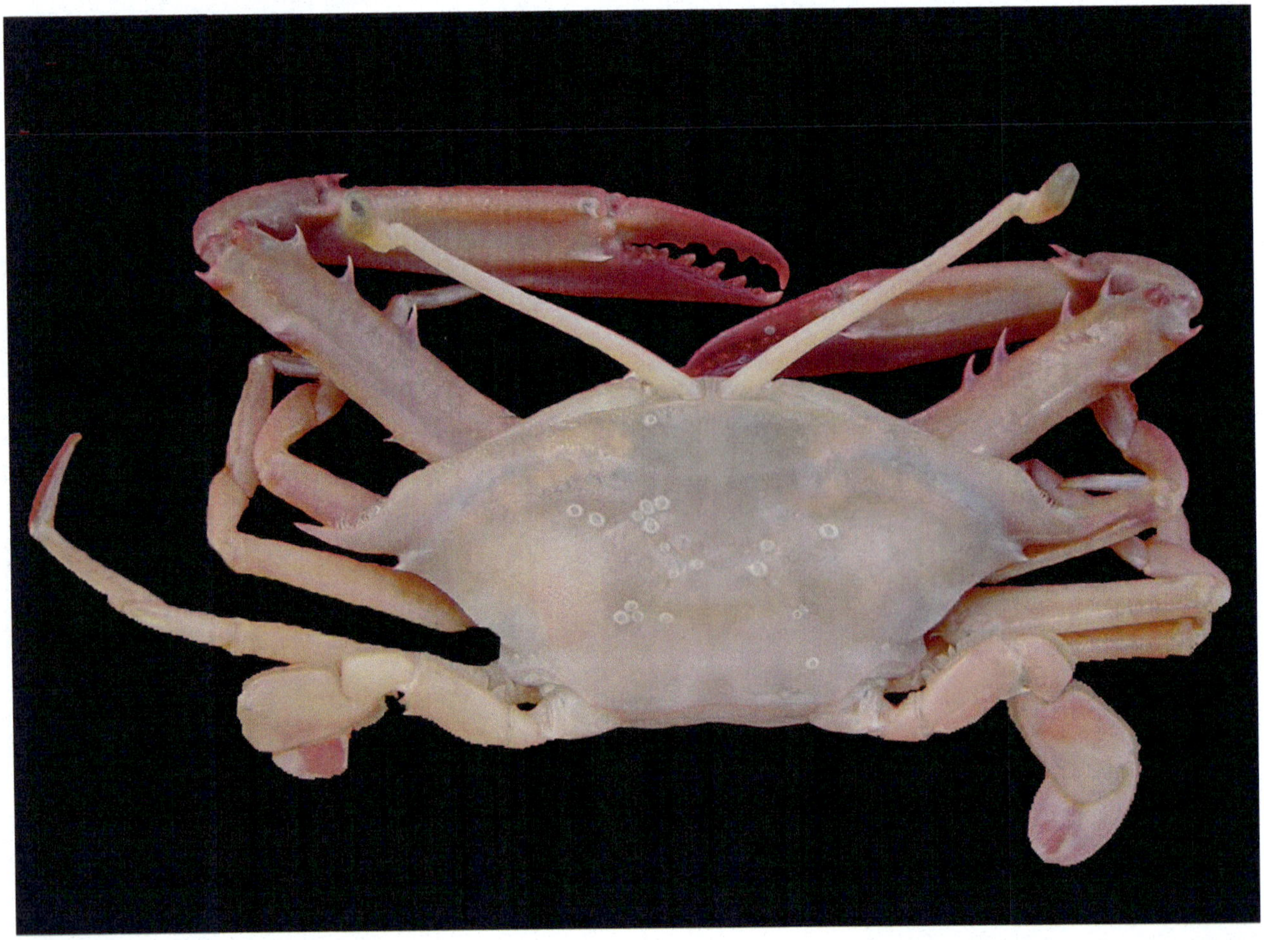

Fig. 20.26 *Podophthalmus vigil* (Fabricius, 1798). Male (ZUTC Brach1038), CL = 34.8, CB = 91.2 mm, Bandar-Abbas, Iran, Persian Gulf, depth 70 m, coll. R. Naderloo (Photo: R. Abdollahi)

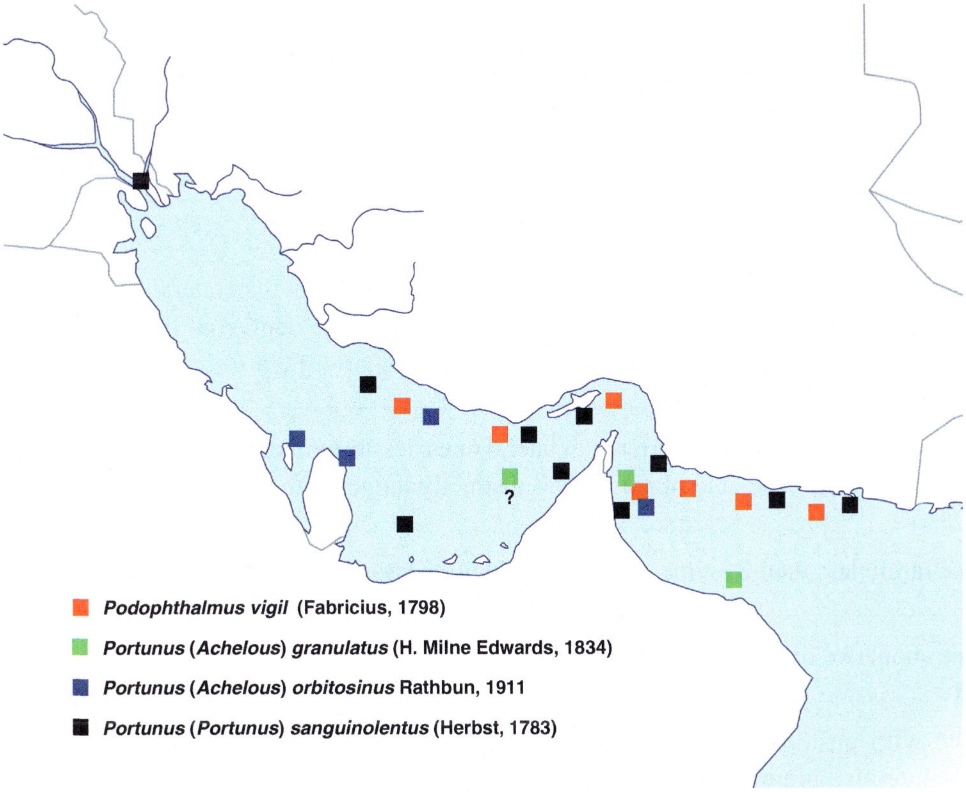

Fig. 20.27 Map showing the recorded localities of *Podophthalmus vigil* (Fabricius, 1798), *Portunus* (*Achelous*) *granulatus* (H. Milne Edwards, 1834), *Portunus* (*Achelous*) *orbitosinus* Rathbun, 1911, and *Portunus* (*Portunus*) *sanguinolentus* (Herbst, 1783)

Genus *Portunus* Weber, 1795

Key to species of the genus *Portunus* in the region

1. Carapace with posterolateral margins forming curve with posterior margin 2
- Carapace with posterolateral margins forming right angle or conspicuous spine with posterior margin ... 5
2. Third maxillipeds with anterolateral angle of merus not producing laterally 3
- Third maxillipeds with anterolateral angle of merus markedly producing laterally 4
3. Third maxillipeds with merus densely setose on external surface; merus of chelipeds without spine on posterodistal angle; dorsal surface of carapace with three large red spots on posterior part .... *Portunus (Portunus) sanguinolentus*
- Third maxillipeds with external surface of merus nearly glabrous; merus of chelipeds with spine on posterodistal angle; dorsal surface of carapace with different color pattern ..... *Portunus (Portunus) segnis*
4. Carapace with last anterolateral tooth slightly larger than others, dorsal surface with patchy granules; manus of chelipeds with longitudinal furrow along entire length of lower margin; male and female abdomen with second somite with dorsally directed spiniform process; G1 long, markedly curved, tapering apically .... *Portunus (Achelous) orbitosinus*

- Carapace with last anterolateral tooth not larger than others, granules covering entire dorsal surface, never grouped in patches; manus of chelipeds without longitudinal furrow along lower margin; male and female abdomen with second somite without spiniform process; G1 short, stout, apically flared .... *Portunus (Achelous) granulatus*

5. Chelipeds with merus having single distal spine on posterior margin, sometimes with small tubercle proximal to spine 6
- Chelipeds with merus having two distal spines on posterior margin 7

6. Median frontal teeth nearly as long as lateral ones, more prominent than laterals; merus of chelipeds with a tubercle proximal to distal spine on posterior margin; carapace depressed on anterior portion, markedly sculptured with beaded granules; last anterolateral tooth shorter than frontal margin; G1 bent laterally on proximal part .... *Portunus (Xiphonectes) tuberculosus*
- Median frontal teeth markedly shorter than lateral ones, less prominent than laterals, lateral teeth triangular, long, markedly sharp; last anterolateral tooth distinctly longer than frontal margin; G1 curved laterally on median part .... *Portunus (Xiphonectes) longispinosus*

7. Carapace rarely less than 2.3 times as broad as long; natatory legs with distal spine on posterior margin of merus .... *Portunus (Xiphonectes) pulchicristatus*
- Carapace about two times as broad as long; natatory legs without distal spine on posterior margin of merus, serrated 8

8. Carapace with posterolateral junction forming a conspicuous spine; median frontal teeth about as large as laterals; granules of median postcardiac granular patch separate, not forming distinct tubercles .... *Portunus (Xiphonectes) hastatoides*
- Carapace with posterolateral junction forming obtuse or right angle; median frontal teeth markedly smaller than laterals; granules of median postcardiac granular patch fused to form distinct tubercles .... *Portunus (Xiphonectes) arabicus*

Portunus (Achelous) granulatus (H. Milne Edwards, 1834)

(Figs. 20.22d, 20.27, and 20.28)

Lupea granulata H. Milne Edwards 1834: 454.
Neptunus granulatus — Alcock 1899a: 45 (?partly).
Neptunus (*Achelous*) *granulatus* — Stephensen 1946: 121.
Portunus (*Achelous*) *granulatus* — Rathbun 1911: 205, pl 15, fig 10.
Portunus granulatus — Guinot 1967a: 257 (in list); Titgen 1982: 250 (in list); Apel and Spiridonov 1998: 286–289, figs 97, 109; Apel 2001: 74.
Portunus (*Achelous*) *granulatus granulatus* — Ng et al. 2008: 151 (in list).

Type locality: Mauritius.

Distribution

Persian Gulf: Alcock (1899a) recorded this species from an unknown locality.

Gulf of Oman: UAE (Apel 2001), Oman (Apel and Spiridonov 1998).

World distribution: Indo-West Pacific: South Africa, Mauritius, Seychelles, Red Sea Somlia, southern Oman, Persian Gulf, Gulf of Oman, Pakistan, Sri Lanka, China, Japan, Indonesia, Australia, Fiji, Tahiti and Kiribati.

Habitat: Subtidal, 16–30 m, substrate sandy.

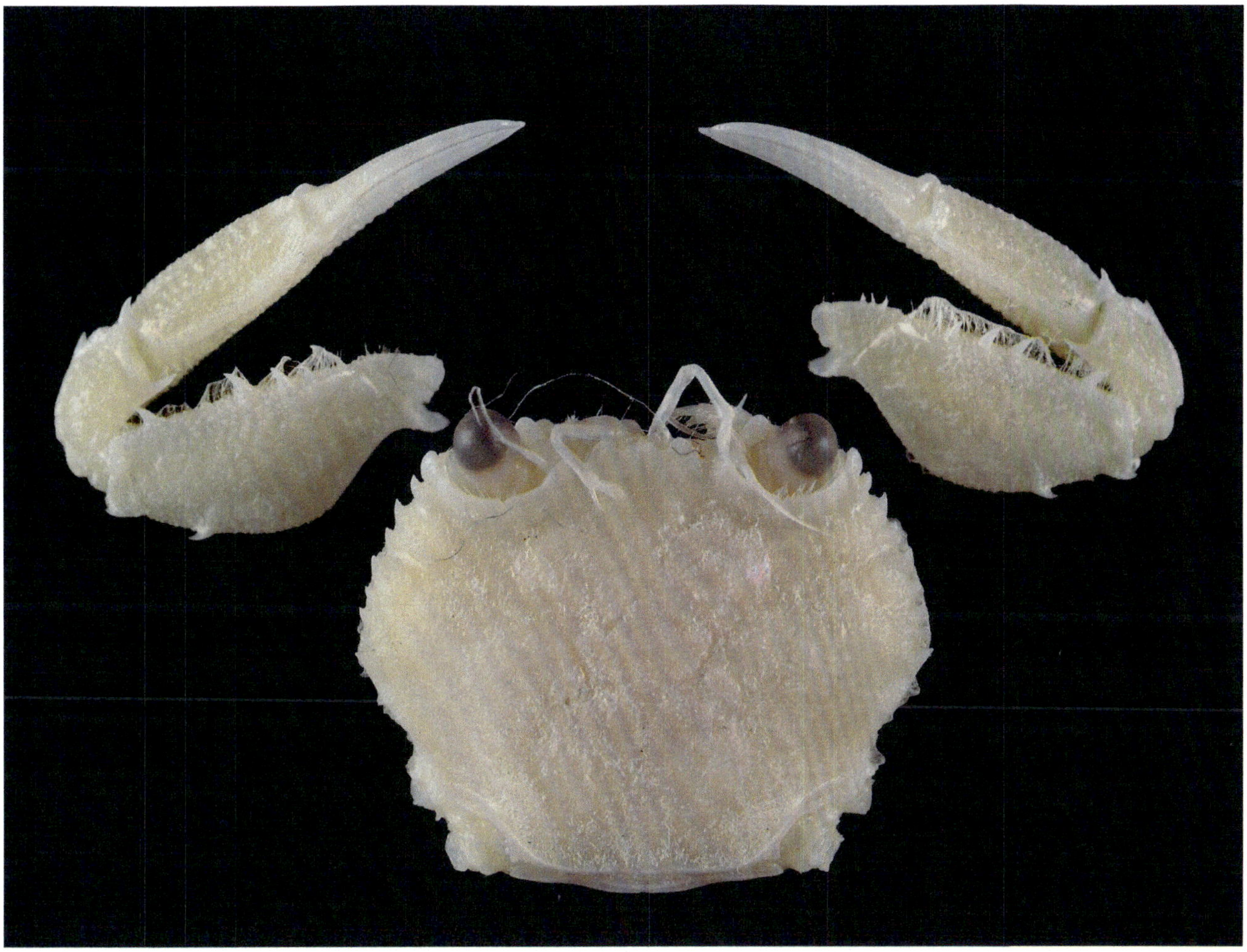

Fig. 20.28 *Portunus* (*Achelous*) *granulatus* (H. Milne Edwards, 1834). Male (SMF 24283), CL = 7.5, CB = 10.1 mm, Sandy Beach Hotel, Al-Aqqa, UAE, Gulf of Oman, depth 4–6 m, 04.07.1995, coll. M. Apel (Photo: S. Träkner)

Portunus (*Achelous*) *orbitosinus* Rathbun, 1911

(Figs. 20.27, 20.29, and 20.30b)

Portunus (*Achelous*) *orbitosinus* Rathbun 1911: 205, pl 15, fig 11.

? *Neptunus granulatus* — Alcock 1899a: 45 (?partly).

Neptunus (*Achelous*) *orbitosinus* — Stephensen 1946: 120.

Portunus orbitosinus — Guinot 1967a: 258 (in list); Titgen 1982: 250 (in list); Apel and Spiridonov 1998: 299, 300, figs 110, 111, 115; Apel 2001: 76.

Portunus (*Achelous*) *orbitosinus* — Ng et al. 2008: 151 (in list).

Type locality: Saint Brandon, Seychelles and Amirante.

Distribution

Persian Gulf: Iran [Stephensen (1946) as *Neptunus* (*Achelous*) *orbitosinus*], Bahrain [Stephensen (1946) as *Neptunus* (*Achelous*) *orbitosinus*], Qatar (Apel 2001). More likely the record of *P. granulatus* by Alcock (1899a) from the Persian Gulf is this species (see Apel 2001).

Gulf of Oman: UAE (Apel and Spiridonov 1998; Apel 2001).

World distribution: Indo-West Pacific: East Africa, Madagascar, Mauritius, Sychelles, Saint Brandon, Amirante, Red Sea, Gulf of Aden, Socotra, Persian Gulf, Gulf of Oman, Thailand, Indonesia, Philippines, Japan, Australia, New Caledonia, Bikini Atoll and Marques Islands.

Habitat: Subtidal, 2–75 m, substrates sandy, muddy, coral rubble and shells.

Fig. 20.29 *Portunus* (*Achelous*) *orbitosinus* Rathbun, 1911. Male (ZMUC CRU936), CL = 19.7, CB = 27.7 mm, NE Bahrain, 26° 50′N, 51° 17′E, depth 33 m, 30.03.1937, coll. G. Thorson (Photo: S. Träkner)

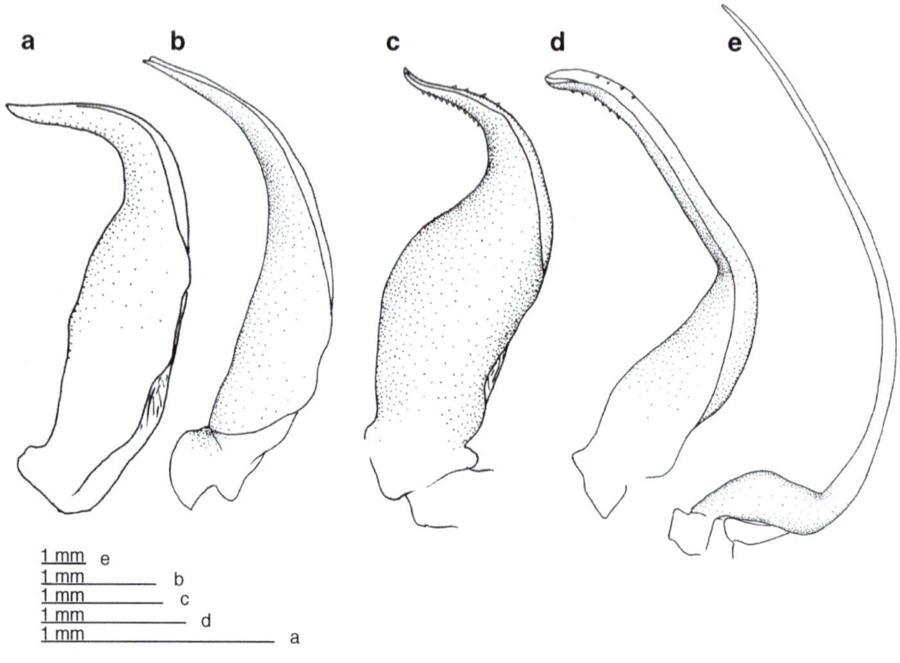

Fig. 20.30 Male G1, ventral view: (**a**) *Portunus* (*Xiphonectes*) *longispinosus* (Dana, 1852) (SMF 24395); (**b**) *Portunus* (*Achelous*) *orbitosinus* Rathbun, 1911 (SMF 38960); (**c**) *Portunus* (*Xiphonectes*) *pulchicristatus* (Gordon, 1931) (MNH 1930.12.2.82/85, syntype); (**d**) *Portunus* (*Xiphonectes*) *tuberculosus* (A. Milne-Edwards, 1861) (MNH 1925.2.25.14/16); (**e**) *Portunus* (*Portunus*) *segnis* (Forskål, 1775) (SNMNH 39).

Portunus (Portunus) sanguinolentus (Herbst, 1783)
(Figs. 20.27 and 20.31)

Cancer sanguinolentus Herbst 1783: 161, pl 8, figs 56, 57.
Neptunus (Neptunus) sanguinolentus — Alcock 1899a: 32; Chopra 1935: 474–476, fig 2; Stephensen 1946: 123, 124, fig 26d.
Neptunus sanguinolentus — Nobili 1906a: 114.
Portunus sanguinolentus — Guinot 1967a: 257 (in list); Hogarth 1989: 104; Carpenter et al. 1997: 38; Apel and Spiridonov 1998: 306, 307, pl 12.
Portunus sanguinolentus sanguinolentus — Stephenson 1972b: 15 (in key), 42, 43; Titgen 1982: 250 (in list).
Portunus (Portunus) sanguinolentus sanguinolentus — Ng et al. 2008: 152 (in list), 156 (in text).
Type locality: India (Tharangambadi).

Distribution

Persian Gulf: Iran (Stephensen 1946; Naderloo and Sari 2007a), Iraq (Apel and Spiridonov 1998; Apel 2001), UAE (Nobili 106a). Chopra (1935) recorded this species from an unknown locality.

Gulf of Oman: Iran (Apel and Spiridonov 1998; Apel 2001; Naderloo et al. 2015), UAE (Apel and Spiridonov 1998; Apel 2001).

World distribution: Indo-West Pacific: East Africa, Red Sea, Persian Gulf, Gulf of Oman, Arabian Sea, Pakistan, India, Sri Lanka, Andaman Sea, Thailand, Philippines, China, Japan Indonesia, Australia, New Zealand, French Polynesia and Hawaii.

Habitat: Subtidal, 3–40 m, substrates sandy and muddy.

Fig. 20.31 *Portunus (Portunus) sanguinolentus* (Herbst, 1783). Male (ZUTC 5072), CL = 45.25, CB = 118.7 mm, Tang, Sistan-Baluchestan, Iran, Gulf of Oman, coll. N. Ghotbeddin (Photo: R. Abdollahi)

Remarks Subcpecies of *Portunus* (*Portunus*) *sanguilolentus sanguilolentus* has beeen proposed by Stephenson (1968) for the population of the species in the Indian Ocean and *Portunus* (*Portunus*) *sanguilolentus hawaiiensis* for the population occuring in the Pacific Ocean. Since that not more work has been done in order to establish the taxonomic postion of these two subspecies. I prefer not use the subspecies here till the clear taxonomic decision being done on this broadly distributed species.

Portunus (*Portunus*) *segnis* (Forskål, 1775)

(Figs. 20.30e, 20.32, 20.33, 20.34, and 20.35)

Cancer segnis Forskål 1775: 18, 91.
Neptunus (*Neptunus*) *pelagicus* — Alcock 1899a: 34; Chopra 1935: 476, 477, fig 3; Stephensen 1946: 124, 125, fig 26e.
Portunus pelagicus — Guinot 1967a: 257 (in list); Pretzmann 1971: 471; Basson et al. 1977: 224, 227, 231, 235, 250, 256; Titgen 1982: 118–120, 250 (in list); Jones 1986a: 161, pl 47; Hogarth 1989: 114; Hywel-Davies 1994: 48, Appen 4; Al-Ghais and Cooper 1996: 423; Bahmani 1997: 2, 6, 8, 9, 12; Cooper 1997: 166; Hornby 1997: 75; Carpenter et al. 1997: 38; Hornby 1997: 15; Apel and Spiridonov 1998: 300–303, pls 10, 11; Al-Khayat and Jones 1999: 58, 61; Apel 2001: 76.
Portunus segnis — Lai et al. 2010: 215–218, figs 6b, 7b, 11, 12, 13, 14, 20b, 21b, 22b, 23b, 24b; Naderloo and Türkay 2012: 39; Naderloo et al. 2013: 449, tab 1.
Portunus (*Portunus*) *segnis* — Ng et al. 2008: 152 (in list), 156 (in text).
Type locality: Red Sea (Egypt).

Distribution.

Persian Gulf: Iran [Stephensen (1946) as *P. pelagicus*; Bahmani (1997) as *P. pelagicus*; Naderloo and Sari (2007a) as *P. pelagicus*; Hosseini (2009) as *P. pelagicus*; Naderloo and Türkay 2012; Naderloo et al. 2013], Kuwait [Jones (1986a) as *Portunus pelagicus*], Saudi Arabia [Basson et al. (1977) as *P. pelagicus*; Apel and Spiridonov (1998) as *P. pelagicus*; Apel (2001) as *P. pelagicus*], Bahrain [Apel and Spiridonov (1998) as *P. pelagicus*; Apel (2001) as *P. pelagicus*], Qatar [Al-Khayat and Jones (1999) as *P. pelagicus*], UAE [Nobili (1906a) as *P. pelagicus*; Titgen (1982) as *P. pelagicus*; Al-Ghais and Cooper (1996) as *P. pelagicus*; Cooper (1997) as *P. pelagicus*; Apel and Spiridonov (1998) as *P. pelagicus*; Apel (2001) as *P. pelagicus*; Lai et al. 2010].

Gulf of Oman: Iran (Ghotbeddin et al. 2012b; Naderloo et al. 2015), Oman [Hogarth (1989, 1994) as *P. Pelagicus*].

World distribution: Western Indian Ocean and Mediterranean: South Africa, East Africa, Red Sea, Gulf of Aden, Socotra, southern Oman, Persian Gulf, Gulf of Oman, Pakistan, west of Indian sub-continent, Mediterranean (Lessepsian migrant).

Habitat: Intertidal to 65 m, substrate mangroves, sandy, rocky/cobble.

Remarks Lai et al. (2010) revised *Portunus pelagicus* (Linnaeus, 1758) species complex using morphological and genetic data. They recognized four species within broadly distributed *Portunus pelagicus* senso lato, which are relatively geographically distinct. According to their data, *Portunus segnis* (Forskål, 1775) is the species distributing in the western Indian Ocean including the Persian Gulf and the Gulf of Oman. Whereas *P. pelagicus* is restricted to Southeast and East Asia.

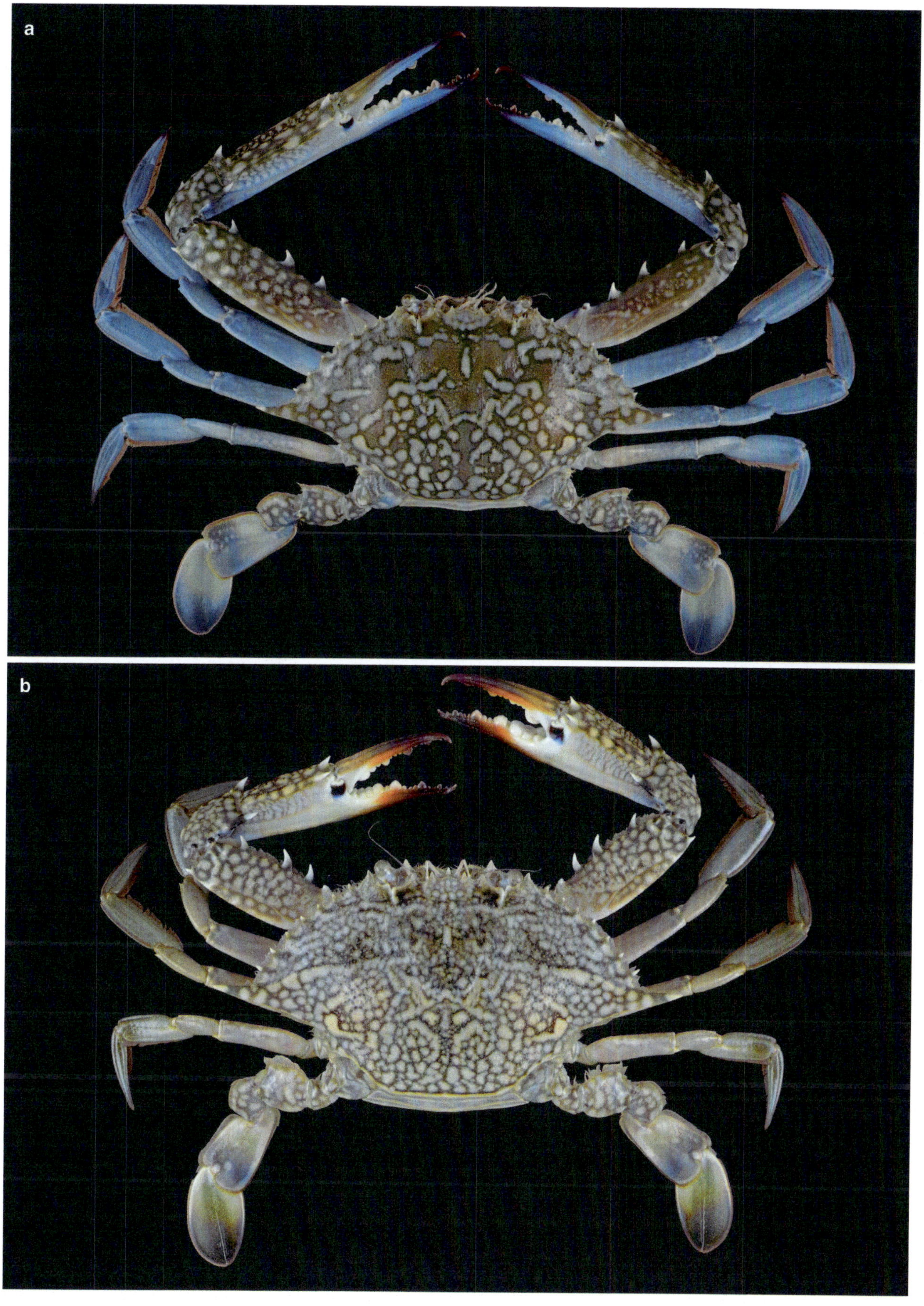

Fig. 20.32 *Portunus* (*Portunus*) *segnis* (Forskål, 1775). (ZUTC 5560), (**a**) Male, CL = 63.20 , CB = 139.00 mm; (**b**) female, CL = 55.00 , CB = 120.50 mm, between Bandar-Abbas and Hormuz Island, Iran, Persian Gulf, 17.08.2015, coll. A. Sari and A. Dehghan (Photo: R. Abdollahi)

Fig. 20.33 *Portunus* (*Portunus*) *segnis* (Forskål, 1775). The crabs in mating position, Bandar-Abbas, East coast of the city, 22.04.2008 (Photo: R. Naderloo)

Fig. 20.34 *Portunus* (*Portunus*) *segnis* (Forskål, 1775). By-catch of crab by a fisherman, Kani, Qeshm Island, 06.01.2008 (Photo: R. Naderloo)

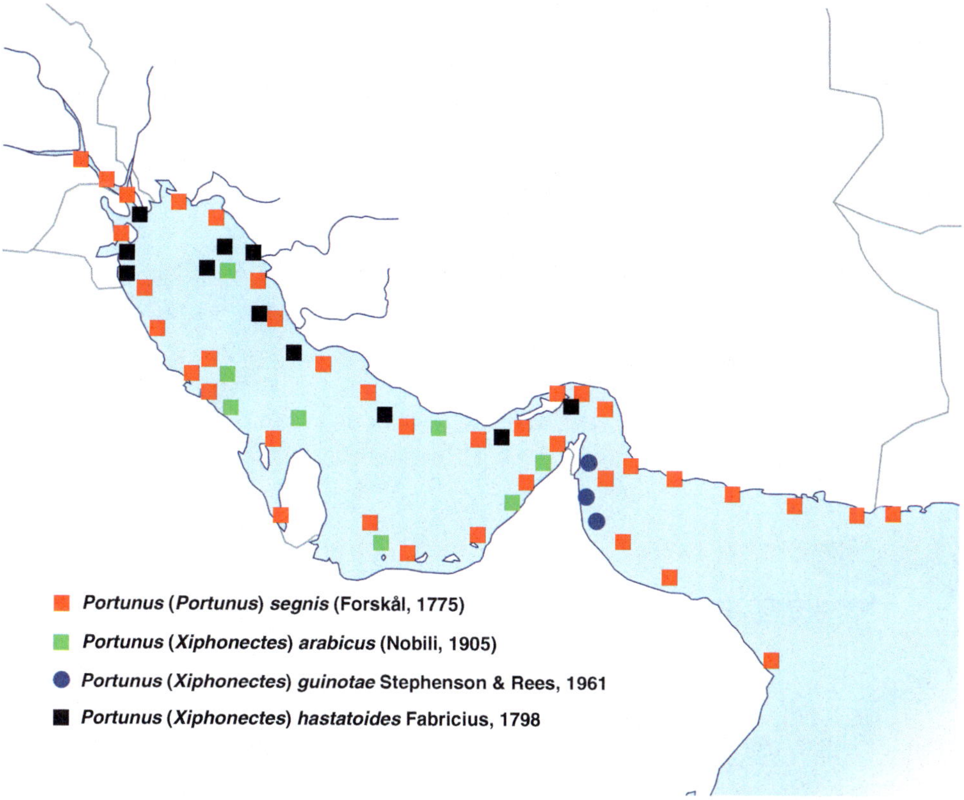

Fig. 20.35 Map showing the recorded localities of *Portunus* (*Portunus*) *segnis* (Forskål, 1775), *Portunus* (*Xiphonectes*) *arabicus* (Nobili, 1905), *Portunus* (*Xiphonectes*) *guinotae* Stephenson and Rees, 1961, and *Portunus* (*Xiphonectes*) *hastatoides* Fabricius, 1798

Portunus (*Xiphonectes*) *arabicus* (Nobili, 1905)

(Figs. 20.22c, 20.35, and 20.36)

Neptunus (*Hellenus*) *arabicus* Nobili 1905a: 163.
Portunus acerbiterminalis Stephenson and Rees 1967a: 14–16, fig 1, pl 1b.
Neptunus (*Hellenus*) *arabicus* — Nobili 1906a: 115, pl 5, figs 22, 22a, 1906b: 190 (in key), 191; Stephensen 1946: 121, 122, fig 26a–c.
Neptunus (*Hellenus*) *andersoni* — Alcock 1899a: 39; Stephensen 1946: 122; Titgen 1982: 250 (in list).
Portunus arabicus — Guinot 1967a: 258 (in list); Titgen 1982: 250 (in list); Cooper 1997: 166, 167; Apel and Spiridonov 1998: 281–284, figs 95, 98, 99, 106; Apel 2001: 74.
Portunus acerbiterminalis — Stephenson and Rees 1967b: 287; Stephenson 1972b: 15 (in key), 38 (partly: material from Saudi Arabia and East Africa).
? *Portunus* sp. nr. *hastatoides* — Basson et al. 1977: 224, 227, 231, 243, 250.
Portunus (*Xiphonectes*) *arabicus* Ng et al. 2008: 152 (in list).

Type locality: *Neptunus* (*Hellenus*) *arabicus*: Persian Gulf (Arzanah Island, Abu Dhabi, UAE) and Djibouti. Apel and Spiridonov (1998) selected a specimen as lectoype from the Persian Gulf (25° 10′N, 55° 10′E – 24° 55′N, 54° 40′E). — *Portunus acerbiterminalis*: Persian Gulf (Tarut Bay, Saudi Arabia).

Distribution

Persian Gulf: Iran (Stephensen 1946; Naderloo and Sari 2007a), Saudi Arabia [Stephenson and Rees (1967a) as *P. acerbiterminalis*; Basson et al. (1977) as *Portunus* sp. nr. *aastatoides*; Apel and Spiridonov 1998; Apel 2001], Bahrain (Stephensen 1946), UAE (Nobili 1905a, 1906a; Cooper 1997).

Gulf of Oman: No record.

World distribution: Northwestern Indian Ocean: Red Sea, Gulf of Aden, Somalia, Socotra and Persian Gulf.

Habitat: Subtidal, 16–30 m, substrate sandy.

Fig. 20.36 *Portunus* (*Xiphonectes*) *arabicus* (Nobili, 1905). Male (ZUMC CRU935), CL = 13.4, CB = 26.6 mm, N Khark Island, Iran, Persian Gulf, depth 16–17 m, 15.03.1937, coll. G. Thorson (Photo: S. Träkner)

Portunus (*Xiphonectes*) *guinotae* Stephenson and Rees, 1961

(Figs. 20.35, 20.37, and 20.38)

Portunus guinotae Stephenson and Rees 1961: 425, figs 1b, d, g, 2d–f.
Portunus guinotae — Apel and Spiridonov 1998: 290, 291, figs 102–104; Apel 2001: 75.
Portunus (*Xiphonectes*) *guinotae* — Ng et al. 2008: 152 (in list).

Type Locality: Maldives.

Distribution

Persian Gulf: No record.

Gulf of Oman: UAE (Apel and Spiridonov 1998; Apel 2001).

World distribution: Indo-West Pacific: Seychelles, Gulf of Oman, Maldives and Tuamotu.

Habitat: Subtidal, 3–60 m, substrate sandy.

Portunus (*Xiphonectes*) *hastatoides* Fabricius, 1798

(Figs. 20.22e, 20.35, and 20.39)

Portunus hastatoides Fabricius 1798: 368.
Portunus hastatoides — Guinot 1967a: 257 (in list); Basson et al. 1977: 250; Titgen 1982: 250 (in list); Apel and Spiridonov 1998: 293–296, figs 100, 101, 105, 107; Apel 2001: 75; Naderloo and Sari 2007a: 343, tab 1.
Neptunus (*Hellenus*) *hastatoides* — Alcock 1899a: 38, 39; Stephensen 1946: 122, 123.
Neptunus (*Hellenus*) *hastatoides* [sic!] — Chopra 1935: 477, 478, text-fig 4.
Portunus (*Xiphonectes*) *hastatoides* — Ng et al. 2008: 152 (in list).

Type locality: Indian Ocean (possibly Tharangambadi on south coast of India).

Fig. 20.37 *Portunus* (*Xiphonectes*) *guinotae* Stephenson and Rees, 1961. Male (SMF 24319), CL = 6.6, CB = 14.8 mm, Martini Rocks, 1 km S Khor Fakkan, UAE, Gulf of Oman, depth 5–18 m, 29.06.1995, coll. M. Apel (Photo: S. Träkner)

Distribution

Persian Gulf: Iran (Stephensen 1946; Naderloo and Sari 2007a; Apel and Spiridonov 1998; Apel 2001), Kuwait (Apel and Spiridonov 1998; Apel 2001).

Gulf of Oman: No record.

World distribution: Indo-West Pacific: South Africa, Madagascar, East Africa, Seychelles, Socotra, Persian Gulf, India, Sri Lanka, Andamans, Burma, Thailand, Malaysia, Philippines, China, Japan, Indonesia, New Guina and Australia.

Habitat: Subtidal, 5–100 m, substrates sandy and muddy.

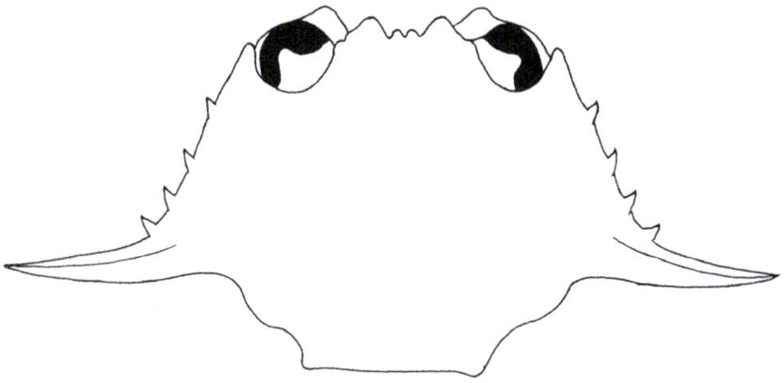

Fig. 20.38 *Portunus* (*Xiphonectes*) *guinotae* Stephenson and Rees, 1961. Drawing of carapace of the species (SMF 24310)

Fig. 20.39 *Portunus* (*Xiphonectes*) *hastatoides* Fabricius, 1798. Male (ZUTC Brach1066), CL = 20.54, CB = 36.87 mm, Iran, Persian Gulf, 2003, coll. R. Naderloo (Photo: R. Abdollahi)

Portunus (*Xiphonectes*) *longispinosus* (Dana, 1852)

(Figs. 20.30a, 20.40, 20.41, and 20.42)

? *Amphitrite longi-spinosa* Dana 1852a: 84.

? *Amphitrite longi-spinosa* — Dana 1852b: 277, pl 17, fig 2.

Neptunus (*Hellenus*) *longispinosus* — Alcock 1899a: 40; Stephensen 1946: 123.

Portunus longispinosus — Guinot 1967a: 258 (in list; partly); Titgen 1982: 250 (in list).

? *Portunus emarginatus* — Basson et al. 1977: 224.

Portunus cf. *longispinosus* — Apel and Spiridonov 1998: 296–298, figs 92, 114; Apel 2001: 276.

Portunus (*Xiphonectes*) *longispinosus* — Ng et al. 2008: 152 (in list).

Type locality: Fiji (Ovalau).

Distribution

Persian Gulf: Saudi Arabia [Basson et al. (1977) as *P. emarginatus*; Apel and Spiridonov 1998; Apel 2001].

Gulf of Oman: UAE (Apel and Spiridonov 1998; Apel 2001).

World distribution: Indo-West Pacific: Madagascar, Islands of the western Indain Ocean, East Africa, Red Sea, Gulf of Aden, Socotra, Persian Gulf, Gulf of Oman, Japan, French Polynesia, Fiji and Hawaii.

Habitat: Subtidal, 3–45 m, substrate sandy near coral reefs.

Fig. 20.40 *Portunus* (*Xiphonectes*) *longispinosus* (Dana, 1852). Male (SMF 22974), CL = 11.84, CB = 17.9 mm, Bab-el-Mandib, Gulf of Aden, depth 40–45 m, 06.03.1995, coll. F.S. Meteor (Photo: S. Träkner)

Fig. 20.41 *Portunus* (*Xiphonectes*) *longispinosus* (Dana, 1852). Drawing of carapace of the species (SMF 24395)

Portunus (*Xiphonectes*) *pulchicristatus* (Gordon, 1931)

(Figs. 20.30c, 20.42, and 20.43)

Neptunus (*Hellenus*) *alcocki* Gordon 1930: 519 [invalid name; homonym].

Neptunus (*Hellenus*) *pulchicristatus* Gordon 1931: 534, figs 8, 10a.

Neptunus (*Hellenus*) *spinipes* — Alcock 1899a: 32 (in key), 39, 40.

Neptunus (*Hellenus*) *pulchicristatus* — Chopra 1935: 479, fig 5.

Portunus pulchicristatus — Stephenson 1972a: 138; Stephenson 1972b: 15 (in key), 42; Tirmizi and Kazmi 1996: 25–27, figs 11, 12a–d; Apel and Spiridonov 1998: 303–305, figs 112, 116; Apel 2001: 76.

Portunus (*Xiphonectes*) *pulchicristatus* — Ng et al. 2008: 152 (in list).

Type locality: Hong Kong.

Distribution

Persian Gulf: Iran (Naderloo and Sari 2007a). Chopra (1935) recorded this species from an unknown locality.

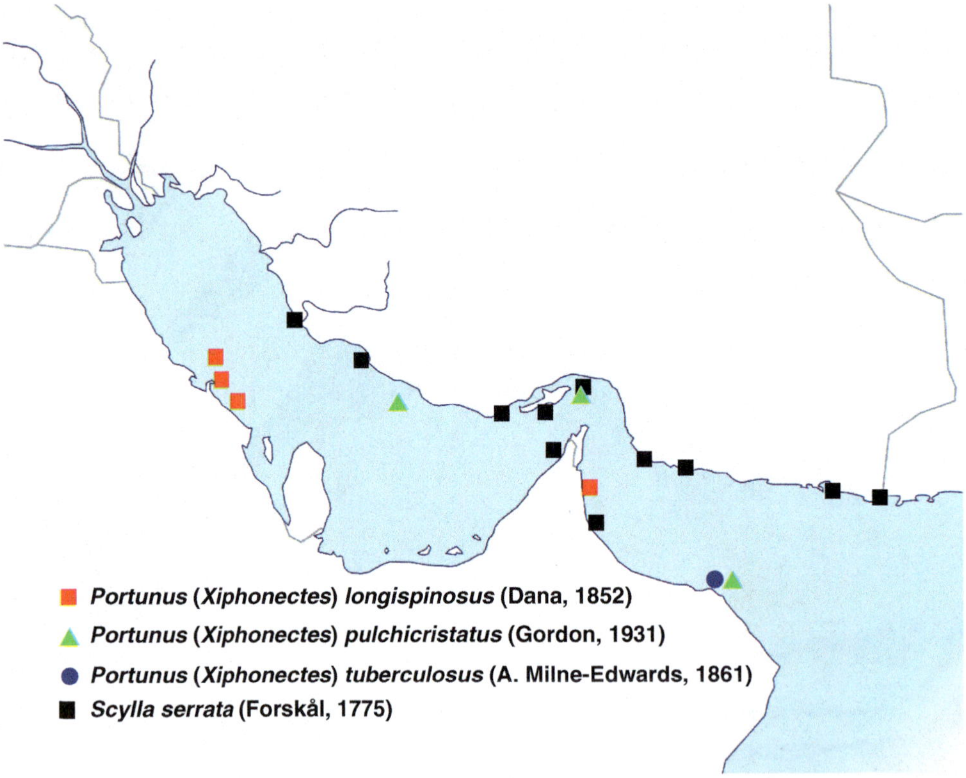

Fig. 20.42 Map showing the recorded localities of *Portunus* (*Xiphonectes*) *longispinosus* (Dana, 1852), *Portunus* (*Xiphonectes*) *pulchicristatus* (Gordon, 1931), *Portunus* (*Xiphonectes*) *tuberculosus* (A. Milne-Edwards, 1861), and *Scylla serrata* (Forskål, 1775)

Fig. 20.43 *Portunus* (*Xiphonectes*) *pulchicristatus* (Gordon, 1931). Female (ZUTC Brach1067), CL = 7.3, CB = 17.08 mm, N Qeshm Island, Iran, Persian Gulf, depth 25 m, 04.03.2003, coll. R. Naderloo (Photo: R. Abdollahi)

Gulf of Oman: Oman [Gordon 1931; Stephenson (1972a, 1972b) as *P. spinipes*].

World distribution: Indo-West Pacific: Persian Gulf, Gulf of Oman, Pakistan, India, Andamans, Burma, Thailand, Indonesia, Philippines, China and Australia.

Habitat: Subtidal, 10–830 m, substrates sandy, muddy and coral reefs.

Portunus (*Xiphonectes*) *tuberculosus* (A. Milne-Edwards, 1861)

(Figs. 20.30d, 20.42, and 20.44)

Neptunus tuberculosus A. Milne-Edwards 1861: 333–334, 339 (in key), pl 31, figs 5, 5a–c. *Neptunus* (*Hellenus*) *tuberculosus* — Alcock 1899a: 42–43; Stephensen 1946: 123. *Portunus tuberculosus* — Guinot 1967a: 257 (in list); Stephenson and Rees 1967a: 52, fig 18, pl 6b; Titgen 1982: 250 (in list); Apel and Spiridonov 1998: 308–310; Apel 2001: 77. *Portunus* (*Xiphonectes*) *tuberculosus* — Ng et al. 2008: 152 (in list).

Type locality: Hawaii.

Distribution

Persian Gulf: Alcock (1899a) recorded this species from an unknown locality.

Gulf of Oman: Oman (Apel and Spiridonov 1998; Apel 2001).

World distribution: Indo-West Pacific: Madagascar, Seychelles, East Africa, Gulf of Aden, Persian Gulf, Gulf of Oman, Maldives, Sri Lanka, Andamans, Burma, Indonesia, Philippines, China, Australia, Palau and Hawaii.

Habitat: Subtidal, 25–115 m, substrates sandy, muddy and coral.

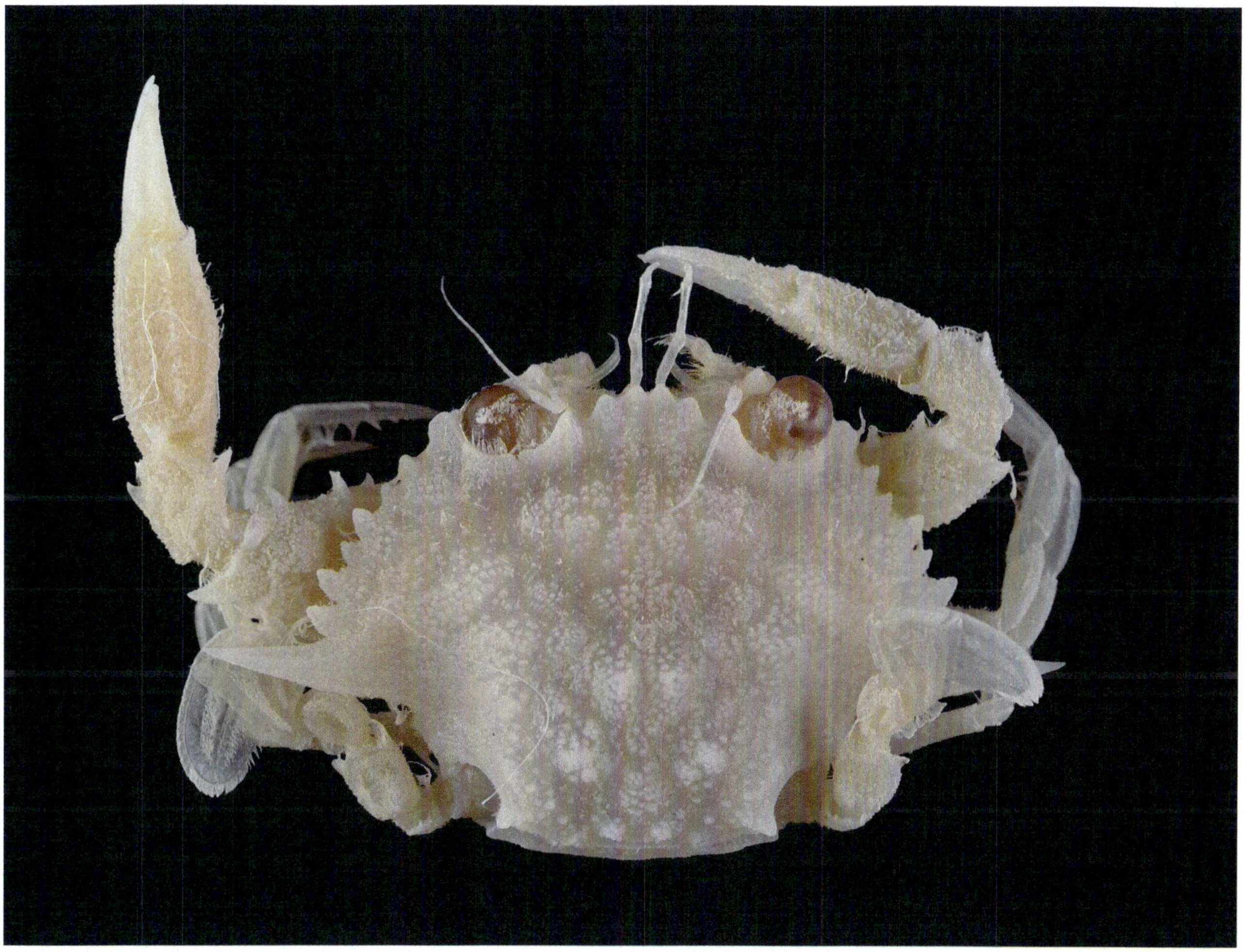

Fig. 20.44 *Portunus* (*Xiphonectes*) *tuberculosus* (A. Milne-Edwards, 1861). Female (SMF 22971), CL = 9.9, CB = 17.9 mm, Bab-el-Mandib, Gulf of Aden, depth 45 m, 06.03.1995, coll. F.S. Meteor (Photo: S. Träkner)

Genus *Scylla* De Haan, 1833

Scylla serrata (Forskål, 1775)

(Figs. 20.42, 20.45, 20.46, and 20.47)

Cancer serratus Forskål 1775: 90.

Scylla serrata — Hogarth 1989: 117; Apel and Spirinonov 1998: 312, 313; Apel 2001: 77; Ng et al. 2008: 153 (in list); Naderloo and Türkay 2012: 39; Naderloo et al. 2013: 449, tab 1.

Type locality: Red Sea (Jeddah, Saudi Arabia).

Distribution

Persian Gulf: Iran (Naderloo and Türkay 2012: 39; Naderloo et al. 2013; Rezaei-Atagholipour et al. 2013), UAE (Hogarth and Beach 2001).

Gulf of Oman: Iran (Rezaei-Atagholipour et al. 2013; Naderloo et al. 2015), UAE (Apel and Spirinonov 1998).

World distribution: Indo-West Pacific: South Africa, East Africa, Mauritius, Red Sea, southern Oman, Persian Gulf, Gulf of Oman, Thailand, Japan, Indonesia, Australia and Tahiti.

Habitat: Mangroves, this species digs large burrows among pneumatophores of mangroves.

Fig. 20.45 *Scylla serrata* (Forskål, 1775). Male (ZUTC 5525), CL = 126, CB = 185 mm, mangroves of Hormuz Island, Iran, Persian Gulf, 16.01.2013, coll. R. Naderloo and H. Salehi (Photo: R. Abdollahi)

Fig. 20.46 *Scylla serrata* (Forskål, 1775). The same specimen in the Hormuz Island, taken by the author (Photo: Farimah Asadi)

Fig. 20.47 *Scylla serrata* (Forskål, 1775) (ZUTC 5525), G1 and G2, ventral view

Genus *Thalamita* Latreille, 1829

Key to species of the genus *Thalamita* in the region

1. Front with two lobes 2
- Front with four or six lobes 6

2. Fourth pair of legs with posterodistal angle of propodus rounded; basal antennal segment smooth or minutely granulate; apical part of G1 strongly curved .... *Thalamita poissonii*
- Fourth pair of legs with posterodistal angle of propodus sharply angular or spiniform; basal antennal segment with distinctly granular or spiny ridge; apical part of G1 not strongly curved 3

3. Basal antennal segment with sharp spines .... *Thalamita quadrilobata*
- Basal antennal segment with granules 4

4. Basal antennal segment with finely granular ridge, shorter or of about the same width as orbit; cardiac ridge on carapace well-developed, granular; G1 straight, stout, getting broad apically .... *Thalamita iranica*
- Basal antennal segment with coarsely granular ridge, distinctly longer than orbit; cardiac ridge on carapace minutely granular, somewhat smooth; G1 slightly curved, slender, tapering apically 5

5. Fourth anterolateral tooth markedly smaller than others; frontal lobes nearly straight; inner surface of manus of chelipeds smooth .... *Thalamita admete*
- Fourth anterolateral tooth not markedly smaller than others; frontal lobes slightly concave; inner surface of manus of chelipeds with granular row on middle portion .... *Thalamita savignyi*

6. Front with four distinct lobes, median frontal lobes distinctly narrower than laterals; posterior margin of propodus of last legs without spinules .... *Thalamita sexlobata*
- Front with six lobes 7

7. Anterolateral margins of carapace with four teeth; dorsal surface of carapace with one pair of gastric ridges .... *Thalamita bandusia*
- Anterolateral margins of carapace with five teeth; dorsal surface of carapace with more than one pair of gastric ridges 8

8. Basal antennal segment with row of spines 9
- Basal antennal segment with row of granules 10

9. Fourth anterolateral tooth of carapace very small; outer surface of cheliped manus with two well-developed costae; mesogastric ridge extending to incision between first and second anterolateral teeth of carapace .... *Thalamita prymna*
- Fourth anterolateral tooth of carapace smaller than others but not rudimentary; outer surface of cheliped manus with one costa near lower margin; mesogastric ridge short .... *Thalamita rubridens*

10. Carapace with very faint ridges on dorsal surface; manus of chelipeds smooth with single costa on outer surface .... *Thalamita crenata*
- Carapace with distinct ridges on anterior part of dorsal surface; manus of chelipeds with two or three costae on outer surface 11

11. Supplementary tooth at base of first anterolateral tooth distinct; cardiac ridge not as distinct as first mesobranchial ridges; G1 with distinct spinules on mesial surface .... *Thalamita loppenthini*
- Supplementary tooth at base of first anterolateral tooth indistinct; cardiac ridge as distinct as mesobranchial ridges; G1 with hardly visible spinules on mesial surface .... *Thalamita indistincta*

Thalamita admete (Herbst, 1803)

(Figs. 20.48, 20.49a, and 20.50)

Cancer admete Herbst 1803: 40, 41, pl 57, fig 1.
Thalamita admete — Alcock 1899a: 74 (in key), 82–84; Yazdani et al. 2008: 1, 2, fig 2; Naderloo and Türkay 2012: 41; Naderloo et al. 2013: 449, tab1.

Type locality: "Ostindien". East India or Indo-Malayan Archipelago.

Distribution

Persian Gulf: Iran (Naderloo and Türkay 2012; Naderloo et al. 2013), Saudi Arabia (Basson et al. 1977), UAE (Titgen 1982). Alcock (1899a) recorded this species from an unknown locality.

Gulf of Oman: Iran (Yazdani et al. 2008; Ghotbeddin et al. 2012b).

World distribution: Indo-West Pacific: East Africa, Madagascar, Réunion, Mauritius, southern Oman, Persian Gulf, Gulf of Oman, Pakistan, India, Thailand, Japan, Philippines, Indonesia, Cocos (Keeling) Islands, Lord Howe Island, Australia, French Polynesia and Hawaii.

Habitat: Intertidal, substrate rocky/cobble.

Fig. 20.48 *Thalamita admete* (Herbst, 1803). Male (ZUTC Brach1279), CL = 16.8, CB = 22.1 mm, south coast of Qeshm Island, Iran, Persian Gulf, 10.01.2008, coll. R. Naderloo (Photo: S. Träkner)

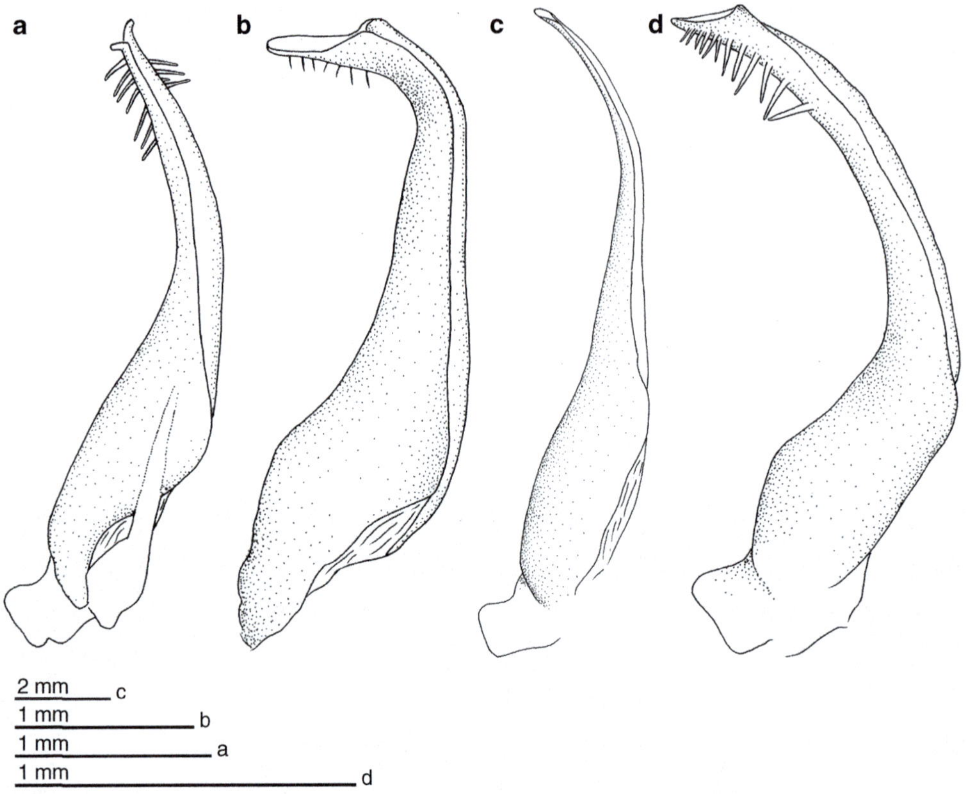

Fig. 20.49 Male G1, ventral view: (**a**) *Thalamita admete* (Herbst, 1803) (SMF 35087); (**b**) *Thalamita bandusia* Nobili, 1905 (SMF 24348); (**c**) *Thalamita crenata* Rüppell, 1830 (SMF 24326); (**d**) *Thalamita indistincta* Apel and Spiridonov, 1998 (SMF 24433)

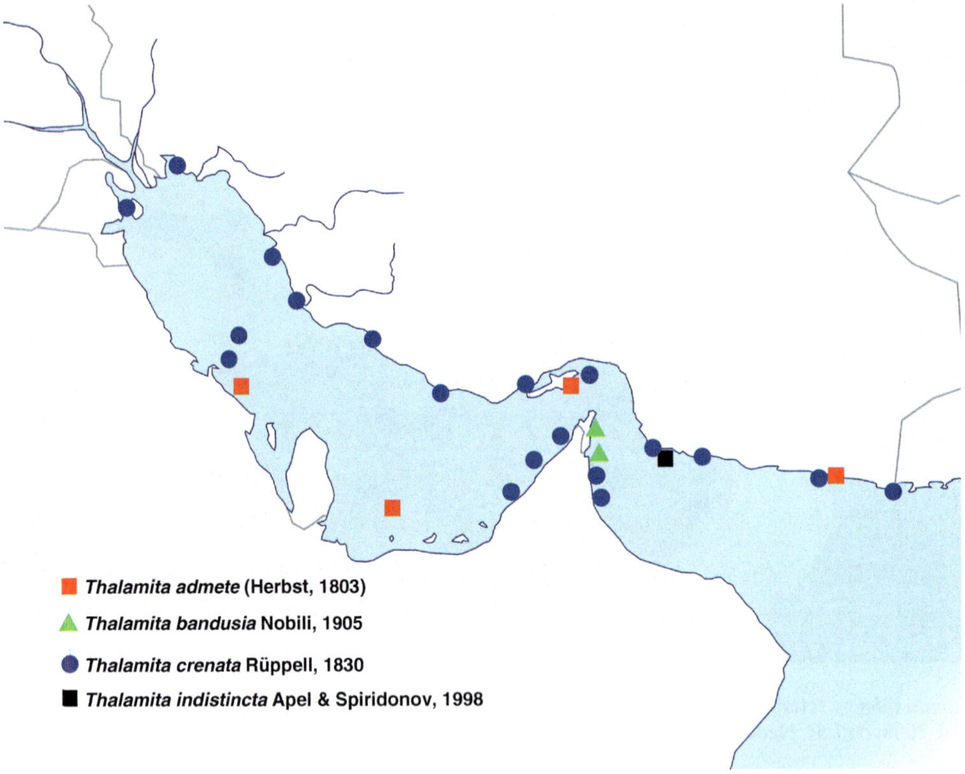

Fig 20.50 Map showing the recorded localities of *Thalamita admete* (Herbst, 1803), *Thalamita bandusia* Nobili, 1905, *Thalamita crenata* Rüppell, 1830, and *Thalamita indistincta* Apel and Spiridonov, 1998

Thalamita bandusia Nobili, 1905
(Figs. 20.49b, 20.50, and 20.51)

Thalamita bandusia Nobili 1905b: 402.
Thalamita bandusia — Nobili 1906b: 211, 212; Apel and Spiridonov 1998: 231–233, figs 43, 45, 46, 60, 61; Apel 2001: 78; Ng et al. 2008: 154 (in list).

Type locality: Bab Al-Mandab and Gulf of Aden.
Distribution
Persian Gulf: No record.
Gulf of Oman: UAE (Apel and Spiridonov 1998).
World distribution: Northwestern Indian Ocean: Red Sea, Gulf of Aden and Gulf of Oman.
Habitat: Subtidal, 10–45 m, substrate rocky/sandy.

Fig. 20.51 *Thalamita bandusia* Nobili, 1905. Male (SMF 24348), CL = 9.5, CB = 14.3 mm, "Martini Rocks", 1 km S Khor Fakkan, UAE, Gulf of Oman, depth 10–18 m, 29.06.1995, coll. M. Apel (Photo: S. Träkner)

Thalamita crenata Rüppell, 1830
(Figs. 20.49c, 20.50, and 20.52)

Thalamita [sic!] *crenata* Rüppell 1830: 6, tab 1, fig 2.
Thalamita crenata — Alcock 1899a: 73 (in key), 76, 77; Nobili 1906a: 119; Basson et al. 1977: 227; Titgen 1982: 122, 123, 250 (in list); Al-Ghais and Cooper 1997: 423, 424; Apel and Spiridonov 1998: 233–235, figs 44, 49, 50, pl 8; Apel 2001: 78; Ng et al. 2008: 154 (in list); Naderloo and Türkay 2012: 41; Naderloo et al. 2013: 449, tab 1.
Thalamita prymna var. *crenata* — Stephensen 1946: 125, 206 (in list).

Type locality: Red Sea (southern part).

Distribution:

Persian Gulf: Iran (Naderloo and Türkay 2012; Naderloo et al. 2013), Saudi Arabia (Basson et al. 1977; Apel and Spiridonov 1998; Apel 2001), UAE (Titgen 1982; Al-Ghais and Cooper 1996; Apel and Spiridonov 1998; Apel 2001).

Gulf of Oman: Iran (Ghotbeddin et al. 2012b; Naderloo et al. 2015), UAE (Apel 2001; Apel and Spiridonov 1998).

World distribution: Indo-West Pacific: South Africa, East Africa, Madagascar, Mauritius, Seychelles, Red Sea, southern Oman, Persian Gulf, Gulf of Oman, Pakistan, India, Andaman Sea, Malaysia, Thailand, Japan, Philippines, Singapore, Cocos (Keeling) Islands, Lord Howe Island, Australia, Samoa, Tuamotu, New Caledonia, French Polynesia and Hawaii.

Habitat: Intertidal, substrates mangroves, rocky, sandy and muddy.

Fig. 20.52 *Thalamita crenata* Rüppell, 1830. Female (SMF 40798), CL = 35.7, CB = 54.9 mm, Mahtabi, Bandar-Khamir, Iran, Persian Gulf, 08.01.2008, coll. M. Türkay and R. Naderloo (Photo: S. Träkner)

Thalamita indistincta Apel and Spiridonov, 1998
(Figs. 20.49d, 20.50, 20.53, and 20.54)

Thalamita indistincta Apel and Spiridonov 1998: 242–249, figs 56–59, 62, 63, 67, 68.
Thalamita (*prymna* (Herbst) var. *picta* Stimpson?) — Stephensen 1946: 125, 126 (partly: ovigerous female).
Thalamita macropus — Spiridonov 1994: 143, fig 7d, e.
Thalamita indistincta — Apel 2001: 78; Ng et al. 2008: 154 (in list).

Type locality: Gulf of Aden.
Distribution:
Persian Gulf: No record.
Gulf of Oman: Iran [Stephensen (1946) as *Thalamita* (*prymna* (Herbst) var. *picta* Stimpson?); Apel and Spiridonov 1998; Apel 2001].
World distribution: Northwestern Indian Ocean: Gulf of Aden and Gulf of Oman.
Habitat: Subtidal, 35–67 m.

Fig. 20.53 *Thalamita indistincta* Apel and Spiridonov, 1998. Male holotype (SMF 24431), CL = 4.4, CB = 5.6 mm, Gulf of Aden, depth 76 m, 16.03.1987, coll. FS "Meteor" (Photo: S. Träkner)

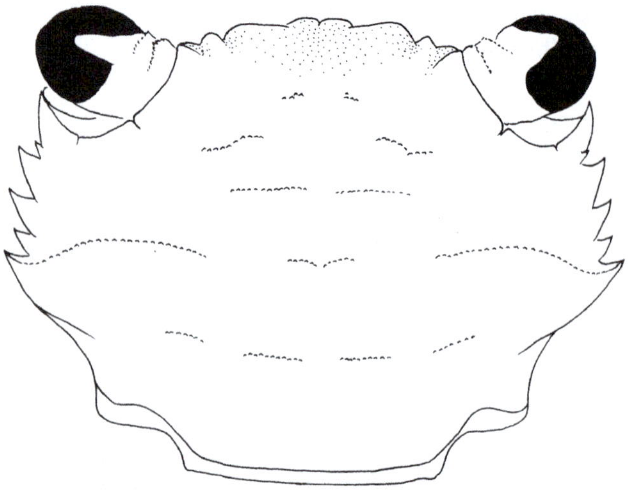

Fig. 20.54 *Thalamita indistincta* Apel and Spiridonov, 1998. Drawing of carapace of the species (SMF 24431)

Thalamita iranica Stephensen, 1946

(Figs. 20.55, 20.56, 20.57a, and 20.58)

Thalamita iranica Stephensen 1946: 128–133, 206 (in list), figs 28–30.

Thalamita iranica — Stephenson 1972a: 149; Titgen 1982: 123, 250 (in list); Apel and Spiridonov 1998: 249, 250, figs 64, 69, 70; Apel 2001: 78; Ng et al. 2008: 154 (in list); Naderloo and Türkay 2012: 41; Naderloo et al. 2013: 449, tab 1.

Type locality: Persian Gulf (Kish Island, Iran).

Distribution

Persian Gulf: Iran (Stephensen 1946; Naderloo and Türkay 2012; Naderloo et al. 2013), Saudi Arabia (Apel and Spiridonov 1998; Apel 2001), Bahrain (Stephensen 1946), UAE (Titgen 1982; Apel and Spiridonov 1998; Apel 2001).

Fig. 20.55 *Thalamita iranica* Stephensen, 1946. Male paratype (ZMUC CRU973), CL = 8.8, CB = 13.7 mm, Iran, Persian Gulf, depth 33 m, 27.03.1937, coll. G. Thorson (Photo: S. Träkner)

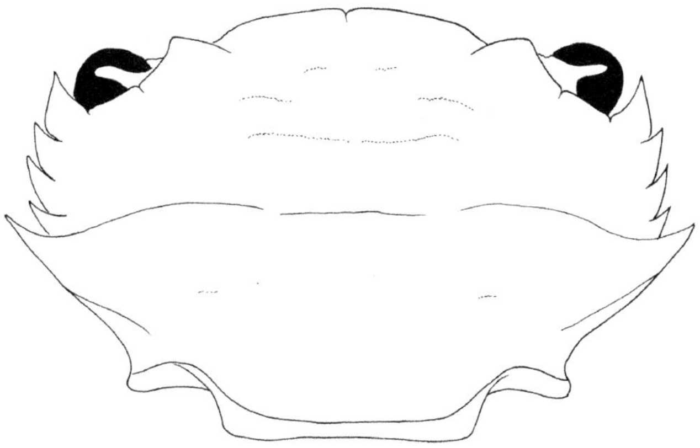

Fig 20.56 *Thalamita iranica* Stephensen, 1946. Drawing of carapace of the same species (ZMUC CRU973)

Gulf of Oman: No record.

World distribution: Northwestern Indian Ocean: Red Sea, Gulf of Aden, Persian Gulf and Gulf of Oman.

Habitat: Intertidal to 58 m, substrates clay, sandy, algae, gravel, oyster bank, cobble, rocky and corals.

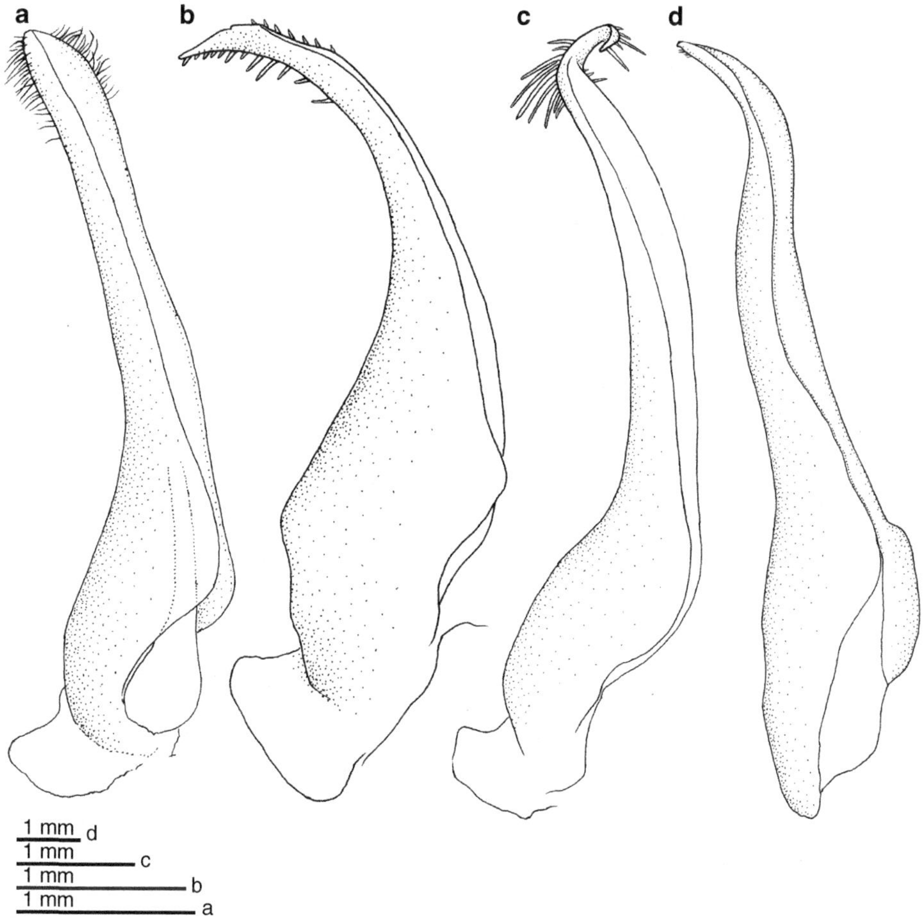

Fig. 20.57 Male G1, ventral view: (**a**) *Thalamita iranica* Stephensen, 1946 (ZMUC CRU813); (**b**) *Thalamita loppenthini* Apel and Spiridonov, 1998 (ZMUC CRU955, holotype); (**c**) *Thalamita poissonii* (Savigny, 1817) (SMF 24339); (**d**) *Thalamita prymna* (Herbst, 1803) (ZUTC Brach1836)

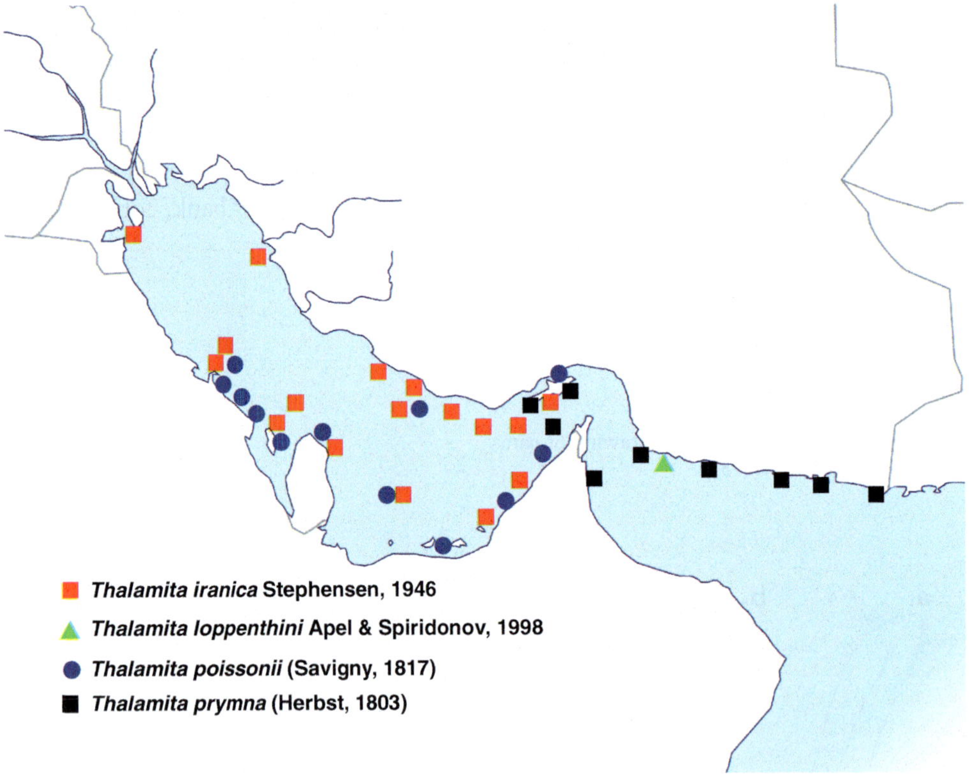

Fig 20.58 Map showing the recorded localities of *Thalamita iranica* Stephensen, 1946, *Thalamita loppenthini* Apel and Spiridonov, 1998, *Thalamita poissonii* (Savigny, 1817), *Thalamita prymna* (Herbst, 1803)

Thalamita loppenthini Apel and Spiridonov, 1998
(Figs. 20.57b, 20.58, and 20.59)

Thalamita loppenthini Apel and Spiridonov 1998: 251–253, figs 65, 66, 71, 72.
Thalamita loppenthini — Apel 2001: 79; Ng et al. 2008: 154 (in list).
Thalamita (*prymna* (Herbst) var. *picta* Stimpson?) — Stephensen 1946: 125, 126 (partly).
Type locality: Gulf of Oman (Jask, Iran).
Distribution

Persian Gulf: No record.

Gulf of Oman: Iran (Apel and Spiridonov 1998; Naderloo et al. 2015).

World distribution: This species is still only known from the Gulf of Oman.

Habitat: Subtidal to 35 m.

Thalamita poissonii (Savigny, 1817)
(Figs. 20.57c, 20.58, and 20.60)

Portunus poissonii Savigny 1817: pl 4, figs 3(1, 2), 5 (1, 2).
Portunus poissonii — Audouin 1826: 84.
Thalamita poissonii — Alcock 1899b: 74 (in key), 81; Nobili 1906a: 120; Nobili 1906b: 206; Stephensen 1946: 133, fig 31; Guinot 1967a: 256 (in list); Titgen 1982: 250 (in list); Vousden 1987: 32, 56; Cooper 1997: 167, fig 13; Apel and Spiridonov 1998: 253–257, figs 73–75; Apel 2001: 79; Ng et al. 2008: 154 (in list); Naderloo and Türkay 2012: 41; Naderloo et al. 2013: 449, tab 1.
Thalamita poissoni — Basson et al. 1977: 224, 231, 250, 256; Al-Ghais and Cooper 1996: 424, 425; Hornby 1997: 75.
Type locality: Red Sea (Gulf of Suez).
Distribution

Persian Gulf: Iran (Naderloo and Türkay 2012; Naderloo et al. 2013), Saudi Arabia (Basson et al. 1977; Apel and Spiridonov 1998; Apel 2001), Bahrain (Stephensen 1946; Vousden 1987), UAE (Nobili 1906a; Al-Ghais and Cooper 1996; Cooper

Fig. 20.59 *Thalamita loppenthini* Apel and Spiridonov, 1998. Holotype male (ZMUC CRU955), CL = 14.42, CB = 22.82 mm, W Jask, Iran, Gulf of Oman, 02.03.1938, coll. B. Loppenthin (Photo: S. Träkner)

Fig. 20.60 *Thalamita poissonii* (Savigny, 1817). Male (SMF 24339), CL = 15.7. CB = 24.3 mm, Jubail Marine Wildlife Santuary, Saudi Arabia, Persian Gulf, 07.12.1991, coll. M. Apel (Photo: S. Träkner)

1996; Hornby 1997; Apel and Spiridonov 1998; Apel 2001). Alcock (1899b) recorded this species from an unknown locality.

Gulf of Oman: No record.

World distribution: East Atlantic, Indo-West Pacific and Mediterranean: East Atlantic, Angola, Canary Islands, South Africa, East Africa, Madagascar, Mauritius, Seychelles, Red Sea, Gulf of Aden, Persian Gulf, western coast of India, Lakshadweep Islands, Sri Lanka and Mediterranean (Lessepsian migrant).

Habitat: Intertidal to 5 m, substrate rocky/cobble.

Remarks Apel and Spiridonov (1998) mentioned some variations in the shape of the frontal lobes, anterolateral teeth of the carapace and pilosity of the carapace. A small male (CL = 9.23, CB = 13.28 mm) collected during the study from the south coast of the Qeshm I., which has a relatively plumose carapace, and the first three anterolateral teeth of the carapace are nearly blunt, whereas specimens examined by Apel and Spiridonov (1998) from the Persian Gulf have acute teeth. The Persian Gulf population of this species has usually the frontal lobe laterally produced (Apel and Spiridonov 1998), while in the present specimen it is indistinct. Apel and Spiridonov (1998) stated that the occurrence of this species in the eastern Indian Ocean is doubtful, since some records from the region (Sakai 1939; Stephenson and Rees 1967a, b) are based on misidentifications.

Thalamita prymna (Herbst, 1803)

(Figs. 20.57d, 20.58, and 20.61)

Cancer prymna Herbst 1803: 41, pl 57, fig 2.
Thalamita prymna — Alcock 1899a: 78; Guinot 1967a: 255 (in list; partly); Tirmizi and Kazmi 1996: 60–63, figs 31, 32, pl 2c; Apel and Spiridonov 1998: 257–260, figs 76, 85; Apel 2001: 79; Yazdani et al. 2008: 2, fig 3; Naderloo and Türkay 2012: 42; Naderloo et al. 2013: 449, tab 1.

Fig. 20.61 *Thalamita prymna* (Herbst, 1803). Male (SMF 49892), CL = 34.1, CB = 42.9 mm, S coast of Qeshm Island, Iran, Persian Gulf, 10.01.2008, coll. M. Türkay and R. Naderloo (Photo: S. Träkner)

not *Thalamita* (*prymna* (Herbst) var. *picta* Stimpson?) — Stephensen 1946: 125, 126 [= *Thalamita loppenthini* Apel and Spiridonov 1998 (partly) and *Thalamita indistincta* Apel and Spiridonov, 1998 (partly)].
not *Thalamita prymna* — Titgen 1982: 124, 250 (in list) [= *Thalamita rubridens* Apel and Spiridonov, 1998].
Type locality: India (southeastern India, Tuticorin).
Distribution

Persian Gulf: Iran (Naderloo and Türkay 2012; Naderloo et al. 2013), UAE (Apel and Spiridonov 1998; Apel 2001).

Gulf of Oman: Iran (Yazdani et al. 2008; Ghotbeddin et al. 2012b), UAE (Apel and Spiridonov 1998; Apel 2001).

World distribution: Indo-West Pacific: South Africa, East Africa, Madagascar, Mauritius, Red Sea, Persian Gulf, Gulf of Oman, Pakistan, India, Andaman Islands, Nicobar Islands, Mergui Archipelago, Malaysia, Thailand, Japan, Philippines, Singapore, Indonesia, Lord Howe Island, Australia, Samoa, Fiji and French Polynesia.

Habitat: Intertidal to 5 m, substrate rocky/cobble.

Thalamita quadrilobata Miers, 1884

(Figs. 20.62, 20.63a, and 20.64)

Thalamita quadrilobata Miers 1884b: 539, 540, pl XLVIII, fig b.
Thalamita quadrilobata — Apel and Spiridonov 1998: 260–263, figs 77–86; Apel 2001: 79; Ng et al. 2008: 155 (in list).
Type locality: Seychelles.

Fig 20.62 *Thalamita quadrilobata* Miers, 1884. Male (SMF 24330), CL = 14.3, CB = 25.1 mm, S Lima, Oman, Gulf of Oman, 29.06.1995, coll. M. Apel (Photo: S. Träkner)

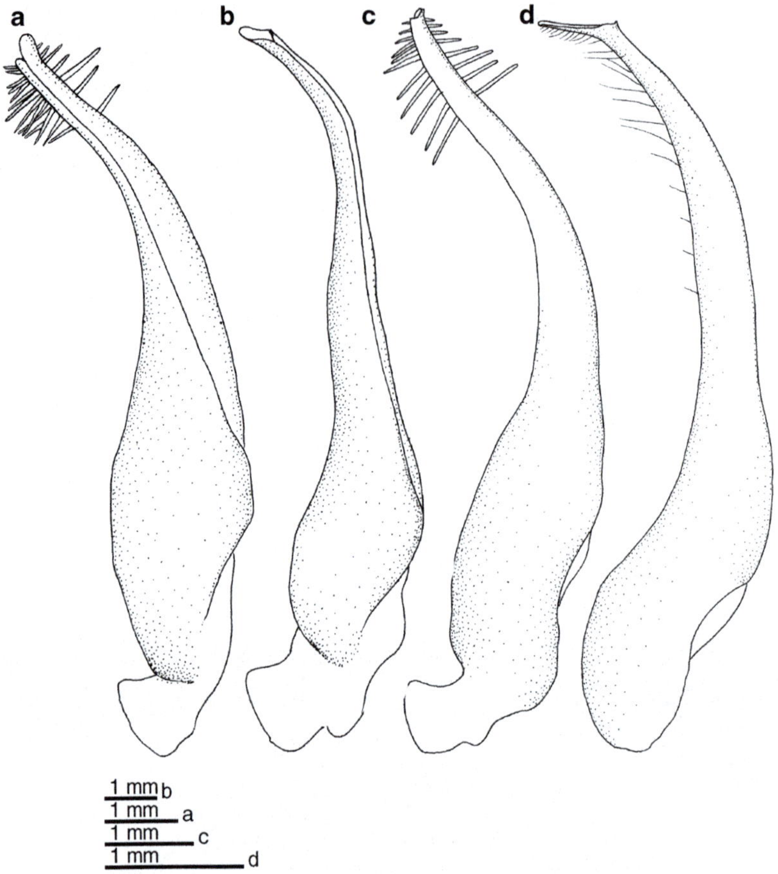

Fig. 20.63 Male G1, ventral view: (**a**) *Thalamita quadrilobata* Miers, 1884 (SMF 24330); (**b**) *Thalamita rubridens* Apel and Spiridonov, 1998 (SMF 24400: holotype); (**c**) *Thalamita savignyi* A. Milne-Edwards, 1861 (ZUTC: Iran); (**d**) *Thalamita sexlobata* Miers, 1886 (ZMUC CRU872)

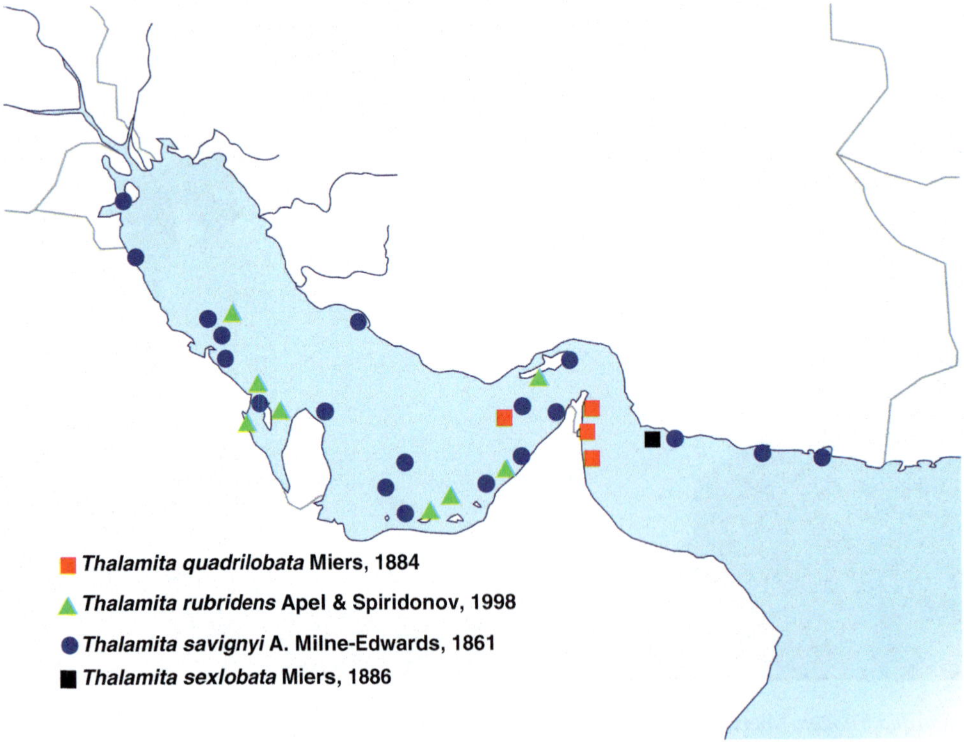

Fig. 20.64 Map showing the recorded localities of *Thalamita quadrilobata* Miers, 1884, *Thalamita rubridens* Apel and Spiridonov, 1998, *Thalamita savignyi* A. Milne-Edwards, 1861, *Thalamita sexlobata* Miers, 1886

Distribution

Persian Gulf: Iran (present study).

Gulf of Oman: UAE (Apel and Spiridonov 1998; Apel 2001).

World distribution: Indo-West Pacific: Madagascar, Mauritius, Seychelles, East Africa, Red Sea, Gulf of Aden, Persian Gulf, Gulf of Oman, Maldives, Andaman Sea, Indonesia, Philippines, Palau, Australia and French Polynesia.

Habitat: Subtidal, 4–25 m.

Thalamita rubridens Apel and Spiridonov, 1998

(Figs. 20.63b, 20.64, and 20.65)

Thalamita rubridens Apel and Spiridonov 1998: 263–266, figs 78–81, 87, 88, pl 9.

Thalamita rubridens — Apel 2001: 80; Ng et al. 2008: 155 (in list); Naderloo and Türkay 2012: 42; Naderloo et al. 2013: 449, tab 1.

Thalamita prymna — Titgen 1982: 263–266, figs 78–81, 87, 88, pl 9.

Type locality: Persian Gulf (Merawwah Island, Abu Dhabi, UAE).

Distribution

Persian Gulf: Iran (Naderloo and Türkay 2012; Naderloo et al. 2013), Saudi Arabia (Apel and Spiridonov 1998), Bahrain (Apel and Spiridonov 1998), UAE (Apel and Spiridonov 1998).

Gulf of Oman: No record.

World distribution: This species is still only known from the Persian Gulf.

Habitat: Intertidal to 9 m, substrate rocky/cobble.

Fig. 20.65 *Thalamita rubridens* Apel and Spiridonov, 1998. Male paratype (SMF 24408), CL = 39.2, CB = 61.1 mm, Abu Ali Island, Saudi Arabia, Persian Gulf, 13.04.1993, coll. M. Apel (Photo: S. Träkner)

Thalamita savignyi A. Milne-Edwards, 1861
(Figs. 20.63c, 20.64, and 20.66)

Thalamita savignyi A. Milne-Edwards 1861: 357, 358, 367 (in key).
Thalamita savignyi — Alcock 1899a: 74 (in key), 84; Apel and Spiridonov 1998: 267–270, figs 48, 82–84; Apel 2001: 80; Ng et al. 2008: 155 (in list); Naderloo and Türkay 2012: 42; Naderloo et al. 2013: 449, tab 1.
Thalamita admete — Alcock 1899a: 82–84 (partly: material from the Persian Gulf).
Thalamita admete var. *savignyi* — Nobili 1906a: 120; Laurie 1915: 440.
Thalamita admete savignyi — Stephensen 1946: 136, 206 (in list).
Thalamita admete — Heller 1861a: 355, 356; Guinot 1967a: 255 (in list; partly: records from the Persian Gulf); Basson et al. 1977: 227, 231, 243; Titgen 1982: 250 (in list).
Thalamita? *admete* — Titgen 1982: 120, 121.
Type locality: Red Sea (Gulf of Suez).
Distribution
Persian Gulf: Iran (Naderloo and Türkay 2012; Naderloo et al. 2013), Saudi Arabia (Apel and Spiridonov 1998; Apel 2001), UAE (Nobili 1906a; Apel and Spiridonov 1998; Apel 2001). Alcock (1899a) recorded this species from an unknown locality.
Gulf of Oman: Iran (Naderloo et al. 2015).
World distribution: Western Indian Ocean: East Africa (Zanzibar), Red Sea, Gulf of Aden, Persian Gulf and Gulf of Oman.
Habitat: Intertidal to 10 m, substrate rocky/cobble.

Fig. 20.66 *Thalamita savignyi* A. Milne-Edwards, 1861. Male (ZUTC Brach1281), CL = 2.87, CB = 40.54 mm, Khalij-Nayband, Iran, Persian Gulf, coll. Kamyab. Photo S. Träkner

Thalamita sexlobata Miers, 1886

(Figs. 20.63d, 20.64, and 20.67)

Thalamita sexlobata Miers 1886: 196, pl 16, figs 2a–c.
Thalamita sexlobata — Alcock 1899a: 87; Stephensen 1946: 136, fig 32c, d; Apel and Spiridonov 1998: 271–273, figs 89, 90, 93; Apel 2001: 80; Ng et al. 2008: 155 (in list).

Type locality: Tonga.

Distribution

Persian Gulf: Alcock (1899a) recorded this species from an unknown locality.

Gulf of Oman: Iran (Stephensen 1946).

World distribution: Indo-West Pacific: Madagascar, Gulf of Aden, Red Sea, Persian Gulf, Gulf of Oman, Andaman Sea, Burma, Indonesia, Japan, Australia, Tonga, Hawaii.

Habitat: Subtidal, 20–80 m, substrates sandy and muddy.

Fig. 20.67 *Thalamita sexlobata* Miers, 1886. Male (ZMUC CRU872), CL = 8.6, CB = 10.3 mm, SW of Jask, Iran, Gulf of Oman, depth 35 m, 29.03.1938, coll. B. Loppenthin (Photo: R. Abdollahi)

Family Xanthidae MacLeay, 1838 (Rubble Crabs, Stone Crabs, Rock Crabs)

21

Diagnosis Carapace subhexagonal to transversely oval, broader than long, regions usually well defined, anterolateral margins convex, confluent with exorbital angle, with four to five denticles or lobes, usually distinctly separated from posterolateral margins. Front relatively narrow, normally with median notch. Orbits small, exorbital lobe small to indistinct, eyestalks short. Antennules folded transversely or obliquely. Antennae with basal antennal segment broad, short, inner angle usually touching front, distal segments entering orbit, antennal flagellum short and slender. Third maxillipeds covering whole buccal cavern, with quadrate-shaped merus. Abdomen of males with tightly fused somites 3–5, abdomen of females with freely moving somites. Genital opening of males coxal. Genital opening of females sternal. G1 various-sized and –shaped; G2 short.

Key to genera of the family Xanthidae in the region

1. Antennules folded obliquely; carapace suncircular to oval, with lateral margins arched; chelipeds markedly unequal; symbiont with hexacorals .... *Cymo*
- Antennules folded transversely; carapace in various forms, mostly xanthoid forms, broader than long; chelipeds equal or subequal 2
2. Walking legs with special dactylus-propodus articulation, markedly formed by lateral extension of proximal margin of dactylus, dactylus-propodus articulation hardly seen in some species, but dactylus is always bifid 3
- Walking legs without special dactylus-propodus articulation; dactylus always ending in simple spine 6
3. Front narrow, narrower than one third of carapace breadth, usually projecting; bilobed with large median fissure; basal antennal segment with anterolateral lobe; dactylus-propodus articulation of walking legs weakly developed .... *Etisus*
- Front broad, broader than one third of carapace breadth; basal antennal segment without anterolateral lobe; dactylus-propodus articulation of walking legs markedly developed 4
4. Basal antennal segment with anterolateral angle long, situated in orbital hiatus; antennal flagellum excluded from orbital hiatus .... *Pilodius*

R. Naderloo, *Atlas of Crabs of the Persian Gulf*, DOI 10.1007/978-3-319-49374-9_21

- Basal antennal segment with anterolateral angle not long; orbital hiatus completely open; antennal flagellum freely directed backwards to orbital hiatus 5

5. Dorsal surface of carapace markedly uneven, tuberculate; anterolateral margin of carapace with two last teeth spiniform; front as broad as posterior margin of carapace; chelipeds relatively short, with tubercles; walking legs with spines along anterior margins of merus and carpus .... *Cyclodius*

- Dorsal surface of carapace relatively smooth, with faint tubercles on anterolateral region; teeth along anterolateral margins of carapace never spiniform; front broader than posterior margin of carapace; chelipeds relatively long, smooth; anterior margins of merus and carpus of walking legs without spine; dark-brown colored species .... *Chlorodiella*

6. Carapace transversely oval, usually much broader than long, with dorsal surface convex, finely granular on lateral surface, regions weekly defined; anterolateral margins entire with small denticles, not distinctly separated from posterolateral margins; front straight, feebly bilobed .... *Neoliomera*

- Carapace xanthoid-shaped, regions well-defined; anterolateral margins convex, markedly separated from the posterolateral margins; posterolateral margins relatively concave 7

7. Distance between exorbital angle and first anterolateral tooth continuing connected to buccal cavern; larger cheliped with differentiated basal tooth on cutting edge of movable finger 8

- Distance between exorbital angle and first anterolateral tooth never connected to buccal cavity; larger cheliped without markedly differentiated basal tooth 9

8. Anterolateral margins of carapace with markedly large pore leading to buccal cavern; posterolateral margins with special depression for situating last walking leg .... *Hepatoporous*

- Distinct distance between exorbital angle and first anterolateral tooth of carapace, with feeble connection to buccal cavern; carapace moderately flat, with transverse rows of fine granules on dorsal surface .... *Medaeops*

9. Carapace and appendages densely granular or tubercular; chelipeds subequal; fingers apically pointed, never spoon-shaped 10

- Carapace and appendages not markedly granular or tubercular; chelipeds subequal to unequal; fingers pointed or spoon-shaped apically 14

10 Posterolateral margins of carapace with deep depression for situating last three pairs of walking legs .... *Psaumis*

- Posterolateral margins of carapace straight or slightly concave 11

11. Thoracic sternum of males with suture between sternites 3 and 4 interrupted medially 12

- Thoracic sternum of males with complete suture between sternites 3 and 4, never interrupted medially 13

12. Carapace and appendages with tubercles tightly fused together, narrow and shallow furrow between regions .... *Actaea*

- Carapace and appendages with tubercles in groups distinctly separated by broad and deep furrows .... *Epiactaea*

13. Carapace with large swollen tubercles covering with small granules, deep and broad furrow between tubercles, dense patch of setae usually in anterior part of tubercles; anterior margin of thoracic sternite 3 smooth .... *Paractaea*

- Carapace with large tubercles covering with small granules, shallow and broad furrow between tubercles, short setae usually in anterior part of tubercles; anterior margin of thoracic sternite 3 granular .... *Gaillardiellus*

14. Walking legs with marked crest along anterior margins, particularly of carpus and propodus; lateral margins of carapace with four teeth or entire without any lobe or tooth 15

- Walking legs without marked crest along anterior margins; lateral margins of carapace with four or more teeth 18

15. Dorsal surface of carapace without any elevation, regions not defined; anterolateral margins of carapace entire, feebly lobate, weekly emarginated .... *Atergatis*

- Dorsal surface of carapace uneven, regions well-defined; anterolateral margins of carapace distinctly emarginated, lobate or dentate 16

16. Walking legs with two crests along anterior margin of carpus, forming special cup-shaped structure .... *Zozymodes*

- Walking legs with one crest along anterior margin of carpus 17

17. Anterolateral margins of carapace with four distinct lobes, last one triangular; dorsal surface of carapace tuberculates, regions well-defined; walking legs with distinct grooves on posterior surface of carpus and propodus .... *Zosimus*

- Anterolateral margins of carapace with a continuous crest, very weekly divided into lobes, last epibracnhial tooth distinct; walking legs without grooves on posterior surface .... *Platypodia*

18. Fingers of chelipeds apically spoon-shaped; anterolateral margins of carapace with four or more teeth 19

- Fingers of chelipeds apically pointed, never spoon-shaped; anterolateral margins of carapace with four or less teeth 20

19. Fingers of chelipeds with large spoon-shaped tip; basal antennal segment long; third maxillipeds with merus broader than long, anterolaterally produced; G1 with short subdistal spines, long apical process, surrounded with mushroom-shaped tubercles .... *Leptodius*

- Fingers of chelipeds with small spoon-shaped tip; basal antennal segment quadrate; third maxillipeds with merus as long as broad; G1 with long subdistal spines, apically bifurcate, without mushroom-shaped tubercles .... *Macromedaeus*

20. Dorsal surface of carapace markedly uneven, granular, regions well defined with deep furrows; anterolateral margins with four distinct teeth; chelipeds with transverse granular crests on upper surface of carpus and palm .... *Paraxanthodes*

- Dorsal surface of carapace smooth or with small granules or punctuate, regions never or weakly defined; anterolateral margins with granules or lobes, never with teeth; chelipeds without any granular crest 21

21. Dorsal surface of carapace completely smooth, without any granules, regions not defined; anterolateral margins of carapace shorter than posterolateral margins; walking legs narrow and long .... *Liagore*

- Dorsal surface of carapace with small granulates or punctuate, with low elevations, regions weakly defined; anterolateral margins of carapace nearly as long as posterolateral margins; walking legs broad and short .... *Xanthias*

Genus *Actaea* De Haan, 1833

Key to species of the genus *Actaea* in the region

- Carapace covered with tubercles, tubercles getting spiniform on lateral margins; appendages entirely coverd with more or less spiniform tubercles .... *Actaea spinosissima*
- Carapace densely covered with conical tubercles; anterolateral margins with tuberculate teeth; appendages with conical tubercles, sometimes spiniform; G1 apical process gently curving laterally .... *Actaea jacquelinae*

Actaea jacquelinae Guinot, 1976

(Figs. 21.1, 21.2, 21.3a and 21.4)

Actaea jacquelinae Guinot 1976: 227, fig 39d, pl 10, figs 3–3b.

Actaea calculosa — Alcock 1898: 152 (partly: material from the Persian Gulf); Chopra and Das 1937: 400 (partly: material from the Persian Gulf); Stephensen 1946: 153; Guinot 1967a: 260 (in list); Titgen 1982: 251 (in list).

Actaea granulata — Nobili 1906a: 127.

Fig. 21.1 *Actaea jacquelinae* Guinot, 1976. Male holotype (MNHN-IU-2014.8622), CL = 12.2, CB = 17.1 mm, SW of Arzanah Island and SO of Zirkuh Island, Abu Dhabi, UAE, Persian Gulf, 1901, coll. Bonnier and Peréz (Photo: S. Träkner)

Actaea savignyi — Stephensen 1946: 151, fig 38c; Guinot 1964: 44; Guinot 1967a: 260 (in list); Evans et al. 1973: 275; Titgen 1982: 125–127, 251 (in list); Jones 1986a: 162, pl 48.

Actaea jacquelinae — Serène 1984: 113, fig 65, pl 14c; Tirmizi and Ghani 1996: 11–13, fig 3; Apel 2001: 81; Naderloo and Sari 2007a; 343, tab 1; Naderloo and Türkay 2012: 43; Naderloo et al. 2013: 449, tab 1.

Type locality: Persian Gulf (Arzanah Island, Abu-Dhabi, UAE).

Distribution

Persian Gulf: Iran [Stephensen (1946) as *A. savignyi*; Apel 2001; Naderloo and Sari 2007a; Naderloo and Türkay 2012; Naderloo et al. 2013], Kuwait [Jones (1986a) as *A. savignyi*], Saudi Arabia (Apel 2001), Bahrain [Stephensen (1946) as *A. savignyi*; Apel 2001], UAE [Nobili (1906a) as *A. granulata*; Evans et al. (1973) and Titgen (1982) as *A. savignyi*; Apel 2001].

Gulf of Oman: Iran (Naderloo et al. 2015).

World distribution: Northern Indian Ocean: Persian Gulf, Gulf of Oman, Pakistan and India.

Habitat: Intertidal to 12.5 m, substrates rocky/cobble and oyster bank.

Remarks Guinot (1976) described this species from Arzanah Island in the Persian Gulf and mentioned the minor differences between *A. jacquelinae* and *A. savignyi*. These two species morphologically are very similar; therefore their records in the different regions of the northwestern Indian Ocean made so much confusion. More material are needed to be examined in order to establish the exact taxonomic position and their geographical distribution in the region.

Fig. 21.2 *Actaea jacquelinae* Guinot, 1976. Female (ZUTC Brach1070), CL = 19.05, CB = 25.64 mm, Iran, Persian Gulf, depth 24 m, 20.12.2003, coll. R. Naderloo (Photo: R. Abdollahi)

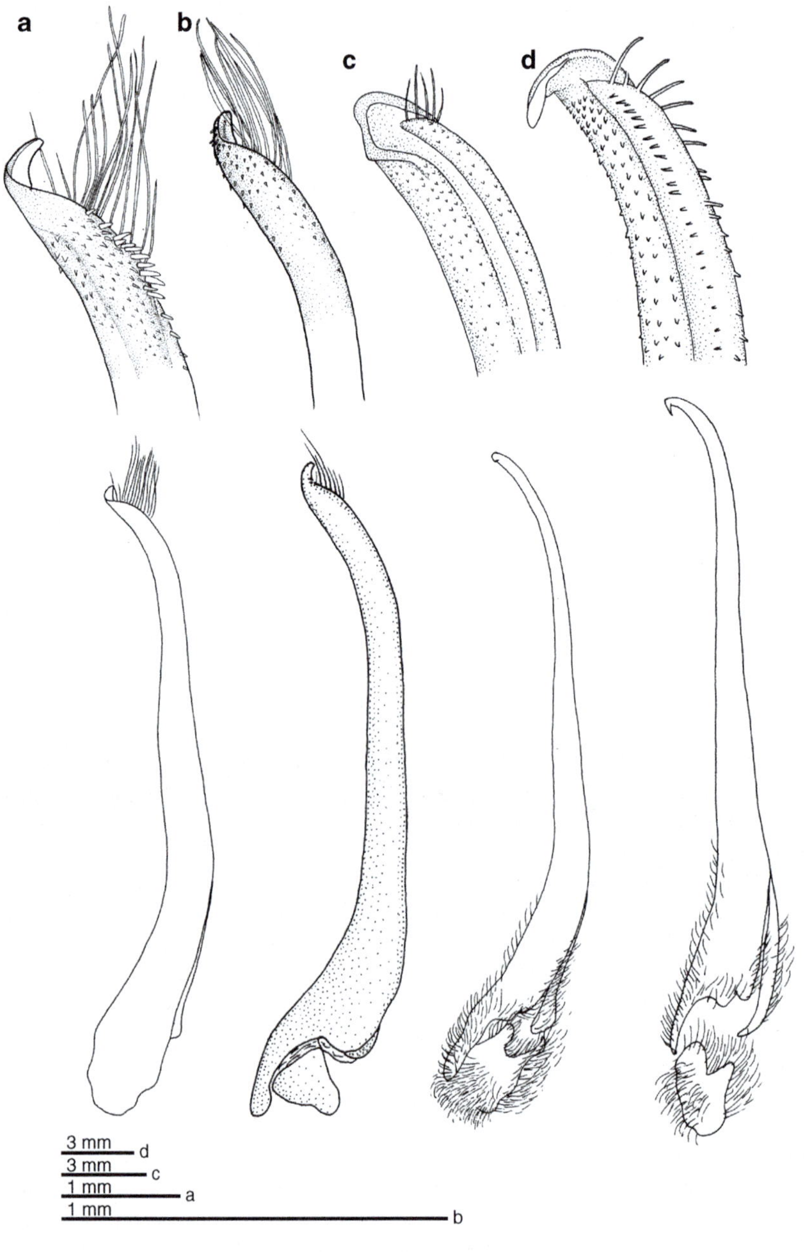

Fig. 21.3 Male G1, ventral view: (**a**) *Actaea jacquelinae* Guinot, 1976 (SMF: UAE); (**b**) *Actaea spinosissima* Borradaile, 1902 (SMF: UAE); (**c**) *Atergatis integerrimus* (Lamarck, 1818) (ZMUC CRU981); (**d**) *Atergatis laevigatus* A. Milne-Edwards, 1865 (SMF: Saudi Arabia)

Actaea spinosissima Borradaile, 1902

(Figs. 21.3b, 21.4, and 21.5)

Actaea spinosissima Borradaile 1902: 256, fig 55.
Actaea spinosissima — Serène 1984: 114, 115, pl 14f; Apel 2001: 82; Ng et al. 2008: 195 (in list).
Type locality: Maldives.

Distribution

Persian Gulf: No record.

Gulf of Oman: UAE (Apel 2001).

World distribution: Indian Ocean: Cargados, Carajos, Socotra, Gulf of Oman, Maldives, Sri Lanka and Indonesia.

Habitat: Shallow subtidal to 20 m, substrates rocky/cobble and oyster bank.

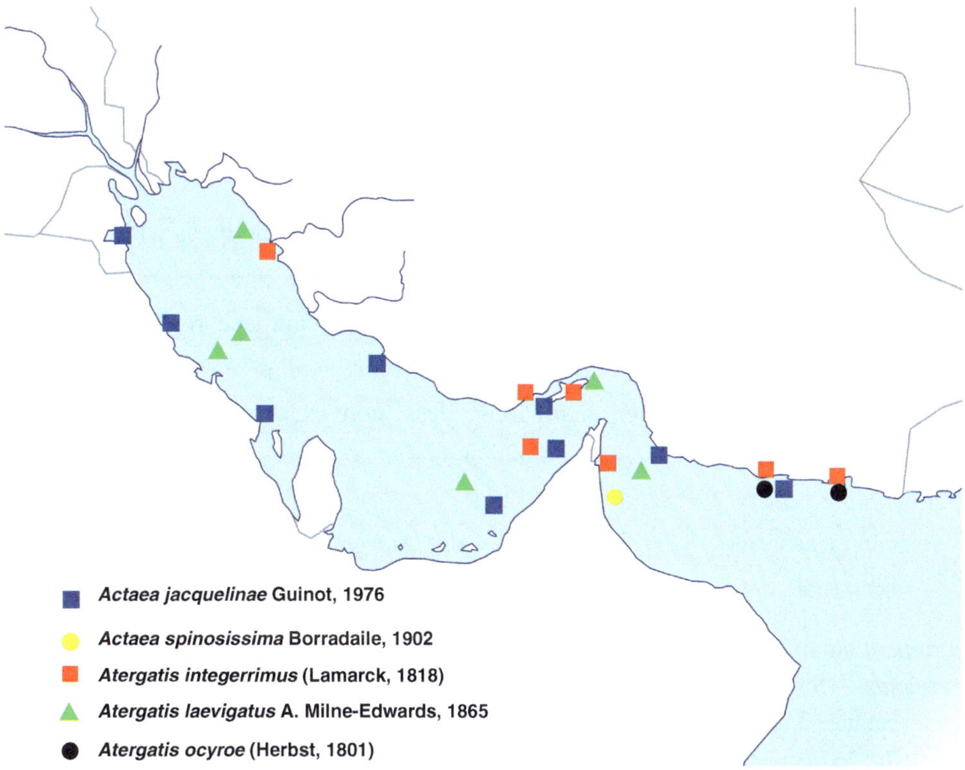

Fig. 21.4 Map showing the recorded localities of *Actaea jacquelinae* Guinot, 1976, *Actaea spinosissima* Borradaile, 1902, *Atergatis integerrimus* (Lamarck, 1818), *Atergatis laevigatus* A. Milne-Edwards, 1865, and *Atergatis ocyroe* (Herbst, 1801)

Fig. 21.5 *Actaea spinosissima* Borradaile, 1902. Male (SMF), CL = 5.9, CB = 7.9 mm, Sandy Beach Hotel, Al-Aqqa, N Khor Fakkan, UAE, Gulf of Oman, depth 3–4 m, 04.07.1995. coll.M. Apel (Photo: S. Träkner)

Genus *Atergatis* De Haan, 1833

Key to species of the genus *Atergatis* in the region

1. Carapace without any epibranchial tooth on lateral margins; chelipeds with blunt crest along upper margin of palm .... *Atergatis laevigatus*

- Carapace with distinct blunt epibranchial tooth on lateral margins; chelipeds with markedly elevated crest along upper margin of palm 2

2. Carapace about 1.4 times broader than long, more convex, carapace regions relatively well defined .... *Atergatis ocyroe*

- Carapace about 1.6 times broader than long, less convex, carapace regions not well defined .... *Atergats integrrimus*

Atergatis integerrimus (Lamarck, 1818)

(Figs. 21.3c, 21.4, and 21.6)

Cancer integerrimus Lamarck 1818: 272.
Atergatis integerrimus — Stephensen 1946: 154 (partly: a male from coral reef near Larak Island, Persian Gulf), fig 39b; Guinot 1967a: 261 (in list; partly); Serène 1984: 151, pl 21f; Tirmizi and Ghani 1996: 16–19, fig 5; Apel 2001: 82; Ng et al. 2008: 205 (in list).

not *Atergatis integerrimus* — Stephensen 1946: 154 (partly: material from south of Bushehr, Iran, Persian Gulf) [= *Atergatis laevigatus* (A. Milne-Edwards, 1865)].

Type locality: Australia.

Distribution

Persian Gulf: Iran (Stephensen 1946; Naderloo et al. 2016).

Gulf of Oman: UAE (Apel 2001).

Fig. 21.6 *Atergatis integerrimus* (Lamarck, 1818). Male (ZUTC 5483), CL = 34.88, CB = 56.21 mm, Park-Qadir, Abu Musa Island, Iran, Persian Gulf, 25.12.2014, coll. S. Ebrahimnejad and R. Abdollahi. (Photo: R. Abdollahi)

World distribution: Indo-West Pacific: Zanzibar, Persian Gulf, Gulf of Oman, Strait of Hormuz, Pakistan, India, Taiwan, Philippines, Japan and Australia.

Habitat: Intertidal to shallow subtidal, substrate sandy.

Atergatis laevigatus A. Milne-Edwards, 1865

(Figs. 21.3d, 21.4, 21.7, and 21.8)

Atergatis laevigatus A. Milne-Edwards 1865b: 241, pl 15, figs 4, 4a.

Atergatis laevigatus — Guinot 1967a: 261 (in list; partly); Serène 1984: 147, 148, pl 21b; Apel 2001: 82; Ng et al. 2008: 205 (in list).

Atergatis roseus — Heller 1861a: 309, 310; Guinot 1967a: 261 (in list; partly); Titgen 1982: 251 (in list); Tirmizi and Ghani 1996: 14–16, fig 4a–f.

Atergatis integerrimus — Stephensen 1946: 154 (partly: material from Station 2, south of Bushehr, Iran, Persian Gulf; not fig 39b).

Type locality: India (Malabar).

Distribution

Persian Gulf: Iran [Heller (1861a) as *A. roseus*; Stephensen (1946) as *A. integerrimus*], Saudi Arabia (Apel 2001), UAE (Apel 2001).

Gulf of Oman: Iran (Naderloo et al. 2015), Oman (Apel 2001).

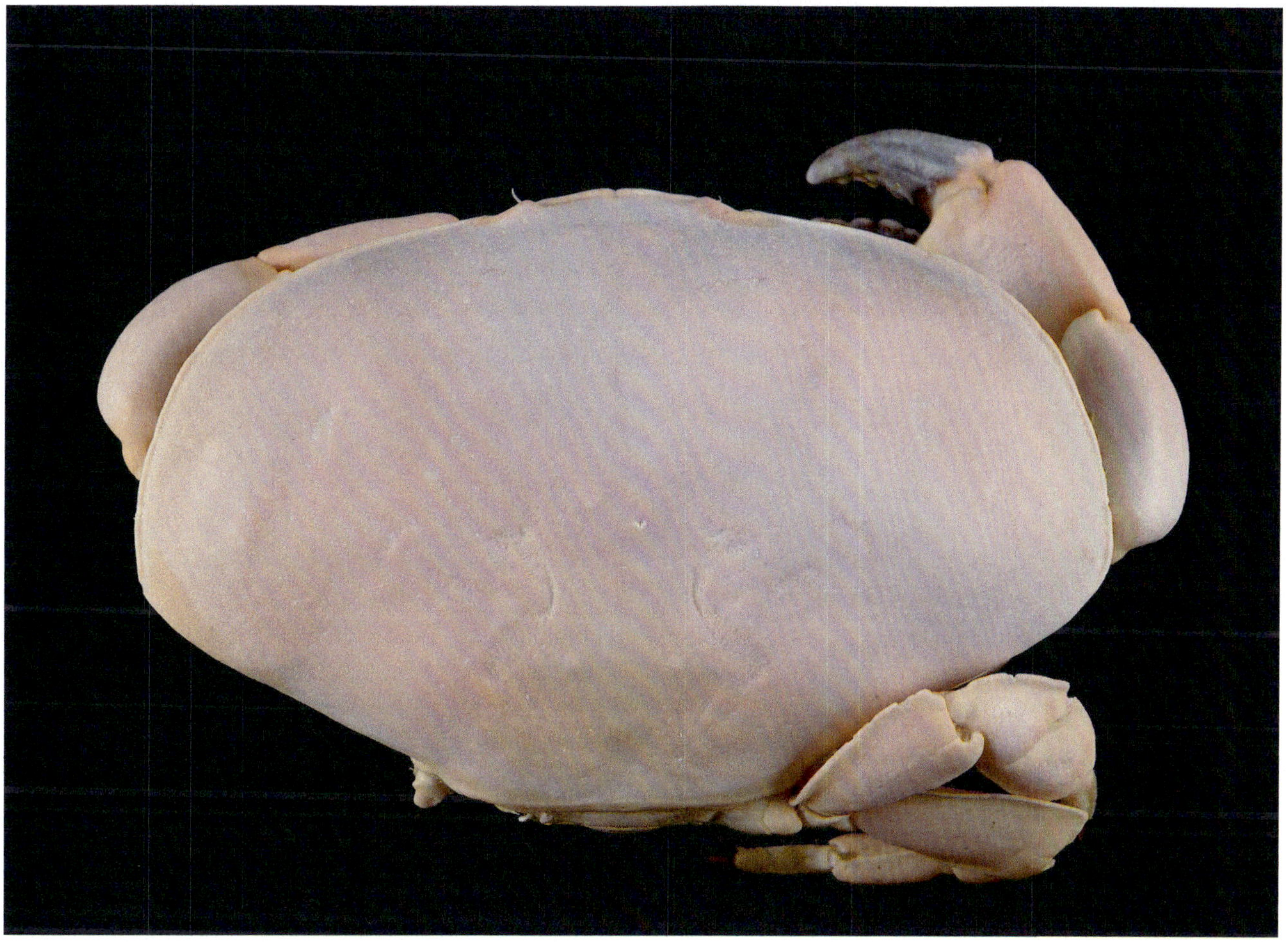

Fig. 21.7 *Atergatis laevigatus* A. Milne-Edwards, 1865. Female (SMF), CL = 53.4, CB = 83.5 mm, Abu Ali, Saudi Arabia, Persian Gulf, 16.05.95, coll. M. Apel (Photo: S. Träkner)

Fig. 21.8 *Atergatis laevigatus* A. Milne-Edwards, 1865. The crab with living colour, Mahtabi, E Bandar-Khamir, Iran, Persian Gulf, 10.01.2008 (Photo: R. Naderloo)

World distribution: Northwestern Indian Ocean: Red Sea, Persian Gulf, Gulf of Oman, Pakistan and southwestern India (Malabar).

Habitat: Intertidal to shallow subtidal, substrate sandy.

Atergatis ocyroe (Herbst, 1801)

(Figs. 21.4, 21.9, and 21.10a)

Cancer ocyroe Herbst 1801:20, pl 54, fig 2.
Atergatis compressipes MacLeay 1838: 59.
Atergatis floridus — Sakai 1999: 32, pl 17c.
Atergatis ocyroe — Ng and Davie 2007: 172–174, figs 1, 2d–h, 3b, 4b, d; Ng et al. 2008: 205 (in list); Ghotbeddin and Naderloo 2014: 3, fig 2c.

Type locality: "Ostindien" (East India or Indo-Malayan Archipelago).

Distribution

Persian Gulf: No record.

Gulf of Oman: Iran (Ghotbeddin and Naderloo 2014; Naderloo et al. 2015).

World distribution: Indian Ocean: South Africa, Gulf of Oman, Pakistan, India and Thailand (Phuket).

Habitat: Shallow subtidal to 38 m.

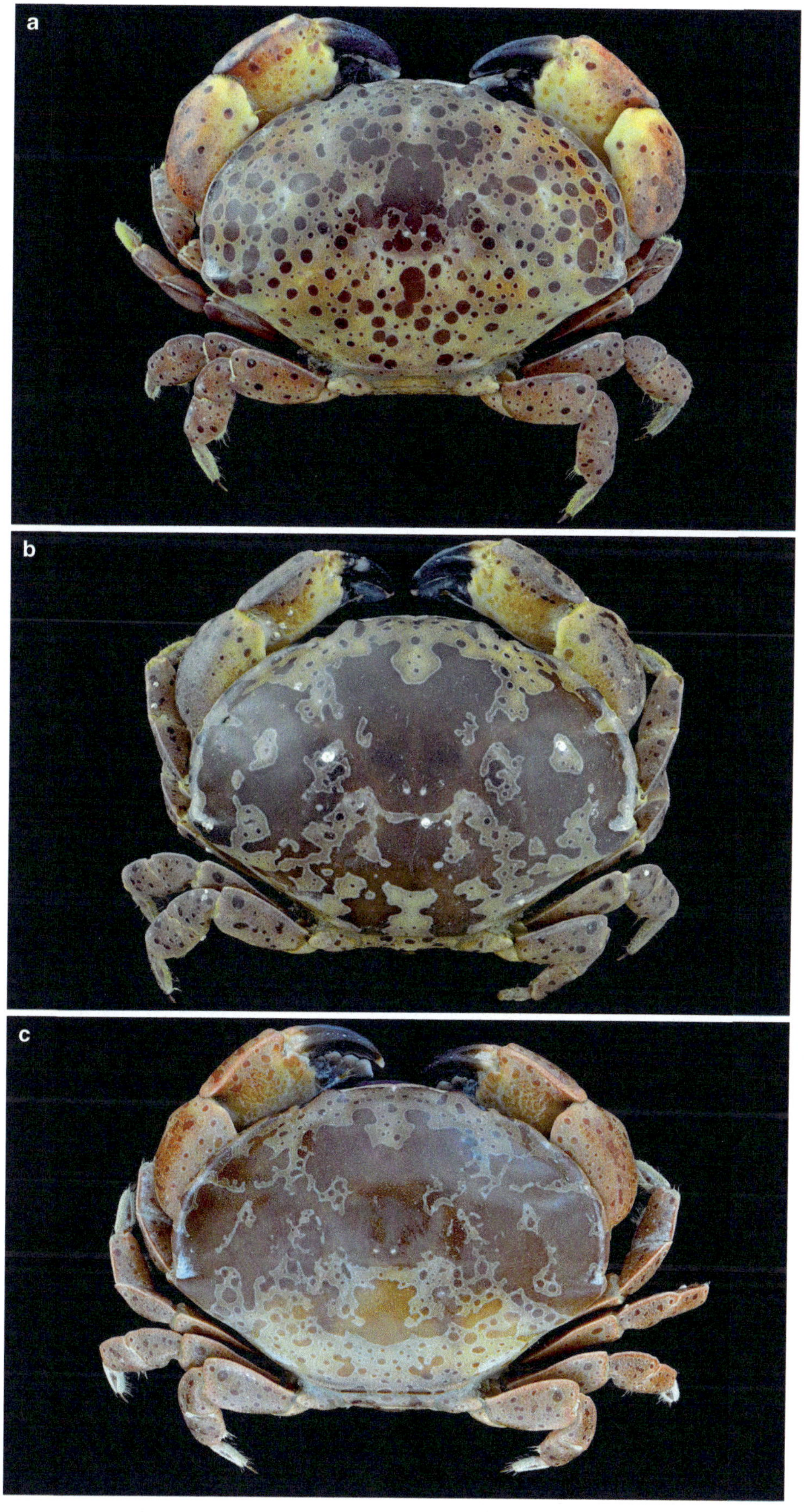

Fig. 21.9 *Atergatis ocyroe* (Herbst, 1801). (ZUTC 5338), Tis, Chabahar, Iran, Gulf of Oman, 18.01.2015, coll. R. Abdollahi and Z. Rajaie: (**a**) male (CL = 33.0, CB = 47.03 mm), (**b**) female (CL = 34.66, CB = 49.01 mm), (**c**) female (CL = 32.75, CB = 47.12 mm) (Photos: R. Abdollahi)

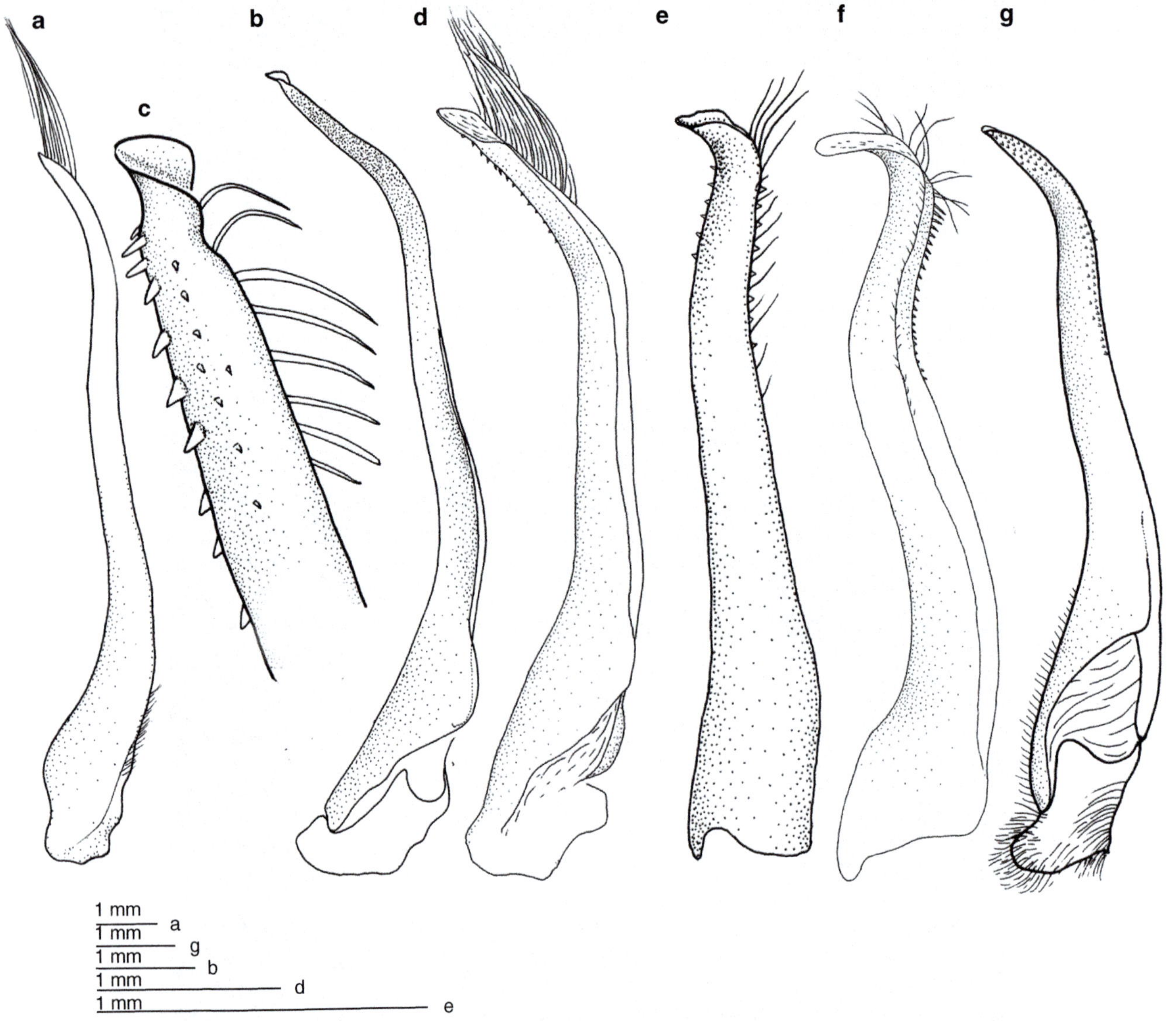

Fig. 21.10 Male G1, ventral view: (**a**) *Atergatis ocyroe* (Herbst, 1801) (ZUTC 5338); (**b**) *Chlorodiella nigra* (Forskål, 1775) (SMF: Saudi Arabia); (**c**) apical part of (**b**); (**d**) *Cyclodius drachi* (Guinot, 1964) (ZUTC 5533); (**e**) *Cymo andreossyi* (Audouin, 1826) (ZUTC 5523); (**f**) *Cymo melanodactylus* Dana, 1852 (ZUTC 5524); (**g**) *Etisus anaglyptus* H. Milne Edwards, 1834 (ZUTC 5527)

Genus *Chlorodiella* Rathbun, 1897

Chlorodiella nigra (Forskål, 1775)

(Figs. 21.10b, c, 21.11, and 21.12)

Cancer niger Forskål 1775: 89.

Chlorodius niger — Nobili 1906a: 128; Stephensen 1946: 156, 157, fig 38d, e; Evans et al. 1973: 275.

Chlorodiella nigra — Guinot 1967a: 262 (in list); Basson et al. 1977: 261; Titgen 1982: 128, 251 (in list); Serène 1984: 258, 259, fig 168, pl 36b; Hogarth 1989: 114; Hogarth 1994: 95; Apel 2001: 83; Ng et al. 2008: 197 (in list).

Type locality: Red Sea (Jeddah, Saudi Arabia).

Distribution

Persian Gulf: Iran (Stephensen 1946; present study), Saudi Arabia (Basson et al. 1977; Apel 2001),

Fig. 21.11 *Chlorodiella nigra* (Forskål, 1775). Male (SMF), CL = 14.6, CB = 25.8 mm, Karan Island, Saudi Arabia, Persian Gulf, depth 2–6 m, 25.05.1995, coll. M. Apel (Photo: S. Träkner)

Bahrain (Apel 2001), Qatar (Ape 2001), UAE (Nobili 1906a; Evans et al. 1973; Apel 2001).

Gulf of Oman: UAE (Apel 2001; present study).

World distribution: Indo-West Pacific: South Africa, Islands in the western Indian Ocean, Gulf of Aden, Red Sea, southern Oman, Persian Gulf, Gulf of Oman, Pakistan, India, Indo-Malayan Archipelago, China, Taiwan, Japan, French Polynesia and Hawaii.

Habitat: Shallow subtidal to 12 m, living on different hard corals (*Acropora* sp., *Pocillopora* sp. and *Porites* sp.).

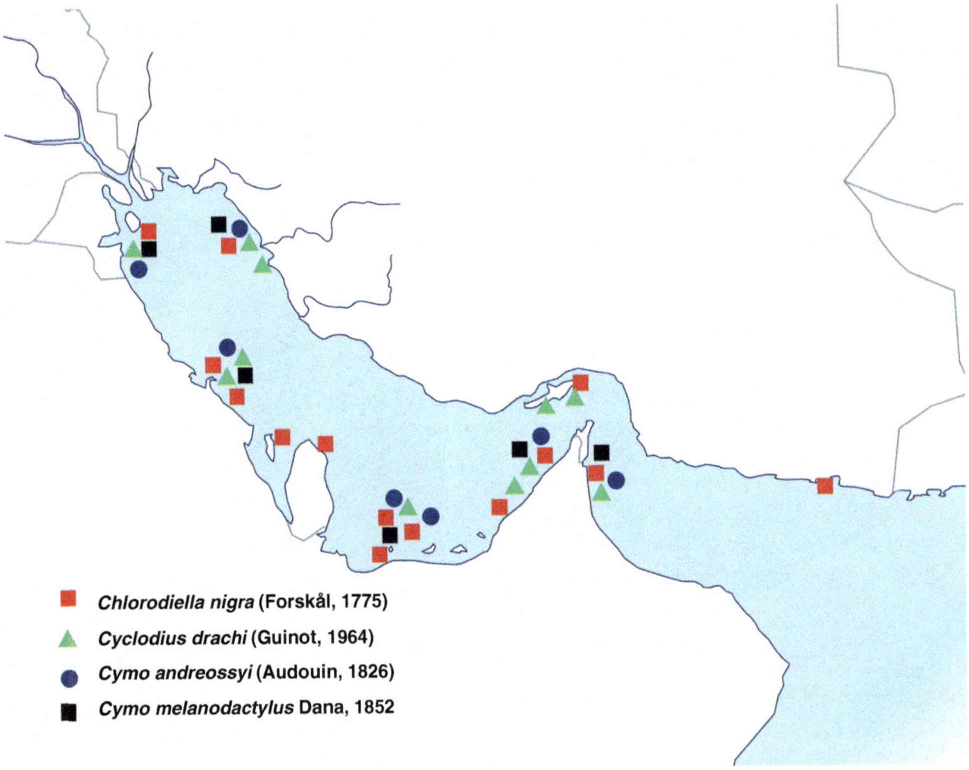

Fig. 21.12 Map showing the recorded localities of *Chlorodiella nigra* (Forskål, 1775), *Cyclodius drachi* (Guinot, 1964), *Cymo andreossyi* (Audouin, 1826), and *Cymo melanodactylus* Dana, 1852

Genus *Cyclodius* Dana, 1851

Cyclodius drachi (Guinot, 1964)

(Figs. 21.10d, 21.12, and 21.13)

Phymodius drachi Guinot 1964: 72, figs 42, 47, pl 3, fig 1, pl 7, fig 1.
Phymodius ungulatus—Nobili 1906a: 129; Stephensen 1946: 157; Titgen 1982: 252 (in list).
Phymodius granulatus — Stephensen 1946: 157, fig 38f; Guinot 1967a: 267 (in list: partly); Titgen 1982: 133, 134; Galil and Vannini 1990: 44; Titgen 1982: 133, 134; Hogarth 1989: 106; Hogarth 1994: 99.
Phymodius drachi—Guinot 1967a: 267 (in list); Serène 1984: 249, 250, fig 154, pl 35d; Apel 2001: 89; Naderloo and Türkay 2012: 45; Naderloo et al. 2013: 449, tab 1.
Cyclodius drachi— Ng et al. 2008: 197 (in list).

Type locality: Madagascar.

Distribution

Persian Gulf: Iran [Stephensen (1946) as *P. granulatus*; Naderloo and Türkay 2012; Naderloo et al. 2013]; Saudi Arabia (Apel 2001), UAE [Nobili (1906a) as *P. ungulatus*; Titgen (1982) as *P. granulatus*; Apel 2001].

Gulf of Oman: Iran (Naderloo et al. 2015), Oman (Hogarth 1989), UAE (Apel 2001).

World distribution: Western Indian Ocean: East Africa, Madagascar, Comoros, Red Sea, Gulf of Aden, Persian Gulf and Gulf of Oman.

Habitat: Intertidal to 3 m, substrates rocky/cobble and coral reefs.

Fig. 21.13 *Cyclodius drachi* (Guinot, 1964). Female (ZUTC), CL = 5.82, CB = 8.70 mm, Jod Village, Sistan-Baluchestan, Iran, Gulf of Oman, 15.11.2005, coll. S. Ebrahimnejad (Photo: R. Abdollahi)

Genus *Cymo* De Haan, 1833

Key to species of the genus *Cymo* in the region

- Chelipeds with light-gray fingers; front with two lobes, one bifurcated spine on each lobe of frontal margin, median notch nearly shallow, round; G1 with apical process long, bent laterally with less than 90 degree .... *Cymo andreossyi*
- Chelipeds with fingers characteristically dark brown, white in apical part; front with two lobes, two bifurcated spines on each lobe of frontal margin, median notch nearly deep; G1 with apical process short, hook-shaped .... *Cymo melanodactylus*

Cymo andreossyi (Audouin, 1826)

(Figs. 21.10e, 21.12, and 21.14)

Pilumnus andreossyi Savigny 1817: pl 5, fig 5.
Pilumnus andreossyi — Audouin 1826: 86.
Cymo andreossyi — Nobili 1906a: 129, 130; Guinot 1967a: 263 (in list); Hogarth 1989: 106; Hogarth 1994: 95; Apel 2001: 43; Ng et al. 2008: 197 (in list); Brösing et al. 2014: 196–200, figs 1, 2a–d, 3a–f. 14a; Ng et al. 2008: 197 (in list).
Cymo andreossyi var. *melanodactyla* — Stephensen 1946: 153, fig 38a, b.
Type locality: Red Sea (Egypt).

Distribution

Persian Gulf: Iran [Stephensen (1946) as *C. andreossyi* var. *melanodactylus*; present study], Saudi Arabia (Apel 2001), UAE (Nobili 1906a; Apel 2001).
Gulf of Oman: UAE (Apel 2001).
World distribution: Indo-West Pacific: South Africa, East Africa, Red Sea, southern Oman, Persian Gulf, Gulf of Oman, Pakistan, Sri Lanka, Andaman Islands, Nicobar Islands, Taiwan, Japan and Australia.

Habitat: Shallow subtidal to 4 m, living on different hard corals (*Acropora* sp., *Pocillopora* sp., and *Porites* sp.).

Fig. 21.14 *Cymo andreossyi* (Audouin, 1826). Male (ZUTC 5523), CL = 12.18, CB = 12.85 mm, Near Ariport, Abu Musa Island, Iran, Persian Gulf, 16.05.2015, coll. R. Naderloo (Photo: R. Abdollahi)

Cymo melanodactylus Dana, 1852
(Figs. 21.10f, 21.12, and 21.15)

Cymo melanodactylus Dana 1852b: 225.
Cymo melanodactylus — Guinot 1967a: 263 (in list); Serène 1984: 34, fig 8, pl 2b; Hogarth 1989: 106, 114; Hogarth 1994: 95; Apel 2001: 43; Ng et al. 2008: 198 (in list); Brösing et al. 2014: 202–207, figs 8a–h, 9a–f, 10a–d, 14c, 15a, b.
Cymo andreossyi var. *melanodactyla* — Nobili 1906a: 129, 130; Titgen 1982: 251 (in list).
not *Cymo andreossyi* var. *melanodactyla* — Stephensen 1946: 153, fig 38a, b.

Type locality: Indonesia (Java).

Distribution

Persian Gulf: Iran (Stephensen 1946; present study), Saudi Arabia (Apel 2001), UAE (Nobili 1906a; Apel 2001).

Fig. 21.15 *Cymo melanodactylus* Dana, 1852. Female (ZUTC 5524), CL = 11.90, CB = 12.81 mm, Park-Qadir, Abu Musa Island, Iran, Persian Gulf, 15.05.2015, coll. R. Naderloo (Photo: R. Abdollahi)

Gulf of Oman: UAE (Apel 2001).

World distribution: Indo-West Pacific: Madagascar, Red Sea, Gulf of Aden, southern Oman, Persian Gulf, Gulf of Oman, Sri Lanka, Andaman Islands, Mergui Archipelago, Taiwan, Japan, Indonesia, Australia, Fiji and Samoa.

Habitat: Shallow subtidal to 10 m, living on different hard corals (*Acropora* sp., and *Pocillopora* sp.).

Genus *Epiactaea* Serène, 1984

Epiactaea margaritifera (Odhner, 1925)

(Figs. 21.16 and 21.17)

Actaea margaritifera Odhner 1925: 48, pl 3, fig 10.

Actaea nodulosa — Henderson 1893: 356; Alcock 1898: 148; Nobili 1906b: 257, pl 10, fig 2.

Actaea margaritifera — Stephensen 1946: 152; Guinot 1967a: 261 (in list); Titgen 1982: 251 (in list).

Epiactaea margaritifera — Serène 1984: 117, fig 70, pl 15b; Hogarth 1989: 106; Hogarth 1994: 95; Apel 2001: 84; Ng et al. 2008: 195 (in list).

Type locality: Gulf of Aden.

Distribution

Persian Gulf: Iran (Stephensen 1946). Alcock (1898) recorded this species as *Actaea nodulosa* from an unknown locality.

Gulf of Oman: No record.

World distribution: Indian Ocean: Gulf of Aden, southern Oman, Persian Gulf, Pakistan, Sri Lanka, Thailand, Indonesia, Singapore and Torres Strait.

Habitat: Subtidal to 33 m.

Fig. 21.16 *Epiactaea margaritifera* (Odhner, 1925). Male (ZMUC CRU901), CL = 6.6, CB = 3.6 mm, Bushehr, Iran, Persian Gulf, 29° 10′N, 50° 30′E, depth 32 m, 30.01.1938, coll. B. Loppenthin (Photo: S. Träkner)

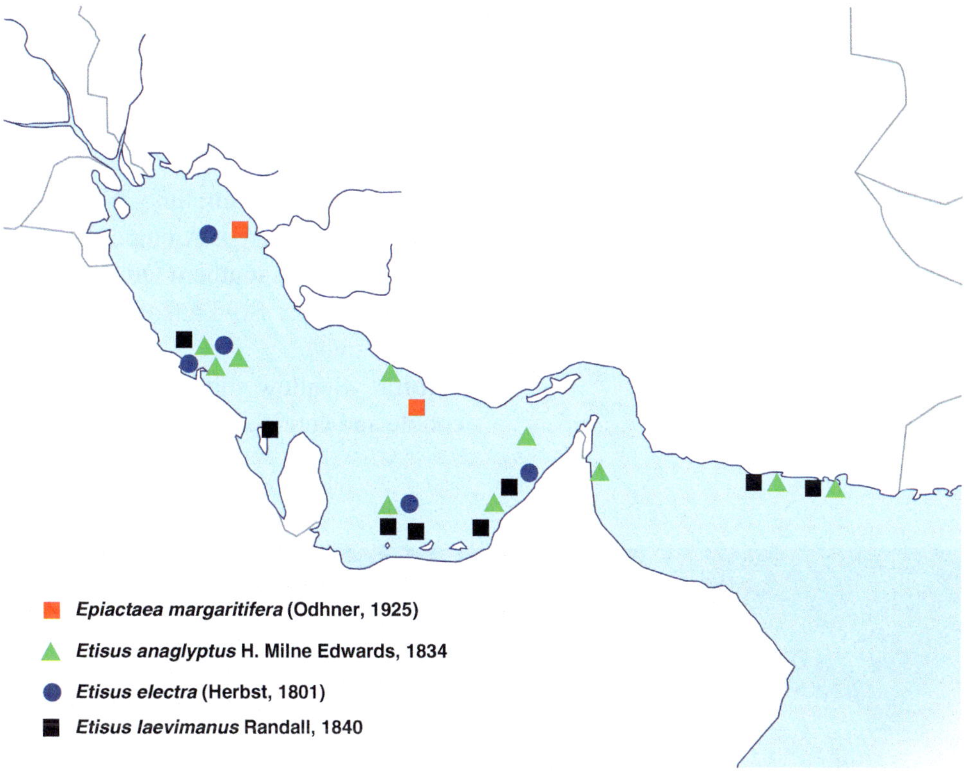

Fig. 21.17 Map showing the recorded localities of *Epiactaea margaritifera* (Odhner, 1925), *Etisus anaglyptus* H. Milne Edwards, 1834, *Etisus electra* (Herbst, 1801), and *Etisus laevimanus* Randall, 1840

Genus *Etisus* H. Milne Edwards, 1834

Key to species of the genus *Etisus* in the region

1. Front with two broad lobes, obliquely straight, broader than orbit; anterolateral margins with broad lobes, apically dentiform; dorsal surface of carapace smooth, regions moderately defined; walking legs with sparse long setae .... *Etisus laevimanus*

- Front with two lobes shorter than orbit, apically concave or straight; anterolateral margins with distinct teeth; dorsal surface of carapace markedly uneven, regions well defined; walking legs with dense long setae 2

2. Frontal lobes apically straight, slightly extended beyond inner orbital angle; walking legs with spines .... *Etisus anaglyptus*

- Frontal lobes apically concave, markedly extended beyond inner orbital angle; walking legs without spines .... *Etisus electra*

***Etisus anaglyptus* H. Milne Edwards, 1834**
(Figs. 21.10g, 21.17, and 21.18)

Etisus anaglyptus H. Milne Edwards 1834: 411, 412.
Etisus anaglyptus — H. Milne Edwards 1837b: pl 11, fig 4; Guinot 1964: 57, fig 33a–c; Guinot 1967a: 263 (in list); Basson et al. 1977: 261; Serène 1984: 218–224 (in key), 228, fig 137, pl 33a; Hogarth 1994: 96; Apel 2001: 84; Ng et al. 2008: 198 (in list).
Etisodes anaglyptus — Alcock 1898: 133; Stephensen 1946: 158.
Type locality: Southeast Asia.

Distribution

Persian Gulf: Iran (Stephensen 1946), Saudi Arabia (Basson et al. 1977; Apel 2001), UAE (Apel 2001).

Gulf of Oman: Iran (Naderloo et al. 2015), UAE (Apel 2001).

World distribution: Indo-West Pacific: Madagascar, Seychelles, Chagos Archipelago, Somalia, Red Sea, Gulf of Aden, southern Oman, Persian Gulf, Gulf of Oman, India, Sri Lanka, Taiwan, Japan, Australia and Samoa.

Habitat: Shallow subtidal to 20 m, substrates rocky/cobble and corals.

Fig. 21.18 *Etisus anaglyptus* H. Milne Edwards, 1834. Female (ZUTC 5527), CL = 20.15, CB = 30.30 mm, Iran, Gulf of Oman, 2010, coll. N. Ghotbeddin (Photo: S. Träkner)

Etisus electra (Herbst, 1801)

(Figs. 21.17, 21.19, and 21.20a)

Cancer electra Herbst 1801: 34, pl 51, fig 6.
Etisodes electra — Nobili 1906a: 122; Stephensen 1946: 158, fig 41a.
Etisus electra — Guinot 1964: 49, 50, figs 21, 22, 28, 30, pl 5, fig 1; Guinot 1967a: 263 (in list); Basson et al. 1977: 224, 261; Serène 1984: 228, fig 138, pl 21d; Hogarth 1989: 106; Hogarth 1994: 96; Sakai 1999: 34, pl 18c; Apel 2001: 84, 85; Ng et al. 2008: 198 (in list).
Etisus frontalis — Titgen 1982: 130, 251 (in list).

Type locality: "Ostindien". East India or Indo-Malayan Archipelago.

Distribution

Persian Gulf: Iran (Apel 2001), Saudi Arabia (Basson et al. 1977; Apel 2001), UAE [Nobili 1906a; Titgen (1982) as *Etisus frontalis*].

Gulf of Oman: No record.

World distribution: Indo-West Pacific: Madagascar, Mauritius, Somalia, Red Sea, southern Oman, Persian Gulf, India, Maldives, Andaman Sea, Nicobar Islands, Taiwan, Philippines, Japan, Australia and Samoa.

Habitat: Intertidal to 10 m, substrates rocky/cobble and dead corals.

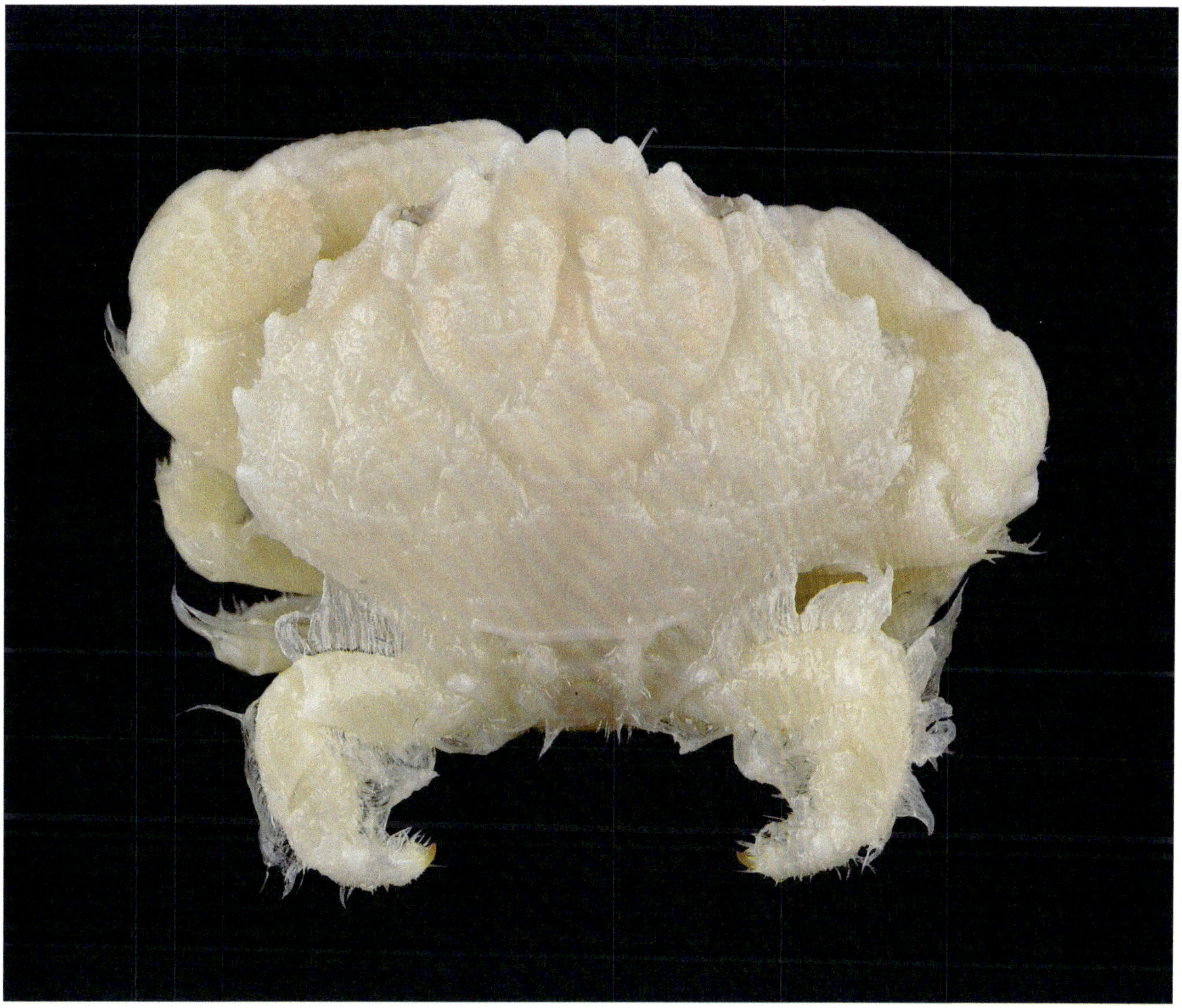

Fig. 21.19 *Etisus electra* (Herbst, 1801). Male (SMF), CL = 21.0, CB = 16.7 mm, Karan Island, Saudi Arabia, Persian Gulf, 27.05.1995, coll. M. Apel (Photo: S. Träkner)

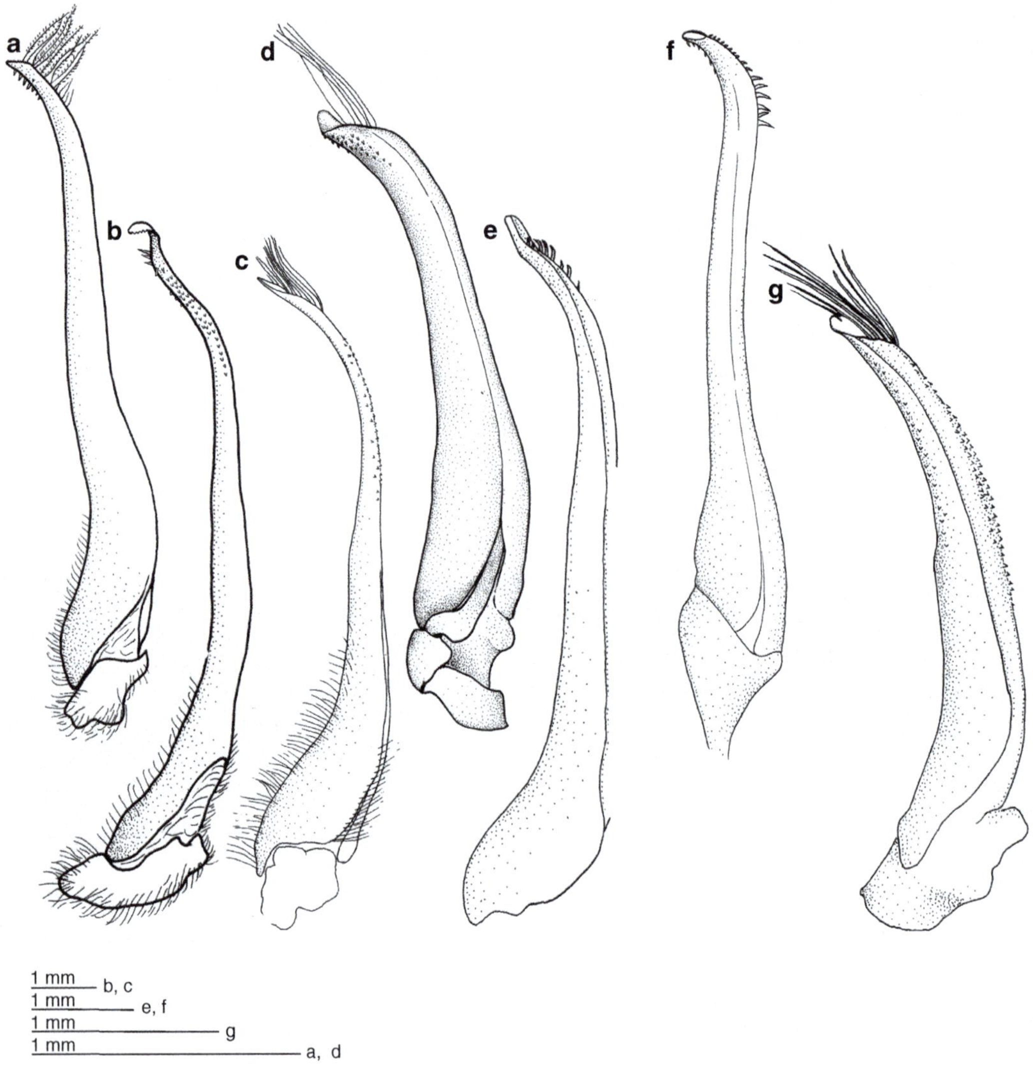

Fig. 21.20 Male G1, ventral view: (**a**) *Etisus electra* (Herbst, 1801) (SMF: Saudi Arabia); (**b**) *Etisus laevimanus* Randall, 1840. Male (SMF: Saudi Arabia); (**c**) *Gaillardiellus rueppelli* (Krauss, 1843) (SMF: Socotra); (**d**) *Hepatoporus guinotae* (Zarenkov, 1971) (SMF: Gulf of Oman, UAE); (**e**) *Leptodius exaratus* (H. Milne Edwards, 1834) (ZUTC 5081); (**f**) *Macromedaeus voeltzkowi* (Lenz, 1905) (ZUTC: Iran, Gulf of Oman); (**g**) *Medaeops neglectus* (Balss, 1922) (ZUTC Brach1311)

Etisus laevimanus Randall, 1840

(Figs. 21.17, 21.20b, and 21.21)

Etisus laevimanus Randall 1840: 115.

Etisus laevimanus — Alcock and Anderson 1894b: 200; Alcock 1898: 131; Nobili 1906a: 121; Guinot 1964: 56; Stephensen 1946: 157; Guinot 1967a: 263 (in list); Basson et al. 1977: 261; Titgen 1982: 251 (in list); Serène 1984: 225, 226, fig 136, pl 33b; Carpenter et al. 1997: 40; Apel 2001: 85; Ng et al. 2008: 198 (in list).

Type locality: Hawaii.

Distribution

Persian Gulf: Saudi Arabia (Basson et al. 1977; Apel 2001), Bahrain (Nobili 1906a; Apel 2001), UAE (Nobili 1906a; Apel 2001). Alcock (1898) recorded this species from an unknown locality.

Gulf of Oman: Iran (Naderloo et al. 2015).

World distribution: Indo-West Pacific: Madagascar, Seychelles, Red Sea, southern Oman, Persian Gulf, Gulf of Oman, India, Maldives, Taiwan, Japan, Australia and Hawaii.

Habitat: Intertidal to 9 m, substrates rocky/cobble and dead corals.

Fig. 21.21 *Etisus laevimanus* Randall, 1840. Male (SMF), CL = 38.4, CB = 60.4 mm, Abu Ali, Saudi Arabia, Persian Gulf, 16.05.1995, coll. M. Apel (Photo: S. Träkner)

Genus *Euxanthus* Dana, 1851

***Euxanthus exsculptus* (Herbst, 1790)**
(Figs. 21.22 and 21.23)

Cancer exsculptus Herbst 1790: 265, pl 21, fig 121.
Cancer melissa Herbst 1801: 7, pl 51, fig 1.
Euxanthus melissa — Alcock 1898: 110.
Euxanthus exsculptus — Guinot-Dumortier 1960: 169, pl 1, fig 4, pl 2, fig 10, pl 6, figs 36, 37, pl 8, figs 42–47; Serène 1984: 86, fig 48, pl 11b; Tirmizi and Ghani 1996: 33–36, figs 11 a–f, 12a–d; Apel 2001: 85; Ng et al. 2008: 199 (in list).

Type locality: *Cancer exsculptus*: Unknown. — *Cancer melissa*: Unknown.

Distribution

Persian Gulf: No record.

Gulf of Oman: UAE (Apel 2001).

World distribution: Indo-West Pacific: Mauritius, Gulf of Oman, Pakistan, Mumbai, Sri Lanka, Nicobar Islands, Andaman Sea, Mergui Archipelago, Indonesia, Philippines, China, Taiwan, Japan, Australia, Fiji, Samoa and New Caledonia.

Habitat: Intertidal to shallow subtidal, substrates rocky/cobble and corals.

Fig. 21.22 *Euxanthus exsculptus* (Herbst, 1790). Male (ZRC 1998.973), CL = 52.1, CB = 34.0 mm, Laguis, N Biutan, Indonesia, 05.1993, coll. PKL Ng (Photo: Jose C.E. Medosa)

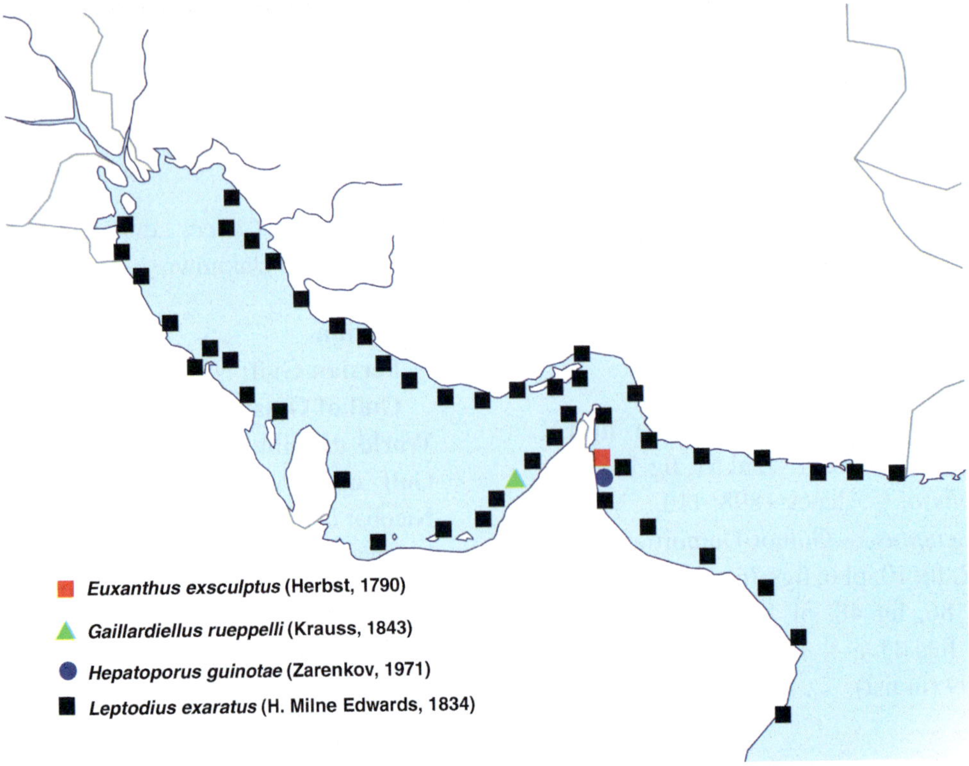

Fig. 21.23 Map showing the recorded localities of *Euxanthus exsculptus* (Herbst, 1790), *Gaillardiellus rueppelli* (Krauss, 1843), *Hepatoporus guinotae* (Zarenkov, 1971), and *Leptodius exaratus* (H. Milne Edwards, 1834)

Genus *Gaillardiellus* Guinot, 1976

***Gaillardiellus rueppelli* (Krauss, 1843)**
(Figs. 21.20c, 21.23, and 21.24)

Cancer (*Aegle*) *rüppelli* Krauss 1843: 28, pl 1, fig 1.
Actaea ruppellii — Alcock 1898: 144; Guiont 1964: 41; Guinot 1967a: 260 (in list); Titgen 1982: 251 (in list).
Actaea rüppelli — Nobili 1906a: 125.
Actaea rüppellii — Stephensen 1946: 150, 151.
Gaillardiellus rueppelli — Guinot 1976: 254, figs 42a, 43a, 44b, pl 16, figs 1, 1a; Serène 1984: 118, 119, fig 71, pl 15f; Apel 2001: 85; Ng et al. 2008: 195 (in list).

Type locality: South Africa (Natal).
Distribution
Persian Gulf: UAE (Nobili 1906a). Alcock (1898) recorded this species from an unknown locality.
Gulf of Oman: No record.
World distribution: Indo-West Pacific: South Africa, East Africa, Madagascar, Mauritius, Réunion, Aldabra, Amirante, Saint Brandon, Persian Gulf, Chagos Archipelago, Sri Lanka, Mergui Archipelago, Strait of Malacca, Vietnam, Japan, Fiji and Samoa.
Habitat: Subtidal, 10–110 m, substrates rocky/cobble and corals.

Fig. 21.24 *Gaillardiellus rueppelli* (Krauss, 1843). Male (SMF), CL = 15.8, CB = 21.9 mm, Socotra, Yemen, depth 17 m, 30.03.1999, coll. M. Apel (Photo: S. Träkner)

Genus *Hepatoporus* Serène, 1984

Hepatoporus guinotae (Zarenkov, 1971)

(Figs. 21.20d, 21.23, and 21.25)

Carpoporus guinotae Zarenkov 1971: 191, fig 86.
Hepatoporus guinotae — Serène 1984: 75, fig 40, pl 10d–f; Apel 2001: 86; Ng et al. 2008: 199 (in list).

Type locality: Red Sea (southern part).

Distribution

Persian Gulf: No record.

Gulf of Oman: UAE (Apel 2001).

World distribution: Western Indian Ocean: Madagascar, Kenya, Red Sea and Gulf of Oman.

Habitat: Subtidal, 18–108 m.

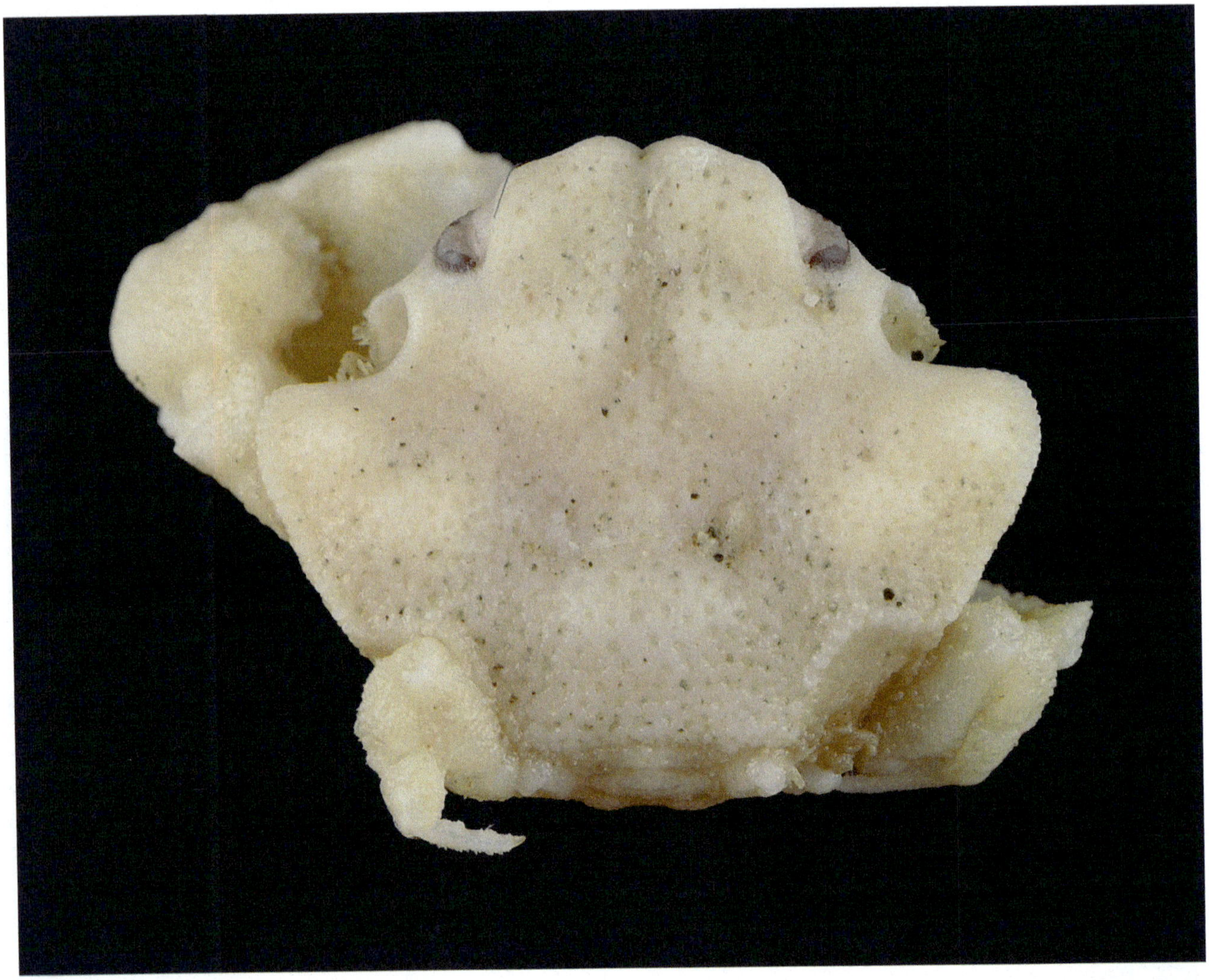

Fig. 21.25 *Hepatoporus guinotae* (Zarenkov, 1971). Male (SMF), CL = 6.5, CB = 9.00 mm, Khor Fakkan, UAE, Gulf of Oman, depth 18–20, 30.06.1995, coll. M. Apel (Photo: S. Träkner)

Genus *Leptodius* A. Milne-Edwards, 1863

Leptodius exaratus (H. Milne Edwards, 1834)

(Figs. 21.20e, 21.23, 21.26, and 21.27)

Chlorodius exaratus H. Milne Edwards 1834: 402;
Chlorodius exaratus — H. Milne Edwards 1849: pl 11, fig 3.
Leptodius exaratus — Nobili 1906a: 121; Guinot 1964: 2, 3, 25–27; Guinot 1967a: 265 (in list); Titgen 1982: 131, 132; Serène 1984: 183, fig 106, pl 26a; Hogarth 1989: 106, 113, 114; Hogarth 1994: 96; Tirmizi and Ghani 1996: 48–50, fig 18; Apel 2001: 86; Ng et al. 2008: 203 (in list); Naderloo and Türkay 2012: 44; Naderloo et al. 2013: 449, tab 1.

Xantho (*Leptodius*) *exaratus* — Alcock 1898: 118 (in key), 119; Jones 1986a: 162, pl 47.

Xantho (= *Leptodius*) *exaratus* — Stephensen 1946: 149, 150, fig 37c.

Xantho exaratus — Basson et al. 1977: 227, 238.

Leptodius (*Xantho*) *exaratus* — Hornby 1997: 15.

Type locality: "Les côtes de l'Inde" (Indian Coast).

Distribution

Persian Gulf: Iran (Stephensen 1946; Naderloo and Türkay 2012; Naderloo et al. 2013), Kuwait (Jones 1986a), Saudi Arabia (Basson et al. 1977; Apel 2001), Bahrain (Nobili 1906a; Stephensen 1946; Apel 2001), UAE (Nobili 1906a; Titgen 1982; Hornby 1997; Apel 2001).

Gulf of Oman: Iran (Naderloo et al. 2015).

World distribution: Indo-West Pacific: South Africa, East Africa, Madagascar, Mauritius, Seychelles, Red Sea, Gulf of Aden, Socotra, southern Oman, Persian Gulf, Gulf of Oman, Pakistan, India, Sri Lanka, Bay of Bengal, Mergui Archipelago, Christmas Island, Middleton and Elisabeth Reefs, Australia and Hawaii.

Habitat: Intertidal, substrate rocky/cobble.

Fig. 21.26 *Leptodius exaratus* (H. Milne Edwards, 1834). Male (SMF 40743), CL = 21.4, CB 29.3 mm, Basaeedu, Qeshm Island, Iran, Persian Gulf, 07.05.2008, coll. R. Naderloo and A. Kazemi (Photo: S. Träkner)

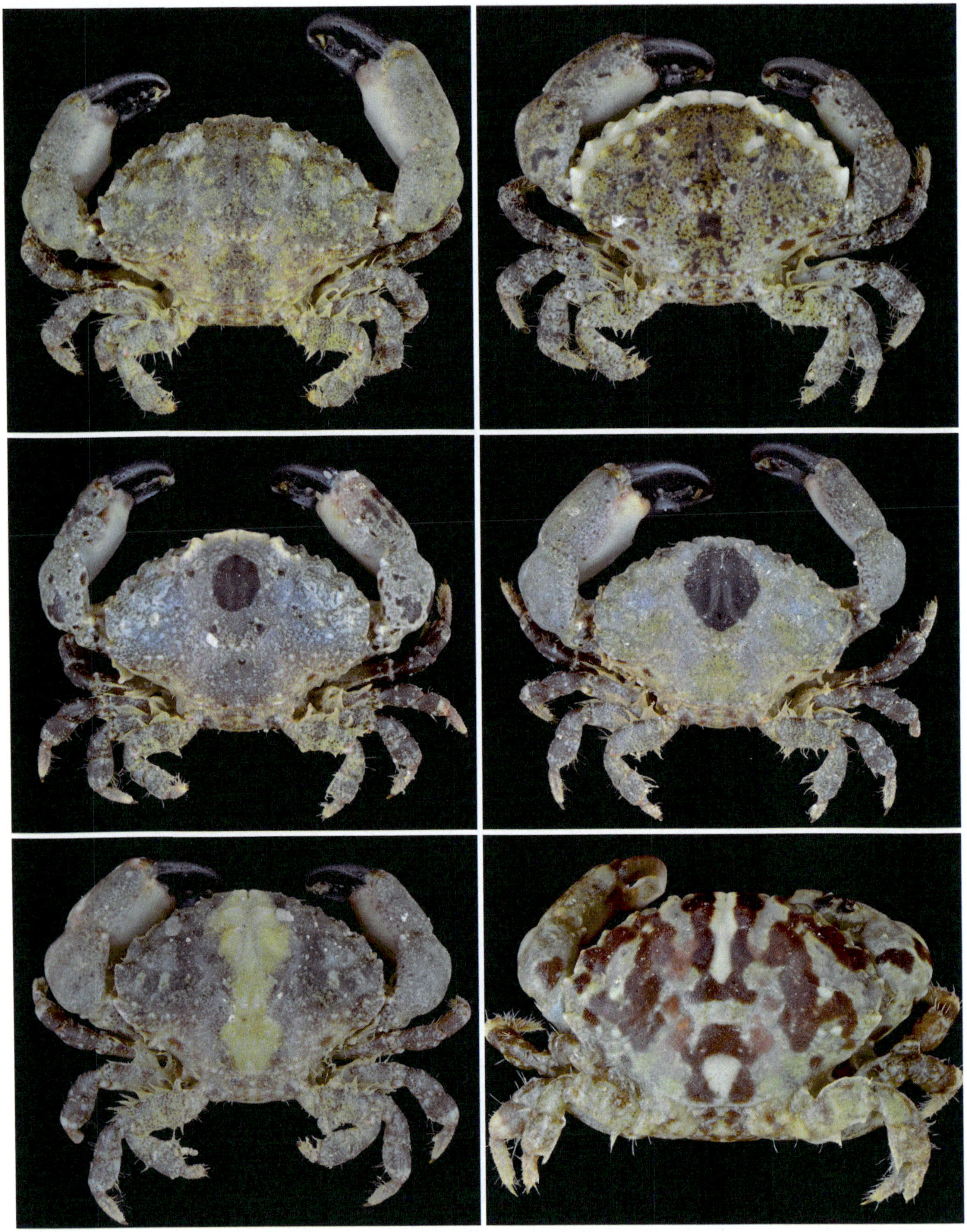

Fig. 21.27 *Leptodius exaratus* (H. Milne Edwards, 1834). The variety color patterns of the species finding in just a one station, Park-Qadir, Abu Musa Island, Iran Persian Gulf (Photo: R. Abdollahi)

Genus *Liagore* De Haan, 1833

Liagore erythematica Guinot, 1971
(Figs. 21.28, 21.29, and 21.30)

Liagore erythematica Guiont 1971a: 1076.
Liagore rubromaculatus — Kemp 1923: 408, 409, pl 10, fig 2; Chopra 1935: 508, fig 16; Stephensen 1946: 154, fig 39a; Titgen 1982: 252 (in list).
Liagore erythematica — Guinot 1971b: 1091–1097, figs 1, 3; Apel 2001: 87; Naderloo and Sari 2007a: 343, tab 1; Ng et al. 2008: 203 (in list).

Type locality: Between Sri Lanka and Kolkata (20° 51′N, 87° 58′E).
Distribution
Persian Gulf: Iran [Stephensen (1946) as *L. rubromaculatus*; Naderloo and Sari 2007a]. Kemp (1923) and Chopra (1935) recorded this species as *L. rubromaculata* from the Persian Gulf (29° 59'N, 50° 03'E).
Gulf of Oman: No record.
World distribution: Northern Indian Ocean: Persian Gulf and Bay of Bengal.
Habitat: Subtidal, 28–33 m, substrate muddy.

Fig. 21.28 *Liagore erythematica* Guinot, 1971. Female (ZUTC Brach1071), CL = 16.5, CB = 22.15 mm, Iran, Persian Gulf, 26° 41′N, 54° 04′E, depth 34 m, 20.12.2003, coll. R. Naderloo (Photo: R. Abdollahi)

Genus *Macromedaeus* Ward, 1942

Key to species of the genus *Macromedaeus* inthe region

1. Anterolateral margins of carapace with four teeth, teeth triangular, interconnected; fingers of chelipeds light brown .... *Macromedaeus voetzkowi*
- Anterolateral margins of carapace with five teeth, teeth tubercle-form, broadly separated 2

Fig. 21.29 *Liagore erythematica* Guinot, 1971. The specimen showing the color pattern of the species, collected from the unknown locality in the Persian Gulf, not preserved

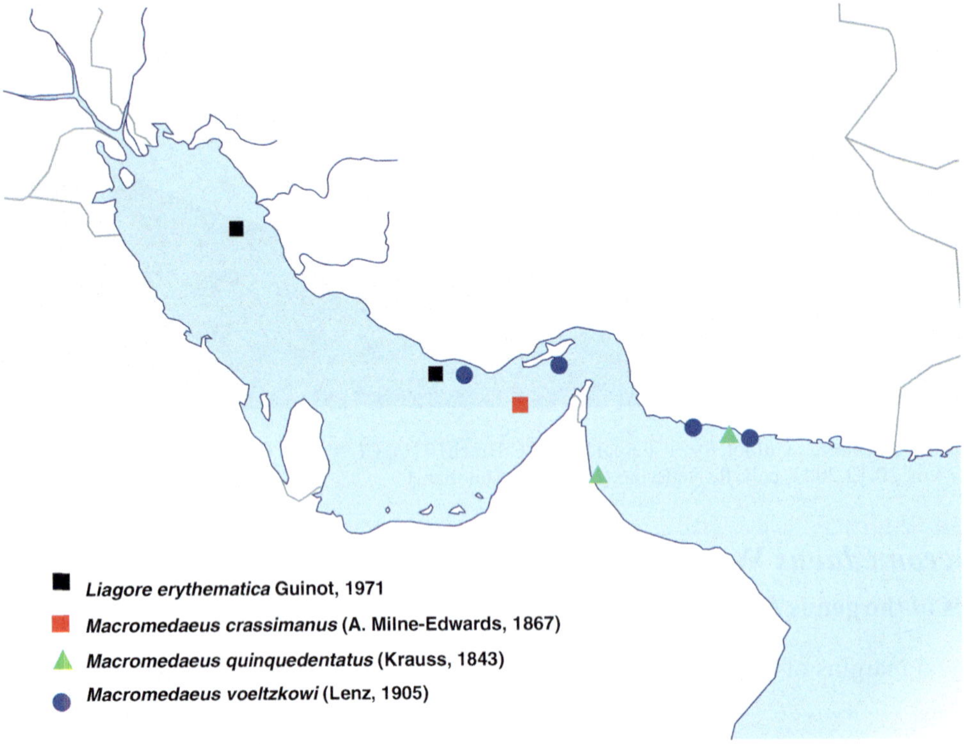

Fig. 21.30 Map showing the recorded localities of*Liagore erythematica* Guinot, 1971, *Macromedaeus crassimanus* (A. Milne-Edwards, 1867), *Macromedaeus quinquedentatus* (Krauss, 1843), and *Macromedaeus voeltzkowi* (Lenz, 1905)

2. Fronto-orbital margin slightly less than half of the greatest breadth of carapace; regions of carapace defined with deep grooves; G1 apical part mesially tuberculate, subdistal process short; fingers of chelipeds dark brown to black .... *Macromedaeus quinquedentatus*

- Fronto-orbital margin distinctly less than half of the greatest breadth of carapace; regions of carapace defined with shallow grooves; G1 apical part without tubercules along mesial surface, subdistal process relatively long; fingers light to dark brown .... *Macromedaeus crassimanus*

Macromedaeus crassimanus (A. Milne-Edwards, 1867)

(Figs. 21.30 and 21.31)

Xantho crassimanus A. Milne-Edwards 1867: 267.
Xantho (*Leptodius*) *crassimanus* — Alcock 1898: 120.
Macromedaeus crassimanus — Guinot 1968: 708; Serène 1984: 177 (in key), 179, fig 103, pl 25b; Ng et al. 2008: 203 (in list).
Type locality: New Caledonia.

Distribution:

Persian Gulf: Iran (Naderloo et al. 2016).

Gulf of Oman: No record.

World distribution: Indo-West Pacific: Red Sea, Persian Gulf, Paksitan, India, Sri Lanka, Andaman Islands, Taiwan, China, Indonesia, Sumatra, Palau, Christmas Islands, Australia and New Caledonia.

Habitat: Intertidal to shallow subtidal, substrate rocky/cobble.

Fig. 21.31 *Macromedaeus crassimanus* (A. Milne-Edwards, 1867). Male (ZUTC 5477), CL = 21.47, CB = 34.36 mm, Park-Qadir, Abu Musa Island, Iran, Persian Gulf, 29.12.2014, coll. S. Ebrahimnejad and R. Abdollahi (Photo: R. Abdollahi)

Macromedaeus quinquedentatus (Krauss, 1843)
(Figs. 21.30 and 21.32)

Xantho quinquedentatus Krauss 1843: 30, pl 1, fig 3a–d.
Xantho (*Leptodius*) *euglyptus* Alcock 1898: 121, 122.
Leptodius euglyptus — Alcock 1899c: pl 36, fig 1.
Macromedaeus quinquedentatus — Serène 1984: 177 (in key), 179–181, fig 104, pl 25d–e; Hogarth 1989: 106, Hogarth 1994: 97, 98; Tirmizi and Ghani 1996: 51–53, fig 19; Apel 2001: 87; Ng et al. 2008: 203 (in list); Ghotbeddin and Naderloo 2014: 3, fig 2b.

Type locality: *Xantho quinquedentatus*: South Africa. — *Xantho* (*Leptodius*) *euglyptus*: Sri Lanka and Mergui Archipelago.

Distribution:

Persian Gulf: No record.

Gulf of Oman: UAE (Apel 2001), Iran (Ghotbeddin and Naderloo 2014; Naderloo et al. 2015).

World distribution: Indo-West Pacific: South Africa, Madagascar, Somalia, Gulf of Aden, southern Oman, Gulf of Oman, Pakistan, India, Sri Lanka, Mergui Archipelago, Indo-Malayan Archipelago and Hawaii.

Habitat: Intertidal, substrate rocky/cobble.

Fig. 21.32 *Macromedaeus quinquedentatus* (Krauss, 1843). Female (ZUTC Brach1309), CL = 21.2, CB = 23.1 mm, Jod Village, Sistan-Baluchestan, Iran, Gulf of Oman, 15.11.2005, coll. R. Naderloo and A. Kazemi (Photo: R. Abdollahi)

Macromedaeus voeltzkowi (Lenz, 1905)

(Figs. 21.20f, 21.30, and 21.33)

Macromedaeus voeltzkowi Lenz 1905: 353, pl 47, figs 6, 6a.
Macromedaeus voeltzkowi — Serène1984: 177 (in key), 178, 179, fig 102, pl 25f; Ng et al. 2008: 203 (in list); Naderloo and Türkay 2012: 44; Ghotbeddin and Naderloo 2014: 2, fig 2a.
Type locality: Zanzibar.

Distribution

Persian Gulf: Iran (Naderloo and Türkay 2012).
Gulf of Oman: Iran (Ghotbeddin et al. 2012a; Ghotbeddin and Naderloo 2014; Naderloo et al. 2015).
World distribution: Western Indian Ocean: South Africa, East Africa, Red Sea, Somalia, Persian Gulf and Gulf of Oman.
Habitat: Intertidal, substrate rocky/cobble.

Fig. 21.33 *Macromedaeus voeltzkowi* (Lenz, 1905). Ovigerous female (SMF 40777), CL = 4.89, CB = 6.79 mm, Jod Village, Sistan–Baluchestan, Iran, Gulf of Oman, 23.10.2006, coll. R. Naderloo and A. Kazemi (Photo: S. Träkner)

Genus *Medaeops* Guinot, 1967

Key to species of the genus *Medaeops* in the region

- Anterolateral margins of carapace with four distinct teeth, second and third ones largest; frontal lobes slightly oblique laterally; merus of walking legs flattened, with granules on posterior surface; medium-sized species .... *Medaeops neglectus*
- Anterolateral margins of carapace with four hardly defined granular lobes, separated by narrow fissures; frontal lobes granular, nearly straight; merus of walking legs cylindrical, posterior surface smooth; large-sized species .... *Medaeops edwardsi*

Medaeops edwardsi Guinot, 1967
(Figs. 21.34 and 21.35)

Medaeops edwardsi Guinot 1967b: 369, figs 33, 42
Medaeops edwardsi — Serène 1984: 91 (in key), 92, fig 53, pl 12e; Ghani and Tirmizi 1992: 43, fig 4; Tirmizi and Ghani 1996: 54–56, fig 20; Ng et al. 2008: 199 (in list).
Type locality: India (Malabar Coast).
Distribution
Persian Gulf: No record.
Gulf of Oman: Iran (Naderloo et al. 2015).
World distribution: Western Indian Ocean: Madagascar, Gulf of Oman and India.
Habitat: Intertidal, substrate rocky/cobble.

Medaeops neglectus (Balss, 1922)
(Figs. 21.20g, 21.35, and 21.36)

Xantho neglectus Balss 1922: 6.
Xantho neglectus — Chopra and Das 1937: 397.
Medaeops neglectus — Guinot 1967b: 367, fig 21, 31, 41; Serène 1984: 91, 92, fig 52, pl 12b Ng et al. 2008: 199 (in list); Mendosa et al. 2009: figs 2d–f, 3e, f, 5c, d; Naderloo and Türkay 2012: 44.
Medaeus granulosus — Jones 1986a: 161, 162, pl 48; Hogarth 1989: 106; Tirmizi and Ghani 1996: 56–59, figs 21, 22; Apel 2001: 87.
Type locality: Red Sea.
Distribution
Persian Gulf: Iran (Naderloo and Türkay 2012).
Gulf of Oman: Iran (Naderloo et al. 2015).

Fig. 21.34 *Medaeops edwardsi* Guinot, 1967. Male (ZUTC 5315), CL = 25.9, CB = 44.11 mm, Jod Village, Sistan–Baluchestan, Iran, Gulf of Oman, 30.12.2013, coll. S. Ebrahimnejad (Photo: R. Abdollahi)

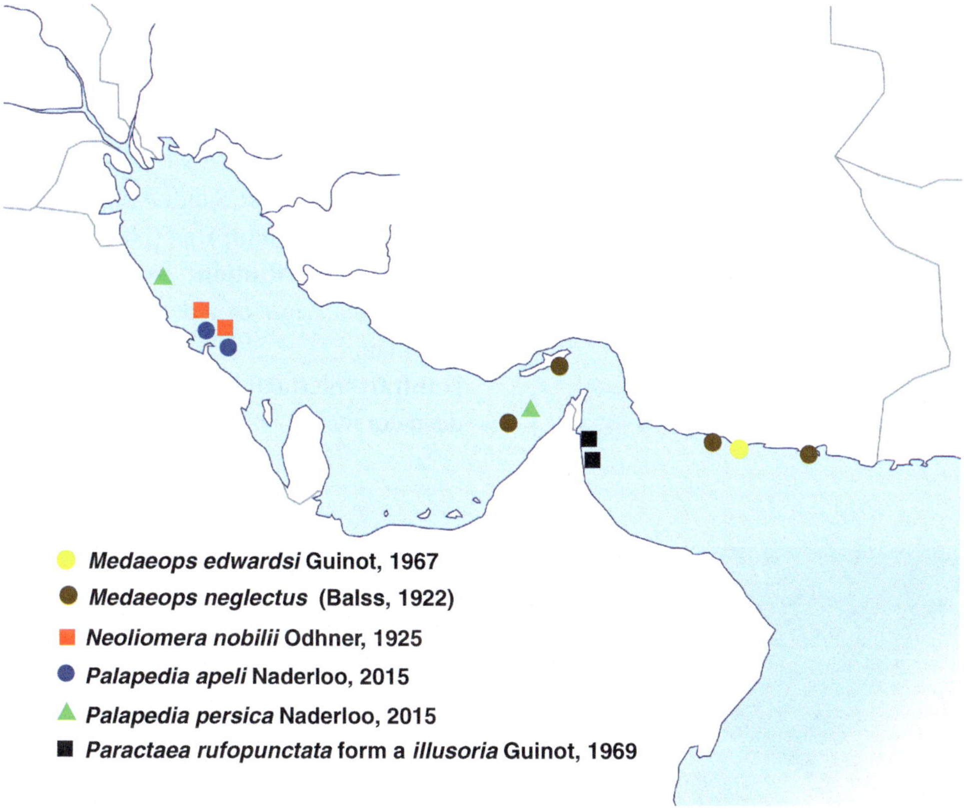

Fig. 21.35 Map showing the recorded localities of *Medaeops edwardsi* Guinot, 1967, *Medaeops neglectus* (Balss, 1922), *Neoliomera nobilii* Odhner, 1925, *Palapedia apeli* Naderloo, 2015, *Palapedia persica* Naderloo, 2015, and *Paractaea rufopunctata* forma *illusoria* Guinot, 1969

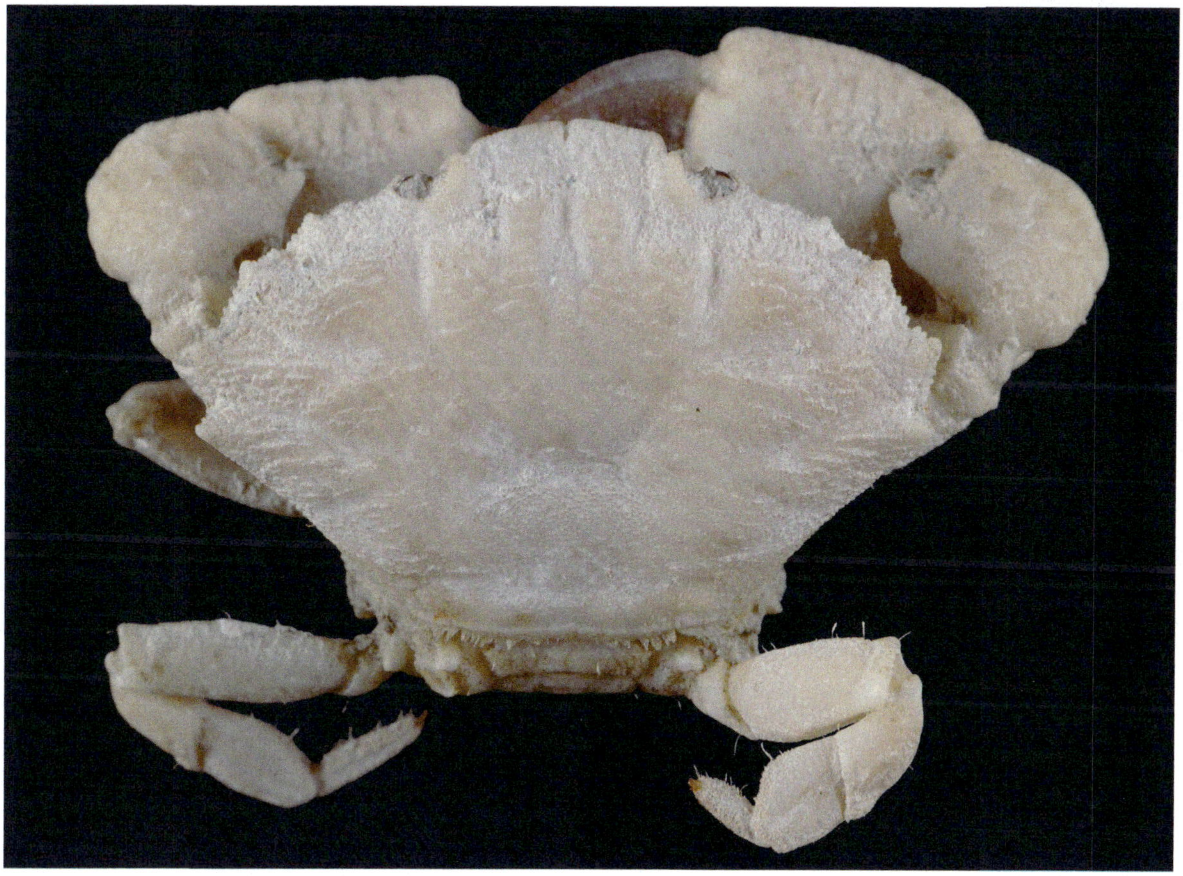

Fig. 21.36 *Medaeops neglectus* (Balss, 1922). Male (ZUTC Brach1311), CL = 23.20, CB = 14.10 mm, Zeytun Park Beach, Qeshm Island, Iran, Persian Gulf, 08.01.2008, coll. R. Naderloo and M. Türkay (Photo: S. Träkner)

World distribution: Western Indian Ocean: East Africa, Madagascar, Mayotte Island, Red Sea, Gulf of Aden, Persian Gulf and Gulf of Oman.

Habitat: Intertidal, substrate rocky/cobble.

Genus *Neoliomera* Odhner, 1925

Neoliomera nobilii Odhner, 1925

(Figs. 21.35, 21.37, and 21.39a)

Neoliomera nobilii Odhner 1925: 30, pl 2, fig 9.
Neoliomera nobilii — Serène 1984: 71, fig 29, pl 9b; Apel 2001: 88; Ng et al. 2008: 201 (in list).
Liomera pubescens var — Nobili 1906b: 222.

Type locality: Perim and Djibouti.

Distribution

Persian Gulf: Saudi Arabia (Apel 2001).

Gulf of Oman: Iran (Naderloo et al. 2015).

World distribution: Northwestern Indian Ocean: Red Sea, Gulf of Aden, Persian Gulf and Gulf of Oman.

Habitat: Intertidal to 8 m, substrates rocky/cobble and dead corals.

Fig. 21.37 *Neoliomera nobilii* Odhner, 1925. Male (SMF), CL = 11.1, CB = 19.1 mm, Karan Island, Saudi Arabia, Persia Gulf, 24.05.1995, coll. M. Apel (Photo: S. Träkner)

Genus *Palapedia* Ng, 1993

Palapedia apeli Naderloo 2015

(Figs. 21.35, 21.38, and 21.39b)

Palapedia apeli Naderloo 2015b: 270–273, figs 2a–b, 4a–f, 5d–f.

Kraussia (*nitida* Stimpson?) — Stephensen 1946: 138, fig 33.

Type locality: Persian Gulf (Bahrain).

Distribution

Persian Gulf: Bahrain (Naderloo 2015b).

Gulf of Oman: No record.

World distribution: This species is still only recorded from the Persian Gulf.

Habitat: Subtidal to 23 m, substrate gravel.

Fig. 21.38 *Palapedia apeli* Naderloo, 2015. Male holotype (ZMUC CRU945), CL = 9.63, CB = 9.91 mm, Bahrain, Persian Gulf, depth 23 m, 27.03.1937, coll. G. Thorson. Photo: S. Träkner (Reproduced from Naderloo (2015b) by permission of Magnolia Press)

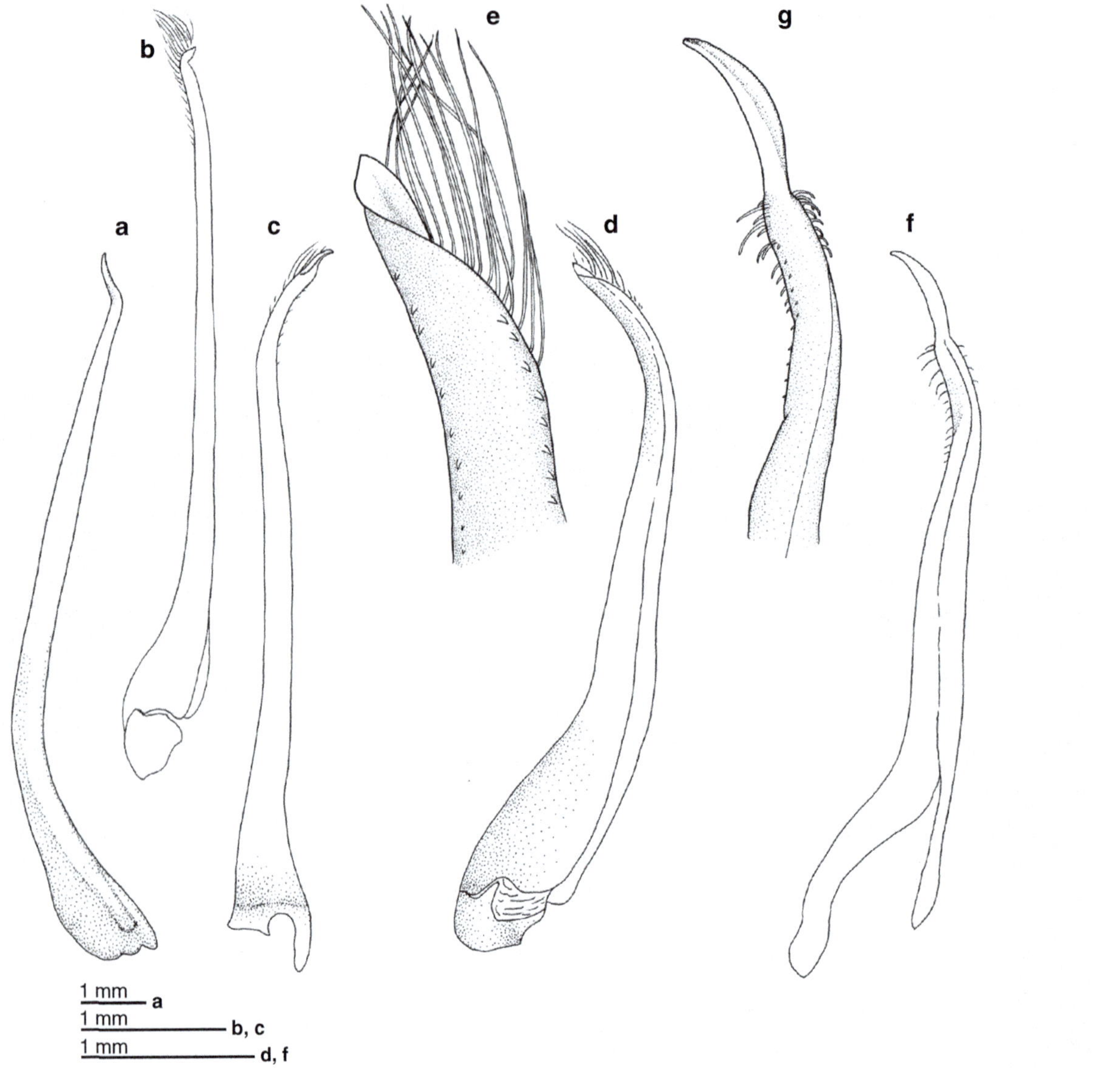

Fig. 21.39 Male G1, ventral view: (**a**) *Neoliomera nobilii* Odhner, 1925 (SMF: Saudi Arabia); (**b**) *Palapedia apeli* Naderloo, 2015 (ZMUC CRU945; holotype); (**c**) *Palapedia persica* Naderloo, 2015 (SMF: paratype); (**d**) *Paractaea rufopunctata* forma *illusoria* Guinot, 1969 (SMF: UAE); (**e**) apical part of (**d**); (**f**) *Pilodius spinipes* Heller, 1861 (ZUTC 5930); (**g**) apical part of (**f**)

Palapedia persica Naderloo 2015

(Figs. 21.35, 21.39c, and 21.40)

Palapedia persica Naderloo 2015b: 266–270, figs 1a–b, 3a–f, 5a–c.

Palapedia n. sp. — Apel 2001: 88 (partly: material from Saudi Arabia).

Type locality: Persian Gulf (Abu Musa Island, Iran).

Distribution

Persian Gulf: Iran (Naderloo 2015b), Saudi Arabia (Naderloo 2015b).

Gulf of Oman: No record.

World distribution: This species is still only recorded from the Persian Gulf.

Habitat: Subtidal, 10–12 m, substrate sandy in corals reefs.

Fig. 21.40 *Palapedia persica* Naderloo, 2015. Female holotype (ZUTC 5484), CL = 15.56, CB = 17.05 mm, Park-Qadir, Abu Musa Island, Iran, Persian Gulf, 13.04.2015, coll. R. Naderloo and A. Sari. Photo: R. Abdollahi (Reproduced from Naderloo (2015b by permission of Magnolia Press)

Genus *Paractaea* Guinot, 1969

Paractaea rufopunctata forma *illusoria*, Guinot 1969

(Figs. 21.35, 21.39d, e, and 21.41)

Paractaea rufopunctata forma *illusoria* Guinot 1969: 247, fig 22.
Paractaea rufopunctata forma *illusoria* — Serène 1984: 125; Ng et al. 2008: 196 (in list).
Paractaea rufopunctata illusoria — Apel 2001: 88, 89.
? *Paractaea rufopunctata* — Hogarth 1989: 106; Hogarth 1994: 98.

Type locality: Gulf of Aden (Aden and Djibouti).

Distribution

Persian Gulf: No record.

Gulf of Oman: UAE (Apel 2001).

World distribution: Northwestern Indian Ocean: Red Sea, Gulf of Aden, Socotra, southern Oman and Gulf of Oman.

Habitat: Shallow subtidal to 20 m, substrate rocky/ cobble.

Fig. 21.41 *Paractaea rufopunctata* forma *illusoria* Guinot, 1969. Male (SMF), CL = 10.3, CB = 14.6 mm, Fujairah, UAE, Gulf of Oman, 27.06.95, coll. M. Apel (Photo: S. Träkner)

Remarks: Guinot (1969), along with description of the genus *Paractaea*, described nine different forms of the species *Paractaea rufopunctata* (H. Milne Edwards, 1834). He described *Paractaea rufopunctata* forma *illusoria* from the Gulf of Aden. Apel (2001) collected several specimens of this species and assigned them to *Paractaea rufopunctata illusoria*. I examined the specimens and herewith support that the Persian Gulf's material are more similar to *Paractaea rufopunctata illusoria*. But this species needs a through taxonomic revision.

Genus *Paraxanthodes* Guinot, 1968

***Paraxanthodes cumatodes* (MacGilchrist, 1905)**
(Figs. 21.42 and 21.44)

Xanthodes cumatodes MacGilchrist 1905: 258.
Xanthodes cumatodes — Alcock and Annandale 1907: pl 79, fig 1.
Xanthias cumatodes — Stephensen 1946: 148; Guinot 1964: 28; Guinot 1967a: 269 (in list); Titgen 1982: 252 (in list).
Paraxanthodes cumatodes — Guinot 1968: 723, fig 60; Serène 1984: 209, pl 30c; Apel 2001: 89; Ng et al. 2008: 204 (in list).

Type locality: Strait of Hormuz (26° 20′N, 53° 54′E).

Distribution

Persian Gulf: MacGilchrist (1905) and Guinot (1968) recorded this species from the Strait of Hormuz.

Gulf of Oman: No record.

World distribution: Indo-West Pacific: Red Sea, Strait of Hormuz, Philippines and New Caledonia.

Habitat: Subtidal, 7–102 m.

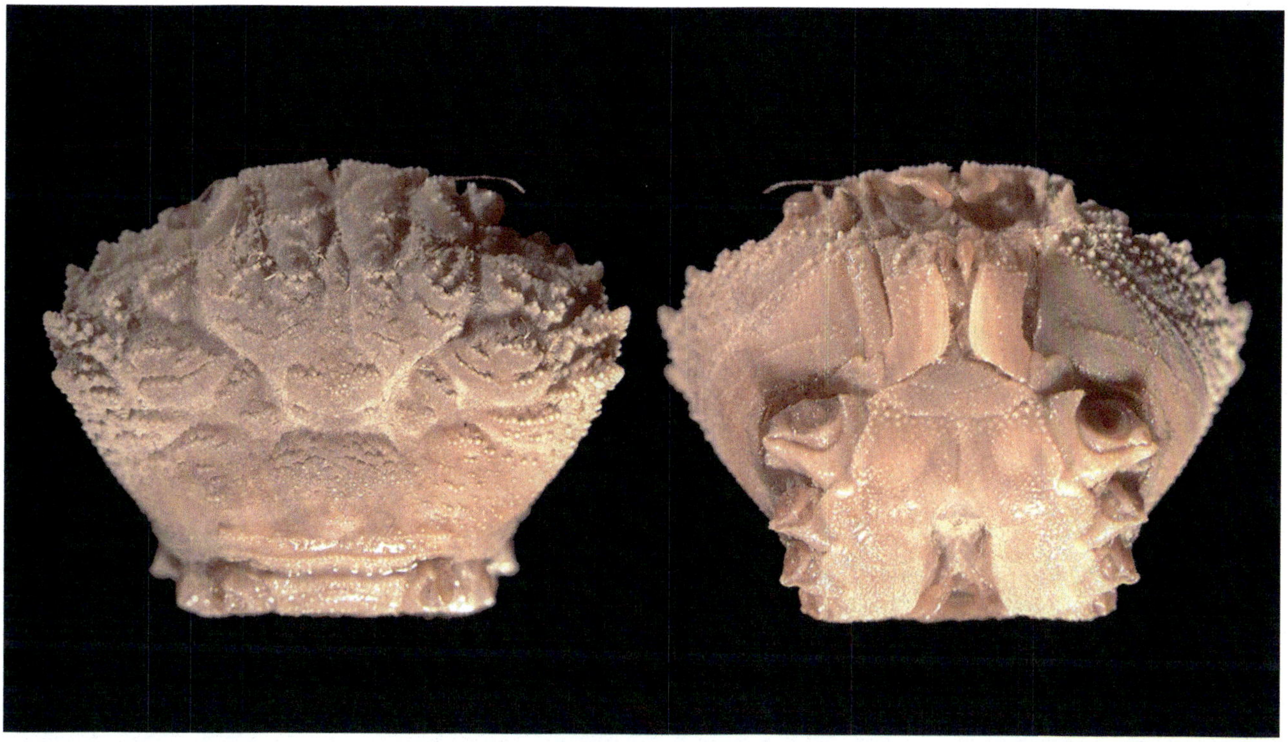

Fig. 21.42 *Paraxanthodes cumatodes* (MacGilchrist, 1905). Male holotype (Calcutta Museum), CL = 8.00, CB = 6.4 mm, Strait of Hormuz, 26° 20′N 53° 54′E

Genus *Pilodius* Dana, 1851

Pilodius spinipes Heller, 1861
(Figs. 21.39f, g, 21.43, and 21.44)

Pilodius spinipes Heller 1861b: 11, 12.
Pilodius spinipes — Heller 1861a: 340, 341, pl 3, fig 22; Titgen 1982: 134, 252 (in list); Serène 1984: 237 (in key), 243, figs 143 g, 148, pl 34a, b; Clark and Galil 1993: 1155–1159, figs 15a–g, 38a, b, 39a, 44a, c; Hogarth 1989: 106; Hogarth 1994: 99; Apel 2001: 89; Ng et al. 2008: 197 (in list).
Type locality: Red Sea.
Distribution
Persian Gulf: Iran (Naderloo et al. 2016), Saudi Arabia (Apel 2001), UAE (Titgen 1982).
Gulf of Oman: UAE (Apel 2001).
World distribution: Indian Ocean: South Africa, East Africa, Madagascar, Mauritius, Seychelles, Aldabra, Red Sea, Gulf of Aden, southern Oman, Persian Gulf, Gulf of Oman, India, Andaman Sea and Mergui Archipelago.
Habitat: Intertidal to 20 m, substrate coral reefs.

Genus *Platypodia* Bell, 1835

Platypodia anaglypta (Heller, 1861)
(Figs. 21.44, 21.45, and 21.46a)

Atergatis anaglyptus Heller 1861b: 6.
Atergatis anaglyptus — Heller 1861a: 312, pl 2, figs 11, 12.
Lophactaea anaglypta — Alcock 1898: 102; Stephensen 1946: 147.
Platypodia anaglypta — Guinot 1967a: 268 (in list); Titgen 1982: 252 (in list); Serène 1984: 158 (in key), 159, fig 92, pl 22a; Hogarth 1989: 106; Hogarth 1994: 99; Apel 2001: 89, 90.
Lophozozymus anaglyptus— Ng et al. 2008: 206 (in list), 210 (in text).
Type locality: Red Sea.
Distribution
Persian Gulf: Iran (Naderloo et al. 2016), Saudi Arabia (Apel 2001). Alcock (1898) recorded this species from an unknown locality.
Gulf of Oman: UAE (Apel 2001).

Fig. 21.43 *Pilodius spinipes* Heller, 1861. Male (ZUTC 5930), CL = 7.66, CB = 11.65 mm, Abu Musa Island, Iran, Persian Gulf, 15.05.2015, coll. R. Naderloo and A. Sari (Photo: R. Abdollahi)

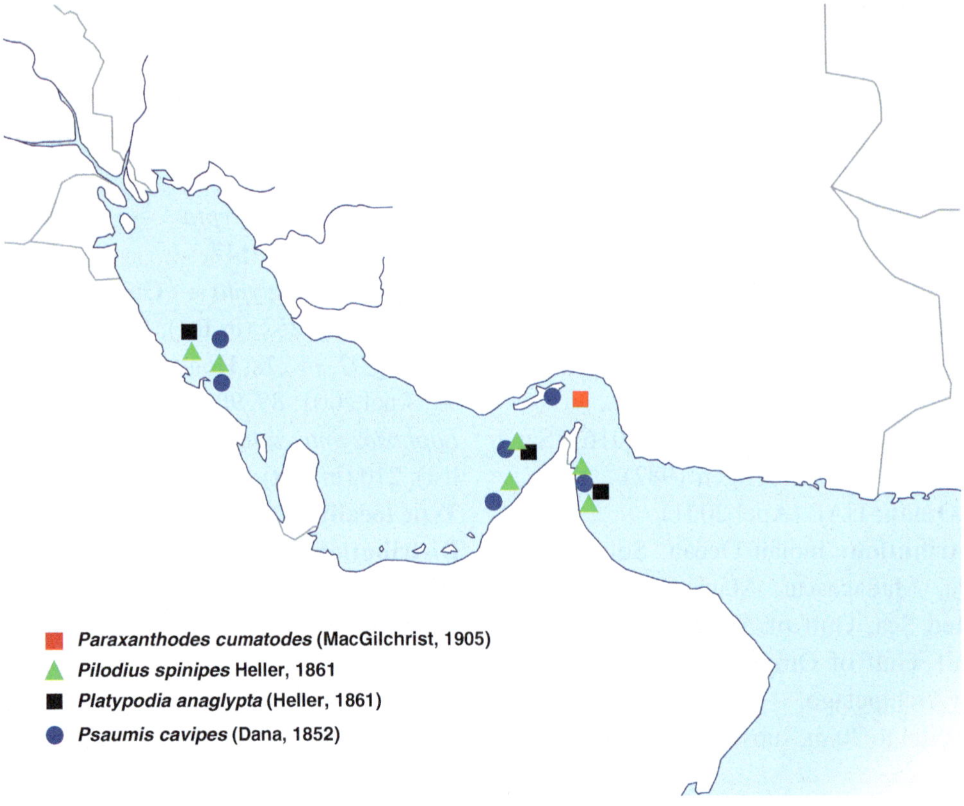

Fig. 21.44 Map showing the recorded localities of *Paraxanthodes cumatodes* (MacGilchrist, 1905), *Pilodius spinipes* Heller, 1861, *Platypodia anaglypta* (Heller, 1861), and *Psaumis cavipes* (Dana, 1852)

Fig. 21.45 *Platypodia anaglypta* (Heller, 1861). Male (SMF), CL = 20.1, CB = 30.0 mm, Abu Ahmed, Fujairah, UAE, Gulf of Oman, 02.07.1995, coll. M. Apel (Photo: S. Träkner)

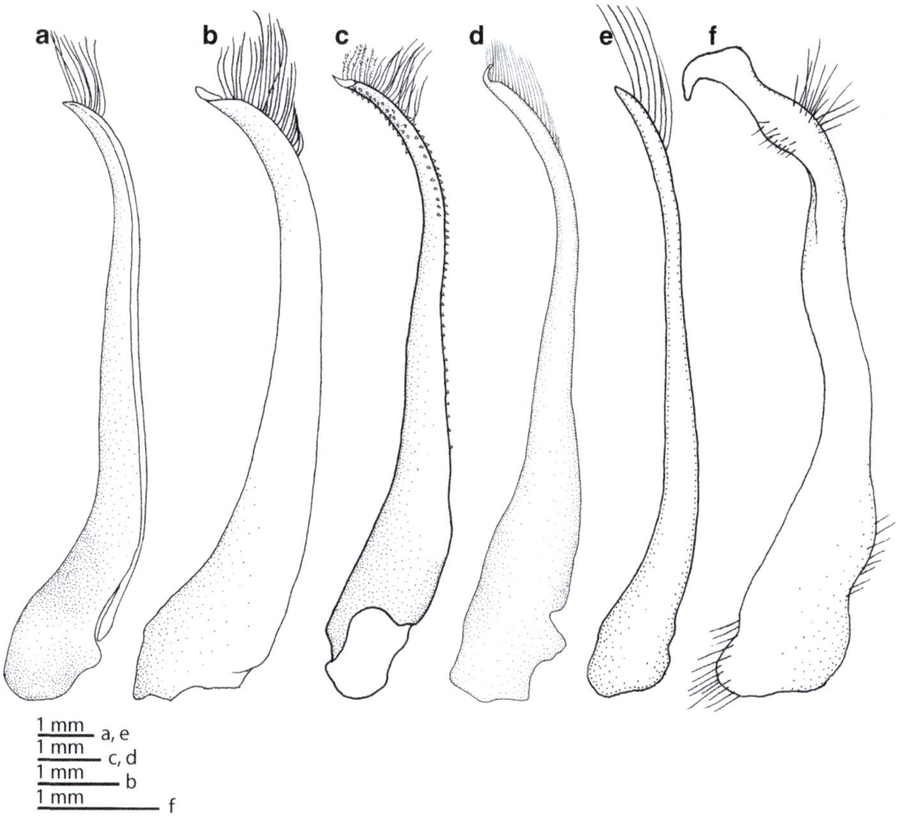

Fig. 21.46 Male G1, ventral view: (**a**) *Platypodia anaglypta* (Heller, 1861) (SMF: UAE); (**b**) *Psaumis cavipes* (Dana, 1852) (SMF: Saudi Arabia); (**c**) *Xanthias punctatus* (H. Milne Edwards, 1834) (SMF: Saudi Arabia); (**d**) *Xanthias sinensis* (A. Milne-Edwards, 1867) (ZUTC: Abu Musa, Persian Gulf); (**e**) *Zosimus aeneus* (Linnaeus, 1758) (ZUTC 5479); (**f**) *Zozymodes xanthoides* (Krauss, 1843) (ZUTC: Persian Gulf)

World distribution: Indo-West Pacific: Madagascar, Comoros, Seychelles, southern Oman, Persian Gulf, Gulf of Oman, Maldives, Sri Lanka, Japan, Australia, New Caledonia, Samoa, Salomon Island, Coetivy and Palau.

Habitat: Shallow subtidal to 10 m, substrates cobble and coral reefs.

Remarks Ng et al. (2008) replaced this species from the genus *Platypodia* Bell, 1835, to the genus *Lophozozymus* A. Milne-Edwards, 1863. They discussed that this species has smooth carapace and appendages, therefore it is different from all members of *Platypodia*. Further, it is close to *Lophozozymus* because of the presence of groove on the last anterolateral teeth and special color pattern with having prominent purple carapace with bright spots. But, I checked some specimens collected from the Persian Gulf which their morphology are completely constant with the genus *Platypodia* given by Serène (1984), therefore I leave it in *Platypodia* at the moment and believed that the taxonomy of the genera *Platypodia* and *Lophozozymus* must be studied in detail. Genetic and morphological study of Xanthidae by Lai et al. (2011) also showed that *P. anaglyptus* is rather different from *Lophozozymus*, therefore better to leave this species in *Platypodia* at the moment.

Genus *Psaumis* Kossmann, 1877

Psaumis cavipes (Dana, 1852)

(Figs. 21.44, 21.46b, and 21.47)

Actaeodes cavipes Dana 1852a: 78.
Cancer fossulatus Girard 1859: 149, pl 4, figs 2–2b.
Psaumis fossulata — Kossmann 1877: 27, pl 1, fig 3, pl 3, fig 10.
Actaea cavipes — Alcock 1898: 147; Stephensen 1946: 151; Guinot 1967a: 260 (in list); Titgen 1982: 251 (in list).
Actaea fossulata — Nobili 1906a: 126.

Fig. 21.47 *Psaumis cavipes* (Dana, 1852). Female (ZUTC 5325), CL = 8.90, CB = 13.78 mm, Dargahan, Qeshm Island, coll. H. Rahimian (Photo: R. Abdollahi)

Psaumis cavipes — Serène 1984: 129, 130, fig 76, pl 18f; Hogarth 1989: 106; Hogarth 1994: 99; Apel 2001: 90; Ng et al. 2008: 196 (in list).

Type locality: *Actaeodes cavipes*: Fiji and Samoa. — *Cancer fossulatus*: Red Sea.

Distribution

Persian Gulf: Iran (Naderloo et al. 2016), UAE [Nobili (1906a) as *Actaea fossulata*; Apel 2001].

Gulf of Oman: UAE (Apel 2001).

World distribution: Indo-West Pacific: Madagascar, Comoros, Mauritius, Amirante, Tanzania, Somalia, Red Sea, Gulf of Aden, southern Oman, Persian Gulf, Gulf of Oman, Pakistan, Chagos Archipelago, Andaman Sea, Vietnam, Taiwan, Fiji and Samoa.

Habitat: Intertidal to 6 m, substrates rocky/cobble and coral reefs.

Genus *Xanthias* Rathbun, 1897

Xanthias punctatus (H. Milne Edwards, 1834)

(Figs. 21.46c, 21.48, and 21.49)

Xantho punctatus H. Milne Edwards 1834: 396, 397.

Xantho punctatus — A. Milne-Edwards 1873: 199, pl 7, fig 6.

Xanthias punctatus — Serène 1984: 197, fig 113, pl 28e; Apel 2001: 90; Ng et al. 2008: 204 (in list).

Type locality: Mauritius.

Distribution

Persian Gulf: Saudi Arabia (Apel 2001).

Gulf of Oman: No record.

World distribution: Indo-West Pacific: Madagascar, Mauritius, Seychelles, Somalia, Red Sea, Persian

Fig. 21.48 *Xanthias punctatus* (H. Milne Edwards, 1834). Male (SMF 29985), CL = 17.1, CB = 28.1 mm, 1 km E Siquara, Socotra, depth 14 m, 23.03.1999, coll. M. Apel (Photo: S. Träkner)

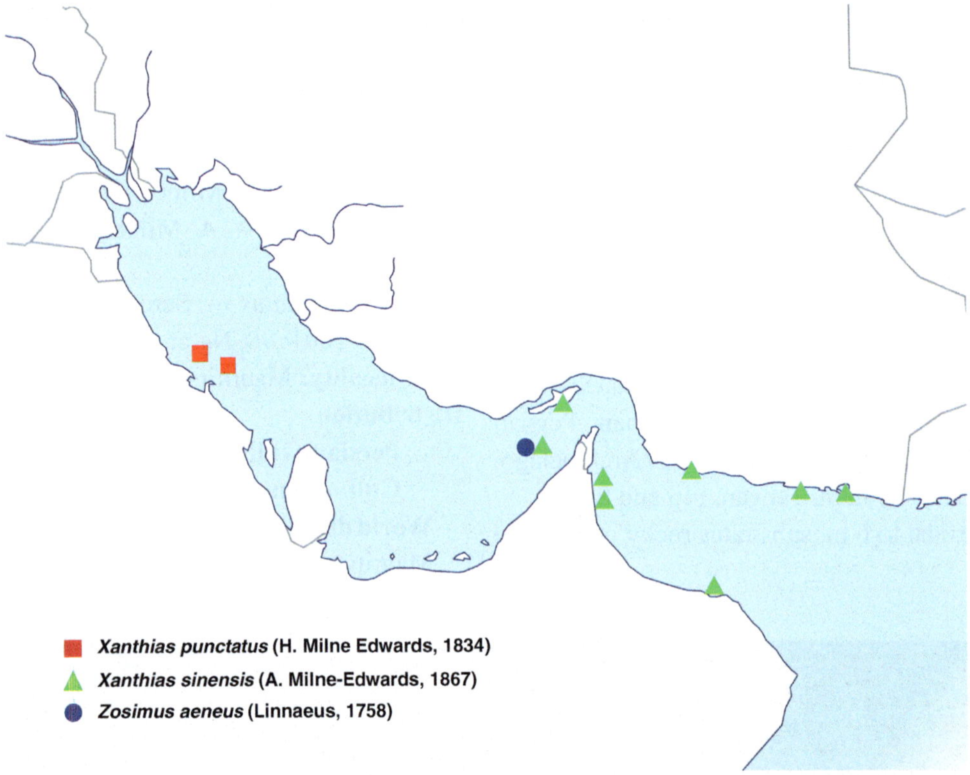

Fig. 21.49 Map showing the recorded localities of *Xanthias punctatus* (H. Milne Edwards, 1834), *Xanthias sinensis* (A. Milne-Edwards, 1867), and *Zosimus aeneus* (Linnaeus, 1758)

Gulf, India, Lakshadweep Islands, Sri Lanka, Indo-Malayan Archipelago, Australia and New Caledonia.

Habitat: Subtidal, 7–15 m, substrates rocky/cobble with dead corals.

Xanthias sinensis (A. Milne-Edwards, 1867)

(Figs. 21.46d, 21.49, and 21.50)

Pseudozius sinensis A. Milne-Edwards 1867: 278.
Lioxantho asperatus Alcock 1898: 92.
Lioxantho asperatus — Alcock and Anderson 1899: pl 36, fig 9; Nobili 1906a: 120.
Xanthias sinensis — Serène 1984: 195, fig 111, pl 28a; Hogarth 1989: 106, 114; Hogarth 1994: 100; Hornby 1997: 16; Apel 2001: 90, 91; Ng et al. 2008: 204 (in list); Naderloo and Türkay 2012: 45; Naderloo et al. 2013: 449, tab 1.

Type locality: *Pseudozius sinensis*:? China. — *Lioxantho asperatus*: Pakistan (Karachi).

Distribution

Persian Gulf: Iran (Naderloo and Türkay 2012; Naderloo et al. 2013).

Gulf of Oman: Iran (Naderloo et al. 2015), UAE (Hornby 1997; Apel 2001), Oman (Serène 1984).

World distribution: Indian Ocean: Somalia, Red Sea, Gulf of Aden, southern Oman, Persian Gulf, Gulf of Oman, Pakistan and China.

Habitat: Intertidal, substrate rocky/cobble.

Fig. 21.50 *Xanthias sinensis* (A. Milne-Edwards, 1867). The specimen is not preserved, Tang, Sistan-Baluchestan, Iran, Gulf of Oman, 24.05.2013, coll. R. Naderloo (Photo: R. Abdollahi)

Genus *Zosimus* Leach, 1818

***Zosimus aeneus* (Linnaeus, 1758)**
(Figs. 21.46e, 21.49, and 21.51)

Cancer aeneus Linnaeus 1758: 630.
Zosimus floridus Herbst 1783.
Zosimus amphotrite Herbst 1801.
Zosimus aeneus — Galil and Vannini 1990: 39; Ng et al. 2008: 206 (in list); Naderloo et al. 2016: 2, fig 3.

Type locality: Unknown.
Distribution
Persian Gulf: Iran (Naderloo et al. 2016).
Gulf of Oman: No record.
World distribution: Western Indian Ocean: South Africa, Madagascar, Réunion, Mayotte, Seychelles, Tanzania, Mozambique, Somalia, Red Sea and Persian Gulf.
Habitat: Intertidal to shallow subtidal, substrate rocky/cobble.

Fig. 21.51 *Zosimus aeneus* (Linnaeus, 1758). Female (ZUTC 5479), CL = 44.10, CB = 64.39 mm, Park-Qadir, Abu Musa Island, Iran, Persian Gulf, 22.12.2014, coll. S. Ebrahimnejad and R. Abdollahi (Photo: R. Abdollahi)

Genus *Zozymodes* Heller, 1861

Key to species of the genus *Zozymodes* in the region

- Anterolateral margins of carapace with four small teeth; transverse sterae of granules on dorsal surface of carapace, prominent on lateral regions; cup-shaped structure markedly prominent on dorsal surface of merus and carpus of walking legs .... *Zozymodes cavipes*
- Anterolateral margins of carapace with four lobes; small granules regularly distributed over dorsal surface of carapace; cup-shaped structure on dorsal surface of merus and carpus of walking legs hardly distinguishable .... *Zozymodes xanthoides*

Zozymodes cavipes (Dana, 1852)

(Figs. 21.52 and 21.53)

Chlorodius cavipes Dana 1852a: 79.
Leptodius cavipes — De Man 1887: 34; Klunzinger 1913: 214; Nobili 1906b: 243.
Zozymodes cavipes —Guinot 1968: 714, fig 36; Serène 1984: 152 (in key), 152, 153, fig 88, pl 19f; Tirmizi and Ghani 1992: 135–138, figs 1a–c, 2a–d; Ng et al. 2008: 206 (in list); Naderloo et al. 2016: 3, fig 4.
Type locality: Unknown.

Distribution

Persian Gulf: Iran (Naderloo et al. 2016).
Gulf of Oman: Iran (Naderloo et al. 2015).
World distribution: Indo-West Pacific: East Africa, Madagascar, Aldabra Islands, Red Sea, Gulf of Aden, Persian Gulf, Gulf of Oman, Pakistan, Andaman Islands, Chagos Archipelago, Mergui Archipelago, Malaysia, Indonesia, Taiwan, China, Japan, Christmas Island and Australia.
Habitat: Intertidal, substrate rocky/cobble.

Fig. 21.52 *Zozymodes cavipes* (Dana, 1852). Female (ZUTC 5526), CL = 9.79, CB = 14.78 mm, Tis, Chabahar, Iran, Gulf of Oman, 15.09.2013, coll. R. Naderloo (Photo: R. Abdollahi)

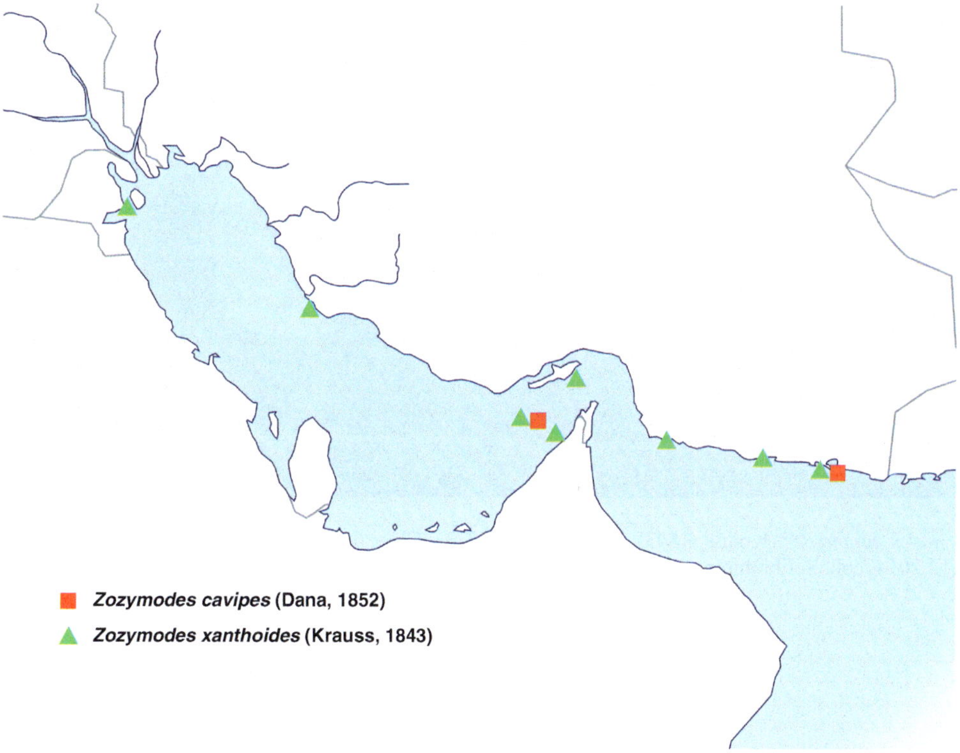

Fig. 21.53 Map showing the recorded localities of *Zozymodes cavipes* (Dana, 1852), and *Zozymodes xanthoides* (Krauss, 1843)

Zozymodes xanthoides (Krauss, 1843)
(Figs. 21.46f, 21.53, and 21.54)

Cancer (*Pilumnus*) *Xanthoides* Krauss 1843: 32, pl 1, fig 6a, b.
Zozymodes carinipes Heller 1861b: 8.
Zozymodes Xanthoides — Stephensen 1946: 156, fig 39c, d; Guinot 1967a: 269 (in list); Titgen 1982: 140, 141, 52 (in list); Serène 1984: 152 (in key), 153, fig 89, pl 19d; Hogarth 1994: 100; Apel 2001: 91; Naderloo and Türkay 2012: 45; Naderloo et al. 2013: 449, tab 1.

Type locality: *Cancer* (*Pilumnus*) *Xanthoides*: South Africa (Natal). — *Zozymodes carinipes*: Red Sea.

Distribution

Persian Gulf: Iran (Stephensen 1946; Naderloo and Türkay 2012; Naderloo et al. 2013), UAE (Titgen 1982).

Gulf of Oman: Iran (Naderloo et al. 2015).

World distribution: Western Indian Ocean: South Africa, Madagascar, Somalia, Red Sea, Gulf of Aden, Socotra, southern Oman, Persian Gulf and Gulf of Oman.

Habitat: Intertidal, substrate rocky/cobble.

Fig. 21.54 *Zozymodes xanthoides* (Krauss, 1843). Male (ZUTC 5773), CL = 6.7, CB = 9.2 mm, Park-Qadir, Abu Musa Island, Iran, Persian Gulf, 25.12.2014, coll. S. Ebrahimnejad and R. Abdollahi (Photo: S. Shahbazi)

Family Pseudoziidae Alcock, 1898 (Stone Crabs)

22

Diagnosis Carapace transversely oval, much broader than long, dorsal surface moderately convex, regions poorly defined; anterolateral margins with broad lobes or very small denticles; junction between anterolateral and posterolateral margins marked with small denticle. Front broad, broader than orbit, straight, weakly bilobed, gently deflexed. Orbits small, eyestalks short. Antennules folded transversely. Antennae with basal antennal segment extending between inner infraorbital angle and lateral frontal margin. Thoracic sternites 1–2 fused, suture not visible, suture of sternites 2–3 visible, suture of sternites 3–4 medially interrupted. Abdomen of males and females with seven freely moving somites. G1 long, slender; G2 short, 1/3 to 1/4 of G1. Genital opening of males coxal. Genital opening of females sternal.

Genus *Pseudozius* Dana, 1851

Pseudozius caystrus (Adams and White, 1849)
(Figs. 22.1, 22.2 and 22.3a)

Panopeus caystrus Adams and White 1849: 42, pl 9, fig. 2.
Pseudozius caystrus — Guinot 1967a: 272 (in list); Serène 1984: 313, figs 242, 243, pl 48b; Hogarth 1989: 106, 113, 114; Hogarth 1994: 102; Ng and Wang 1994: 86, 97, figs 1–8; Tirmizi and Ghani 1996: 73–75, fig 28; Neumann and Spiridonov 1999: 50; Apel 2001: 103; Ng et al. 2008: 180 (in list); Naderloo and Türkay 2012: 44; Khvorov et al. 2012: 1739 (in list); Naderloo et al. 2013: 449, tab 1.

Type locality: "Eastern Seas".
Distribution
Persian Gulf: Iran (Naderloo and Türkay 2012; Naderloo et al. 2013).
Gulf of Oman: Iran (Naderloo et al. 2015), UAE (Apel 2001), Oman (Hogarth 1989).
World distribution: Indo-West Pacific: South Africa, East Africa, Mauritius, Red Sea, Socotra, southern Oman, Persian Gulf, Gulf of Oman, Pakistan, India, Lakshadweep Islands, China, Japan, Philippines, Christmas Island, Cocos (Keeling) Islands, Ashmore Reef, Australia, Tasmania, Marshall Islands and French Polynesia.
Habitat: Intertidal, substrate rocky/cobble.

R. Naderloo, *Atlas of Crabs of the Persian Gulf*, DOI 10.1007/978-3-319-49374-9_22

Fig. 22.1 *Pseudozius caystrus* (Adams and White, 1849). Male (ZUTC 5531), CL = 15.74, CB = 25.52 mm, Qadir Park, Abu Musa Island, Iran, Persian Gulf, 17.05.2015, coll. R. Naderloo (Photo: S. Shahbazi)

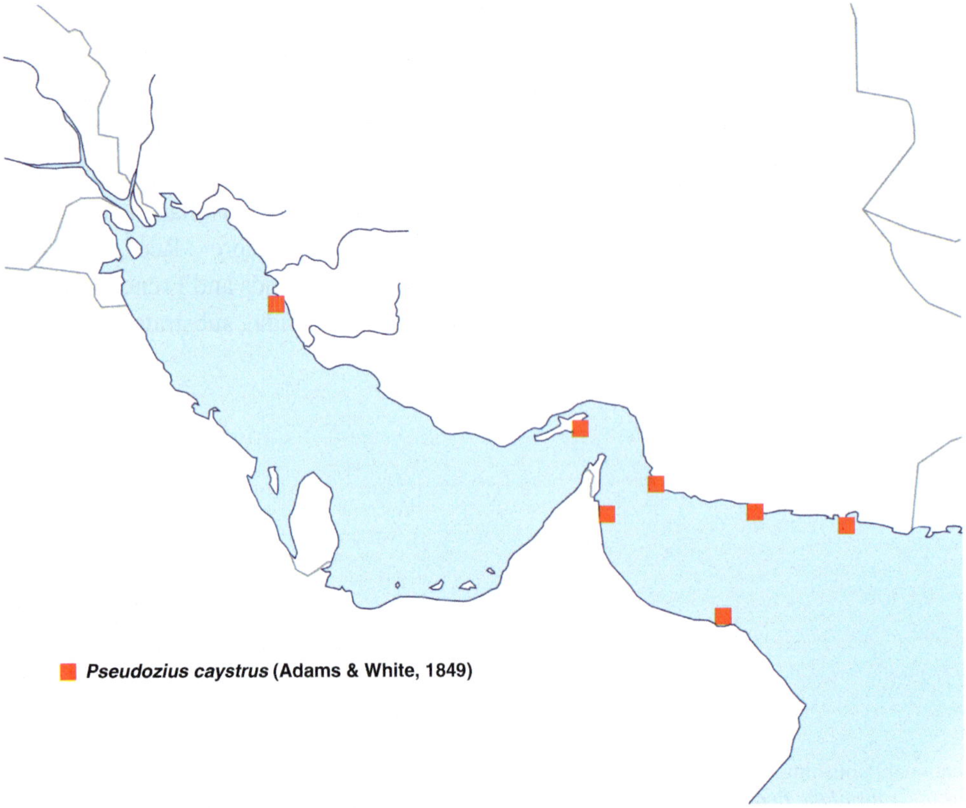

Fig. 22.2 Map showing the recorded localities of *Pseudozius caystrus* (Adams and White, 1849)

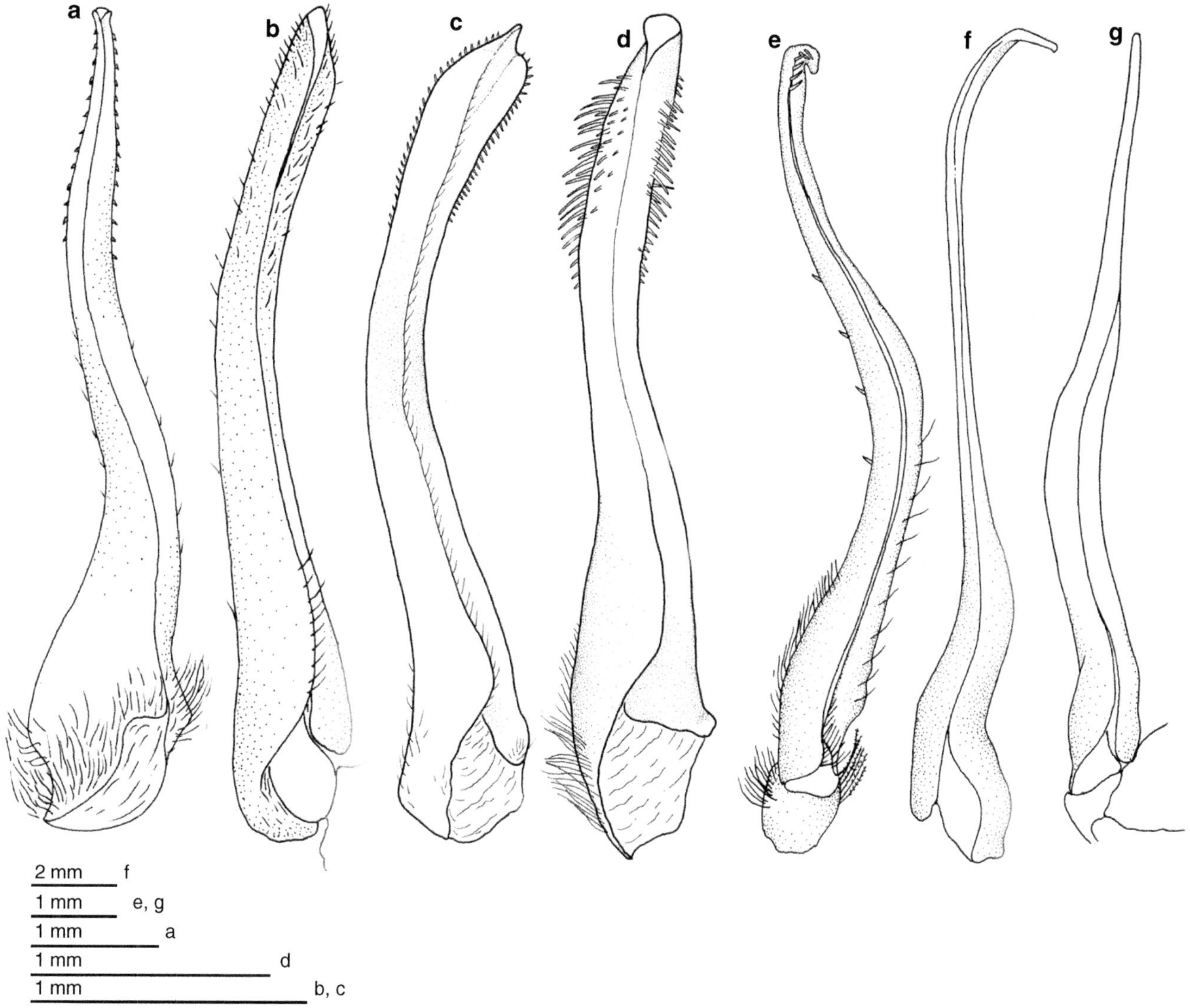

Fig. 22.3 Male G1, ventral view: (**a**) *Pseudozius caystrus* (Adams and White, 1849) (ZUTC: Iran); (**b**) *Trapezia tigrina* Eydoux and Souleyet, 1842 (SMF: Socotra); (**c**) *Trapezia cymodoce* (Herbst, 1801) (SMF: Saudi Arabia); (**d**) *Tetralia cavimana* Heller, 1861 (SMF: Saudi Arabia); (**e**) *Dentoxanthus iranicus* Stephensen, 1946 (India); (**f**) *Galene bispinosa* (Herbst, 1783) (ZMUC CRU891); (**g**) *Halimede tyche* (Herbst, 1801) (SMF: Kuwait)

Family Tetraliidae Castro, Ng and Ahyong, 2004 (Coral Crabs) 23

Diagnosis Carapace hexagonal to longitudinally oval, dorsal surface slightly convex, smooth, polished, regions not defined; anterolateral margins entire, slightly diverging posteriorly, posterolateral margins sharply converging posteriorly; junction between lateral margins not distinct. Front straight, finely denticulate, markedly broad, broader than posterior margin of carapace. Antennules folded transversely. Antennae with basal antennal segment laterally produced, closing orbital hiatus. Orbits about half as broad as front; eyestalks long, slightly extending beyond exorbital tooth, not completely covered by orbit. Third maxillipeds with merus round, markedly smaller than ischium. Chelipeds markedly unequal. Walking legs relatively short, moderately flat. Abdomen of males and females with seven freely moving somites. G1 stout, nearly straight; G2 stout slightly curved, less than as half as length of G1. Genital opening of males coxal. Genital opening of females sternal.

Genus *Tetralia* Dana, 1851

Tetralia cavimana Heller, 1861
(Figs. 22.3d, 23.1, and 23.2)

Tetralia cavimana Heller 1861b: 14.
Tetralia cavimana — Heller 1861a: 353, pl 3, figs 24, 25; Galil 1988: 59 (in key), 59–62, figs 1a, 2a, b; Hogarth 1994: 103; Castro 1999: 100, 101.
Tetralia glaberrima — Nobili 1906a: 143; Stephensen 1946: 161, fig 42c, d; Guinot 1967a: 275 (in list; partly); Evans et al. 1973: 275; Basson et al. 1977: 162; Titgen 1982: 139, 140, 252 (in list); Hogarth 1989: 106; Apel 2001: 84; Ng et al. 2008: 184.

Type locality: Red Sea.
Distribution
Persian Gulf: Iran [Stephensen (1946) as *T. glaberrima*; present study], Saudi Arabia [Basson et al. (1977) as *T. glaberrima*; Apel 2001], UAE [Nobili 1906a; Evans et al. (1973) and Titgen (1982) as *T. glaberrima*; Apel 2001].
Gulf of Oman: UAE (Apel 2001).
World distribution: Western Indian Ocean: Kenya, Red Sea, Gulf of Aden, southern Oman, Persian Gulf and Gulf of Oman.
Habitat: Subtidal, 3–14 m, substrate coral reefs, particularly symbiont with hard coral *Acropora* sp.

R. Naderloo, *Atlas of Crabs of the Persian Gulf*, DOI 10.1007/978-3-319-49374-9_23

Fig. 23.1 *Tetralia cavimana* Heller, 1861. Male (SMF), CL =10.1, CB = 10.2 mm, Abu Ali, Saudi Arabia, Persian Gulf, 13.04.1993, coll. M. Apel (Photo: S. Träkner)

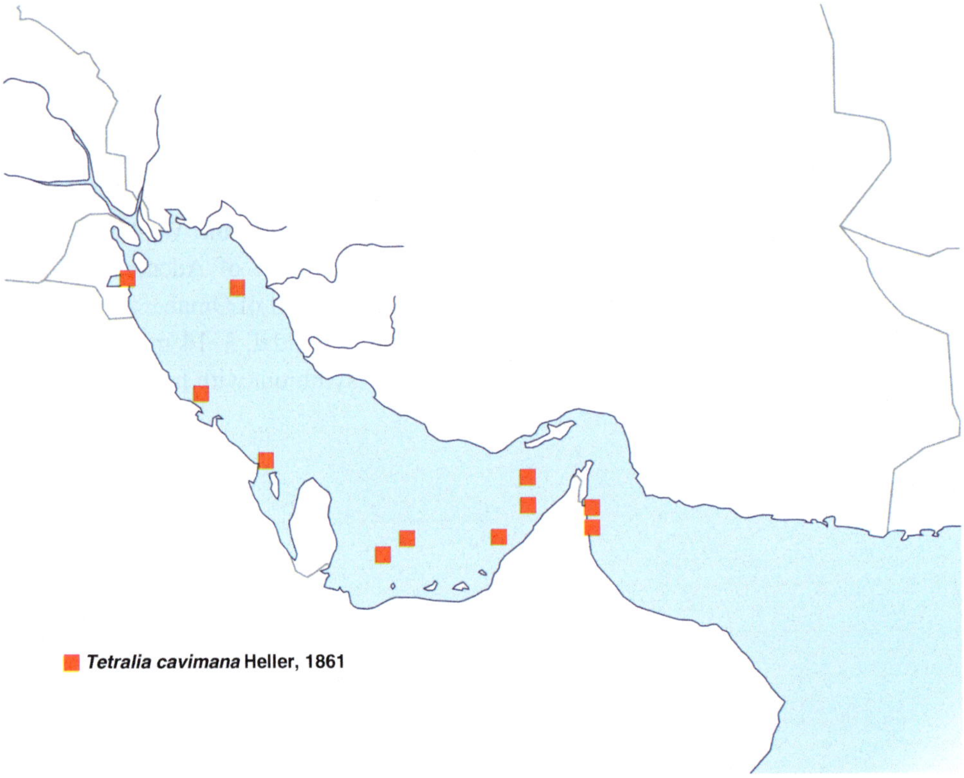

Fig. 23.2 Map showing the recorded localities of *Tetralia cavimana* Heller, 1861

Family Trapeziidae Miers, 1886 (Coral Crabs)

24

Diagnosis Carapace trapezoidal, hexagonal, octagonal, transversely oval; dorsal surface moderately convex, smooth, polished; anterolateral margins entire, without denticles, posterolateral margins gently converging posteriorly, junction between lateral margins marked with epibranchial spine. Front straight, denticulate, markedly broad, distinctly broader than posterior margin of carapace. Orbits small, about as half as front breadth, eyestalks relatively large, not fully covered by orbits. Antennules folded transversely. Antennae with slender basal antennal segment, not expanded anteriorly. Third maxillipeds with merus quadrate, shorter than ischium. Chelipeds equal, slightly unequal. Walking legs moderately long, slightly flattened. Abdomen of males with somites 3–5 fused, sutures hardly visible. Abdomen of females with seven freely moving somites. G1 relatively slender, nearly straight; G2 stout, nearly straight, less than as half as length of G1. Genital opening of males coxal. Genital opening of females sternal.

Key to species of the family Trapeziidae in the region

1. Front slightly broader than posterior margin of carapace, with V or U shaped median notch, with two distinct lobes; G1 with distal spines or setae, symbionts of soft or black corals 2

- Front distinctly broader than posterior margin of carapace, without distinct lobe, dentate; symbionts of scleractinian corals of family Pocilloporidae 3

2. Suture between second and third thoracic sternites complete; posterior margin of dactylus of walking legs with spines; living on corals like alcyonaceans, gorgonians and antipatharians .... *Quadrella coronata*

- Suture between second and third thoracic sternites not complete; posterior margin of dactylus of walking legs without spines; living on antipatharian corals .... *Quadrella reticulata*

3. Chelipeds with dense patch of setae on dorsal margin; upper surface of carapace without special pattern .... *Trapezia cymodose*

- Chelipeds without patch of setae on dorsal margin; carapace and appendages with red spots on upper surface .... *Trapezia tigrina*

R. Naderloo, *Atlas of Crabs of the Persian Gulf*, DOI 10.1007/978-3-319-49374-9_24

Genus *Quadrella* Dana, 1851

Quadrella coronata Dana, 1852
(Figs. 24.1 and 24.2)

Quadrella coronata Dana 1852a: 84.
Quadrella coronata — Dana 1852b: 266; Dana 1855:pl 16, figs 5a–d; Serène 1984: 289, fig 195, pl 41f; Galil 1986: 282–284, 291 (in key), figs 3e, 4c–e; Castro 1999: 95, 96; Apel 2001: 94; Ng et al. 2008: 185 (in list).
Type locality: Sulu Sea or Balabac Strait (between Borneo and Philippines).

Distribution
Persian Gulf: No record.
Gulf of Oman: UAE (Castro 1999; Apel 2001).
World distribution: Indo-West Pacific: South Africa, East Africa, Madagascar, Comoros, Mozambique, Amirante, Seychelles, Red Sea, Gulf of Oman, India, Sri Lanka, Maldives, Indo-Malayan Archipelago, Philippines, China, Japan and Australia.
Habitat: Subtidal, 12–540 m, living in association with soft and black corals.

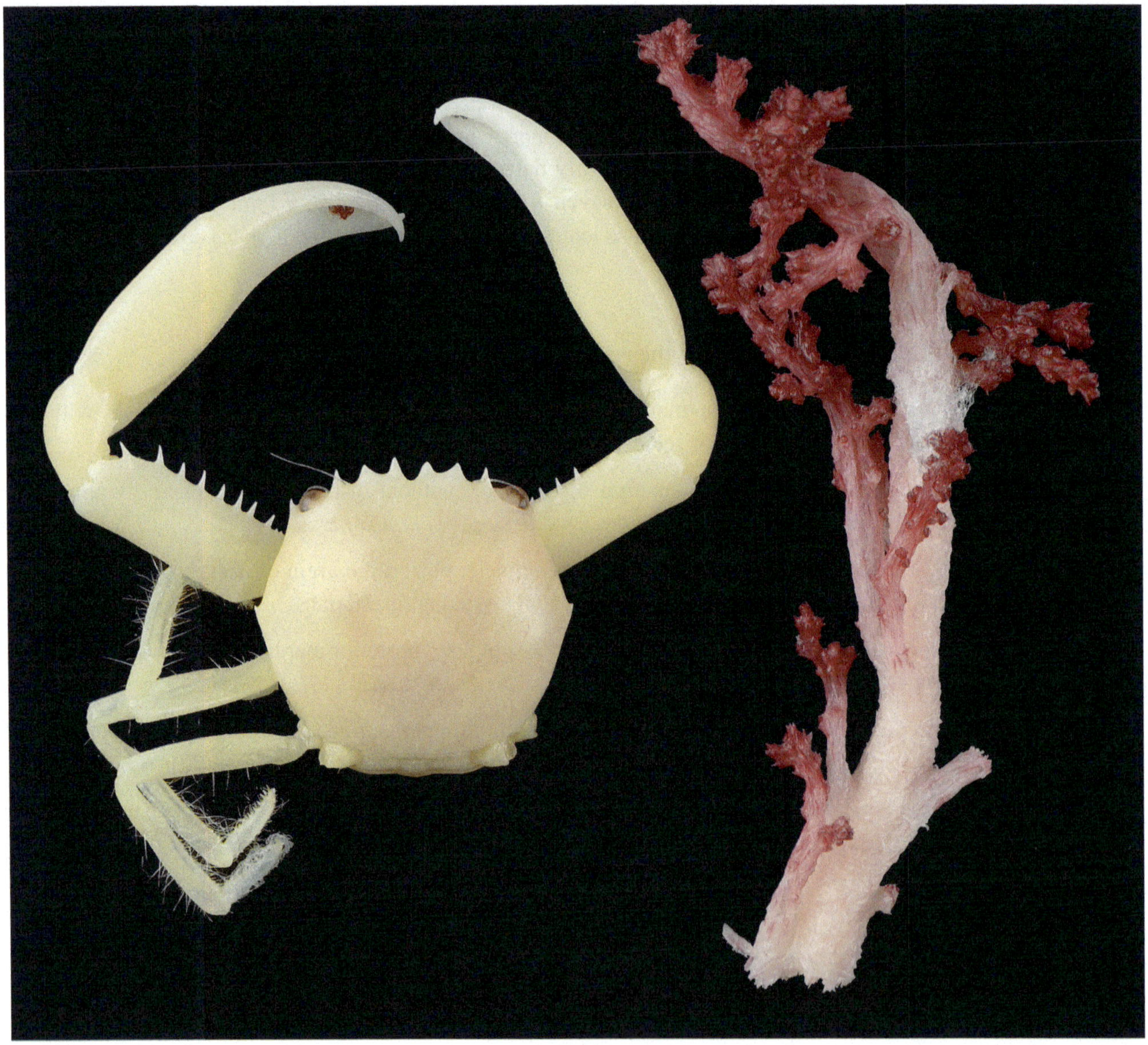

Fig. 24.1 *Quadrella coronata* Dana, 1852. Female (SMF 24083), CL = 24.8, CB = 23.3 mm, SO of Khor Fakkan, Coral Gardens, on soft coral *Siphonogorgia* sp., depth 24 m, 25° 21′N, 56° 23′E, 03.07.1995, coll. M. Apel (Photo: S. Träkner)

Quadrella reticulata Alcock, 1898
(Figs. 24.2 and 24.3)

Quadrella coronata var. *reticulata* Alcock 1898.
Quadrella reticulata — Serène 1984: 286, 287; Castro 1999: 38; Castro et al. 2004: 56; Naderloo and Sari 2005b: 67, figs 2, 3; Naderloo and Sari 2007a: 343, tab1; Ng et al. 2008: 185 (in list).
Type locality: Sri Lanka.
Distribution

Persian Gulf: Iran (Naderloo and Sari 2005b; Naderloo and Sari 2007a).
Gulf of Oman: No record.
World distribution: Indo-West Pacific: Red Sea, Persian Gulf, India, Sri Lanka, Andaman Sea, Indonesia, South China Sea, Philippines and Japan.

Habitat: Subtidal, 18–82 m, living in association with black corals.

Genus *Trapezia* Latreille, 1828

Trapezia cymodoce (Herbst, 1801)
(Figs. 22.2c, 24.2, and 24.4)

Cancer cymodoce Herbst 1801: 22, pl 51, fig 5.
Trapezia cymodoce — Stephensen 1946: 161, 162, fig 42c, d; Guinot 1967a: 276 (in list); Basson et al. 1977: 262; Galil 1987: 160 (in key), 161–163, fig 1; Hogarth 1989: 106, 114; Hogarth 1994: 102; Castro 1999: 104–107; Apel 2001: 84; Ng et al. 2008: 186 (in list).
Type locality: "Ostindien". East India or Indo-Malayan Archipelago.
Distribution

Persian Gulf: Iran (Stephensen 1946), Saudi Arabia (Basson et al. 1977; Apel 2001), UAE (Apel 2001).
Gulf of Oman: UAE (Apel 2001).

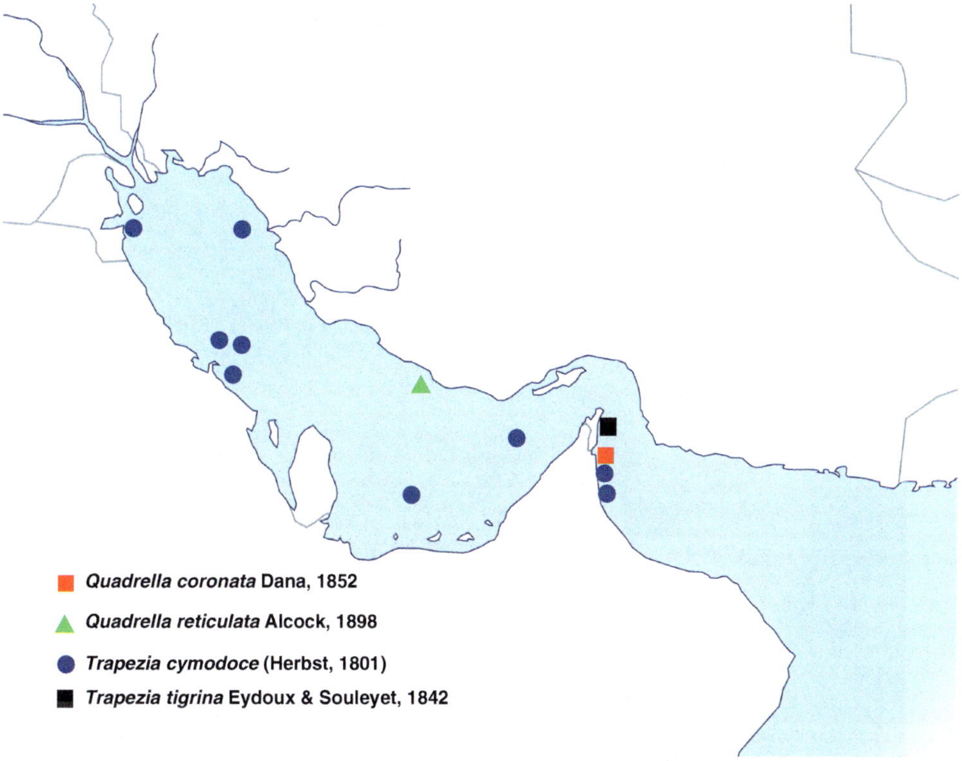

Fig. 24.2 Map showing the recorded localities of *Quadrella coronata* Dana, 1852, *Quadrella reticulata* Alcock, 1898, *Trapezia cymodoce* (Herbst, 1801), and *Trapezia tigrina* Eydoux and Souleyet, 1842

Fig. 24.3 *Quadrella reticulata* Alcock, 1898. Female (ZUTC Brach1072), CL = 10.8, CB = 12. 7 mm, Iran, Persian Gulf, 53° 37′N, 26° 04′E, 23.12.2003, coll. R. Naderloo (Photo: R. Abdollahi)

Fig. 24.4 *Trapezia cymodoce* (Herbst, 1801). Male (SMF), CL = 10.2, CB = 10.4 mm, Karan Island, Saudi Arabia, Persian Gulf, 07.05. 1995, coll. M. Apel (Photo: S. Träkner)

Fig. 24.5 *Trapezia tigrina* Eydoux and Souleyet, 1842. Female (SMF MAP136), CL = 12.1, CB = 14.2 mm, Rhyi Di Hamri, Socotra, Yemen, depth 7–9 m, from living *Pocillopra* sp., 19.03.1999, coll. M. Apel (Photo: S. Träkner)

World distribution: Western Indian Ocean: South Africa, East Africa, Madagascar, Mauritius, Seychelles, Red Sea, Gulf of Aden, southern Oman, Persian Gulf and Gulf of Oman.

Habitat: Subtidal, 3–13 m, living on coral, particularly on *Pocilopora* sp.

Trapezia tigrina Eydoux and Souleyet, 1842

(Figs. 22.2b, 24.2, and 24.5)

Trapezia tigrina Eydoux and Souleyet 1842: 232, pl 2, fig 4.

Trapezia tigrina — Serène 1984: 275, 276, fig 182, pl 39c, d; Galil 1987: 11–13, fig 5; Hogarth 1989: 106; Hogarth 1994: 103; Castro 1999: 115, 116; Apel 2001: 85; Ng et al. 2008: 186 (in list).

Type locality: Red Sea.

Distribution

Persian Gulf: No record.

Gulf of Oman: Oman (Castro 1999; Apel 2001).

World distribution: Indo-West Pacific: Madagascar, Mozambique, Mauritius, Amirante, Seychelles, Tanzania, Kenya, Somalia, Red Sea, Gulf of Aden, southern Oman, Gulf of Oman, Maldives, Chagos Archipelago, Nicobar Islands, Indo-Malayan Archipelago, Philippines, Tuamotu and Marshall Islands.

Habitat: Subtidal, living on corals, particularly on *Pocilopora* sp. Apel (2001) collected this species from *Pocilopora damicornis* in 3–25 m in the Gulf of Oman.

Family Galenidae Alcock, 1898

25

Diagnosis Carapace subquadrate to subpentagonal, slightly broader than long, markedly thick; anterolateral and posterolateral margins nearly parallel, distinct junction between them marked with tooth or small tubercle. Front straight or slightly produced, bilobed, much shorter than posterior margin of carapace. Orbits small, not fully covering eyes; eyestalks short. Antennules folded transversely. Antennae with long basal antennal segment, not reaching to frontal margin. Third maxillipeds with merus quadrate, much shorter than ischium. Chelipeds robust, subequal to unequal. Walking legs relatively slender, last legs without any modification. Abdomen of males and females with seven freely moving somites. G1 slender, long, nearly straight; G2 much shorter than G1. Genital opening of males coxal. Genital opening of females sternal.

Key to species of the family Galenidae in the region

1. Dorsal surface of carapace relatively smooth; anterolateral margins with two denticles; front relatively straight, not extending beyond orbit, with small median denticles.................................... *Galene bispinosa*
- Dorsal surface of carapace relatively granulate; front bilobed; distinctly extending beyond....................... 2

2. Dorsal surface of carapace with finely granulated elevations; lateral margins with large epibranchial spine; front distinctly broader than orbit, with weak medial notch, continuous with inner orbital angle; upper margin of palm and fingers with distinct keel.. *Dentoxanthus iranicus*
- Dorsal surface of carapace with large tubercles; anterolateral margins with large tubercles; front as long as orbit, with distinct medial notch, seperated from orbital denticle; upper margin of palm and fingers tuberculate.. *Halimede tyche*

Genus *Dentoxanthus* Stephensen, 1946

Dentoxanthus iranicus Stephensen, 1946

(Figs. 25.1, 25.2, and 22.2e)

Dentoxanthus iranicus Stephensen 1946: 164, 165, fig 43.
Dentoxanthus iranicus — Tirmizi and Serène 1971: 29; Tirmizi and Kazmi 1982: 310–313, figs 2, 3; Števcic and Ng 1988: 1, 2; Apel 2001: 97; Ng et al. 2008: 137, 2015: 1, figs 1–3.

Type locality: Gulf of Oman (Ras-Meidani, Iran).

Distribution

Persian Gulf: No record.

Gulf of Oman: Iran (Stephensen 1946).

World distribution: Northern Indian Ocean: Gulf of Oman, Pakistan and India.

Habitat: Subtidal to 10 m, substrate hard clay with sand.

R. Naderloo, *Atlas of Crabs of the Persian Gulf*, DOI 10.1007/978-3-319-49374-9_25

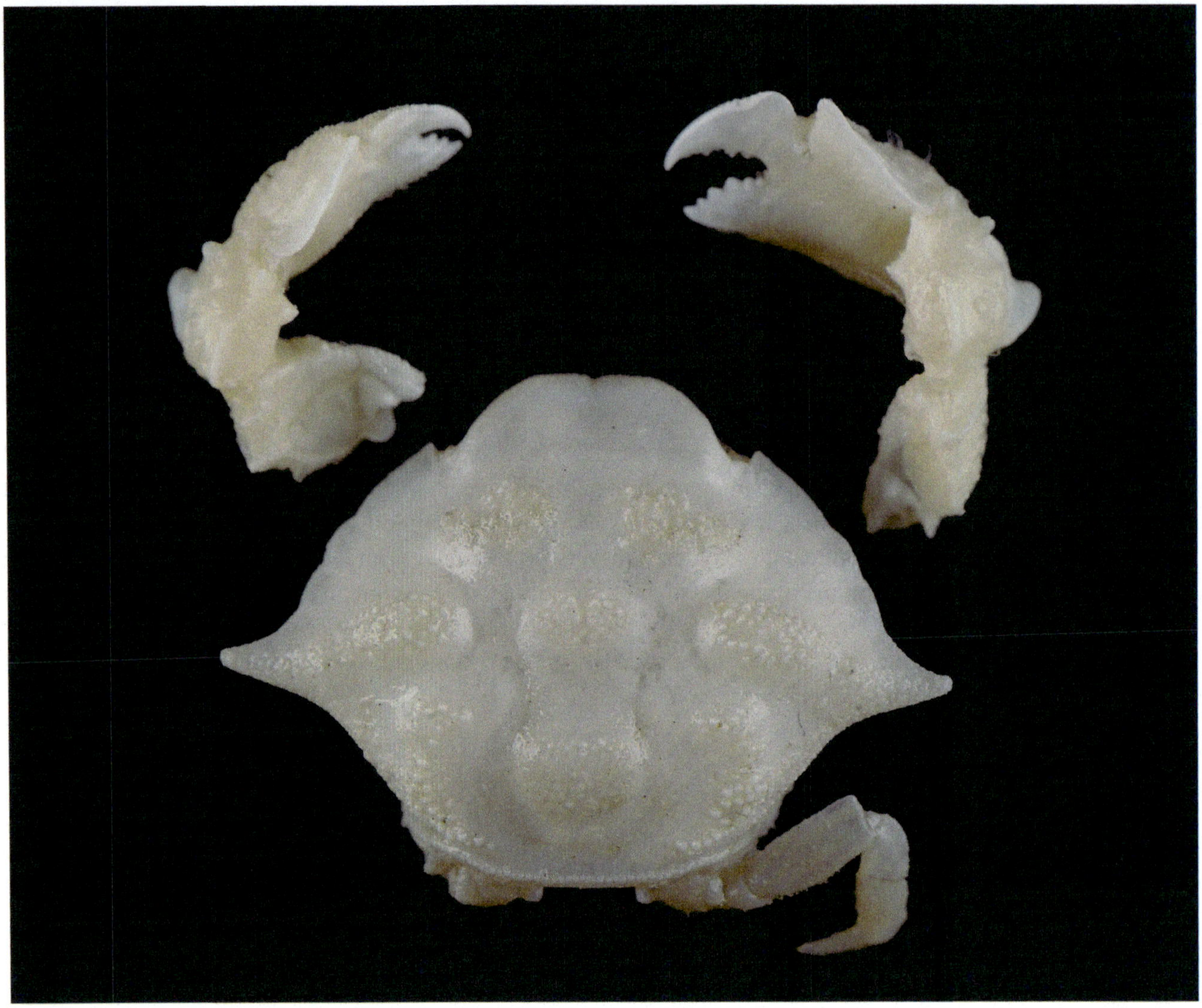

Fig. 25.1 *Dentoxanthus iranicus* Stephensen, 1946. Female holotype (ZMUC CRU975), CL = 7.00, CB = 10.05 mm, Ras-Meydani, Iran, Gulf of Oman, 01.04.1938, coll. B. Loppenthin (Photo: S. Träkner)

Genus *Galene* De Haan, 1833

Galene bispinosa (Herbst, 1783)

(Figs. 22.2f, 25.2, and 25.3)

Cancer bispinosus Herbst 1783: 144, pl 6, fig 45.
Cancer bispinosus — Herbst 1801: 11, pl 54, fig 1.
Galene bispinosa — Chopra 1935: 509, 510, 17a–c; Stephensen 1946: 153, 154; Titgen 1982: 252 (in list); Dai and Yang 1991: 303, 304, pl 39 (2), fig 158 (2); Sakai 1999: 36, pl 19c; Apel 2001: 102; Ng et al. 2008: 137 (in list); Valinasab et al. 2012: 439–441, fig 2.

Type locality: "Ostindien". East India or Indo-Malayan Archipelago.

Distribution

Persian Gulf: Iran (Stephensen 1946).

Gulf of Oman: Iran (Valinasab et al. 2012; Naderloo et al. 2015).

World distribution: Indo-West Pacific: Persian Gulf, Gulf of Oman, India, Bay of Bengal, Vietnam, China, Taiwan, Japan and Australia.

Habitat: Subtidal to 26 m, substrates sandy and muddy.

Genus *Halimede* De Haan, 1835

Halimede tyche (Herbst, 1801)

(Figs. 22.2g, 25.2, and 25.4)

Cancer tyche Herbst 1801: 35, 36, pl 52, fig 3.
Halimede hendersoni Nobili 1905a: 62.
Halimede hendersoni — Nobili 1906a: 123, pl 6, fig 1.
Halimede ochtodes — Stephensen 1946: 158, fig 41b–d; Basson et al. 1977: 243, 250; Ghani and Tirmizi 1992: 41–43, fig 3a–g; Tirmizi and Ghani 1996: 37–40, figs 13, 14.
Halimede tyche — Sakai 1999: 35, pl 18 h; Galil 2000: 329, 330, fig 3a, b; Apel 2001: 98; Ng et al. 2008: 138 (in list); Naderloo and Sari 2007a: 344, tab 1.
Halimede sp. — Evans et al. 1973: 275.

Type locality: *Cancer tyche*: "Ostindien". East India or Indo-Malayan Archipelago.— *Halimede hendersoni*: Persian Gulf (Arzanah Island, Abu Dhabi, UAE).

Distribution

Persian Gulf: Iran [Stephensen (1946) as *H. ochtodes*; Naderloo and Sari 2007a], Kuwait (Apel 2001), Saudi Arabia [Basson et al. (1977) as *H. ochtodes*; Apel 2001], UAE [Nobili (1905a, 1906a) as *H. hendersoni*; Evans et al. (1973) as *Halimede* sp.].

Gulf of Oman: No record.

World distribution: Indo-West Pacific and Mediterranean: Red Sea, Persian Gulf, Pakistan, India, Thailand, Taiwan, Japan, Australia and Mediterranean (Lessepsian migrant).

Habitat: Subtidal, 2–30 m, substrates sandy and muddy.

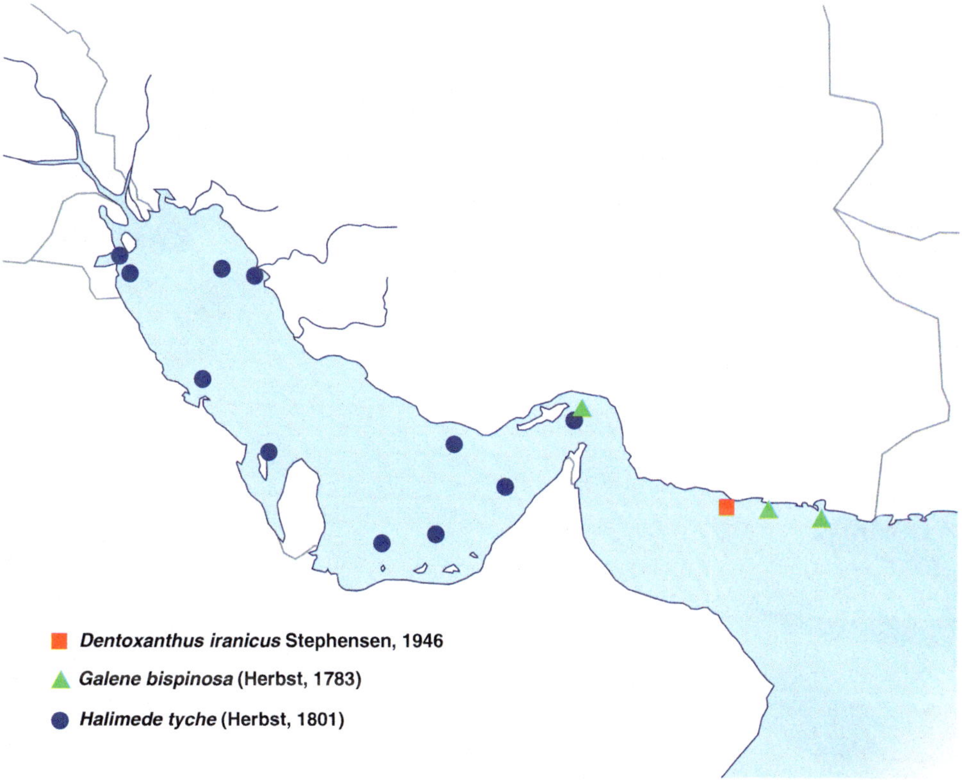

Fig. 25.2 Map showing the recorded localities of *Dentoxanthus iranicus* Stephensen, 1946, *Galene bispinosa* (Herbst, 1783), and *Halimede tyche* (Herbst, 1801)

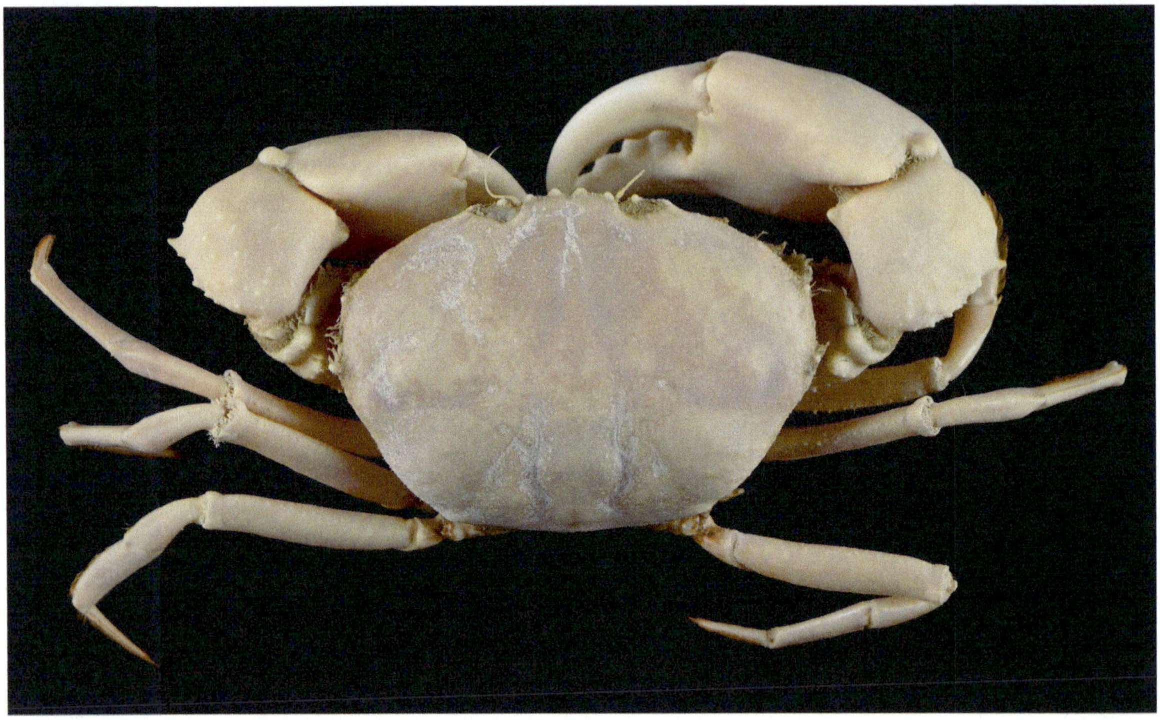

Fig. 25.3 *Galene bispinosa* (Herbst, 1783). Male (ZMUC CRU891), CL = 40.2, CB = 80.8 mm, S Hormuz Island, Iran, Persian Gulf, depth 26 m, 22.04.1937, coll. G. Thorson (Photo: S. Träkner)

Fig. 25.4 *Halimede tyche* (Herbst, 1801). Female (SNMNH 17), CL = 17.9, CB = 23.2 mm, Ras az-Zawr, Saudi Arabia, depth 2–5 m, 17.05.1995, coll. M. Apel (Photo: S. Träkner)

Family Pilumnidae Samouelle, 1819 (Hairy Crabs)

26

Diagnosis Carapace hexagonal, transversely rectangular or ovate; dorsal surface moderately convex, smooth, with granules or spines, usually with long setae, sometimes without setae, regions relatively well defined; anterolateral margins usually with three teeth, spines or lobes, junction between anterolateral and posterolateral margins demarcated with tubercles or spines. Front shorter than posterior margin of carapace, often bilobed with distinct median notch, not or slightly deflexed. Orbits slightly shorter than frontal margin, with or without fissures on supraorbital margin; eyestalks moderately long, not fully covered by orbit. Antennules folded transversely. Antennae with short flagellum. Abdomen of males and females with seven freely moving somites, male abdomen never reaching to anterior margin of cheliped coxa. G1 slender, distinctly S-shaped; G2 very short, stout, comma-shaped. Genital opening of males coxal. Genital opening of females sternal.

Key to genera of the family Pilumnidae in the region

1. Carapace pyriform, distincly longer than broad; front markedly produced .... *Gonatonotus*
- Carapace quadrate, hexagonal or transversly oval, broader than long, or as long as broad; front not markedly produced 2
2. Carapace with anterolateral margins markedly arched; no distinct junction between anterolateral and posterolateral margins; anterolateral margins with lobes, never with spines; upper surface of carapace smooth 3
- Carapace with distinct junction between anterolateral and posterolateral margins; anterolateral margins with lobes, teeth or spines; upper surface of carapace smooth or granular 8
3. Upper surface of carapace and appendages markedly covered with long setae; anterolateral margins not markedly arched .... *Heteropilumnus*
- Margins of carapace and appendages only with long setae; anterolateral margins arched 4
4. Eyes distinctly formed, cornea always pigmented .... *Paraselwynia*
- Eyes indistinct or nearly indistinct; if distinct, first abdominal somite of males and females exremely long, covering abdominal sternum 5
5. Carapace much broader than long; posterolateral margins of carapace running parallel posteriorly; front relatively broad 6
- Carapace slighly broader than long; posterolateral margins of carapace slightly diverging posteriorly; front distinctly narrow 7

R. Naderloo, *Atlas of Crabs of the Persian Gulf*, DOI 10.1007/978-3-319-49374-9_26

6. Male abdomen with first somite extremely broad, filling distance between last walking legs .... *Typhlocarcinops*

- Male abdomen with first somite broader than other somites, but never filling the distance between last walking legs .... *Typhlocarcinus*

7. First antennae with thickly setose flagellum; third maxillipeds with merus produced anterolaterally .... *Mertonia*

- First antennae with naked flagellum; third maxillipeds with merus not produced anterolaterally .... *Xenophthalmodes*

8. Upper surface of carapace completely glabrous, anterolateral margins with small denticles; front broadly bilobed, straight; living in muddy intertidal .... *Eurycarcinus*

- Upper surface of carapace with short or long setae, with distinct spines, teeth or lobes; front in different forms; living in rocky intertidal or subtida 9

9. Upper surface of carapace and appendages markedly covered with long setae; anterolateral margins with spines; distinct lobule between frontal lobe and inner orbital angle; telson extended beyond half of fused sternites 3–4 .... *Pilumnus*

- Upper surface of carapace and appendages without long setae, or with scarce long setae; anterolateral margins without spines or not all teeth spine-shaped; distance between front and inner orbital margin without lobule or with small lobule; telson never reached half of fused sternites 3–4 10

10. Upper surface of carapace with fine velvet form setae; posterolateral margins of carapace markedly concave .... *Actumnus*

- Upper surface of carapace without fine velvet-form setae, scarcely covered with long setae; posterolateral margins of carapace not concave 11

11. Anterolateral margins of carapace with relatively low teeth; large and small chelipeds markedly tuberculate on upper and outer surfaces; transverse suture between sternites 1 and 2 in males indistinct; third and fourth walking legs with distinct teeth along posterior margin of ischium and merus .... *Cryptopilumnus*

- Anterolateral margins of carapace with well grown teeth; at least large chelipeds distinctly smooth without any tooth; distinct transverse suture between sternites 1 and 2 in males; walking legs without teeth on posterior margin of ischium and carpus 12

12. Posterior surface of carapace with granular ridges, serrated by setae, regions moderately defined; front broadly bilobed, with distinct lateral lobules; infraorbital margin with blunt tooth in internal portion, easily visible from dorsal surface of carapace .... *Pilumnopeus*

- Posterior surface of carapace relatively smooth, without setae, regions weakly defined; front broadly bilobed, without lateral lobules; infraorbital margin relatively flat, without blunt tooth visible from dorsal surface of carapace .... *Heteropanope*

Genus *Actumnus* Dana, 1851

Key to species of the genus *Actumnus* in the region

1. Carapace relatively smooth on dorsal surface, covered with dense short setae, regions not markedly defined, front faintly bilobed, not markedly produced 2
- Carapace relatively uneven on dorsal surface, regions well defined, not entirely covered with dense short setae, front markedly bilobed, produced 3
2. Carapace regions not defined; anterolateral margins with broad lobes; posterolateral margins as long as anterolateral margins; upper surface of palm and fingers with small granules .... *Actumnus simplex*

   Carapace regions weekly defined; anterolateral margins with broad denticles; posterolateral margins distinctly longer than anterolateral margins; upper surface of palm and fingers with large granules .... *Actumnus setifer*
3. Carapace with two lines of regular granules parallely running from frontal region to gastric and cardiac regions, making a cross-liked pattern .... *Actumnus margarodes*
- Carapace markedly uneven, regions well defined 4
4. Anterolateral margins with three blunt teeth, first one large and low, last one largest; chelipeds with line of granules on upper margin, small granules scarcely on outer surface .... *Actumnus tesselatus*
- Anterolateral margins with spiniform teeth, supplemented with additional small spines or granules; elevations on carapace additionally granulated 5
5. Anterolateral teeth with additional spines; granules on carapace dentiform; furrows between regions with setae .... *Actumnus obesus*
- Anterolateral teeth with additional granules; granules on carapace small, not dentiform; no setae on furrows of carapace .... *Actumnus asper*

Actumnus asper (Rüppell, 1830)
(Figs. 26.1 and 26.2)

Xantho asper Rüppell 1830: 21, pl 4, fig 8.
Actumnus bonnieri Nobili 1905a: 163.
Actumnus bonnieri — Nobili 1906a: 132, pl 6, fig 32.
Actumnus asper — Stephensen 1946: 141, 142, 206, fig 35c, d; Guinot 1964: 98–100, fig 55a, b, 57, pl 11(2); Guinot 1967a: 272 (in list); Basson et al. 1977: 224, 250, 261; Titgen 1982: 127, 128, 251 (in list); Apel 2001: 95: Ng et al. 2008: 139 (in list).
Type locality: *Actumnus asper*: Red Sea (Jeddah). — *Actumnus bonnieri*: Persian Gulf (Abu Dhabi, UAE).

Distribution

Persian Gulf: Iran (Stephensen 1946), Bahrain (Stephensen 1946), Saudi Arabia (Basson et al. 1977; Apel 2001), UAE [Nobili (1905a) as *Actumnus bonnieri*; Titgen 1982].

Gulf of Oman: No record.

World distribution: Indo-West Pacific: Seychelles, Red Sea, Gulf of Aden, Persian Gulf, India (Gulf of Mannar), Sri Lanka, Andaman Sea, Indo Malayan Archipelago, South China Sea and Tuamotu.

Habitat: Subtidal, 20–30 m, substrates shell, gravel and oyster bank.

Fig. 26.1 *Actumnus asper* (Rüppell, 1830). Male (ZMUC CRU914), CL = 9. 00, CB = 10.25 mm, Bahrain, Persian Gulf, 16.03.1938, coll. B. Loppenthin (Photo: S. Träkner)

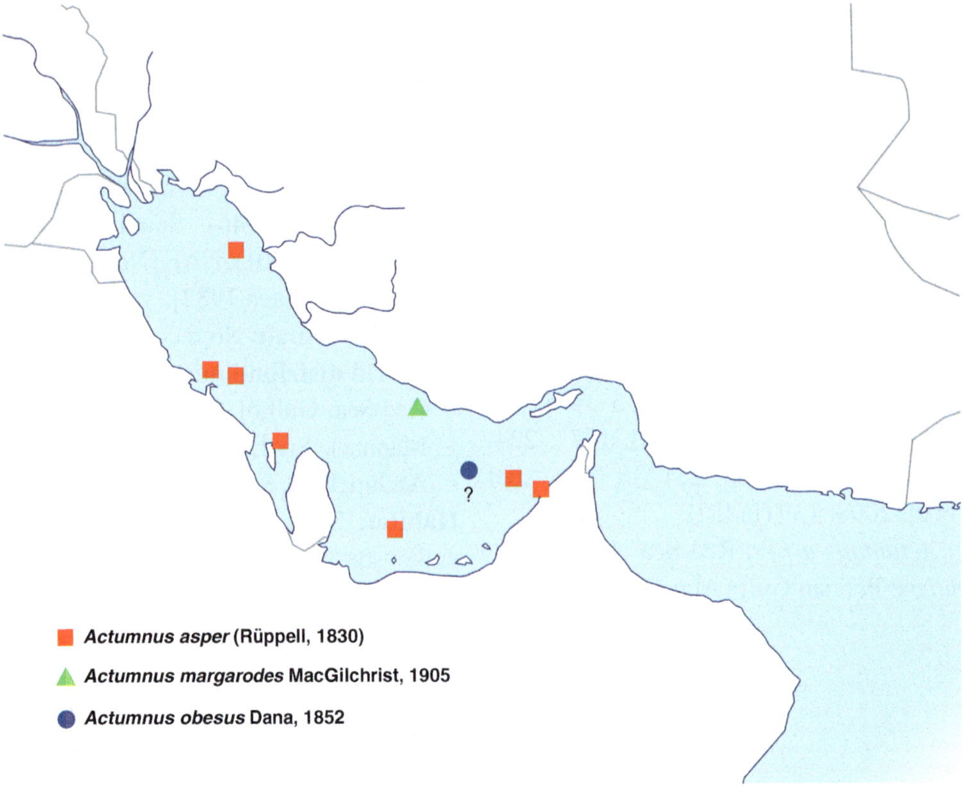

Fig. 26.2 Map showing the recorded localities of *Actumnus asper* (Rüppell, 1830), *Actumnus margarodes* MacGilchrist, 1905, and *Actumnus obesus* Dana, 1852

Actumnus margarodes MacGilchrist, 1905
(Figs. 26.2 and 26.3)

Actumnus margarodes MacGilchrist 1905: 260.
Actumnus margarodes — Alcock and MacGilchrist 1905: pl 76, fig 3; Apel 2001: 95; Ng et al. 2008: 139 (in list).
Type locality: Persian Gulf (Kish Island, Iran, 26° 20′N, 53° 54′E).
Distribution
Persian Gulf: Iran (MacGilchrist 1905).
Gulf of Oman: No record.
World distribution: This species is still only recorded from the Persian Gulf.
Habitat: Unknown.

Actumnus obesus Dana, 1852
(Figs. 26.2 and 26.4)

Actumnus obesus Dana 1852a: 82.
Actumnus obesus — Dana 1852b: 244, pl 14, fig 3a, b; Balss 1933: 37; Stephensen 1946: 142, 143; Guinot 1967a: 272 (in list); Takeda and Miyake 1969: 107–109; Titgen 1982: 251 (in list); Apel 2001: 96; Ng et al. 2008: 139 (in list).
Actumnus setifer — Alcock 1898: 202 (partly).
Type locality: Hawaii.
Distribution
Persian Gulf: Alcock (1898) as *A. setifer* (De Haan, 1835) and Balss (1933) recorded this species from unknown localities.

Fig. 26.3 *Actumnus margarodes* MacGilchrist, 1905. Male holotype (Calcutta Museum), CL = 6. 2, CB = 5.1 mm, Kish Island, Iran, Persian Gulf, 26° 20′N, 53° 54′E. Appendages are detatched

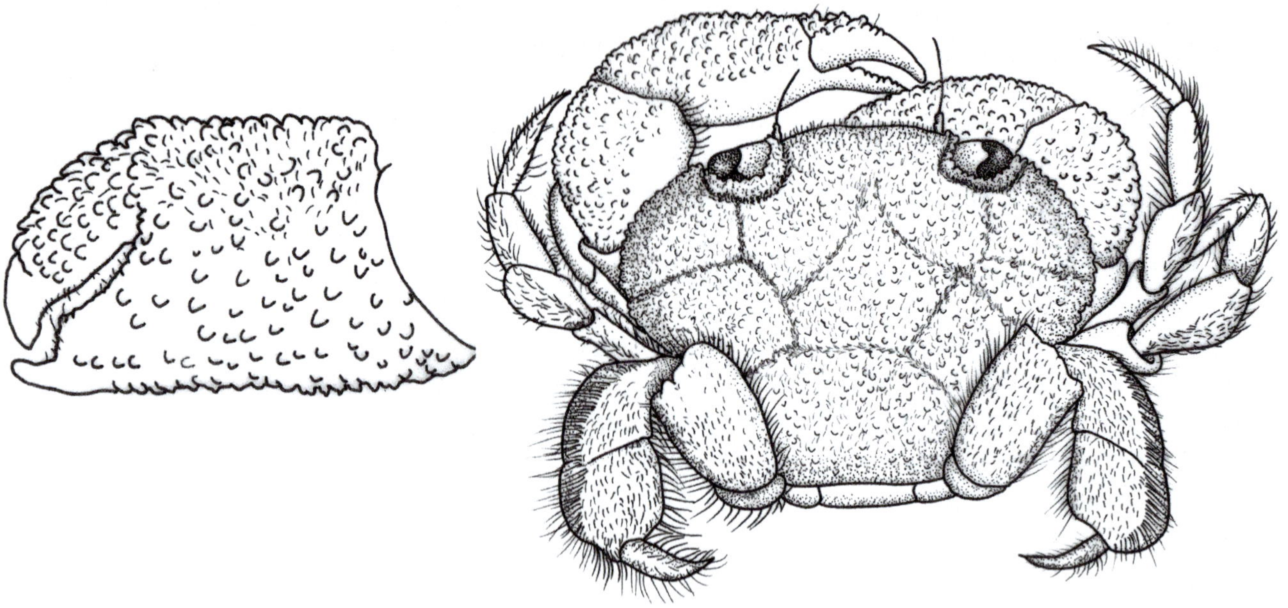

Fig. 26.4 *Actumnus obesus* Dana, 1852. Drawings were provided using original illustratiobs of Dana (1852b: pl 14, figs 3, 3a)

Gulf of Oman: No record.

World distribution: Indo-West Pacific: Madagascar, Amirante, Red Sea, Persian Gulf, Andaman Sea, Australia, New Guinea, Samoa, Fiji and Hawaii.

Habitat: Unknown.

Remarks Balss (1933) mentioned that some specimens of material identified by Alcock (1898) as *A. setifer* (De Haan, 1835) are clearly assigned to *A. obesus*. It is not clear if all specimens of Alcock's (1898) material are *A. obesus* or not.

Actumnus setifer (De Haan, 1835)

(Figs. 26.5 and 26.6)

Cancer (*Pilumnus*) *setifer* De Haan 1835: 50.
Actumnus tomentosus Dana 1852b: 243, 244.
Cancer (*Xantho*) *setifer* — De Haan 1835: pl 3, fig 3.
Actumnus tomentosus — Alcock 1898: 202; Nobili 1906a: 132.
Actumnus setifer — Alcock 1898: 202 (partly); Stephensen 1946: 143; Takeda and Miyake 1969: 115–118, fig 9d–f; Titgen 1982: 251 (in list); Hogarth 1989: 106; Hogarth 1994: 100; Apel 2001: 96; Ng et al. 2008: 139 (in list).
Actumnus setifer setifer — Guinot 1967a: 272 (in list).

Type locality: *Cancer* (*Pilumnus*) *setifer*: Japan. — *Actumnus tomentosus*: Tahiti or Upolu.

Distribution

Persian Gulf: Iran (Stephensen 1946), Bahrain (Stephensen 1946), UAE [Nobili (1906a) as *A. tomentosus*]. Alcock (1898) recorded this species from an unknown locality.

Gulf of Oman: No record.

World distribution: Indo-West Pacific: East Africa, Mauritius, Seychelles, Red Sea, southern Oman, Persian Gulf, India, Sri Lanka, Bay of Bengal, Mergui Archipelago, Andaman Islands, Indo-Malaysian Archipelago, China, Japan, Australia, New Caledonia and Tahiti.

Habitat: Subtidal, 20–49 m, substrates sandy, shell and gravel.

Remarks Rathbun (1911) described a species of the genus, namely *Actumnus amirantensis* Rathbun, 1911, from the western Indian Ocean and discussed the difference between his new taxa with *Actumnus setifer setifer*. The difference provided by Rathbun (1911) are more likely valid (Ng, pers comm), therefore it is possible that all old records of the species be assigned to *A. amirantensis* (see Naderloo and Ng 2011). The genus needs to be revised in order to put the species in their proper taxonomic position.

Fig. 26.5 *Actumnus setifer* (De Haan, 1835). Female (ZMUC CRU913), CL = 10.2, CB = 10.65 mm, Bushehr, Iran, Persian Gulf, depth 49 m, 14.03.1937, coll. G. Thorson (Photo: S. Träkner)

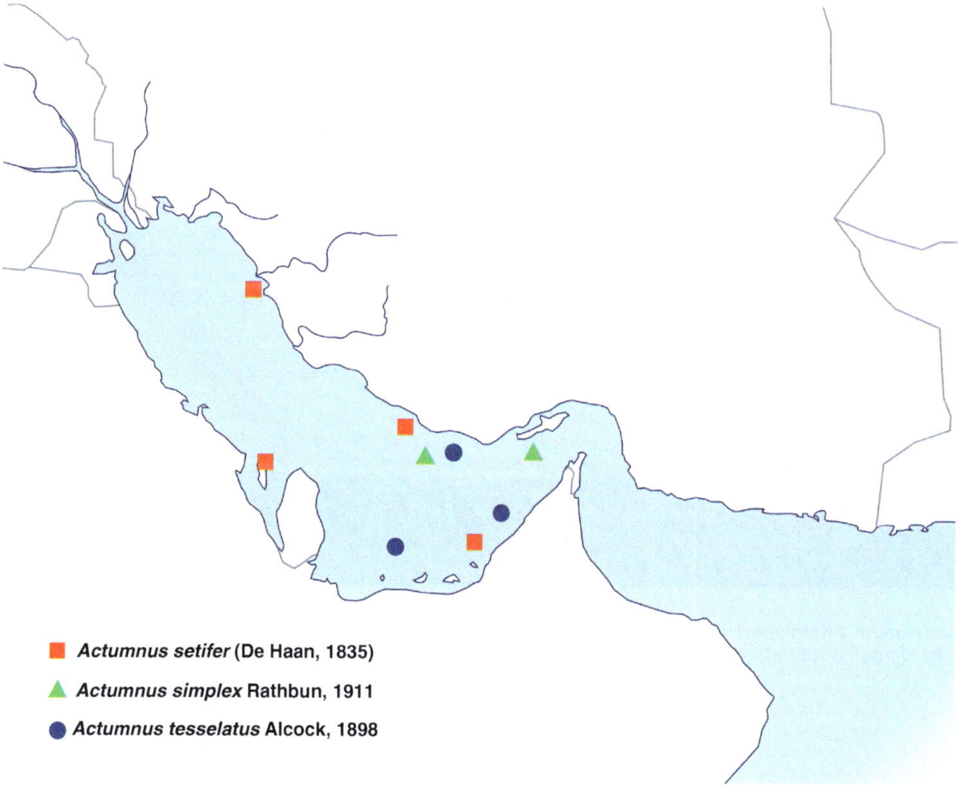

Fig. 26.6 Map showing the recorded localities of *Actumnus setifer* (De Haan, 1835), *Actumnus simplex* Rathbun, 1911, and *Actumnus tesselatus* Alcock, 1898

Actumnus simplex Rathbun, 1911
(Figs. 26.6 and 26.7)

Actumnus simplex Rathbun 1911: 232, 233, pl 16, figs 10, 11.
Actumnus simplex — Ng et al. 2008: 180 (in list); Naderloo and Ng 2011: 1596, figs 1–3.
Type locality: Amirante Island.
Distribution
Persian Gulf: Iran (Naderloo and Ng 2011).
Gulf of Oman: No record.
World distribution: Western Indian Ocean: Amirante Island and Persian Gulf.
Habitat: Subtidal, 40–45 m, substrates sandy and muddy.

Actumnus tesselatus Alcock, 1898
(Figs. 26.6 and 26.8)

Actumnus tesselatus Alcock 1898: 205.
Actumnus tesselatus — Alcock and Anderson 1899: pl 37, fig 9; Nobili 1906a: 132; Stephensen 1946: 142; Titgen 1982: 251 (in list); Apel 2001: 96; Ng et al. 2008: 140 (in list).
Type locality: Persian Gulf.
Distribution
Persian Gulf: Iran (Stephensen 1946), UAE (Nobili 1906a). Alcock (1898) described this species from the Persian Gulf without giving an exact locality.
Gulf of Oman: No record.
World distribution: This species is still only recorded from the Persian Gulf.

Fig. 26.7 *Actumnus simplex* Rathbun, 1911. Female (ZUTC Brach1284), CL = 8.90, CB = 11.03 mm, Iran, Persian Gulf, depth 40–50 m, 09.02.2004, coll. R. Naderloo (Photo: S. Träkner)

Fig. 26.8 *Actumnus tesselatus* Alcock, 1898. Female (ZMUC CRU911), CL = 1.65, CB = 2.25 mm, Kish Island, Iran, Persian Gulf, 08.04.1937, coll. G. Thorson (Photo: S. Träkner)

Habitat: Stephensen (1946) recorded the present species from a sandy beach in Kish Island, while Nobili's (1906a) material came from Arzanah Island (UAE) from an oyster bank at depth of 20–30 m.

Genus *Cryptopilumnus* Hsueh, Huang and Ng, 2009

Cryptopilumnus pereiodontus (Davie and Ghani, 1993)

(Figs. 26.9, 26.10a and 26.11)

Pilumnopeus pereiodontus Davie and Ghani 1993: 61–65, figs 1, 2.

Pilumnopeus pereiodontus — Ng et al. 2008: 141 (in list).

Cryptopilumnus pereiodontus — Hsueh et al. 2009: 328, tab 1; Naderloo et al. 2016: 3, fig 5.

Type locality: Pakistan (Charna Island).

Distribution

Persian Gulf: Iran (Naderloo et al. 2016).

Gulf of Oman: Iran (Naderloo et al. 2015).

World distribution: Northern Indian Ocean: Persian Gulf, Gulf of Oman and Pakistan.

Habitat: Intertidal to shallow subtidal, substrate rocky/cobble.

Fig. 26.9 *Cryptopilumnus pereiodontus* (Davie and Ghani, 1993). Female (ZUTC 5480), CL = 3.76, CB = 4.90 mm, Park-Qadir, Abu Musa Island, Iran, Persian Gulf, 31.12.2014, coll. S. Ebrahimnejad and R. Abdollahi (Photo: S. Shahbazi)

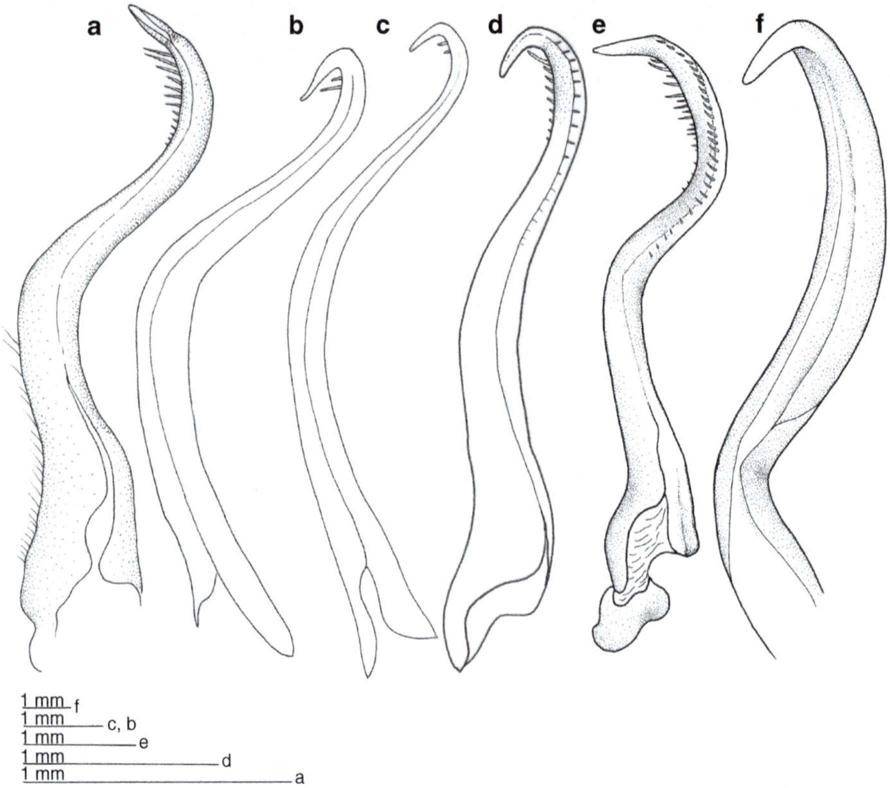

Fig. 26.10 Male G1, ventral view: (**a**) *Cryptopilumnus pereiodontus* (Davie and Ghani, 1993) (ZUTC 5535); (**b**) *Eurycarcinus integrifrons* De Man, 1879 (ZUTC Brach1174); (**c**) *Eurycarcinus orientalis* A. Milne-Edwards, 1867 (ZUTC Brach1173); (**d**) *Heteropanope glabra* Stimpson, 1858 (ZUTC: Qeshm, Tula); (**e**) *Heteropilumnus trichophoroides* De Man, 1895 (ZUTC 1298); (**f**) *Mertonia lanka* Laurie, 1906 (ZMUC CRU862)

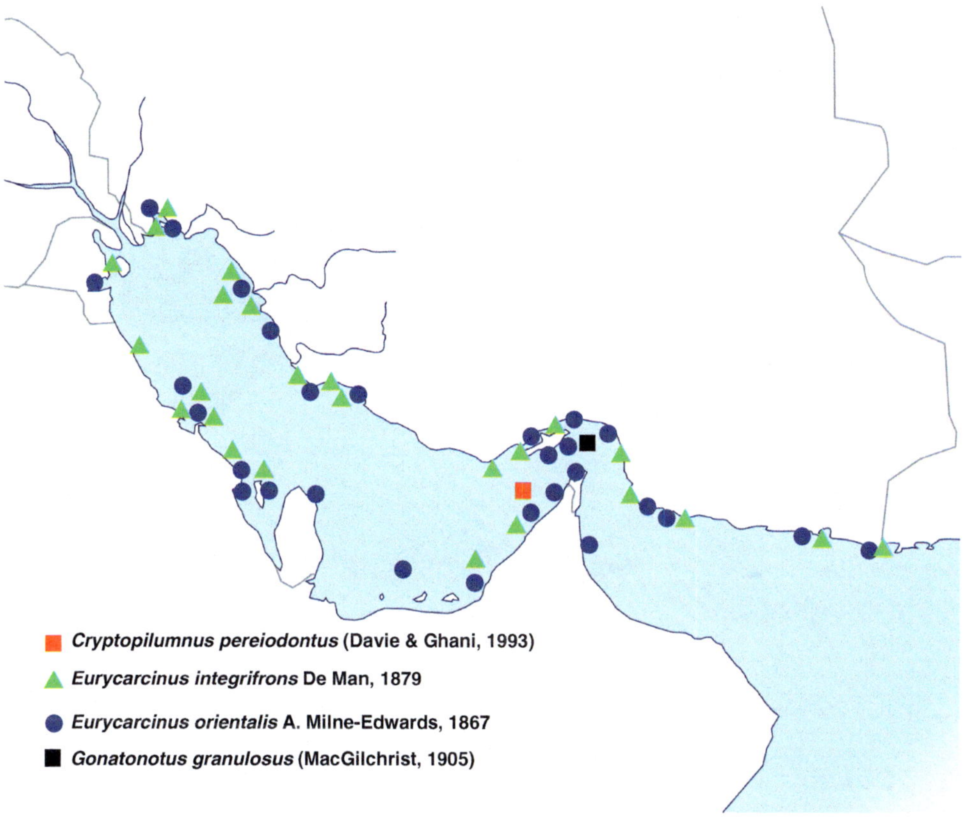

Fig. 26.11 Map showing the recorded localities of *Cryptopilumnus pereiodontus* (Davie and Ghani, 1993), *Eurycarcinus integrifrons* De Man, 1879, *Eurycarcinus orientalis* A. Milne-Edwards, 1867, and *Gonatonotus granulosus* (MacGilchrist, 1905)

Genus *Eurycarcinus* A. Milne-Edwards, 1867

Key to species of the genus *Eurycarcinus* in the region

- Carapace distinctly oval, markedly broader than long (CB/CL = 1.5–1.6), posterolateral margins sharply converging posteriroly .... *Eurycarcinus orientalis*
- Carapace nearly subquadrate, slightly broader than long (CB/CL = 1.2–1.3), posterolateral margins gently converging posteriroly .... *Eurycarcinus integrifrons*

Eurycarcinus integrifrons De Man, 1879

(Figs. 26.10b, 26.11 and 26.12)

Eurycarcinus integrifrons De Man 1879: 55, 56.
Eurycarcinus integrifrons — Apel 1994b: 415, 433, 434; Apel 1996b: 331; Al-Ghais and Cooper 1996: 425, 426; Apel 2001: 97; Naderloo and Sari 2007a: 344, tab 1; Naderloo and Türkay 2012: 36; Naderloo et al. 2013: 449, tab 1, 456, tab 2.
Eurycarcinus orientalis — Alcock 1898: 210, 211 (partly); Chhapgar 1957a: 436, 437, pl 11d–f; Tirmizi et al. 1986: 8–10, fig 3a–d; Tirmizi and Ghani 1996: 30–32, fig 10; Hornby 1997: 16 (partly).
Litocheira (*amoyensis* Gordon?) — Stephensen 1946: 169–171, fig 46; Titgen 1982: 252 (in list).
Eurycarcinus sp. (? *integrifrons*) — Apel and Türkay 1992: 194, 195, 204, 205.
Eurycarcinus integifrons — Ng et al. 2008: 140 (in list) [typing error].

Fig. 26.12 *Eurycarcinus integrifrons* De Man, 1879. Coast of Desalination Plant, Qeshm Island, Iran, Persian Gulf, 05.01.2008 (Photo: A. Sari)

Type locality: Unknown (possibly India).

Distribution

Persian Gulf: Iran [Stephensen (1946) as *Litocheira* (*amoyensis* Gordon?); Apel 2001; Naderloo and Sari 2007a; Naderloo and Türkay 2012; Naderloo et al. 2013], Saudi Arabia (Apel 1994b, 1996b), UAE (Al-Ghais and Cooper 1996).

Gulf of Oman: Iran (Naderloo et al. 2015).

World distribution: Northwestern Indian Ocean: Gulf of Aden, Persian Gulf, Gulf of Oman, Pakistan and India.

Habitat: Intertidal to 32 m, substrates muddy and mangroves.

Eurycarcinus orientalis A. Milne-Edwards, 1867

(Figs. 26.10c, 26.11 and 26.13)

Eurycarcinus orientalis A. Milne-Edwards 1867: 277.
Eurycarcinus orientalis — Jones 1986a: 162, pl 47; Vousden 1987: 36, tabs 4, 7; Apel and Türkay 1992: 194, 204, 205; Ismail and Ahmed 1993: 158; Apel 1994a: 43, 44; Apel 1994b: 411, 412, 433, 434; Hywel-Davies 1994: 37, 48; Apel 1996b: 330, 331; Al-Khayat and Jones 1996: 806, fig 5; Tirmizi and Ghani 1996: 30–32, fig 10; Cooper 1997: 168–170, figs 4, 5, 15; Hornby 1997: 16 (partly); Al-Khayat and Jones 1999: 58, 61; Apel 2001: 97, 98; Ng et al. 2008: 140 (in list).
Eurycarcinus grandidieri — Alcock 1898: 211, 212.
Eurycarcinus sp. — Basson et al. 1977: 58, 228 (in list), fig 38; Titgen 1982: 131.

Fig. 26.13 *Eurycarcinus orientalis* A. Milne-Edwards, 1867. The species in two different colour, Tiab, Iran, Gulf of Oman, 26.12.2012 (Photo: R. Naderloo)

not *Eurycarcinus orientalis* — Alcock 1898: 210, 211; Chhapgar 1957a: 436, 437, pl 11, d–f; Tirmizi et al. 1986: 8–10, fig 3a–d; Tirmizi and Ghani 1996: 30–32, fig 10 [= *Eurycarcinus integrifrons* De Man, 1879].

Type locality: India (Mumbai).

Distribution

Persian Gulf: Iran (Naderloo and Sari 2007a; Naderloo and Türkay 2012: Naderloo et al. 2013), Kuwait (Jones 1986a), Saudi Arabia (Basson et al. 1977; Apel 1994a, 1994b), Qatar (Al-Khayat and Jones 1999; Apel 2001), Bahrain (Vausden 1987; Apel 2001), UAE (Titgen 1982; Cooper 1997; Hornby 1997; Apel 2001).

Gulf of Oman: Iran (Naderloo et al. 2015).

World distribution: Indian Ocean: Gulf of Aden, Persian Gulf, Gulf of Oman, Pakistan, India and Thailand.

Habitat: Intertidal, substrates muddy and mangroves.

Genus *Gonatonotus* White, 1847

Gonatonotus granulosus (MacGilchrist, 1905)

(Figs. 26.11, and 26.14)

Eumedonus granulosus MacGilchrist 1905: 253, 254.
Eumedonus granulosus — Alcock and Annandale 1907: pl 77, fig 2; Stephensen 1946: 114; Guinot 1967a: 299 (in list); Titgen 1982: 249 (in list); Apel 2001: 68.
Gonatonotus granulosus — Ng et al. 2008: 139 (in list).

Type locality: Strait of Hormuz ("Investigator" Stations 291 and 296).

Distribution

Persian Gulf: MacGilchrist (1905) described this species from the Strait of Hormuz.

Gulf of Oman: No record.

World distribution: Indo-West Pacific: South Africa, Mozambique, Zanzibar, Mauritius, Red Sea, Persian Gulf, Amirante Island and Japan.

Habitat: Subtidal, substrates muddy and sandy.

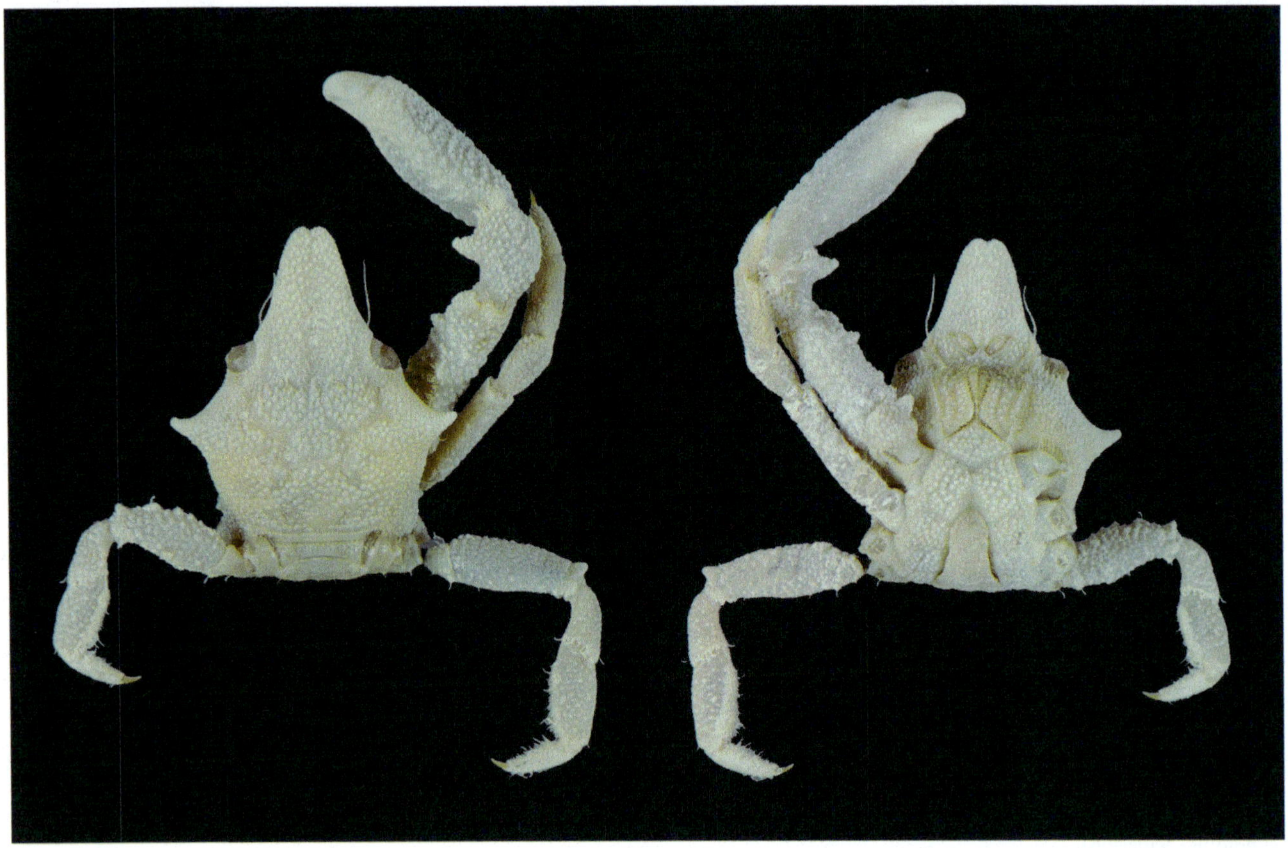

Fig. 26.14 *Gonatonotus granulosus* (MacGilchrist, 1905). Male holotype (Calcutta Museum), Strait of Hormuz, "Investigator" , stations 291 and 296

Genus *Heteropanope* Stimpson, 1858

Heteropanope glabra Stimpson, 1858

(Figs. 26.10d, 26.15 and 26.16)

Heteropanope glabra Stimpson 1858a: 35.
Heteropanope glabra Stimpson — 1907: 63, pl 8, fig 1.
Eurycarcinus maculatus Alcock 1898: 212.
Heteropanope glabra — Davie 1989: 130–134, figs 1a–j, 2; Ng et al. 2008: 140 (in list); Naderloo and Türkay 2012: 37; Naderloo et al. 2013: 449, tab 1.

Type locality: Hong Kong.

Distribution

Persian Gulf: Iran (Naderloo and Türkay 2012: Naderloo et al. 2013).

Gulf of Oman: Iran (Naderloo et al. 2015).

World distribution: Indo-West Pacific: East Africa, Persian Gulf, Gulf of Oman, Pakistan, Andaman Sea, Thailand (Phuket), Hong Kong, Taiwan, Mergui Archipelago, Singapore, Australia and New Caledonia.

Habitat: Intertidal, substrates muddy, mangroves and rocky/cobble.

Genus *Heteropilumnus* De Man, 1895

Heteropilumnus trichophoroides De Man, 1895

(Figs. 26.10e, 26.16 and 26.17)

Heteropilumnus trichophoroides De Man 1895: 549–552, pl 13, fig 8.

Fig. 26.15 *Heteropanope glabra* Stimpson, 1858. Male (SMF 38716), CL = 10.71, CB = 7.04 mm, Zeytun Park Beach, Qeshm Island, Iran, Persian Gulf, 08.01.2008, coll. M. Türkay and R. Naderloo (Photo: S. Träkner)

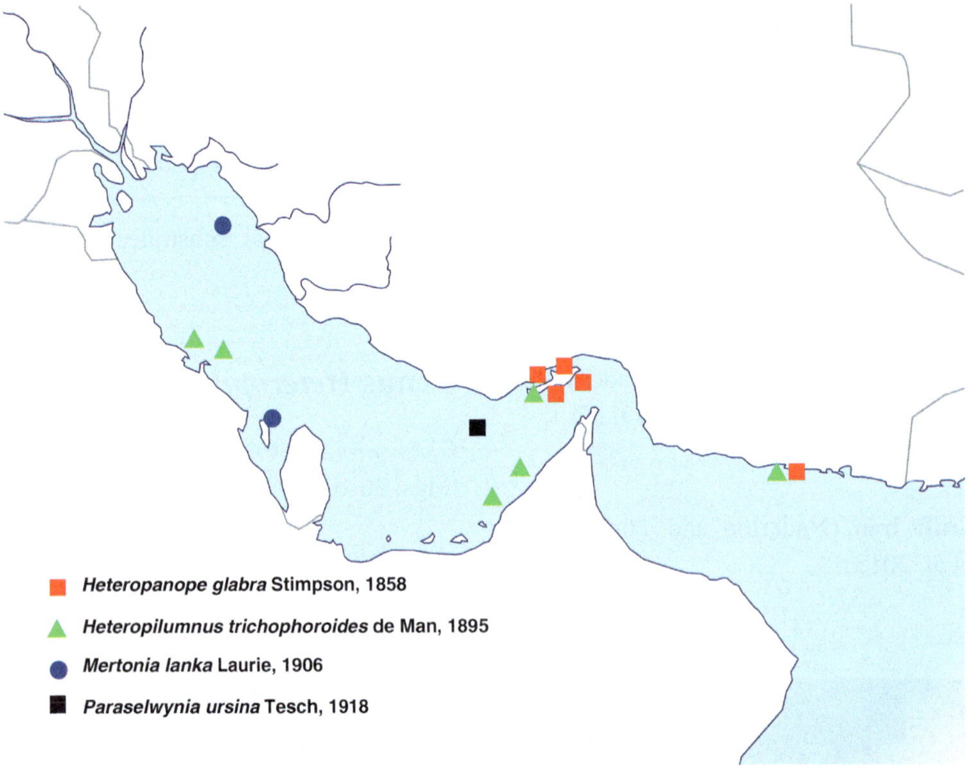

Fig. 26.16 Map showing the recorded localities of *Heteropanope glabra* Stimpson, 1858, *Heteropilumnus trichophoroides* De Man, 1895, *Mertonia lanka* Laurie, 1906, and *Paraselwynia ursina* Tesch, 1918

Fig. 26.17 *Heteropilumnus trichophoroides* De Man, 1895. Female (ZUTC Brach1298), CL = 20.0, CB = 15.8 mm, S Qeshm Island, Iran, Persian Gulf, 11.12.2008, coll. M. Türkay and R. Naderloo (Photo: S. Träkner)

Pilumnus trichophoroides — Nobili 1906a: 134.
Heteropilumnus trichophoroides — Stephensen 1946: 143, 144; Basson et al. 1977: 261, 265; Tirmizi and Ghani 1996: 44–46, figs 16, 17; Apel 2001: 98; Ng et al. 2008: 140 (in list); Naderloo and Türkay 2012: 37: Naderloo et al. 2013: 449, tab 1.
Type locality: Indonesia (Sulawesi, Celebes).
Distribution

Persian Gulf: Iran (Naderloo and Türkay 2012; Naderloo et al. 2013), Saudi Arabia (Basson et al. 1977), UAE (Nobili 1906a).

Gulf of Oman: Iran (Naderloo et al. 2015).

World distribution: Indo-West Pacific: Seychelles, Red Sea, Persian Gulf, Gulf of Oman, Pakistan, Chagos Archipelago, Indian Archipelago, China, Japan, Indonesia and Australia.

Habitat: Intertidal, substrate rocky/cobble.

Genus *Mertonia* Laurie, 1906

Mertonia lanka Laurie, 1906

(Figs. 26.10f, 26.16 and 26.18)

Mertonia lanka Laurie 1906: 424, pl 1, fig 1.
Mertonia lanka — Stephensen 1946: 180, fig 51; Serène 1964: 234, fig 13, pl 21b; Titgen 1982: 252 (in list); Apel 2001: 103; Ng et al. 2008: 143 (in list).
Type locality: India (Gulf of Mannar) and Sri Lanka.
Distribution

Persian Gulf: Iran (Stephensen 1946), Bahrain (Stephensen 1946).

Gulf of Oman: No record.

World distribution: Indo-West Pacific: Persian Gulf, India, Sri Lanka, Gulf of Thailand, Indonesia and Japan.

Habitat: Subtidal to 25 m, substrates muddy, shells and gravel.

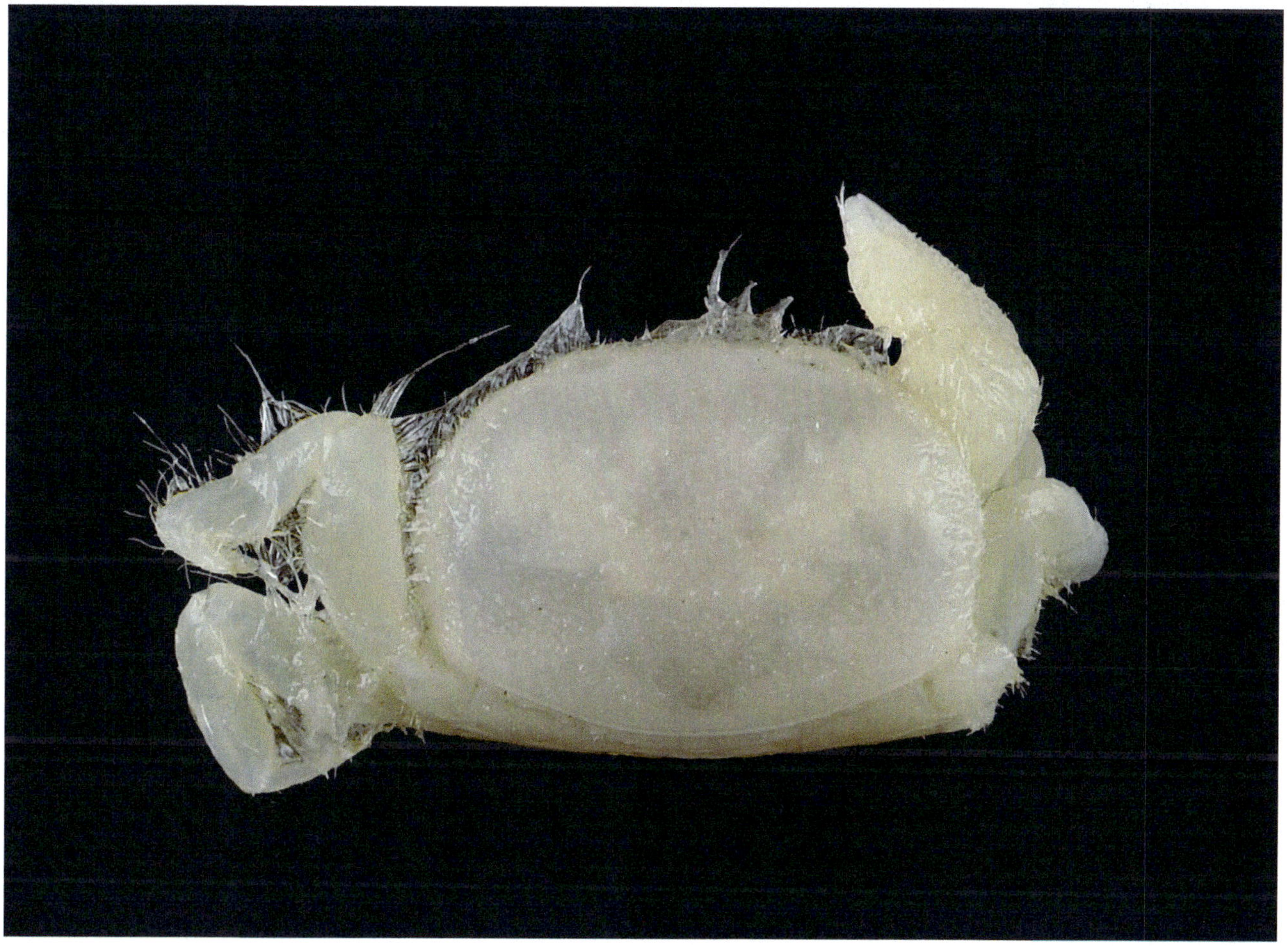

Fig. 26.18 *Mertonia lanka* Laurie, 1906. Male (ZMUC CRU862), CL = 3.3, CB = 4.8 mm, Bahrain, Persian Gulf, depth 23 m, 27.03.1937, coll. G. Thorson (Photo: S. Träkner)

Fig. 26.19 *Paraselwynia ursina* Tesch, 1918. Female (ZMUC CRU964), CL = 4.44, CB = 6.45 mm, Farur Island, Iran, Persian Gulf, depth 6–31 m, 22.03.1938, coll. B. Loppenthin (Photo by S. Träkner)

Genus *Paraselwynia* Tesch, 1918

Paraselwynia ursina Tesch, 1918

(Figs. 26.16, 26.19 and 26.20)

Paraselwynia ursina Tesch 1918: 222, pl 14, fig 2.
Paraselwynia ursina — Stephensen 1946: 173; Titgen 1982: 252 (in list); Apel 2001: 103; Ng et al. 2008: 143 (in list).

Type locality: Indonesia (Kei Island).

Distribution

Persian Gulf: Iran (Stephensen 1946), Bahrain (Stephensen 1946).

Gulf of Oman: No record.

World distribution: Indo-West Pacific: Persian Gulf and Indonesia.

Habitat: Unknown.

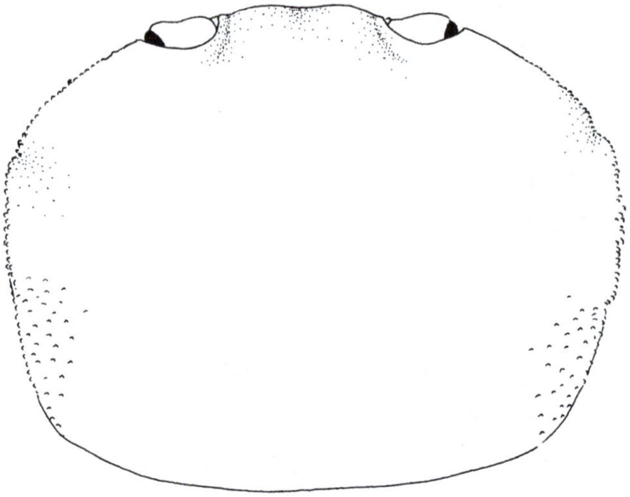

Fig. 26.20 *Paraselwynia ursina* Tesch, 1918. Drawing of carapace of the species

Genus *Pilumnopeus* A. Milne-Edwards, 1867

Pilumnopeus convexus (Maccagno, 1936)

(Figs. 26.21, 26.22a and 26.23)

Heteropanope convexa Maccagno 1936: 176, 177.
Pilumnopeus vauquelini — Stephensen 1946: 141, fig 35a, b; Guinot 1967a: 275 (in list; partly); Basson et al. 1977: 228, 231, 243; Titgen 1982: 252 (in list); Hornby 1997: 15; Naderloo and Türkay 2012: 37.
Pilumnopeus convexa — Davie 1989: 142, 143.
Pilumnopeus convexus — Cooper 1997: 171–173, figs 6, 16; Apel 2001: 98; Ng et al. 2008: 141 (in list); Naderloo and Türkay 2012: 37: Naderloo et al. 2013: 449, tab 1.
Type locality: Red Sea (Eritrea, Asab).

Distribution

Persian Gulf: Iran (Naderloo and Türkay 2012: Naderloo et al. 2013), Saudi Arabia (Apel 2001), Bahrain (Apel 2001), UAE (Cooper 1997; Apel 2001).

Gulf of Oman: Iran (Naderloo et al. 2015).

World distribution: Western Indian Ocean: South Africa, Red Sea, Gulf of Aden, Persian Gulf and Gulf of Oman.

Habitat: Intertidal, substrates rocky/cobble, oyster bank, muddy with shells and mangroves.

Remarks: A second species of the genus, namely *Pilumnopeus vaquelini* (Audouin, 1826) has been frequently recorded from the region (Stephensen 1946; Guinot 1967a; Basson et al. 1977; Titgen 1982; Hornby 1997; Naderloo and Türkay 2012). Apel (2001) exam-

Fig. 26.21 *Pilumnopeus convexus* (Maccagno, 1936). Male (SMF 38505), CL = 12.2, CB = 18.1 mm, Zeytun Park Beach, Qeshm Island, Iran, Persian Gulf, 28.03.1937, coll. M. Türkay and R. Naderloo (Photo: S. Träkner)

ined material of Stephensen (1946) and mentioned that all his records of *P. vauquelini* is misidentification of *P. convexus*. He cited four main differences between these two species, of which three are dealing with carapace morphology. But, the most reliable character for separating these two species is G1 which has the same morphology in all here examined specimens. I examined huge number of specimens of *Pilumnopeus* in the Persian Gulf and the Gulf of Oman, but at last could not convincingly separate *P. convexus* from *P. vaquelini*. Therefore, I hesitantly agree with Apel (2001) that all records of *P. vauquelini* from the region can be recognized as *P. convexus*. There is definitely a necessary need for taxonomic revision of *Pilumnopeus* together with its closed genera, *Heteropanope* and *Benthopanope* in the region. Ghory et al. (2013) studied the taxonomic problem of *P. convexus* and briefly discussed on the differences between *P. convexus* and *P. vaquelini*, but still the latter species needs to be re-examined and redescribed in the region (Naderloo and Türkay 2012).

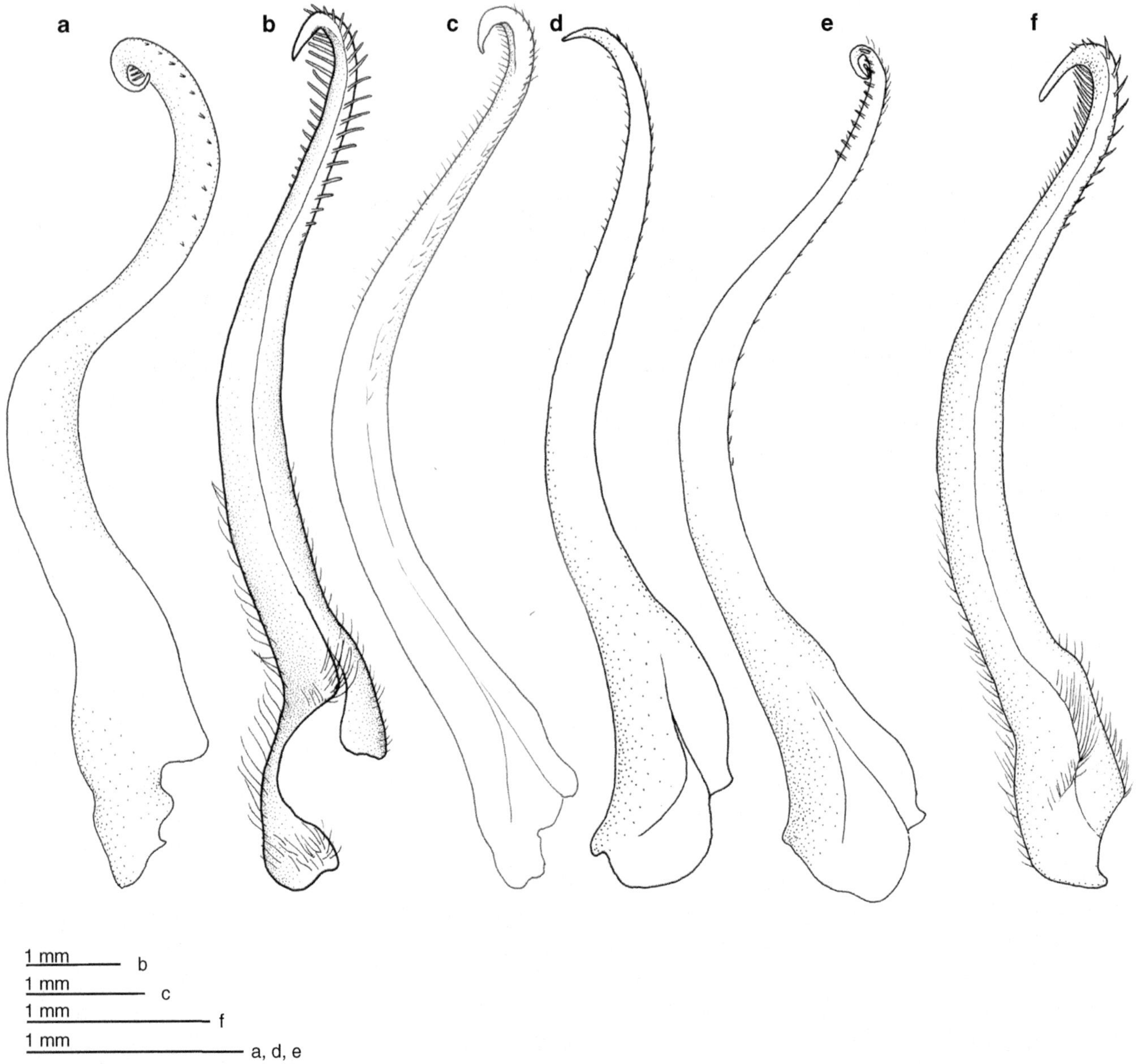

Fig. 26.22 Male G1, ventral view: (**a**) *Pilumnopeus convexus* (Maccagno, 1936) (ZUTC Brach1125); (**b**) *Pilumnus incanus* (Forskål, 1775) (ZUTC 1351); (**c**) *Pilumnus longicornis* Hilgendorf, 1878 (ZUTC Brach1111); (**d**) *Pilumnus minutus* De Haan, 1835 (ZUTC Brach1076); (**e**) *Pilumnus savignyi* Heller, 1861 (ZUTC Brach1078); (**f**) *Pilumnus vespertilio* (Fabricius, 1793) (ZUTC Brach1778)

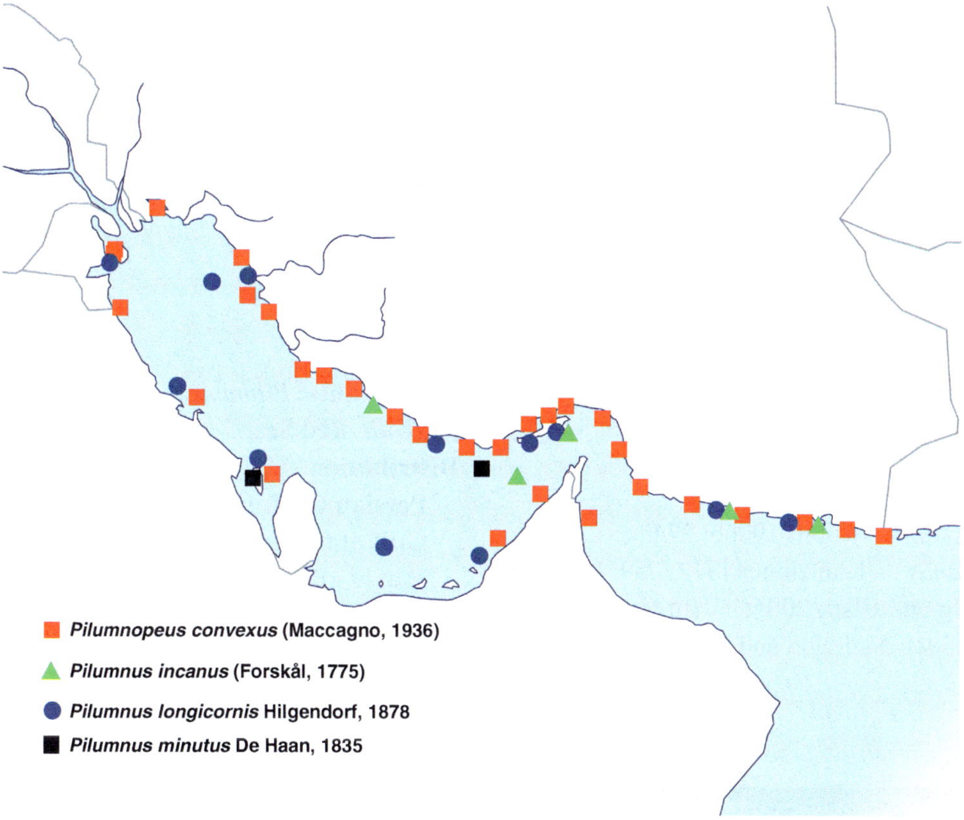

Fig. 26.23 Map showing the recorded localities of *Pilumnopeus convexus* (Maccagno, 1936), *Pilumnus incanus* (Forskål, 1775), *Pilumnus longicornis* Hilgendorf, 1878, and *Pilumnus minutus* De Haan, 1835

Genus *Pilumnus* Leach, 1816

Key to species of the genus *Pilumnus* in the region

1. Carapace and appendages thickly covered with long setae, lateral margins not visible .... *Pilumnus vespertilio*

- Carapace and appendages may with long setae, but never thick, lateral margins visible 2

2. Dorsal surface of carapace with distinct granules .... *Pilumnus incanus*

- Dorsal surface of carapace without granules, may with small pits 3

3. Exorbital angle with small tooth, three to five small teeth between exorbital angle and second anterolateral tooth, second tooth of anterolateral margin with three to four small denticles; dorsal surface of carapace covered with relatively thick short setae; walking legs with spines on anterior margin of merus; medium-sized species .... *Pilumnus longicornis*

- Exorbital angle with well-developed spine-shaped tooth, no tooth between exorbital angle and second anterolateral tooth; dorsal surface without thick short setae; walking legs without spines on anterior margin of merus; small-sized species 4

4. Walking legs with distinct distal spine on posterior margin of carpus; infraorbital margin without spine on inner part; no spine on hepatic region; G1 with apical process bent laterally in about 90 degree .... *Pilumnus minutus*

- Walking legs without distal spine on posterior margin of carpus; infraorbital margin with distinct spine on inner part; distinct hepatic spine visible from dorsal view between exorbital angle and second anterolateral teeth; G1 with markedly hook-shaped apical process 5

5. Larger chelipeds with large spines covering outer surface of palm .... *Pilumnus propinquus*

- Larger chelipeds with palm mostly smooth on outer surface .... *Pilumnus savignyi*

Pilumnus incanus (Forskål, 1775)

(Fig. 26.22b, 26.23 and 26.24)

Cancer incanus Forskål 1775: 92.
Pilumnus forskalii — Nobili 1906b: 274.
Pilumnus incanus — Klunzinger 1913: 259, fig 16, tab 7, figs a–d; Ng and Clark 2005: 11 (in text); Ng et al. 2008: 141 (in list); Naderloo and Türkay 2012: 38.

Type locality: *Pilumnus incanus*: Red Sea. — *Pilumnus forskalii*: Red Sea.

Distribution

Persian Gulf: Iran (Naderloo and Türkay 2012).
Gulf of Oman: Iran (Naderloo et al. 2015).
World distribution: Northwestern Indian Ocean: Red Sea, Persian Gulf and Gulf of Oman.

Habitat: Intertidal, substrate rocky/cobble.

Fig. 26.24 *Pilumnus incanus* (Forskål, 1775). Male (ZUTC Brach1778), CL = 15.78, CB = 21.65 mm, Iran, Gulf of Oman, coll. N. Ghotbeddin (Photo: R. Abdollahi)

Pilumnus longicornis Hilgendorf, 1878
(Figs. 26.22c, 26.23 and 26.25)

Pilumnus longicornis Hilgendorf 1878: 794, pl 1, figs 8, 9.
Pilumnus longicornis — Nobili 1906a: 135; Balss 1933: 15; Stephensen 1946: 144, 145, fig 36a; Guinot 1964: 94, 95; Guinot 1967a: 274 (in list); Basson et al. 1977: 227, 237, 243, 250, 256, 262, 265; Jones 1986a: 163, pl 48; Tirmizi and Ghani 1996: 68–70, fig 26; Hornby 1997: 15; Apel 2001: 99; Naderloo and Sari 2007a: 344, tab 1; Ng et al. 2008: 142 (in list); Naderloo and Türkay 2012: 37; Naderloo et al. 2013: 449, tab 1.
Type locality: Mozambique (Inhambane).

Distribution

Persian Gulf: Iran (Stephensen 1946; Naderloo and Sari 2007a; Naderloo and Türkay 2012; Naderloo et al. 2013), Kuwait (Jones 1986a), Saudi Arabia (Basson et al. 1977), Bahrain (Stephensen 1946), UAE (Hornby 1997).

Gulf of Oman: Iran (Naderloo et al. 2015).

World distribution: Indo-West Pacific: South Africa, East Africa, Madagascar, Mauritius, Seychelles, Amirante, Red Sea, Persian Gulf, Gulf of Oman, Pakistan, India, Sri Lanka, Diego Garcia, Saint Brandon, Saya de Malha Bank, Mergui Archipelago, Taiwan, Philippines, Celebes, Java, Singapore, Australia and New Zealand.

Habitat: Intertidal to 33 m, substrate rocky/cobble with weeds.

Fig. 26.25 *Pilumnus longicornis* Hilgendorf, 1878. Male (SMF), CL = 19.8, CB = 27.2 mm, Ras Al-Khaima, UAE, Persian Gulf, 13.07.1995, coll. M. Apel (Photo: S. Träkner)

Pilumnus minutus De Haan, 1835
(Figs. 26.22d, 26.23 and 26.26)

Cancer (*Pilumnus*) *minutus* De Haan 1835: 50, pl 3, fig 2.
Pilumnus hirsutus Stimpson 1858a: 37.
Pilumnus minutus — Takeda and Miyake 1968: 40, 41, fig 9d–e; Titgen 1982: 252 (in list); Apel 2001: 99; Naderloo and Sari 2007a: 344, tab 1; Ng et al. 2008: 142 (in list).
Pilumnus hirsutus — Stimpson 1907: 69, fig 9; Alcock 1898: 197; Stephensen 1946: 146, fig 36d–f.
Type locality: *Cancer* (*Pilumnus*) *minutus*: Japan. — *Pilumnus hirsutus*: North China Sea.
Distribution
Persian Gulf: Iran (Naderloo and Sari 2007a), Bahrain [Stephensen (1946) as *Pilumnus hirsutus*].
Gulf of Oman: No record.
World distribution: Unknown. Because of the taxonomic complexity, the exact records of the species is not clear.
Habitat: Shallow subtidal to 26 m, substrate rocky/cobble.

Pilumnus propinquus Nobili, 1905
(Figs. 26.27 and 26.28)

Pilumnus propinquus Nobili 1905a: 163.
Pilumnus propinquus — Nobili 1906a: 140–142; Nobili 1906b: 277, pl 10, fig 7; Balss 1933: 12; Stephensen 1946: 147, 206 (in list); Guinot 1964: 3 (in list), 7, 95, 97, fig 56a, b; Guinot 1967a: 274 (in list); Titgen 1982: 252 (in list); Hogarth 1989: 106; Hogarth 1994: 101; Apel 2001: 100; Ng et al. 2008: 142 (in list).
Pilumnus? *propinquus* — Titgen 1982: 137.

Fig. 26.26 *Pilumnus minutus* De Haan, 1835. Male (ZUMC CRU956), CL = 11.10, CB =14.9 mm, North of Bahrain, Persian Gulf, depth 23 m, 27.03.1937, coll. G. Thorson (Photo: S. Träkner)

Fig. 26.27 *Pilumnus propinquus* Nobili, 1905. Syntype male (NMHN B.12958), CL = 6.00, CB = 8.1 mm, NO of Arzanah Island, UAE, Persian Gulf, 1901, coll. Bonnier and Perez (Photo: S. Träkner)

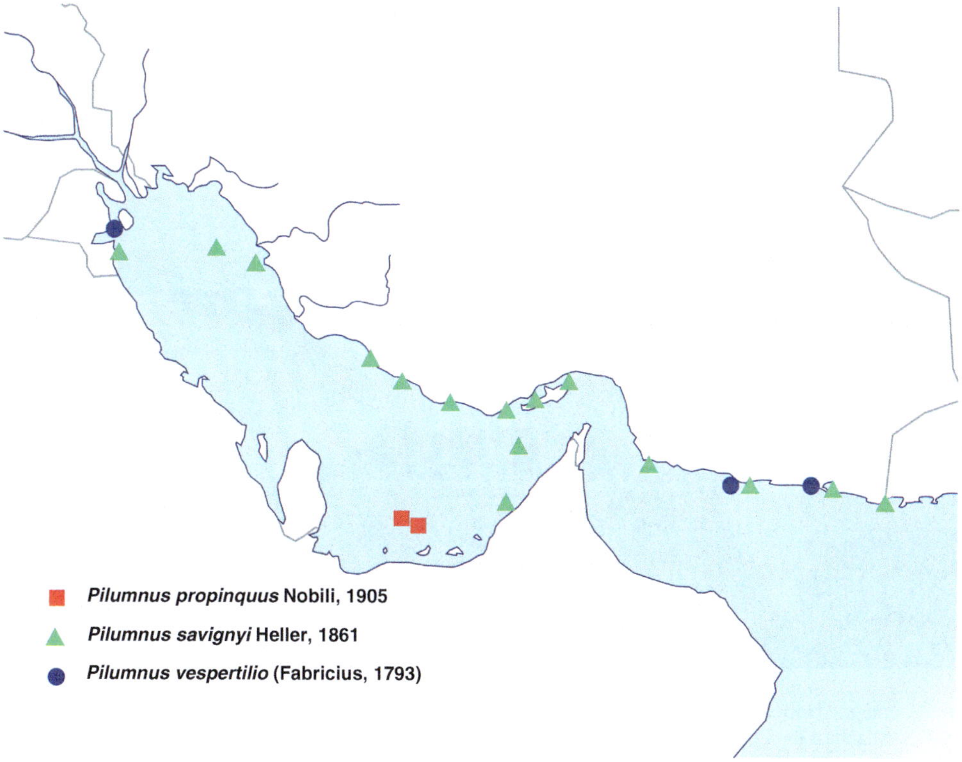

Fig. 26.28 Map showing the recorded localities of *Pilumnus propinquus* Nobili, 1905, *Pilumnus savignyi* Heller, 1861, and *Pilumnus vespertilio* (Fabricius, 1793)

Type locality: Persian Gulf (Arzanah Island, Abu-Dhabi, UAE) and Red Sea.

Distribution

Persian Gulf: UAE (Nobili 1905a).

Gulf of Oman: No record.

World distribution: Northwestern Indian Ocean: Aldabra, Red Sea, southern Oman and Persian Gulf.

Habitat: Unknown.

Remarks: Nobili (1905a) described this species based on the material from the Persian Gulf and the Red Sea. I examined the type series of the species (3 males, 4 females and 3 juveniles) collected from the Arzanah Island in the Persian Gulf in 1901 and deposited in the MNHN (MNHN-IU-2014-8625). Overall morphology of the species is generally very similar to *P. savignyi* in particular regarding the apical part of the G1, having small spine on hepatic margin of carapace, characteristics of front with having two semiquadrate lobes, and overall morphology of the carapace. The only clear difference is seen in the spination of chelipeds. Those of *P. propinquus* have large and more prominent spines on upper and outer surfaces of carpus and palm. The spines on the outer surface of palm of larger chelipeds are distinct and covering the whole surfaces. But, palm of larger cheliped in *P. savignyi* has small spines on proximal and lower parts of outer surface and getting fully smooth on upper surface and margins. *Pilumnus savignyi* is a more common species of the genus finding in different rocky and rubble shores, while no single specimen of *P. propinquus* was collected since its description in 1905. Here I prefer to have them in two distinct species as far as detailed taxonomic studies to be done with examination of more material from both species in the region.

Pilumnus savignyi Heller, 1861

(Figs. 26.22e, 26.28 and 26.29)

Pilumnus savignyi Heller 1861a: 345.
Pilumnus savignyi — Nobili 1906a: 138; Balss 1933: 20; Guinot 1967a: 274 (in list); Titgen 1982: 252 (in list); Jones 1986a: 163, pl 48; Hogarth 1989: 106, 114; Hogarth 1994: 101; Apel 2001: 100; Naderloo and Sari

Fig. 26.29 *Pilumnus savignyi* Heller, 1861. Male (ZUTC Brach1077), CL = 14.60, CB = 19.92 mm, Iran, Persian Gulf, 23.12.2004, coll. R. Naderloo (Photo: S. Shahbazi)

2007a: 344, tab 1; Ng et al. 2008: 142 (in list); Naderloo and Türkay 2012: 37; Naderloo et al. 2013: 449, tab 1.
Pilumnus (*savignyi* Heller?) — Stephensen 1946: 145, 146, fig 36b, c.

Type locality: Red Sea.

Distribution

Persian Gulf: Iran (Stephensen 1946; Naderloo and Sari 2007a; Naderloo and Türkay 2012; Naderloo et al. 2013), Kuwait (Jones 1986a).

Gulf of Oman: Iran (Naderloo et al. 2015).

World distribution: Western Indian Ocean: East Africa, Aldabra, Red Sea, southern Oman, Persian Gulf and Gulf of Oman.

Habitat: Intertidal to 22 m, substrate rocky/cobble.

Pilumnus vespertilio (Fabricius, 1793)

(Figs. 26.22f, 26.28 and 26.30)

Cancer vespertilio Fabricius 1793: 463.
Pilumnus vespertilio — Alcock 1898: 191 (in key), 192, 193; Jones 1986a: 162, pl 48; Hogarth 1989: 106; Hogarth 1994: 101; Tirmizi and Ghani 1996: 70–72, fig 72; Hornby 1997: 15; Apel 2001: 100; Ng et al. 2008: 142 (in list).

Type locality: India (exact locality unknown).

Distribution

Persian Gulf: Kuwait (Jones 1986a).

Gulf of Oman: Iran (Naderloo et al. 2015), UAE (Hornby 1997).

World distribution: Indo-West Pacific: South Africa, East Africa, Mauritius, Seychelles, Red Sea, Gulf of Aden, southern Oman, Persian Gulf, Gulf of Oman, Pakistan, India, Maldives, Andaman Sea, Mergui Archipelego, Thailand, Indo-Malayan Archipelago, Taiwan, Japan, Australia, New Caledonia, Melanesia and Hawaii.

Habitat: Intertidal to 22 m, substrate rocky/cobble.

Fig. 26.30 *Pilumnus vespertilio* (Fabricius, 1793). Female (SMF 38743), CL = 9.1, CB = 12.3 mm, Jod Village, Sistan-Baluchestan, Iran, Gulf of Oman, 15.11.2005, coll. R. Naderloo (Photo: S. Träkner)

Fig. 26.31 *Typhlocarcinops stephenseni* Serène, 1964. Male (ZUMC CRU957), CL = 5.2, CB = 7.4 mm, NO of Kharg Island, Iran, Persian Gulf, depth 22 m, 04.03.1937, coll. G. Thorson (Photo: S. Träkner)

Genus *Typhlocarcinops* Rathbun, 1909

Typhlocarcinops stephenseni Serène, 1964

(Figs. 26.31, 26.32, 26.33a and 26.34)

Typhlocarcinops stephenseni Serène 1964: 223–226, fig 8, pl 19b.

Typhlocarcinops stephenseni — Apel 2001: 103; Naderloo and Sari 2007a: 344, tab 1; Ng et al. 2008: 144 (in list).

Typhlocarcinus nudus —Alcock 1900: 322; Stephensen 1946: 173–176; Titgen 1982: 253 (in list).

Type locality: Serène (1964) described this species based on a specimen from material of Alcock's (1900) *Typhlocarcinus nudus*. Alcock's (1900) material was

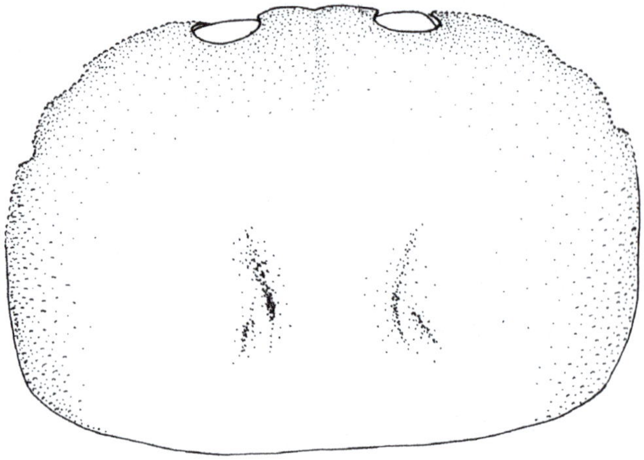

Fig. 26.32 *Typhlocarcinops stephenseni* Serène, 1964. Drawing of carapace of the species

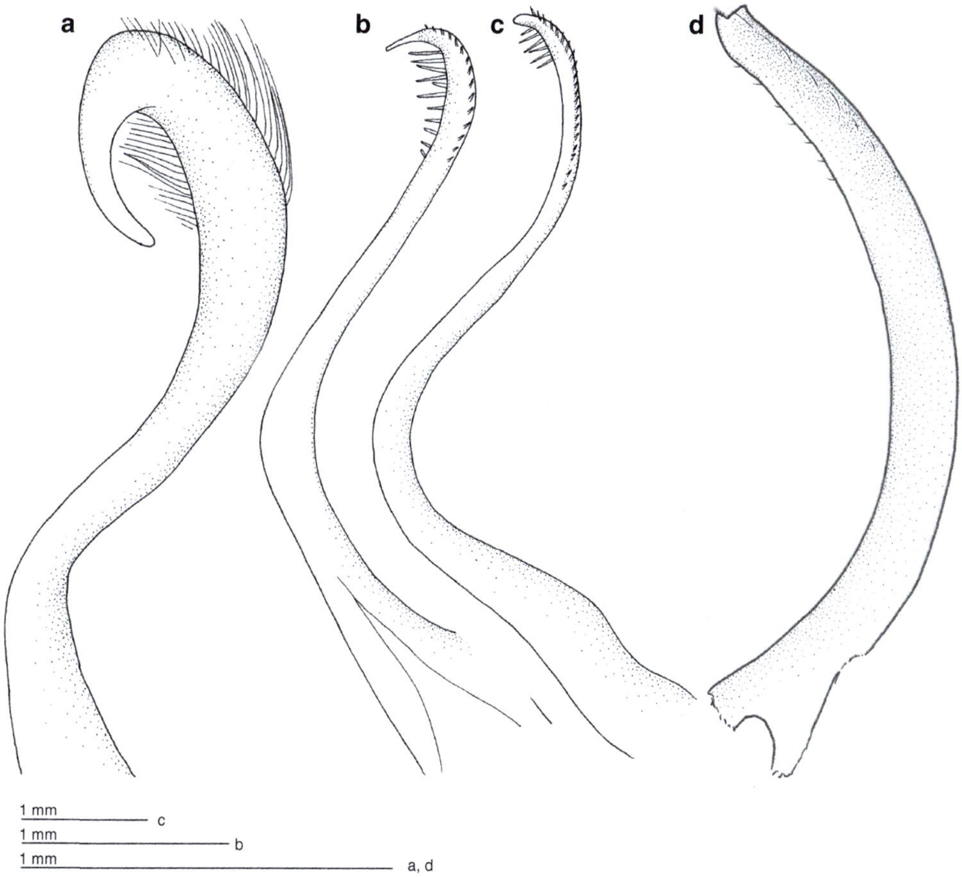

Fig. 26.33 Male G1, ventral view: (**a**) *Typhlocarcinops stephenseni* Serène, 1964 (ZUMC CRU957); (**b**) *Typhlocarcinus dentatus* Stephensen, 1946 (ZUMC CRU974); (**c**) *Typhlocarcinus rubidus* Alcock, 1900 (ZUMC CRU958); (**d**) *Xenophthalmodes dolichophallus* Tesch, 1918 (ZMUC CRU963)

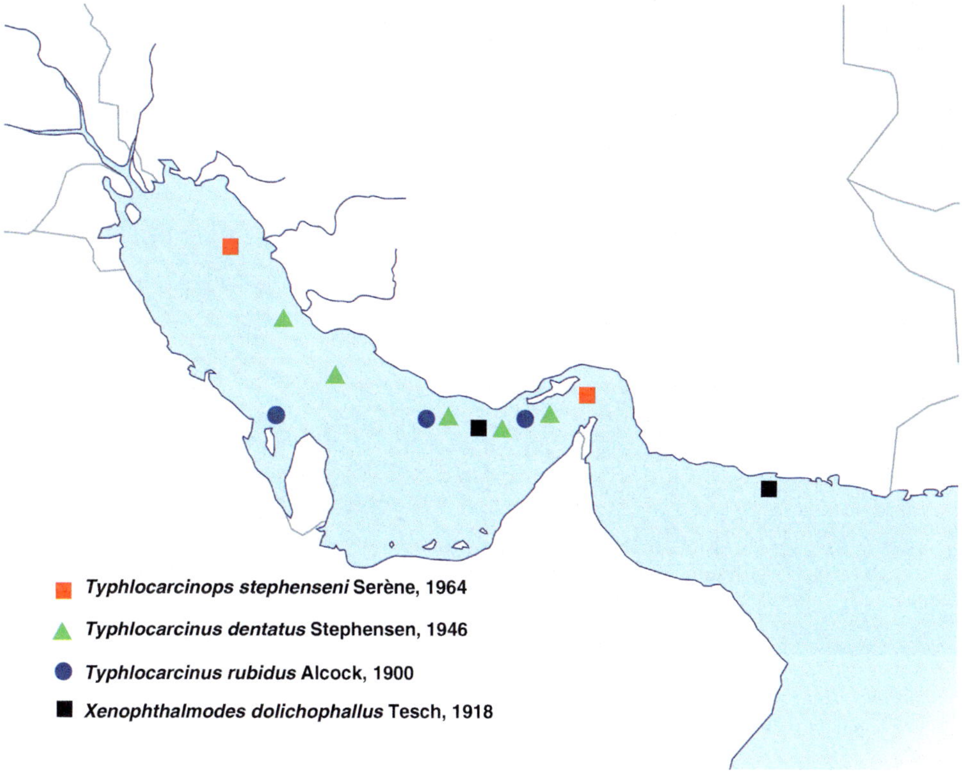

Fig. 26.34 Map showing the recorded localities of *Typhlocarcinops stephenseni* Serène, 1964, *Typhlocarcinus dentatus* Stephensen, 1946, *Typhlocarcinus rubidus* Alcock, 1900, and *Xenophthalmodes dolichophallus* Tesch, 1918

Genus *Typhlocarcinus* Stimpson, 1858

Key to species of the genus *Typhlocarcinus* in the region

- Carapace about 1.4 as broad as long; anterolateral margins of carapace with small denticles; male abdomen with somite 1 narrow, markedly long, slightly less than two times as broad as somite 2 .... *Typhlocarcinus rubidus*
- Carapace about 1.2 as broad as long; anterolateral margins with three rounded teeth, median one largest; male abdomen with somite 1 not markedly narrow and long, about 1.2 times as broad as somite 2 .... *Typhlocarcinus dentatus*

Typhlocarcinus dentatus Stephensen, 1946

(Figs. 26.33b, 26.34, 26.35 and 26.36)

Typhlocarcinus dentatus Stephensen 1946: 177, 178, figs 49i–k, 50.
Typhlocarcinus dentatus — Serène 1964: 207, 208, fig 3, pl 17b; Titgen 1982: 253 (in list); Apel 2001: 103; Naderloo and Sari 2007a: 344, tab 1; Ng et al. 2008: 144 (in list).
Type locality: Persian Gulf (Iran).

Distribution

Persian Gulf: Iran (Stephensen 1946; Naderloo and Sari 2007a).

Gulf of Oman: No record.

World distribution: This species is still only recorded from the Persian Gulf.

Habitat: Subtidal, 7.5–83 m, substrates sandy, muddy and gravel.

Fig. 26.35 *Typhlocarcinus dentatus* Stephensen, 1946. Male (ZUMC CRU969), CL = 6.8, CB = 8.9 mm, Qyanat, Iran, Persian Gulf, depth 7.5 m, 05.04.1937, coll. G. Thorson (Photo: S. Träkner)

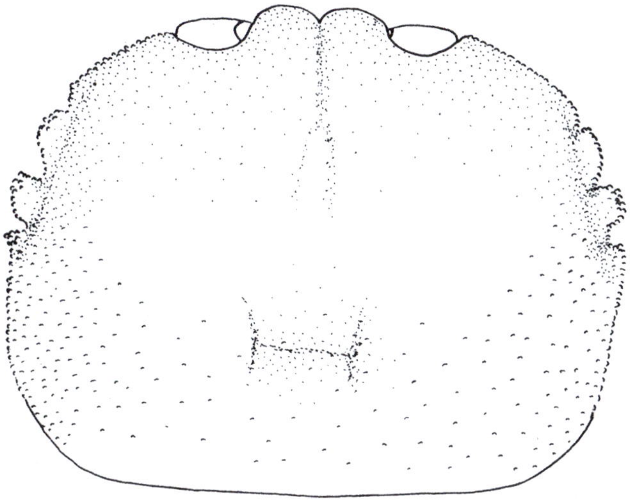

Fig. 26.36 *Typhlocarcinus dentatus* Stephensen, 1946. Drawing of carapace of the species

Typhlocarcinus rubidus Alcock, 1900
(Figs. 26.33c, 26.34, 26.37 and 26.38)

Typhlocarcinus rubidus Alcock 1900: 323.
Typhlocarcinus rubidus—Tesch 1918: 207; Stephensen 1946: 173 (in text); Serène 1964: 212–214, fig 5, pl 18a; Apel 2001: 103; Naderloo and Sari 2007a: 344, tab 1; Ng et al. 2008: 144 (in list).
Typhlocarcinus (*villosus* Stimpson?) — Stephensen 1946: 176, 177; Titgen 1982: 253 (in list).

Type locality: Bay of Bengal.

Distribution

Persian Gulf: Iran [Stephensen (1946) as *Typhlocarcinus* (*villosus* Stimpson?); Serène 1964; Naderloo and Sari 2007a].

Gulf of Oman: No record.

World distribution: Northwestern Indian Ocean: Red Sea, Persian Gulf and Bay of Bengal.

Habitat: Subtidal to 20 m, substrates sandy, muddy and gravel.

Fig. 26.37 *Typhlocarcinus rubidus* Alcock, 1900. Male (ZUMC CRU958), CL = 5.2, CB = 6.8 mm, NO of Bahrain, Persian Gulf, depth 20 m, 31.03.1937, coll. G. Thorson (Photo: S. Träkner)

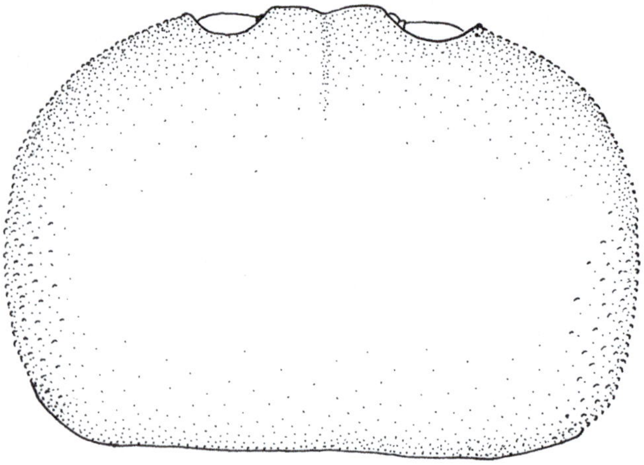

Fig. 26.38 *Typhlocarcinus rubidus* Alcock, 1900. Drawing of carapace of the species

Genus *Xenophthalmodes* Richters, 1880

Xenophthalmodes dolichophallus Tesch, 1918
(Figs. 26.33d, 26.34, 26.39 and 26.40)

Xenophthalmodes dolichophallus Tesch 1918: 216, pl 14, fig 1.
Xenophthalmodes dolichophallus — Stephensen 1946: 178, 179, fig 47c–e; Guinot 1979: 168, fig 37a–c; Apel 2001: 104; Naderloo and Sari 2007a: 344, tab 1; Ng et al. 2008: 144 (in list).

Type locality: Indonesia (Java, Flores and Sulawesi).

Distribution

Persian Gulf: Iran (Naderloo and Sari 2007a).
Gulf of Oman: Iran (Stephensen 1946).
World distribution: Indian Ocean: Persian Gulf, Gulf of Oman and Indonesia.

Habitat: Subtidal, 14 m, substrate sandy.

Fig. 26.39 *Xenophthalmodes dolichophallus* Tesch, 1918. Male (ZMUC CRU963), CL = 4.2, CB = 5.7 mm, 25° 25′N, 59° 28′E, depth 14 m, 01.04.1938, coll. B. Loppenthin (Photo: S. Träkner)

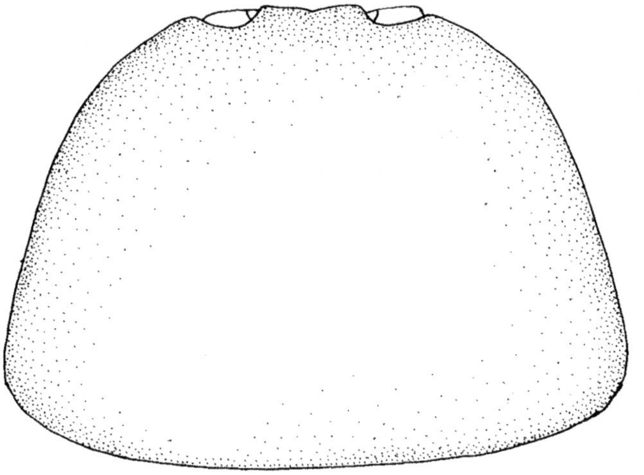

Fig. 26.40 *Xenophthalmodes dolichophallus* Tesch, 1918. Drawing of carapace of the species

Family Hexapodidae Miers, 1886 (Six-Legged Crabs)

27

Diagnosis Carapace transversely rectangular to trapezoidal, markedly broader than long, dorsal surface slightly convex, smooth, regions not well defined. Anterolateral margins entire, usually without dentation, diverging posteriorly, posterior margin markedly broad. Front narrow, straight, gently deflexed. Orbits small, slightly smaller than front, eyes small, not fully covered with orbit dorsally, cornea not broader than stalk. Antennules folded obliquely or transversely. Antennae small, basal segment short. Third maxillipeds not fully covering buccal cavern. Chelipeds robust. Three pairs of walking legs, well developed, last legs characteristically lost. Abdominal sternite 8 markedly reduced, concealed under carapace. Abdomen of males with variable number of somites fused, first two somites reduced, not filling distance between coxae of last walking legs. Abdomen of females with seven free-moving somites. G1 slender, nearly straight; G2 smaller than G1. Genital opening of males coxosternal. Genital opening of females sternal.

Genus *Hexapus* De Haan, 1835

Hexapus sexpes (Fabricius, 1798)
(Figs. 27.1, 27.3a, and 27.4)

Cancer sexpes Fabricius 1798.
Hexapus (*Hexapus*) *estuarinus* Sankarankutty 1975: 1, figs 1, 2.
Hexapus stephenseni Serène and Soh 1976: 24, 25.
Cancer sexpes — Manning and Holthuis 1981: 172.
Hexapus sexpes — Nobili 1906a: 146; Stephensen 1946: 182–185, fig 53a–e; Manning and Holthuis 1981: 172; Manning 1982: 157–161, figs 1a–f, 2; Apel 2001: 101; Ng et al. 2008: 86 (in list).

Type locality: *Cancer sexpes*: Southeast India. — *Hexapus estuarinus*: Southwest India (Cochin). — *Hexapus stephenseni*: Andaman Sea.

Distribution

Persian Gulf: Iran [Stephensen 1946; Serène and Soh (1976) as *Hexapus stephenseni*], Qatar (Apel 2001), UAE (Nobili 1906a).

Gulf of Oman: No record.

World distribution: Indo-West Pacific: Persian Gulf, India, Thailand, Indonesia and Japan.

Habitat: Subtidal to 7 m, substrates muddy and sandy.

R. Naderloo, *Atlas of Crabs of the Persian Gulf*, DOI 10.1007/978-3-319-49374-9_27

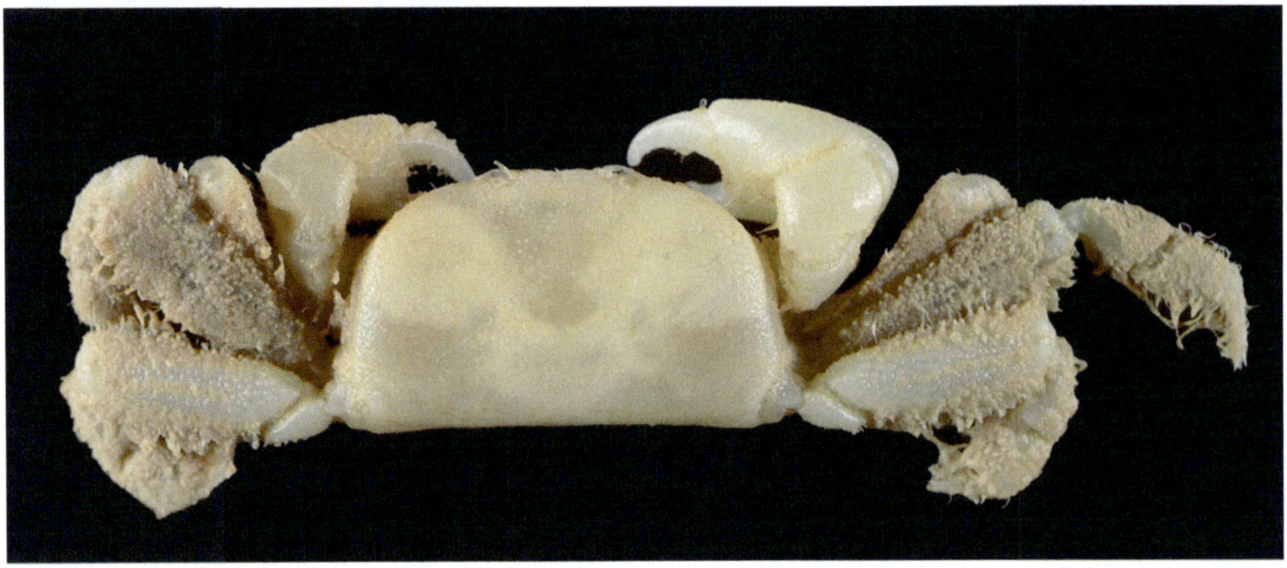

Fig. 27.1 *Hexapus sexpes* (Fabricius, 1798). Male (ZMUC CRU863), CL = 3.8, CB = 6.1 mm, Bushehr, Iran, Persian Gulf, depth 7 m, 18.03.1937, coll. G. Thorson (Photo: S. Träkner)

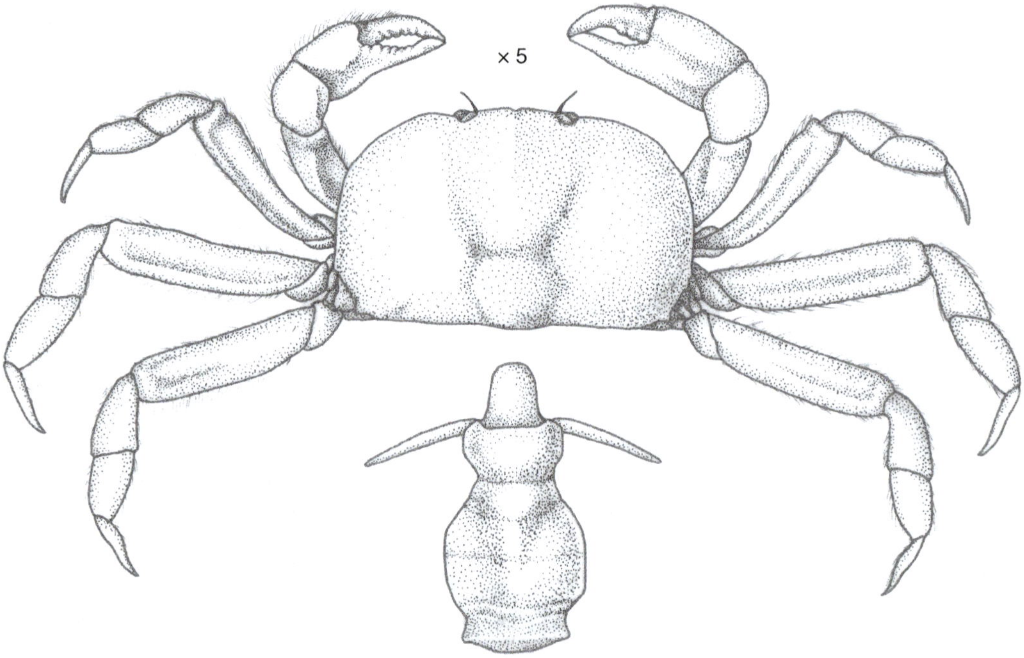

Fig. 27.2 *Lambdophallus sexpes* Alcock, 1900. Drawing of the crab from dorsal view and male abdomen, depicted by S. Pazoki using the original drawing of Alcock and Mcardle (1903)

Genus *Lambdophallus* Alcock, 1900

Lambdophallus sexpes Alcock, 1900

(Figs. 27.2, 27.3b, and 27.4)

Lambdophallus sexpes Alcock 1900: 330, 331.
Lambdophallus sexpes — Alcock and Mcardle 1903: pl 63, figs 1, 1a, 1b; Tesch 1918: 239; Stephensen 1946: 181, 182, fig 52a–c; Titgen 1982: 252 (in list); Apel 2001: 101; Ng et al. 2008: 86 (in list).

Type locality: Bay of Bengal.

Distribution

Persian Gulf: Iran (Stephensen 1946; Apel 2001).

Gulf of Oman: No record.

World distribution: Northern Indian Ocean: Persian Gulf and Bay of Bengal.

Habitat: Subtidal to 70 m, substrate muddy.

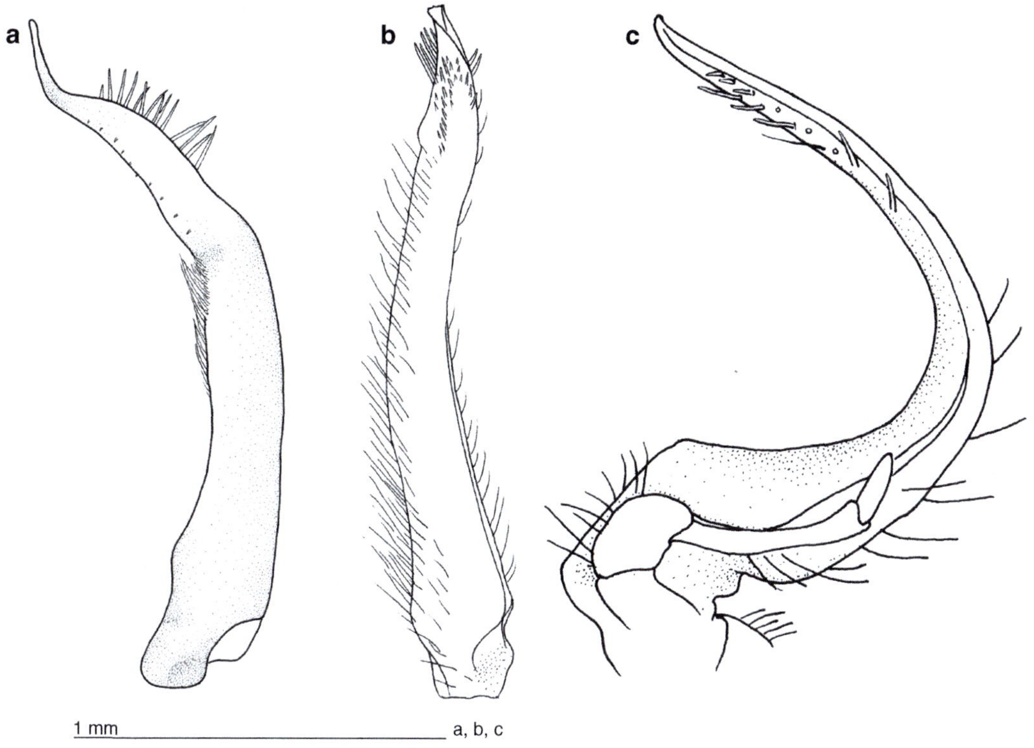

Fig. 27.3 Male G1, ventral view: (**a**) *Hexapus sexpes* (Fabricius, 1798) (ZMUC CRU863); (**b**) *Lambdophallus sexpes* Alcock, 1900 (Modified from Stephensen 1946); (**c**) *Xenophthalmus wolffi* Takeda and Miyake, 1970

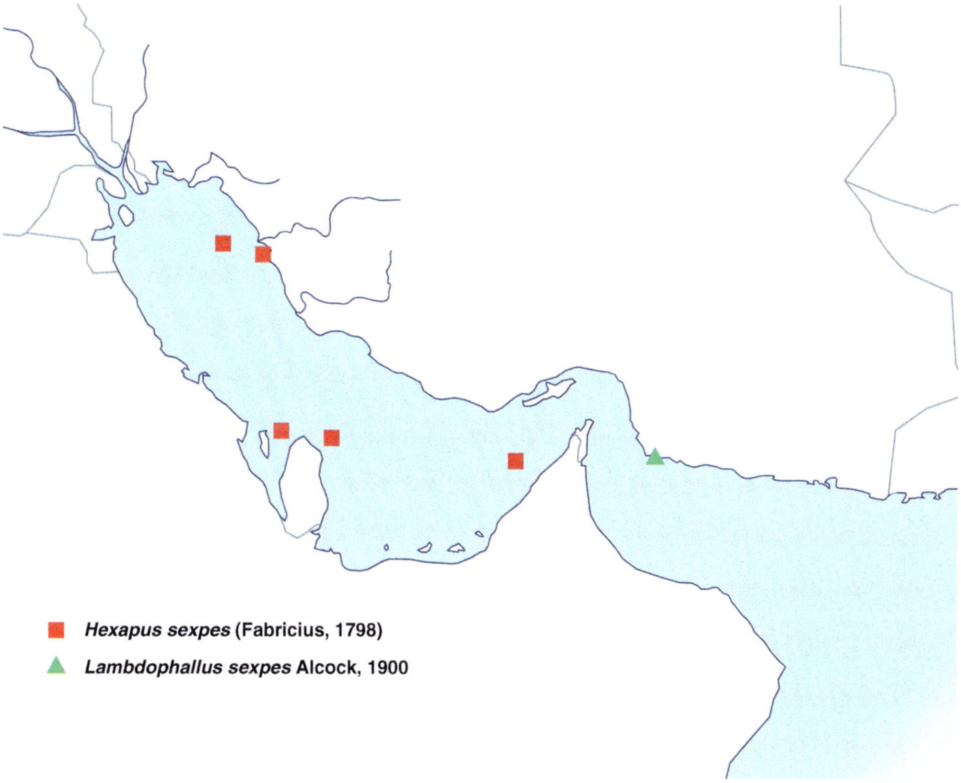

Fig. 27.4 Map showing the recorded localities of *Hexapus sexpes* (Fabricius, 1798), *Lambdophallus sexpes* Alcock, 1900

Family Xenophthalmidae Stimpson, 1858

28

Diagnosis Carapace subtrapezoid, thick, dorsal surface slightly convex, smooth, regions not well defined; lateral margins of carapace entire, without any dentation, diverging posteriorly, no distinct junction between anterolateral and posterolateral margins. Front narrow, deflexed. Orbits characteristically longitudinal on dorsal surface of carapace, narrow, eyestalks relatively long, situated in orbits. Antennules folded transversely. Antennae directed dorsally on carapace, situating along inner side of eyes. Third maxillipeds without gap between, fully covering buccal cavern. Chelipeds nearly equal, relatively small. Walking legs well developed, flattened, margins serrated with long setae, unequal in size and shapes, last legs much smaller. Abdomen of males and females with seven freely moving somites. Genital opening of males coxal. Genital opening of females sternal.

Genus *Xenophthalmus* White, 1846

Xenophthalmus wolffi Takeda and Miyake, 1970

(Figs. 27.3c, 28.1, and 28.2)

Xenophthalmus wolffi Takeda and Miyake 1970: 11–18, figs 1c–h, 2f–i.
Xenophthalmus wolffi — Titgen 1982: 253 (in list); Ghani and Tirmizi 1995b: 653–655; Apel 2001: 105; Ng et al. 2008: 245 (in list).
Xenophthalmus pinnotheroides — Stephensen 1946: 186, 187, fig 54a–c.
? *Xenophthalmus* sp. 2 — Basson et al. 1977: 243.

Type locality: Strait of Hormuz.
Distribution
Persian Gulf: Iran [Stephensen (1946) as *X. pinnotheroides*; Takeda and Miyake 1970], Saudi Arabia [Basson et al. (1977) as *Xenophthalmus* sp.].
Gulf of Oman: No record.
World distribution: Northern Indian Ocean: Persian Gulf and Pakistan.
Habitat: Intertidal to 70 m, substrate muddy.

R. Naderloo, *Atlas of Crabs of the Persian Gulf*, DOI 10.1007/978-3-319-49374-9_28

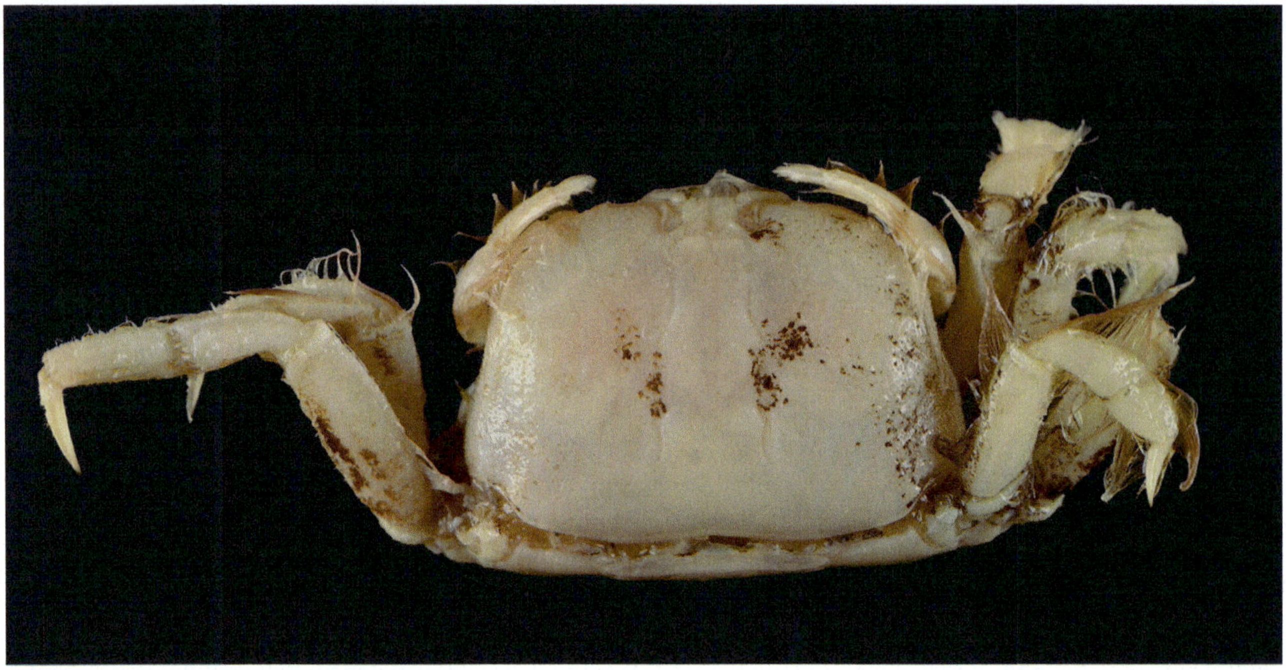

Fig. 28.1 *Xenophthalmus wolffi* Takeda and Miyake, 1970. Female (SMF), CL = 8.5, CB = 10.4 mm, N Jubail, Saudi Arabia, Persian Gulf, 03.12.1992, coll. M. Richmond (Photo: S. Träkner)

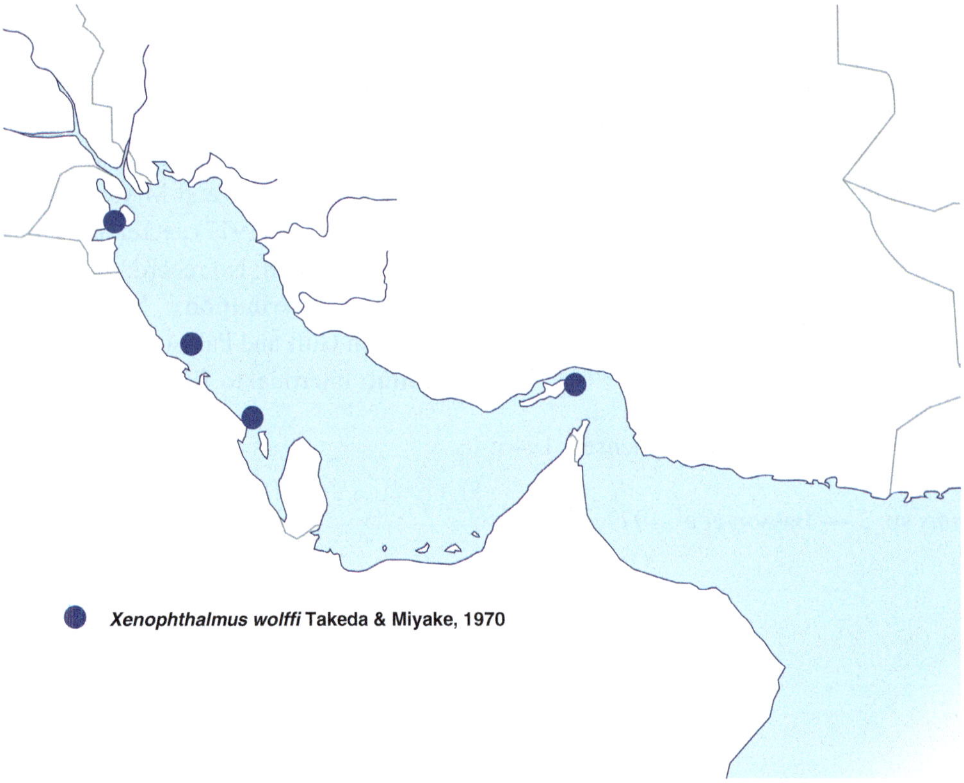

Fig. 28.2 Map showing the recorded localities of *Xenophthalmus wolffi* Takeda and Miyake, 1970

Family Palicidae Bouvier, 1898 (Stilt Crabs)

29

Diagnosis Carapace subquadrate to subcircular, dorsal surface of carapace flat, markedly uneven, with high elevations and tubercles, anterolateral margins with teeth. Front with two to four lobes. Orbits deep, broad, supraorbital and infraorbital margins usually with lobes or denticles, eyestalks relatively long, usually with tubercles. Antennules folded transversely. Antennae with basal segment rectangular or variously expanded, flagellum long. Third maxillipeds small, not fully covering buccal cavern. Chelipeds small to moderate in size, equal or unequal. Walking legs slender, last one markedly reduced, but similar to other ones. Abdomen of males and females with seven free-moving somites. G1 nearly robust, twisted; G2 shorter than G1. Genital opening of males coxal. Genital opening of females sternal.

Genus *Neopalicus* Moosa and Serène, 1981

Neopalicus jukesii (White, 1847)
(Figs. 29.1, 29.2, and 29.3)

Cymopolia jukesii White 1847c: 358, pl 2, fig 1.
Cymopolia carinipes Paul'son 1875: 73, pl 9, figs 4–4c.
Palicus jukesi — Alcock 1900: 451; Rathbun 1911: 240, pl 19, fig 9b.
Neopalicus jukesi — Moosa and Serène 1981: 41 (in key), 42–44, figs 7a, b, pl 2c.
Neopalicus jukesii — Castro 2000b: 554–558, figs 39b, 40b, c, 41c, 49, 61c; Apel 2001: Ng et al. 2008: 127 (in list).
Type locality: *Cymopolia jukesii*: Australia (Sir Charles Hardy's Islands, Torres Strait). — *Cymopolia carinipes*: Red Sea.
Distribution
Persian Gulf: Bahrain (Apel 2001).
Gulf of Oman: No record.
World distribution: Indo-West Pacific: Madagascar, Red Sea, Gulf of Aden, Persian Gulf, Maldives, Sri Lanka, Andaman Sea, Indonesia, Philippines, Australia and New Caledonia.

Habitat: Subtidal, 18–101 m, substrates coarse sand with fragmented shells.

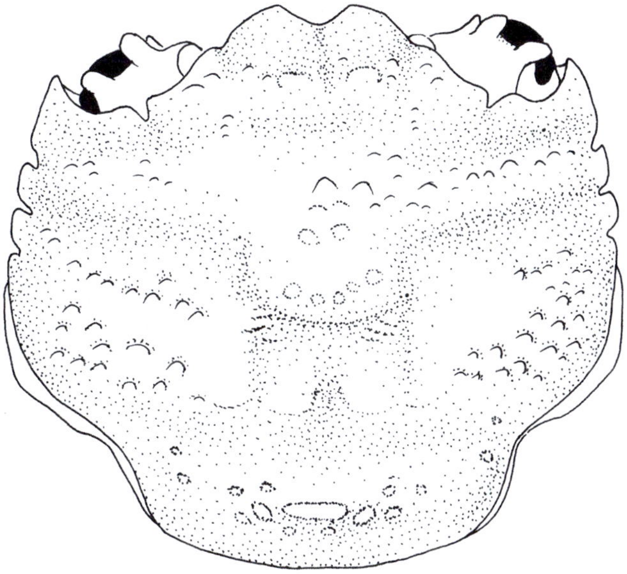

Fig. 29.1 *Neopalicus jukesii* (White, 1847). Drawing of carapace of the species

R. Naderloo, *Atlas of Crabs of the Persian Gulf*, DOI 10.1007/978-3-319-49374-9_29

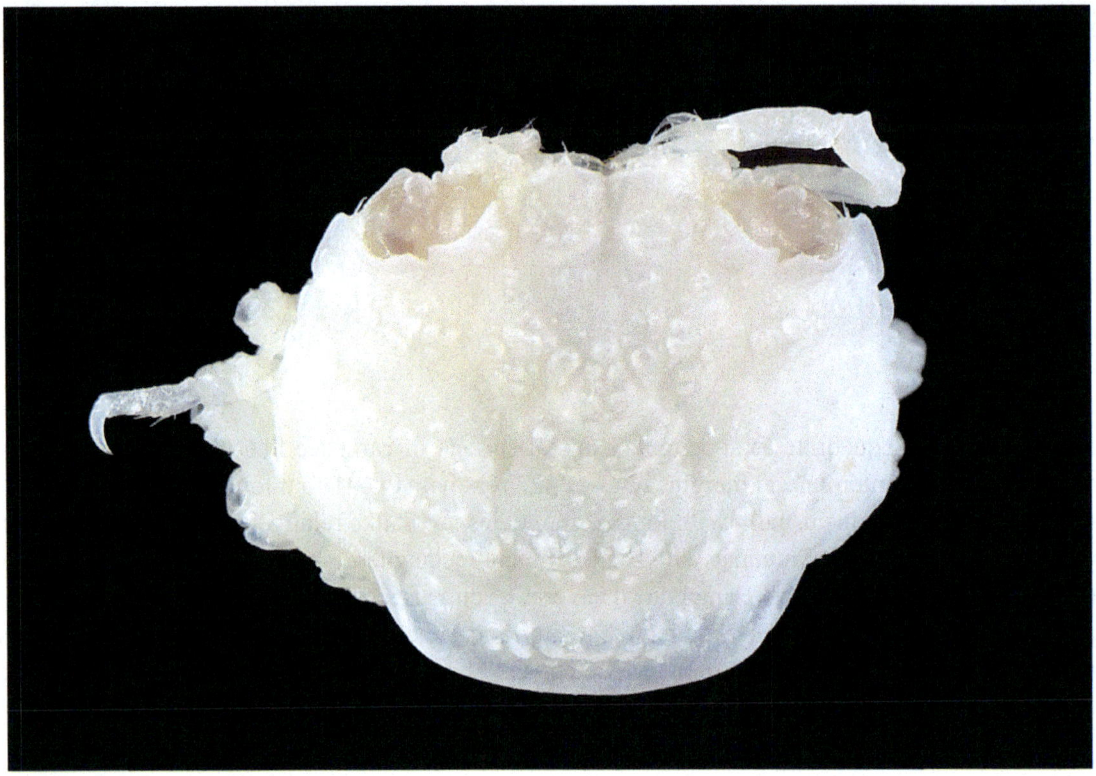

Fig. 29.2 *Neopalicus jukesii* (White, 1847). Juvenile female (ZMUC CRU813), CL = 3.9, CB = 5.5 mm, Bahrain, Persian Gulf, depth 23 m, 27.03.1937, coll. G. Thorson (Photo: S. Träkner)

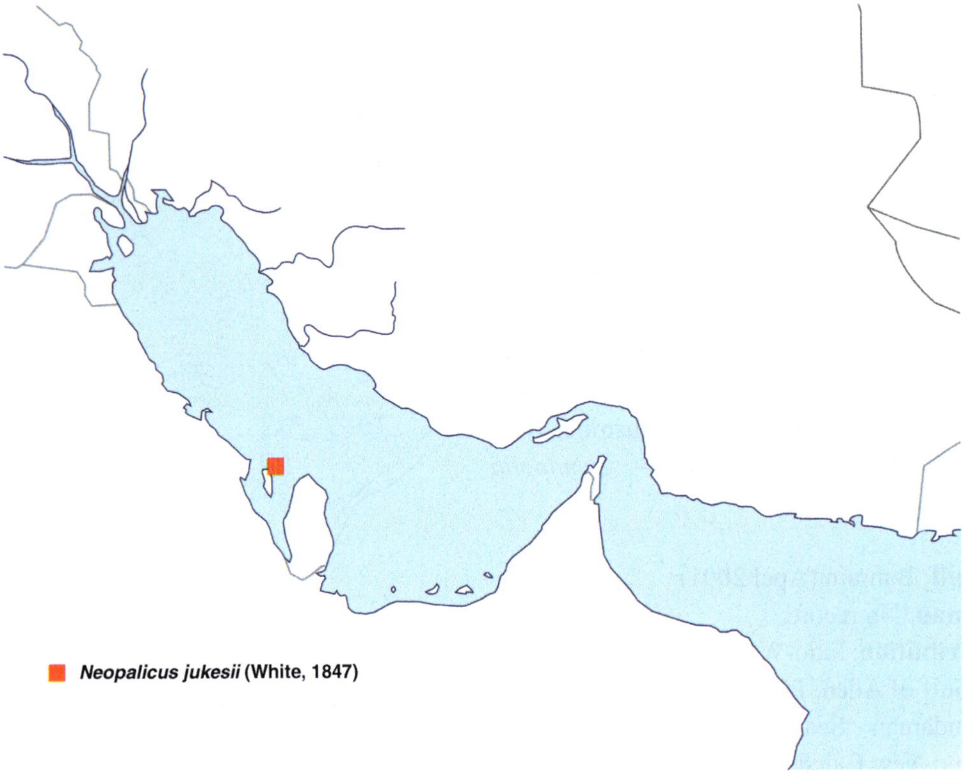

Fig. 29.3 Map showing the recorded localities of *Neopalicus jukesii* (White, 1847)

Part III

Section Eubrachyura, Subsection Thoracotremata Guinot, 1977

Family Grapsidae MacLeay, 1838 (Shore Crabs, Talon Crabs)

30

Diagnosis Carapace quadrate to subcircular, dorsal surface moderately flat, usually with transversal ridges; lateral margins entire or with one dentation behind exorbital angle, converging posteriorly, junction of anterolateral and posterolateral margins not distinct. Front broad, strongly deflexed, broadly bilobed. Orbits well developed, shorter than front, eyestalks moderately long. Antennules folded transversely. Antennae with basal segment bent, flagellum short. Third maxillipeds with broad rhomboidal gap between, mandibles visible when third maxillipeds closed, no longitudinal setose ridge on outer surface of ischium and merus. Chelipeds medium-sized, merus usually with developed spines on anterodistal margin, fingers apically spoon-shaped. Walking legs robust, distinctly flattened, usually armed with spines and bristles. Abdomen of males and females with seven freely moving somites. G1 robust, nearly straight, usually with dense setae on apical part; G2 much smaller than G1. Genital opening of males and females sternal.

Key to genera of the family Grapsidae in the region

- Lateral margins of carapace straight, entire, without postorbital tooth; frontal region smooth, without granules .... *Metopograpsus*
- Lateral margins of carapace arched, with postorbital tooth; frontal region with granules .... *Grapsus*

Genus *Grapsus* Lamarck, 1801

Key to species of the genus *Grapsus* in the region

- Lateral margins of carapace markedly arched; outer infraorbital angle with acute tooth; third maxillipeds with long merus, slightly shorter than ischium; male abdomen with somite 5 of about 2.3 times as broad as long; G1 slightly curved; large sized species .... *Grapsus albolineatus*
- Lateral margins of carapace very slightly arched, nearly straight; outer infraorbital angle with finely denticulate process; third maxillipeds with short merus, markedly shorter than ischium; male abdomen with somite 5 of about 2.8 times as broad as long; G1 relatively straight; medium sized species .... *Grapsus granulosus*

R. Naderloo, *Atlas of Crabs of the Persian Gulf*, DOI 10.1007/978-3-319-49374-9_30

Grapsus albolineatus Latreille in Milbert, 1812
(Figs. 30.1, 30.2, 30.3, 30.5, and 30.10a, b)

Grapsus albolineatus Latreille in Milbert, 1812: 274.
Grapsus albolineatus — Vannini and Valmori 1981: 64, figs 1d, 2d, 3d, 4d; Crosnier 1965: 13, figs 8, 11; Holthuis 1977: 145; Titgen 1982: 142, 253 (in list); Tirmizi and Ghani 1996: 138, fig 53; Carpenter et al. 1997: 40; Apel 2001: 116; Ng et al. 2008: 216 (in list); Naderloo 2011: 2–5, figs 1a–e, 5a, b; Naderloo and Türkay 2012: 45; Naderloo et al. 2013: 350, tab 1.
Grapsus maculatus subsp. *tenuicrustatus* — Stephensen 1946: 194, 210 (in list), fig 59a–d.
Grapsus tenuicrustatus — Pretzmann 1971: 479; Basson et al. 1977: 47, 145, 227; Titgen 1982: 253 (in list).
Type locality: Mauritius.

Distribution
Persian Gulf: Iran [Stephensen (1946) as *Grapsus maculatus* subsp. *tenuicrustatus*; Pretzmann (1971) as *G. tenuicrustatus*; Naderloo 2011; Naderloo and Türkay 2012; Naderloo et al. 2013], Saudi Arabia [Basson et al. (1977) as *G. tenuicrustatus*; Apel 2001], UAE (Titgen 1982; Apel 2001).
Gulf of Oman: Iran (Naderloo 2011; Naderloo et al. 2015), UAE (Apel 2001).
World distribution: Indo-West Pacific: East Africa, Mauritius, Somalia, Socotra, Red Sea, Gulf of Aden, southern Oman, Persian Gulf, Gulf of Oman, Pakistan, India, Sri Lanka, Bay of Bengal, Nicobar Islands, Andaman Sea, Mergui Archipelago, China, Japan, Indonesia, Singapore, Cocos (Keeling) Islands, Australia and Hawaii.
Habitat: Intertidal, substrate rocky.

Fig. 30.1 *Grapsus albolineatus* Latreille in Milbert, 1812. Male (ZUTC Brach1175), CL = 51.00, CB = 54.6 mm, Portuguese castle, Chabahar, Iran, Gulf of Oman, 17.11.2005, coll. R. Naderloo and A. Kazemi (Photo: S. Träkner. Reproduced from Naderloo (2011) by permission of Magnolia Press)

Fig. 30.2 *Grapsus albolineatus* Latreille in Milbert, 1812. Kuwei, Qeshm Island, Iran, Persian Gulf, 05.01.2008 (Photo: R. Naderloo)

Fig. 30.3 *Gapsus albolineatus* Latreille in Milbert, 1812. Guwatr, Sistan-Baluchestan, Iran, Gulf of Oman. Photo: Reza Naderloo

***Grapsus granulosus* H. Milne Edwards, 1853**
(Figs. 30.4, 30.5, and 30.10c, d)

Grapsus granulosus H. Milne Edwards 1853: 169.
Grapsus granulosus — Crosnier 1965: 13, figs 9, 12; Holthuis 1977: 148; Vannini and Valmori 1981: figs 1e, 2e, 3e; Apel 2001: 117; Zaouali et al. 2007: 2, fig 1; Ng et al. 2008: 216 (in list), 218; Naderloo 2011: 5–8, figs 2a–f, 5c, d; Naderloo and Türkay 2012: 45; Naderloo et al. 2013: 450, tab 1.
Type locality: Red Sea.

Distribution
Persian Gulf: Iran (Naderloo 2011; Naderloo and Türkay 2012; Naderloo et al. 2013), Saudi Arabia (Apel 2001), UAE (Apel 2001).
Gulf of Oman: Iran (Naderloo 2011; Naderloo et al. 2015), UAE (Apel 2001).
World distribution: Northwestern Indian Ocean and Mediterranean: Red Sea, Gulf of Aden, Persian Gulf, Gulf of Oman and southern Mediterranean (Lessepsian Migrant).
Habitat: Intertidal, substrate rocky/cobble.

Fig. 30.4 *Grapsus granulosus* H. Milne Edwards, 1853. Male (ZUTC Brach1183), CL = 51.00, CB = 54.6 mm, Jod Village, Sistan-Baluchestan, Iran, Gulf of Oman, 25.10.2006, coll. R. Naderloo and A. Kazemi (Photo: S. Träkner. Reproduced from Naderloo (2011) by permission of Magnolia Press)

Genus *Metopograpsus* H. Milne Edwards, 1853

Key to species of the genus *Metopograpsus* in the region

- Carapace with anterolateral margins distinctly converging posteriorly; inner infraorbital margin with conspicuous keel on outer surface; male abdomen with somite 3 slightly longer than somite 4; G1 with small apical process, sharply directed outwards; genital opening of females with operculum distincly elevated, finger shaped .... *Metopograpsus messor*
- Carapace with anterolateral margins slightly converging posteriorly, nearly straight; inner infraorbital margin with a week keel on outer surface; male abdomen with somite 3 slightly shorter than somite 4; G1 with apical process markedly large, gently directed outwards; genital opening of females with operculum located at bottom of depression, with round tip .... *Metopograpsus thukuhar*

Fig. 30.5 Map showing the distribution of *Grapsus albolineatus* Latreille in Milbert, 1812, and *Grapsus granulosus* H. Milne Edwards, 1853

Metopograpsus messor (Forskål, 1775)

(Figs. 30.6, 30.7, 30.10e, and 30.11)

Cancer messor Forskål 1775: 88.

Metopograpsus messor — Nobili 1906a: 155; Stephensen 1946: 195, fig 59e–g; Crosnier 1965: 23, figs 19, 26; Evans et al. 1973: 276; Basson et al. 1977: 228, 235; Titgen 1982: 144, 254 (in list); Jones 1986a: 160, pl 47; Vousden 1987: 32, 36, tab 4, 5, 7; Hogarth 1989: 104; Apel 1994a: 43, 1996: 329, 2001: 118; Al-Khayat and Jones 1996: 797, 800, 803, 812, figs 3, 4, 1999: 58, 61; Al-Ghais and Cooper 1996: 411, fig 2; Cooper 1997: 165; Hornby 1997: 14; Apel and Türkay 1999: 132; Apel 2001: 118; Ng et al. 2008: 217 (in list); Hosseini 2009: 39, 41; Naderloo 2011: 8–11, figs 3a–g, 5e; Naderloo and Türkay 2012: 46; Naderloo et al. 2013: 350, tab 1.

Type locality: Red Sea.

Distribution

Persian Gulf: Iran (Stephensen 1946; Hosseini 2009; Naderloo 2011; Naderloo and Türkay 2012; Naderloo et al. 2013), Kuwait (Jones 1986a), Saudi Arabia (Basson et al. 1977; Apel 1994a, 1996), Bahrain (Nobili 1906a; Stephensen 1946; Vousden 1987), Qatar (Al-Khayat and Jones 1999), UAE (Titgen 1982; Al-Ghais and Cooper 1996; Cooper 1997; Hornby 1997).

Gulf of Oman: Iran (Naderloo 2011; Naderloo et al. 2015), UAE (Apel 2001).

World distribution: Western Indian Ocean: East Africa, Madagascar, Red Sea, southern Oman, Persian Gulf, Gulf of Oman, Pakistan, India and Andaman Islands.

Habitat: Intertidal, substrates rocky, rocky jetties, cobble and mangroves.

Fig. 30.6 *Metopograpsus messor* (Forskål, 1775). Male (SMF 38455), CL = 23.06, CB = 28.60 mm, mangroves behind Airport, Bandar-Abbas, Iran, Persian Gulf, 07.01.2008, coll. R. Naderloo (Photo: S. Träkner. Reproduced from Naderloo (2011) by permission of Magnolia Press)

Fig. 30.7 *Metopograpsus messor* (Forskål, 1775). Mahtabi, Bandar-Khamir, Iran, Persian Gulf, 24.04.2008 (Photo: R. Naderloo)

Metopograpsus thukuhar (Owen, 1839)

(Figs. 30.8, 30.9, 30.10f, e, and 30.11)

Grapsus thukuhar Owen 1839: 80, pl 24, fig 3.
Metopograpsus thukuhar — H. Milne Edwards 1853: 165; Crosnier 1965: 23 (in key), 25, figs 20, 27; Holthuis 1977: 157, 158; Al-Ghais and Cooper 1996: 413, fig 3; Tirmizi and Ghani 1996: 140, fig 54; Hornby 1997: 14; Apel and Türkay 1999: 132; Apel 2001: 118; Ng et al. 2008: 217 (in list); Naderloo 2011: 11–15, figs 4a–g, 5f; Naderloo and Türkay 2012: 46; Naderloo et al. 2013: 350, tab 1.

Type locality: Hawaii.

Distribution

an (Naderloo 2011; Naderloo and Türkay 2012; Naderloo et **Persian Gulf**: Iral. 2013), UAE (Al-Ghais and Cooper 1996; Hornby 1997).

Gulf of Oman: Iran (Naderloo 2011; Naderloo et al. 2015), UAE (Al-Ghais and Cooper 1996; Hornby 1997; Apel 2001).

World distribution: Indo-West Pacific: South Africa, East Africa, Madagascar, Seychelles, Red Sea, Gulf of Aden, Persian Gulf, Gulf of Oman, Pakistan, India, Vietnam, China, Japan, Cocos (Keeling) Islands, Australia, Tahiti and Hawaii.

Habitat: Intertidal, between pneomatophores and trees of mangroves.

Fig. 30.8 *Metopograpsus thukuhar* (Owen, 1839). Male (SMF 38444), CL = 23.7, CB = 28.3 mm, mangroves of S coast of Qeshm Island, Iran, Persian Gulf, 09.01.2008, coll. M. Türkay and R. Naderloo (Photo: S. Träkner. Reproduced from Naderloo (2011) by permission of Magnolia Press)

Fig. 30.9 *Metopograpsus thukuhar* (Owen, 1839). Guwatr, Sistan-Baluchestan, Iran, Gulf of Oman, 2012 (Photo: R. Naderloo)

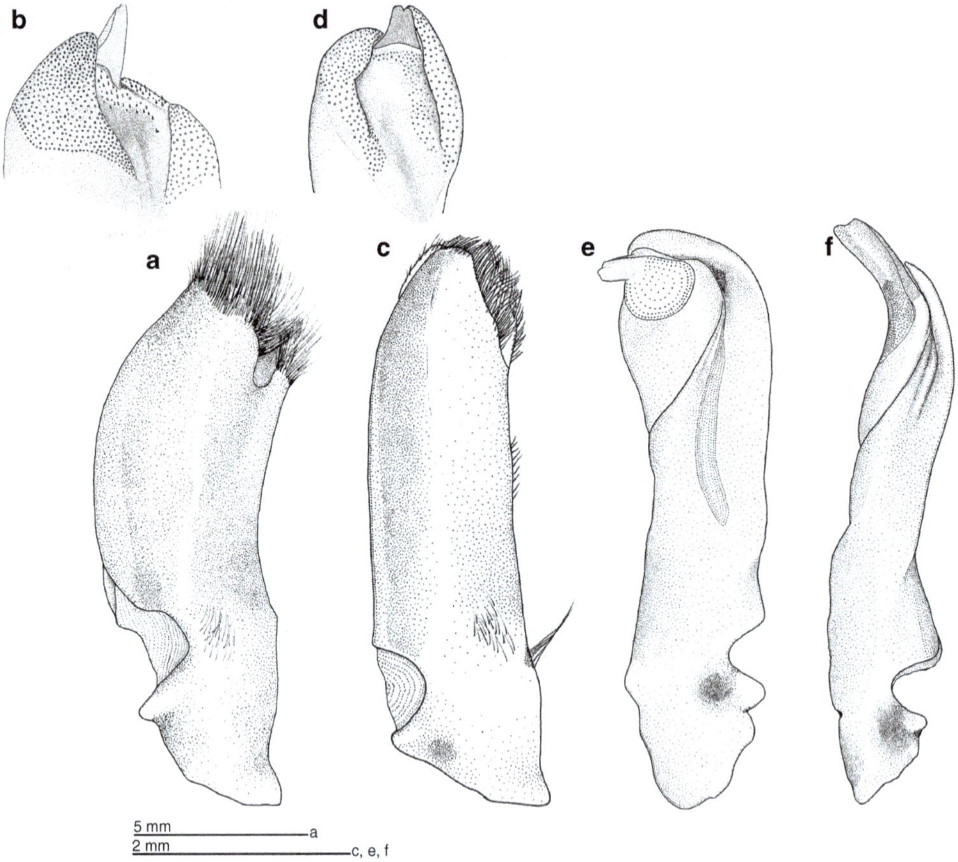

Fig. 30.10 Male G1, ventral view: (**a**) *Grapsus albolineatus* Latreille in Milbert, 1812. (ZUTC Brach1175); (**b**) apical part of (**a**); (**c**) *Grapsus granulosus* H. Milne Edwards, 1853 (ZUTC Brach1183; (**d**) apical part of (**c**); (**e**) *Metopograpsus messor* (Forskål, 1775) (ZUTC Brach11305); (**f**) *Metopograpsus thukuhar* (Owen, 1839) (ZUTC Brach1162) (Reproduced from Naderloo (2011) by permission of Magnolia Press)

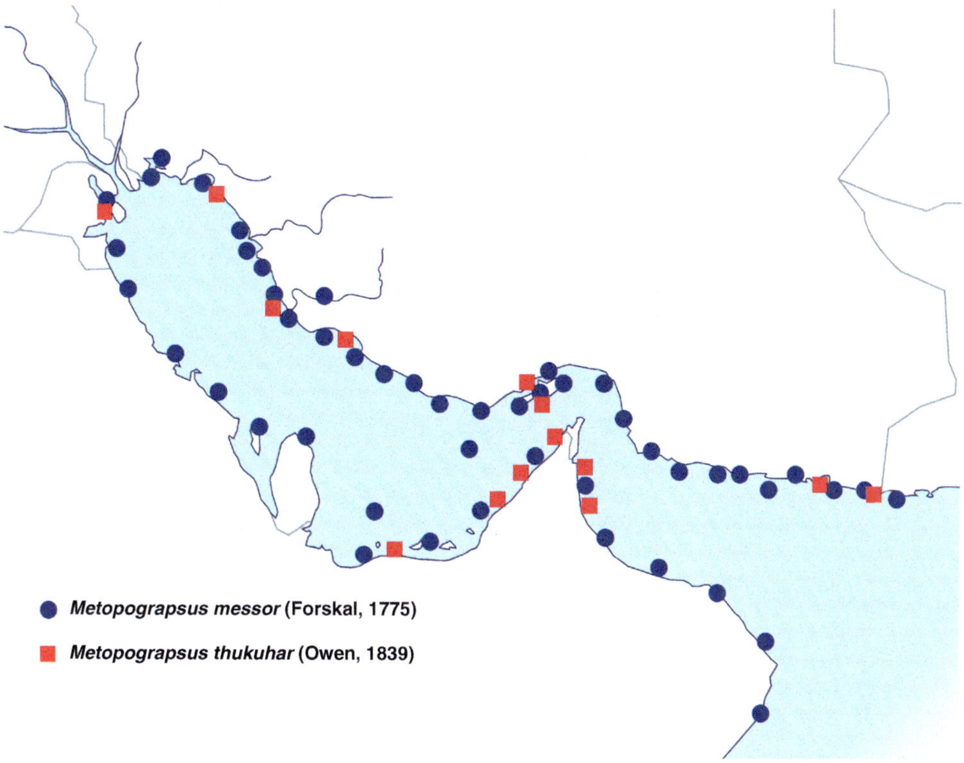

Fig. 30.11 Map showing the recorded localities of *Metopograpsus messor* (Forskål, 1775), and *Metopograpsus thukuhar* (Owen, 1839)

Family Sesarmidae Dana, 1851 (Marsh Crabs)

31

Diagnosis Carapace quadrate to subquadrate, moderately thick, dorsal surface relatively flat, regions more or less defined with shallow grooves, no junction between anterolateral and posterolateral margins, lateral margins entire or with one dentation behind exorbital angle; pterygostomian region fully covered with network of short hooked setae. Front broad, strongly deflexed, as broad as posterior margin of carapace or even broader. Orbits well developed, shorter than front, infraorbital margin running towards buccal cavern, eyestalks moderately long. Antennules folded transversely. Antennae with basal segment large, laterally expanded, flagellum short, often excluded from orbital hiatus. Third maxillipeds with broad rhomboidal gap between, mandibles visible when third maxillipeds closed, oblique setose ridge on outer surface of ischium and merus. Chelipeds robust, subequal, merus usually with at least one spine on anterodistal margin, fingers apically spoon-shaped. Walking legs robust, distinctly flattened, usually armed with spines and bristles. Abdomen of males and females with seven freely moving somites. G1 robust, nearly straight, usually with dense setae on apical part; G2 much smaller than G1. Genital opening of males and females sternal.

Key to genera of the family Sesarmidae in the region

1. Chelipeds of males and females with setal patch on outer surface of palm, longitudinal lines of granules beneath setal patch; posterodistal margin of walking legs denticulate; small-size species .... *Nanosesarma*
- Chelipeds of males and females without setal patch, granules on outer surface of palm not in line; posterodistal margin of walking legs not denticulate; medium-size species 2
2. Chelipeds of males without pectinated crests on upper surface of palm; upper surface of male chelipeds without large tubercles; species of marshlands .... *Chiromantes*
- Chelipeds of males with two or three pectinated crests on upper surface of palm; large tubercles along upper surface of male chelipeds 3
3. Anterolateral margins of carapace with dentation .... *Perisesarma*
- Anterolateral margins of carapace entire, without dentation .... *Parasesarma*

R. Naderloo, *Atlas of Crabs of the Persian Gulf*, DOI 10.1007/978-3-319-49374-9_31

Genus *Chiromantes* Gistel, 1848

Chiromantes boulengeri (Calman, 1920)

(Figs. 31.1, 31.2, 31.6, and 31.11a)

Sesarma boulengeri Calman 1920: 63, fig a.
Chiromantes boulengeri — Apel and Türkay 1999: 133; Apel 2001: 116; Ng et al. 2008: 220 (in list), 223; Naderloo and Schubart 2009: 63, figs 2, 3; Naderloo 2011: 15–17, figs 6a–f, 7a, b; Naderloo and Türkay 2012: 47.
Type locality: Persian Gulf (Basra, Shat al-Arab River, Iraq).

Distribution

Persian Gulf: Iran (Naderloo and Schubart 2009; Naderloo 2011; Naderloo and Türkay 2012), Iraq (Calman 1920; Apel and Türkay 1999), Kuwait (Apel 2001).

Gulf of Oman: No record.

World distribution: Endemic: North Estuarine System of the Persian Gulf (Shat Al-Arab, Arvandroud and Bahmanshir rivers).

Habitat: Muddy bank of the rivers, mostly burrowing in reedbeds.

Fig. 31.1 *Chiromantes boulengeri* (Calman, 1920). Male (SMF 33818), Pole-Tanki Abolhassan, Bahmanshir River, Abadan, Iran, Persian Gulf, 20.05.2008, coll. R. Naderloo and A. Kazemi (Photo: S. Träkner. Reproduced from Naderloo (2011) by permission of Magnolia Press)

Genus *Nanosesarma* Tweedie, 1950

Key to species of the genus *Nanosesarma* in the region

- Male chelipeds with granules along whole upper surface of movable finger; male telson elongate, about 1.6 as long as broad; dorsal surface of carapace with scarse setae; walking legs without denticles on posterior margin of merus .... *Nanosesarma sarii*
- Male chelipeds with granules only on proximal part of upper surface of movable finger; male telson not remarkably elongate, very slightly longer than broad; dorsal surface of carapace with dense long setae; walking legs with distinct denticle medially on posterior margin of merus ...... *Nanosesarma jousseaumei*

Fig. 31.2 *Chiromantes boulengeri* (Calman, 1920). The species in its habitat in Bahmanshir River, Iran (Photo: Ebrahim Gholinejad)

Nanosesarma jousseaumei (Nobili, 1906)

(Figs. 31.3, 31.5c, d, and 31.6)

Sesarma (*Sesarma*) *jousseaumei* Nobili 1905b: 411.
Sesarma (*Sesarma*) *jousseaumei* — 1906b: 323, pl 8, fig 9; Lauri 1915: 416; Guinot 1967a: 283.
Nanosesarma jousseaumei — Holthuis 1977: 172; Ng et al. 2008: 221 (in list), 224; Naderloo and Türkay 2009: 2918, figs 1, 4; Naderloo 2011: 22–26, figs 9b, 10a–e, 11b, 12b; Naderloo and Türkay 2012: 47; Naderloo et al. 2013: 450, tab 1, 458, tab 3.
Nanosesarma minutum — Crosnier 1965: 70, figs 109–115; Vannini and Valmori 1981: 82, fig 15.

Type locality: Gulf of Aden (Djibouti and Obock).
Distribution:

Persian Gulf: Iran (Naderloo and Türkay 2009; Naderloo 2011; Naderloo and Türkay 2012; Naderloo et al. 2013).

Gulf of Oman: Iran (Naderloo et al. 2015).

World distribution: Western Indian Ocean: East Africa (Tanzania), Madagascar, Gulf of Aden, Persian Gulf and Gulf of Oman.

Habitat: Intertidal, substrate rocky/cobble.

Fig. 31.3 *Nanosesarma jousseaumei* (Nobili, 1906). Male (ZUTC Brach1161), CL = 4.8, CB = 4.77 mm, mangroves of Khor-Khalasi, Jask, Iran, Gulf of Oman, 21.11.2005, coll. R. Naderloo and A. Kazemi (Photo: S. Träkner. Reproduced from Naderloo (2011) by permission of Magnolia Press)

Nanosesarma sarii Naderloo and Türkay, 2009

(Figs. 31.4, 31.5a, b, 31.6, and 31.11b)

Nanosesarma sarii Naderloo and Türkay 2009: 2912, figs 1, 2, 3.

Nanosesarma sarii — Naderloo 2011: 18–22, figs 8a–g, 9a, 11a, 12a; Naderloo and Türkay 2012: 47; Naderloo et al. 2013: 450, tab 1, 456, tab 2, 458, tab 3.

Nanosesarma (Nanosesarma) minimum — Tirmizi and Ghani 1996: 159, figs 61.

Nanosesarma minutum — Jones 1986a: 160, pl 46; Apel 2001: 118, 119.

Type locality: Persian Gulf (Iran, Bandar-Abbas).

Distribution

Persian Gulf: Iran (Naderloo and Türkay 2009; Naderloo 2011; Naderloo and Türkay 2012; Naderloo et al. 2013), Kuwait [Jones (1986a) and Apel (2001) as *Nanosesarma minutum*].

Gulf of Oman: Iran (Naderloo 2011; Naderloo et al. 2015).

World distribution: Northern Indian Ocean: Persian Gulf, Gulf of Oman and Pakistan.

Habitat: Intertidal, substrates rocky/cobble, rocky jetties, oyster banks and mangroves.

Fig. 31.4 *Nanosesarma sarii* Naderloo and Türkay, 2009. Male holotype (SMF 34376), CL = 6.12 mm, CB = 7.15 mm, behind Airport, Bandar-Abbas, Iran, Persian Gulf, 07.01.2008, coll. M. Türkay and R. Naderloo (Photo: S. Träkner. Reproduced from Naderloo (2011) by permission of Magnolia Press)

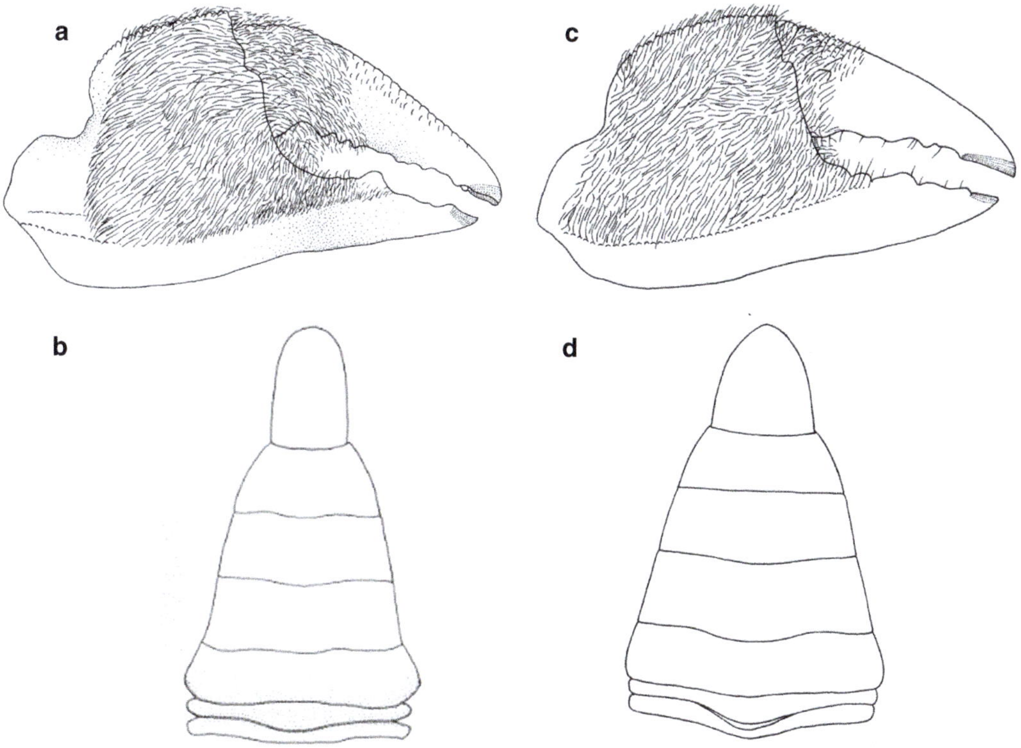

Fig. 31.5 *Nanosesarma sarii* Naderloo and Türkay, 2009: (**a**) right cheliped, outer view, (**b**) male abdomen; *Nanosesarma jousseaumei* (Nobili, 1906): (**c**) left cheliped, outer view, (**d**) male abdomen

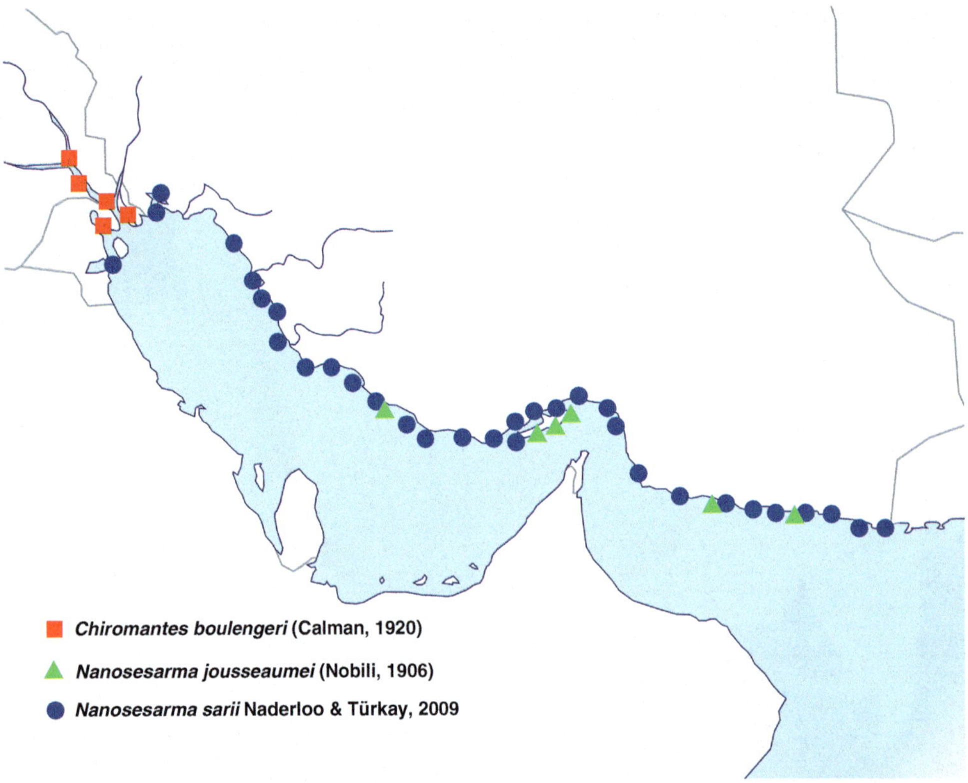

Fig. 31.6 Map showing the recorded localities of *Chiromantes boulengeri* (Calman, 1920), *Nanosesarma jousseaumei* (Nobili, 1906), and *Nanosesarma sarii* Naderloo and Türkay, 2009

Genus *Parasesarma* De Man, 1895

Parasesarma persicum Naderloo and Schubart, 2010

(Figs. 31.7, 31.8, 31.10, and 31.11c)

Parasesarma persicum Naderloo and Schubart 2010: 35, figs 1, 2, 3, 5a, 6.
Parasesarma persicum — Naderloo 2011: 26–29, figs 7c–d, 13a–f; Naser 2011: 186, figs 2, 3; Naderloo and Türkay 2012: 47; Naderloo et al. 2013: 450, tab 1, 456, tab 2, 458, tab 3.
Sesarma plicata — Clayton 1986: 86.
Sesarma (Parasesarma) plicata — Jones 1986a: 160, pl 46.
Sesarma (Parasesarma) plicatum — Al-Ghais and Cooper 1996: 414, fig. 4.
Parasesarma plicatum — Tirmizi and Ghani 1996: 151, fig. 58; Apel and Türkay 1999: 133; Apel 2001: 119.

Type locality: Persian Gulf (Khalij-Nayband, Iran).

Distribution

Persian Gulf: Iran [Naderloo and Schubart (2009) as *P. plicatum*; Naderloo and Schubart 2010; Naderloo 2011; Naderloo and Türkay 2012; Naderloo et al. 2013], Iraq (Naser 2011), Kuwait [Jones (1986a) as *P. plicatum*], UAE [Al-Ghais and Cooper (1996) and Apel (2001) as *P. plicatum*].

Gulf of Oman: Iran (Naderloo 2011; Naderloo et al. 2015).

World distribution: Northern Indian Ocean: Persian Gulf, Gulf of Oman and Pakistan.

Habitat: Intertidal, substrates mangroves, marshlands, muddy banks of lower reaches of Shat Al-Arab and Bahmanshir rivers.

Fig. 31.7 *Parasesarma persicum* Naderloo and Schubart, 2010. Male (SMF 38458), CL = 24.34, CB = 30.16 mm, Nayband, Iran, Persian Gulf, 26.04.2008, coll. R. Naderloo. Photo: S. Träkner (Reproduced from Naderloo (2011) by permission of Magnolia Press)

Fig. 31.8 *Parasesarma persicum* Naderloo and Schubart, 2010. Species in its habitat in Khalij-Nayband, Iran, Persian Gulf (Photo: R. Naderloo)

Genus *Perisesarma* De Man, 1895

Perisesarma guttatum (A. Milne-Edwards, 1869)

(Figs. 31.9, 31.10, and 31.11d)

Sesarma guttatum A. Milne-Edwards 1869: 26.
Sesarma (*Chiromantes*) *guttatum* — Crosnier 1965: 68, figs 94, 97, 106.
Sesarma (*Perisesarma*) *guttatum* — Holthuis 1977: 171.
Sesarma guttatum —Vannini and Volmori 1981: 87, figs 17d, 18d, 19c.
Perisesarma guttatum — Hywel-Davies 1994: 29, 37, 48, app. 4; Apel 2001: 119, 120; Gillikin and Schubart 2004: 440, fig 4a, b; Naderloo 2011: 29–31, fig 7e–f, 14a–e; Naderloo and Türkay 2012: 48; Naderloo et al. 2013: 450, tab 1, 456, tab 2.

Type locality: Zanzibar.

Distribution

Persian Gulf: No record.

Gulf of Oman: UAE (Apel 2001), Oman (Hywel-Davies 1994).

World distribution: Western Indian Ocean: South Africa, Madagascar, Tanzania, Kenya, Somalia, Red Sea and southwestern Gulf of Oman.

Habitat: Intertidal, substrate mangroves.

Fig. 31.9 *Perisesarma guttatum* (A. Milne-Edwards, 1869). Male (SMF 29337), CL = 18.70, CB = 22.49 mm, S Malindi, Watamu, Minda Greek, Kenya, 30° 28′N, 49° 11′E, 09–11.1998, coll. D.P. Gillikin and A. Verheyden (Photo: S. Träkner. Reproduced from Naderloo (2011) by permission of Magnolia Press)

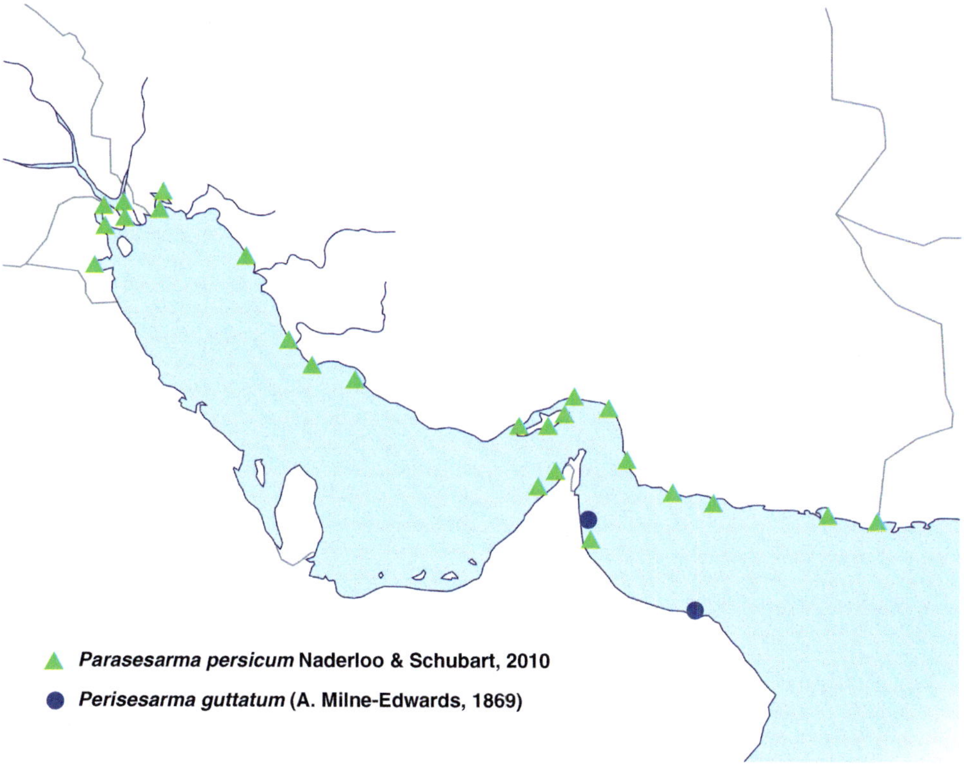

Fig. 31.10 Map showing the recorded localities of *Parasesarma persicum* Naderloo and Schubart, 2010, and *Perisesarma guttatum* (A. Milne-Edwards, 1869)

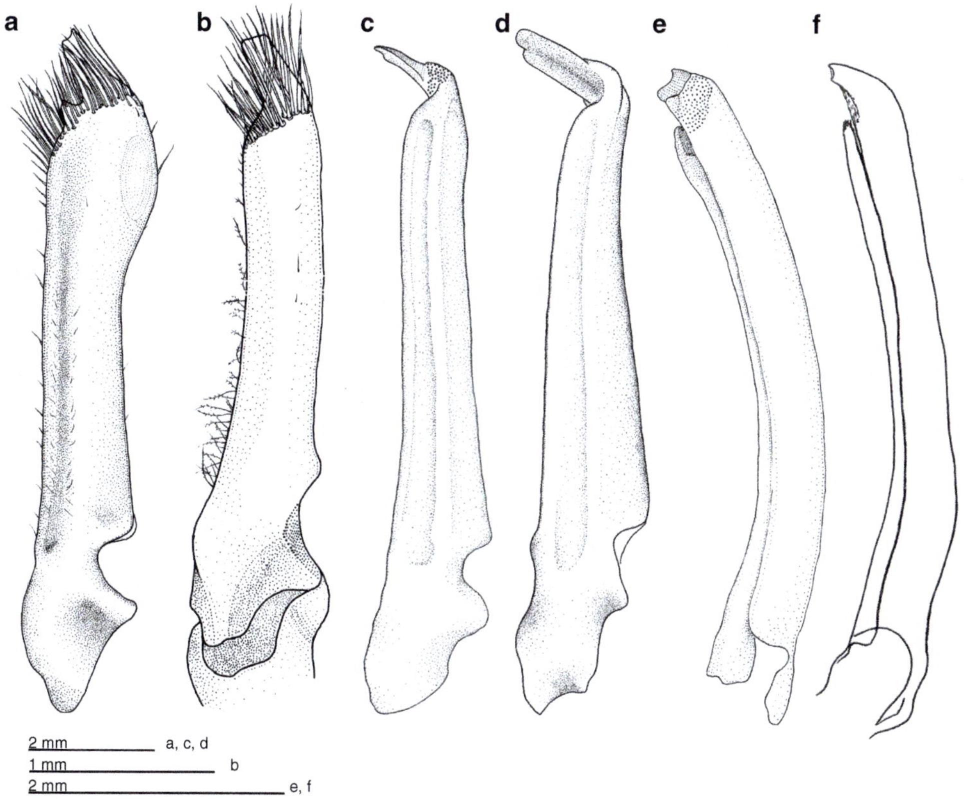

Fig. 31.11 Male G1, ventral view: (**a**) *Chiromantes boulengeri* (Calman, 1920) (SMF 33818); (**b**) *Nanosesarma sarii* Naderloo and Türkay, 2009 (SMF 34376); (**c**) *Parasesarma persicum* Naderloo and Schubart, 2010 (ZUTC Brach1251: holotype); (**d**) *Perisesarma guttatum* (A. Milne-Edwards, 1869) (SMF 29337); (**e**) *Metaplax indica* H. Milne Edwards, 1852 (ZUTC Brach1160); (**f**) *Thalassograpsus harpax* (Hilgendorf, 1892) (SMF 38447)

Family Varunidae H. Milne Edwards, 1853

32

Diagnosis Carapace subquadrate to subcircular, dorsal surface flat to slightly convex, smooth, without transverse ridges; anterolateral margins gently arched, often with two lobes or truncated teeth behind exorbital angle, junction between anterolateral and posterolateral margins more or less distinct. Front straight, mostly bilobed, slightly shorter than posterior margin of carapace. Orbits broad, as broad as or slightly shorter than front; eyestalks moderately long. Pterygostomian region without network of short setae. Antennules folded obliquely. Antennae well developed, with relatively long flagellum. Third maxillipeds with moderate rhomboidal gap between, mandibles hardly visible, longitudinal setose ridge on outer surface of ischium and merus. Chelipeds robust, subequal to unequal, palm characteristically massive. Walking legs without spines along dactylus. Abdomen of males and females with seven freely moving somites; abdomen of females usually rounded. G1 robust, slightly curved, setal patch on apical part; G2 smaller than G1. Genital opening of males and females sternal.

Key to the genera of the family Varunidae in the region

1. Carapace relatively round; front dentate; chelipeds with dense patch of setae on outer surface of palm, euryhaline species finding in rivers .... *Eriocheir*
- Carapace relatively rectangular or quadrate; front lobate; chelipeds without patch of setae on outer surface of palm 2
2. Carapace transversely rectangular; chelipeds of adult males long, slender, nearly equal; inner surface of chelipeds palm without setae; crab of muddy habitats .... *Metaplax*
- Carapace relatively quadrate; chelipeds of males short, robust 3
3. Carapace with anterolateral teeth apically pointed, spiniform; chelipeds unequal in males and equal in females, inner surface of palm with dense setal patch; crab living in rocky/cobble habitats .... *Thalassograpsus*
- Carapace with anterolateral teeth apically blunt, not spiniform; chelipeds equal in males and unequal in females, inner surface of palm smooth, without dense setal patch; crab living in mangroves .... *Pseudohelice*

R. Naderloo, *Atlas of Crabs of the Persian Gulf*, DOI 10.1007/978-3-319-49374-9_32

Genus *Eriocheir* De Haan, 1835

Eriocheir hepuensis Dai, 1991

(Figs. 32.1 and 32.6)

Eriocheir japonica hepuensis Dai 1991: 61, figs 1–11.
Eriocheir hepuensis — Guo et al. 1997: 460; Ng et al. 1998: 493; Ng et al. 1999: 154; Sun et al. 2003: 592; Chu et al. 2003: 738; Tang et al. 2003: 309; Ng et al. 2008: 228 (in list); Naser et al. 2012: figs 4, 5b, 6b, 7b, 8, 9, 10; Naderloo 2014: 1, figs 2–4.
Eriocheir sinensis — Clark et al. 2006: 1–4, figs 1–6; Naser et al. 2011: 120.

Type locality: China (Hepu, Guangxi Province, southern China).

Distribution

Persian Gulf: Iran (Naderloo 2014), Iraq [Clark (2006) as *E. sinensis*; Naser et al. (2011) as *E. sinensis*; Naser et al. 2012], Kuwait [Apel and Bishop (2006) as *E. sinensis*].

Gulf of Oman: No record.

World distribution: Indo-West Pacific: Persian Gulf (Shat Al-Arab, Arvandroud and Bahmanshir rivers), China (Hepu, Aotou, Tongan, Minjiang, Feiyunjiang River, Ouijiang River, Chiagjiang River, Hu and Wang) and Taiwan (Tamshui).

Habitat: Muddy bed of rivers.

Fig. 32.1 *Eriochier hepuensis* Dai, 1991. Female (ZUTC 5235), CL = 28.22, CB = 31.00 mm, Pole-Tanki Abolhassan, Bahmanshir River, Abadan, Iran, Persian Gulf, 20.01.2013, coll. R. Naderloo (Photo: R. Abdollahi)

Genus *Metaplax* H. Milne Edwards, 1852

Metaplax indica H. Milne Edwards, 1852

(Figs. 31.11e, 32.2, 32.3, and 32.6)

Metaplax indicus H. Milne Edwards 1852: 161, 1854: 165, pl 11, fig 2.

Metaplax indicus — Alcock 1900: 431; Apel 1994a: 43, pl 3; Apel 1996: 330, 2001: 117, 118; Apel and Türkay 1999: 133.

Metaplax indica — Clayton 1986: 87, fig 3; Jones 1986a: 161, pl 47; Tirmizi and Ghani 1996: 161, fig 62; Al-Khayat and Jones 1999: 58, 61; Ng et al. 2008: 226 (in list); Naderloo 2011: 31–34, figs 15a–e, 18c, d; Naderloo and Türkay 2012: 48; Naderloo et al. 2013: 450, tab 1, 456, tab 2.

Metaplax indicus occidentalis — Pretzmann 1971: 478, pls 1–3, figs 1–4; Titgen 1982: 254 (in list).

Type locality: Indian Ocean.

Distribution

Persian Gulf: Iran (Pretzmann 1971, Naderloo 2011; Naderloo and Türkay 2012; Naderloo et al. 2013), Kuwait (Clayton 1986; Jones 1986a), Saudi Arabia (Apel 1994a, 1996a, b), Qatar (Al-Khayat and Jones 1999).

Gulf of Oman: Iran (Naderloo 2011; Naderloo et al. 2015).

World distribution: Northern Indian Ocean: Persian Gulf, Gulf of Oman, Pakistan and India.

Habitat: Intertidal, substrates muddy flats and mangroves.

Fig. 32.2 *Metaplax indica* H. Milne Edwards, 1852. Male (ZUTC Brach1160), CL = 16.6, CB = 24.9 mm, Shah-Abdollah, Bushehr, Iran, Persian Gulf, 14.04.2006, coll. R. Naderloo and A. Kazemi (Photo: S. Träkner. Reproduced from Naderloo (2011) by permission of Magnolia Press)

Fig. 32.3 *Metaplax indica* H. Milne Edwards, 1852. The species with live color in its habitat, Saudi Arabia, Persian Gulf (Photo: Michael Apel)

Genus *Pseudohelice* Sakai, Türkay and Yang, 2006

Pseudohelice subquadrata **(Dana, 1851)**
(Figs. 32.4 and 32.6)

Chasmagnathus subquadratus Dana 1851: 251.
Helice leachii Hess 1865: 153.
Helice leachii — Vannini and Valmori 1981: 90, fig 17e; Hywell-Davies 1994: 29, 37, 48; Apel 2001: 117.
Pseudohelice (*Pseudohelice*) *subquadrata* — Sakai et al. 2006: 37–46, figs 50–59.
Pseudohelice quadrata — Ng et al. 2008: 227 (in list) [typing error].

Type locality: Type: Possibly Australia, New South Wales (Dana 1851). — Neotype: Austraila, Sydnsy, Sailors Bay (Sakai et al. 2006).

Distribution

Persian Gulf: No record.

Gulf of Oman: Oman (Hywell-Davies 1994).

World distribution: Indo-West Pacific: Mauritius, Kenya, Seychelles, Red Sea, Gulf of Oman, Thailand, Taiwan, China, Japan, Philippines, Indonesia, Caroline Islands, Australia, New Caledonia, Fiji, Samoa and French Polynesia.

Habitat: Intertidal, substrate mangroves.

Fig. 32.4 *Pseudohelice subquadrata* (Dana, 1851). Female (SMF 48471), CL = 14.9, CB = 17.4 mm, Khor Qarya, Socotra, 23.03.1999, coll. M. Apel (Photo: S. Träkner)

Genus *Thalassograpsus* Tweedie, 1950

Thalassograpsus harpax (Hilgendorf, 1892)
(Figs. 31.11f, 32.5, and 32.6)

Brachynotus harpax Hilgendorf 1892: 38.
Brachynotus harpax — Nobili 1906b: 320, 321.
Thalassograpsus harpax — Tweedie 1950: 134, fig 4; Guinot 1967a: 286; Hartnoll 1975: 307, 310, 317, 322 (in list); Holthuis 1977: 169, 170; Hogarth 1994: 105; Davie and Ng 2007: 258, 259, figs 1, 2, 3; Ng et al. 2008: 227 (in list); Naderloo 2011: 34–36, figs 16a–e, 18a, b; Naderloo and Türkay 2012: 48; Naderloo et al. 2013: 450, tab 1.

Type locality: Gulf of Aden.

Distribution

Persian Gulf: Iran (Naderloo 2011; Naderloo and Türkay 2012; Naderloo et al. 2013).

Gulf of Oman: Oman (Hogarth 1989).

World distribution: Indo-West Pacific: Gulf of Aden, southern Oman, Persian Gulf, Gulf of Oman, China, Taiwan, Japan, Philippines, Singapore, Cocos (Keeling) Islands and Samoa.

Habitat: Intertidal, substrate rocky/cobble.

Fig. 32.5 *Thalassograpsus harpax* (Hilgendorf, 1892). (SMF 38446–38453), CL = 9.85–12.70, CB = 10.50–14.40 mm, Qeshm Island, Iran, Persian Gulf, coll. R. Naderloo. Photo: S. Träkner. (**a**, **b**) dorsal view of females with different colour; (**c**) dorsal view of a male, (**d**) the same male from ventral view, with one cheliped large; (**e**) male with two chelipeds large; (**f**) same species in its habitat with gravel substrate

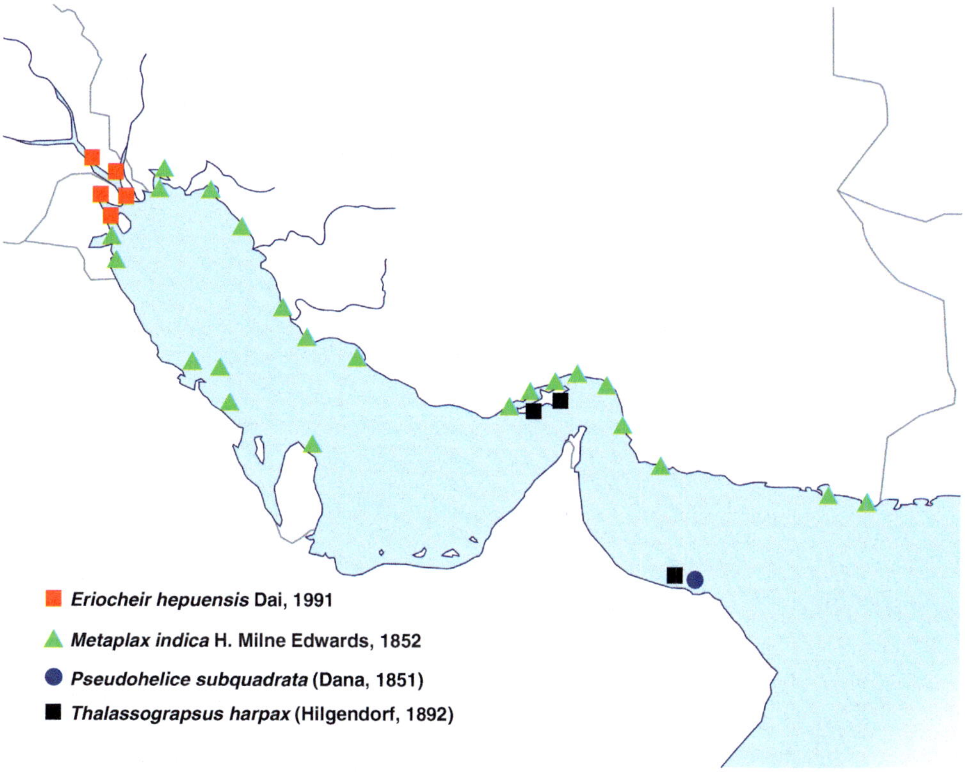

Fig. 32.6 Map showing the recorded localities of *Eriochier hepuensis* Dai, 1991, *Metaplax indica* H. Milne Edwards, 1852, *Pseudohelice subquadrata* (Dana, 1851), and *Thalassograpsus harpax* (Hilgendorf, 1892)

Family Plagusiidae Dana, 1851 (Rafting Crabs)

33

Diagnosis Carapace subcircular, posterior surface relatively flat, sometimes slightly convex, usually with short setose sterae or tense short setae, lateral margins markedly arched, anterolateral margins with several spines. Front dentate, divided into three lobes by markedly deep notches. Orbits well developed, deep, slightly shorter than front. Pterygostomian ridge serrated with setae. Antennules with deep sockets, obliquely folded longitudinally. Antennae with developed basal segment, distal margin sinuous, flagellum short. Third maxillipeds without distinct rhomboidal gap between, mandibles invisible when maxillipeds closed. Abdomen of males and females with somites 3–5 or 3–6 fused. G1 robust, long, straight, with particular apical process surrounded by setae; G2 shorter than G1. Genital opening of males and females sternal.

Key to the species of the family Plagusiidae in the region

- Dosral surface of carapace with granule-shaped tubercles surronding with short setae; front apically triangular, without spines, distincly broader than inner orbital angle; chelipeds without any setae on inner surface of merus and palm; walking legs without spines along anterior margin of merus .... *Plagusia squamosa*

- Dorsal surface of carapace without such tubercles, surface evenly covered with short setae; front appically with spine, as narrow as or narrower than inner orbital angle; chelipeds with dense setal patch on inner surface of merus and palm; walking legs with several large spines along anterior margin of merus .... *Percnon affine*

Genus *Percnon* Gistel, 1848

***Percnon affine* (H. Milne Edwards, 1853)**
(Figs. 33.1 and 33.3)

Acanthopus affine H. Milne Edwards 1853: 180.
Acanthopus pilmanus A. Milne-Edwards 1873: 300, pl 14, fig 5.
Percnon affinis — Nobili 1907: 406, 407.
Percnon affine — Crosnier 1965: 83 (in key), 86–88; figs 132, 133, 137, 142, 147, 148, pl VII, fig 1; McLay 1989: 106–109, figs 1, 2; Naderloo and Fatemi 2015: 1–3, figs 2–5.

Type locality: Hawaiian Islands.

Distribution

Persian Gulf: No record.

Gulf of Oman: Iran (Naderloo and Fatemi 2015).

World distribution: Indo-West Pacific: Madagascar, Gulf of Oman, Cocos (Keeling) Islands, New Caledonia, Hawaii, Tahiti, Fiji and Niue Islands.

Habitat: Intertidal to shallow subtidal, substrate rocky.

R. Naderloo, *Atlas of Crabs of the Persian Gulf*, DOI 10.1007/978-3-319-49374-9_33

Fig. 33.1 *Percnon affine* (H. Milne Edwards, 1853). Male (ZUTC 5314), CL = 35.57, CB = 31.88 mm, Chabahar, Iran, Gulf of Oman, 01.2014, coll. Y. Fatemi (Photo: R. Abdollahi)

Genus *Plagusia* Latreille, 1804

Plagusia squamosa (Herbst, 1790)

(Figs. 33.2 and 33.3)

Cancer squamosus Herbst 1790: 260, 261, pl 90, fig 113.
Plagusia tuberculata Lamarck 1818: 247.
Plagusia squamosa — H. Milne Edwards 1837a: 94; Nobili 1906b: 324; Schubart and Ng 2000: 327, fig 3; Ng et al. 2008: 218 (in list); Naderloo 2011: 36–40, figs 17a–f, 18e–f.
Plagusia tuberculata — Holthuis 1977: 176; Vannini and Valmori 1981: 92, figs 25a, 26a, 27; Hogarth 1989: 113; Tirmizi and Ghani 1996: 148–150, fig 57; Apel 2001: 120.
Plagusia depressa var. *squamosa* — Alcock 1900: 438.

Type locality: *Plagusia squamosa*: "Ostindien". East India or Indo-Malayan Archipelago. — *Plagusia tuberculata*: Mauritius.

Distribution

Persian Gulf: No record.

Gulf of Oman: Iran (Naderloo 2011; Ghotbeddin et al. 2012a; Naderloo et al. 2015), UAE [Apel (2001) as *P. tuberculata*].

World distribution: Indo-West Pacific, Mediterranean, and Eastern Pacific: South Africa, Madagascar, Comoros, Mauritius, East Africa, Somalia, Red Sea, Gulf of Aden, southern Oman, Gulf of Oman, Pakistan, India, China, Japan, Australia, New Zealand, Hawaii, Mediterranean and Eastern Pacific.

Habitat: Intertidal, substrate rocky.

Fig. 33.2 *Plagusia squamosa* (Herbst, 1790). Male (ZUTC 5565), CL = 19.91, CB = 22.18 mm, Chabahar, Iran, Gulf of Oman, 02.2014, coll. Y. Fatemi (Photo: R. Abdollahi)

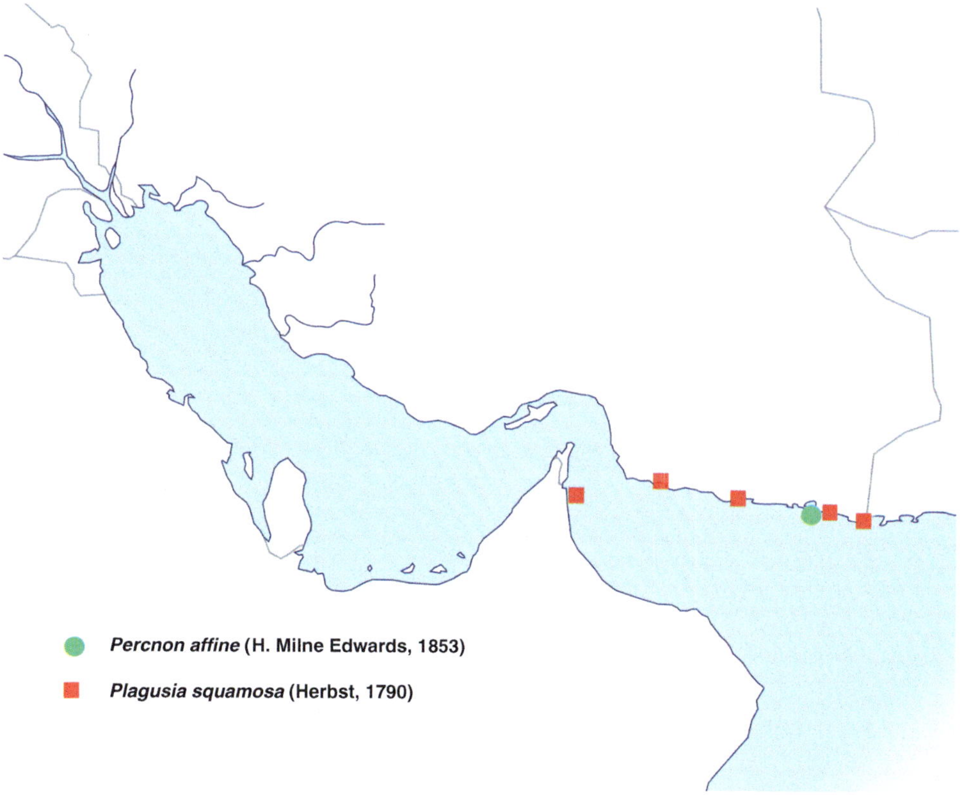

Fig. 33.3 Map showing the recorded localities of *Percnon affine* (H. Milne Edwards, 1853), and *Plagusia squamosa* (Herbst, 1790)

Family Camptandriidae Stimpson, 1858

34

Diagnosis Carapace subquadrate to subovate, broader than long, posterior surface smooth, moderately convex; lateral margins slightly to moderately arched, no distinct junction between anterolateral and posterolateral margins, dorsolateral margins not distinct. Front broad, straight, nearly entire. Orbits long, about as broad as front, margins serrated with setae; eyestalks slender, long with long plumy setae. Antennules folded transversely, slightly obliquely. Antennae with short basal segment, nearly reaching to frontal margin. Third maxillipeds with narrow gap betwenn, merus as long as or longer than ischium. Chelipeds nearly equal, margin serrated with long setae, fingers long, usually twice of palm length, fingers sharply blade-shaped apically. Walking legs long, robust, setose, less setose in females, third one largest. Abdomen of males and females with seven freely moving somites, or some somites partially fused. G1 long, characteristically bent apically to mesial direction, with robust subapical spines; G2 much shorter than G1. Genital opening of males and females sternal.

Key to genera of the family Camptandriidae in the region

1. Carapace transversely oval, markedly broader than long, about 1.5 times as broad as long, or broader; lateral margins non-parallel 2
- Carapace subquadrate, very slightly broader than long; lateral margins straight, parallel 4
2. Carapace about 1.5 times as broad as long, dorsal surface slightly convex, with dense long setae; front entire from dorsal view; anterolateral margins of carapace longer than posterolateral margins, straight; chelipeds without sexual dimorphism .... *Opusia*
- Cararapce more than 1.5 times as broad as long, dorsal surface distinctly convex, without setae; front slightly bilobed from dorsal view; anterolateral margins of carapace distinctly shorter than posterolateral margins; chelipeds with sexual dimorphism 3
3. Abdomen of males with somites 5–7 freely moving, sutures between somites 2–4 complete; lateral angle of front touching inner infraorbital anlge, then closing orbit; posteromedian tooth of epistum broadly triangular, lateral margins concave; palm of chelipeds quadriangular .... *Manningis*
- Abdomen of males with somites 5–6 with limited movability, sutures between somites 2–4 incomplete; lateral angle of front not touching inner infraorbital anlge; posteromedian tooth of epistum sharply triangular, lateral margins convex; palm of chelipeds round .... *Serenella*
4. Carapace about 1.3 times as broad as long; carapace regions relatively well-defined by scattered groups of granules; anterolateral margins with small tooth behind exorbital tooth; chelipeds without sexual dimorphism .... *Nasima*
- Carapace about 1.4 times as broad as long; carapace regions weakly defined by small granules; anterolateral margins entire, without tooth; chelipeds with sexual dimorphism .... *Leptochryseus*

R. Naderloo, *Atlas of Crabs of the Persian Gulf*, DOI 10.1007/978-3-319-49374-9_34

Genus *Leptochryseus* Al-Khayat and Jones, 1996

Leptochryseus kuwaitensis (Jones and Clayton, 1983)

(Figs. 34.1, 34.2, 34.4, and 34.10a)

Cleistostoma kuwaitense Jones and Clayton 1983: 185, fig 2.
Cleistostoma kuwaitense — Clayton 1986: 86–88, fig 3; Jones 1986a: 158, pl 47; Manning 1991: 299, 300; Ismail and Ahmed 1993: 158, 160, 161.
Leptochryseus kuwaitense — Al-Khayat and Jones 1996: 798, 799; Apel and Türkay 1999: 135; Apel 2001: 108; Ng et al. 2008: 233 (in list); Naderloo and Türkay 2012: 48.
Leptochryseus kuwaitensis — Ng et al. 2009: 21–25, figs 7, 8, 9b, 10b, 11g–l.
Type locality: Persian Gulf (Kuwait).
Distribution
Persian Gulf: Iran (Naderloo and Türkay 2012), Kuwait (Jones and Clayton 1983; Clayton 1986; Jones 1986a; Apel 2001).
Gulf of Oman: Iran (Naderloo et al. 2015), UAE (Ismail and Ahmed 1993).
World distribution: Northern Indian Ocean: Persian Gulf and Gulf of Oman.
Habitat: High intertidal, substrate muddy.

Fig. 34.1 *Leptochryseus kuwaitensis* (Jones and Clayton, 1983). (SMF 40498), CL = 24.5, CB = 31.6 mm, behind Bavirat Police State, 15 km E Bandar-Deylam, Iran, Persian Gulf, 23.05.2008, coll. R. Naderloo, A. Kazemi and H. Salehi (Photo: S. Träkner)

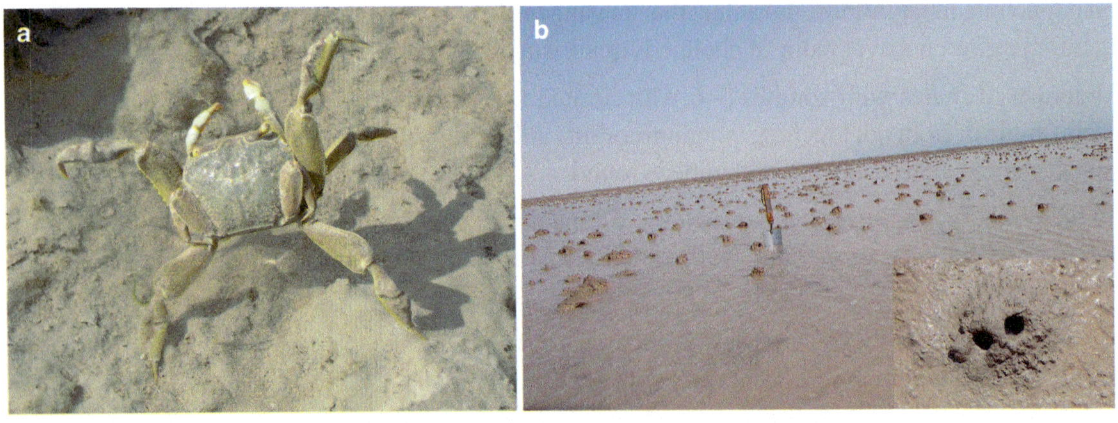

Fig. 34.2 *Leptochryseus kuwaitensis* (Jones and Clayton, 1983). (**a**) The crab in a courtship display, Kuwait, 29.10.2007, Photo: Zainab Sattari; (**b**) the habitat and burrow structure of the crab, Hendijan, Iran, 23.05.2008 (Photo: R. Naderloo)

Genus *Manningis* Al-Khayat and Jones, 1996

Manningis arabicum (Jones and Clayton, 1983) (Figs. 34.3, 34.4, and 34.10b)

Paracleistostoma arabicum Jones and Clayton 1983: 190–192, fig 4.
Paracleistostoma arabicum — Jones 1986a: 158, pl 46; Manning 1991: 299, 300; Apel and Türkay 1992: 191, 204, 205; Apel 1994a: 42–45, figs 1, 2; Apel 1996b: 42–45, figs 1, 2.
Paracleistostoma sp. — Basson et al. 1977: 228.
Manningis arabicum — Al-Khayat and Jones 1996: 797, 798, 800, 809, 810, 812, fig 7; Al-Khayat and Jones 1999: 58, 61; Apel and Türkay 1999: 135; Apel 2001: 111; Ng et al. 2008: 233 (in list); Ng et al. 2009: 10, 11, figs 1b, 4; Naderloo and Türkay 2012: 49; Naderloo et al. 2013: 450, tab 1, 456, tab 2.
Type locality: Persian Gulf (Kuwait).
Distribution

Persian Gulf: Iran (Naderloo and Türkay 2012; Naderloo et al. 2013), Iraq (Ng et al. 2009), Kuwait (Jones and Clayton 1983; Jones 1986a), Saudi Arabia (Basson et al. 1977; Apel 1994a, 1996b, 2001), Qatar (Al-Khayat and Jones 1996, 1999).

Gulf of Oman: No record.

World distribution: Endemic: Northeastern Persian Gulf including entire coast of Iran, Iraq, Kuwait, distributed southward to Saudi Arabia and Gulf of Qatar.

Habitat: Intertidal, substrates muddy and mangroves.

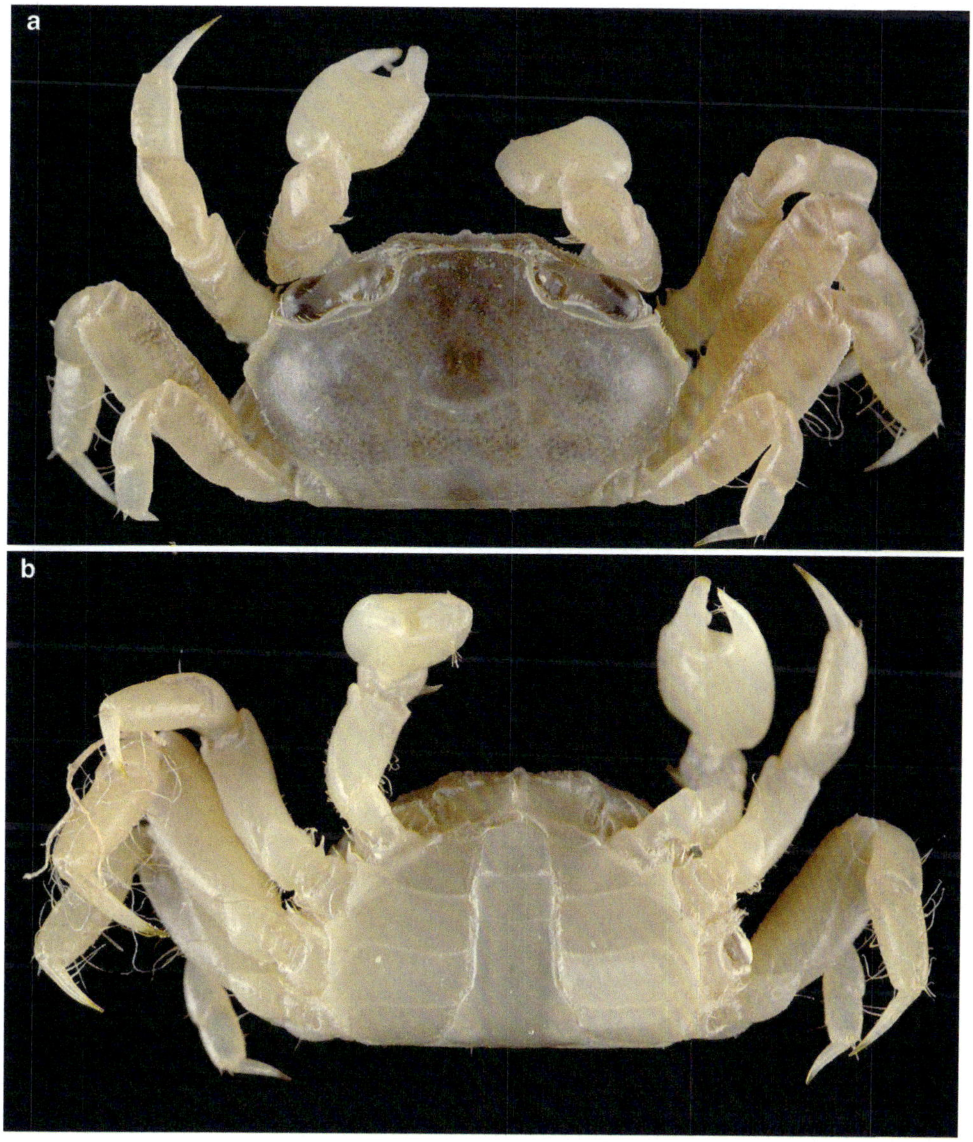

Fig. 34.3 *Manningis arabicum* (Jones and Clayton, 1983). Male (SMF 40510), CL = 7.50, CB = 4.80 mm, mangroves of Dargahan, Qeshm Island, Iran, Persian Gulf, 13.01.2008, coll. M. Türkay and R. Naderloo, (**a**) dorsal view, (**b**) ventral view (Photos: S. Träkner)

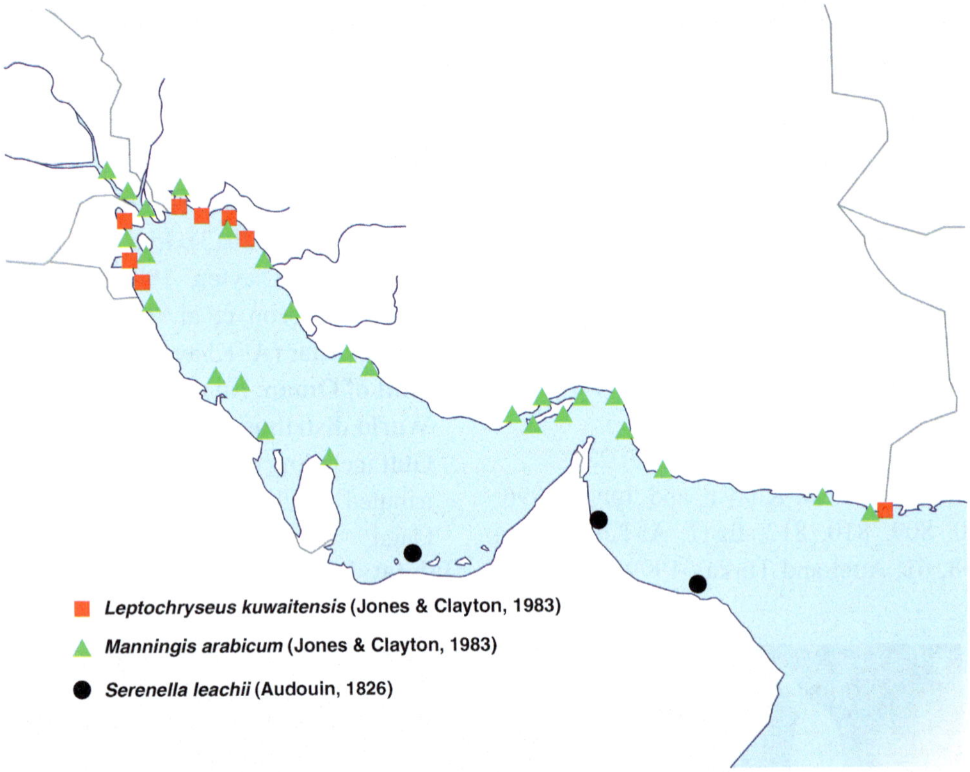

Fig. 34.4 Map showing the distribution of *Leptochryseus kuwaitensis* (Jones and Clayton, 1983), *Manningis arabicum* (Jones and Clayton, 1983), and *Serenella leachii* (Audouin, 1826)

Genus *Nasima* Manning, 1991

Nasima dotilliformis (Alcock, 1900)

(Figs. 34.5, 34.6, 34.8, and 34.10c)

Clistostoma [sic!] *dotilliforme* Alcock 1900: 373.
Clistostoma [sic!] *dotilliforme* — Alcock and Mcardle 1903: pl 64, fig 1.
Cleistostoma dotilliforme — Stephensen 1946: 194; Basson et al. 1977: 53, 56, 228, fig 32; Titgen 1982: 253 (in list); Jones and Clayton 1983: 188, 189, fig 3; Clayton 1986: 87, 88, fig 3; Jones 1986a: 158, pl 45; Vousden 1987: 36; Apel 1994a: 42–45, pl 1.
Nasima dotilliformis — Manning 1991: 304, 305, fig 4; Snowden and Clayton 1995: 84–92; Apel 1996b: 328, 330–337, figs 1, 3, 4; Al-Khayat and Jones 1996: 797, 800, 806, 808, 809, 812, fig 6; Tirmizi and Ghani 1996: 92–94, fig 35; Hornby 1997: 15; Al-Khayat and Jones 1999: 58, 61; Apel and Türkay 1999: 135; Ng et al. 2008: 233 (in list); Ng et al. 2009: 15, 16, figs 5, 6, 10c; Naderloo and Türkay 2012: 49; Naderloo et al. 2013: 450, tab 1, 456, tab 2.
Nasima dotilliforme — Apel 2001: 111.
Nasima dotilleformis [sic!] — Hywel-Davies 1994: 29, 37, 48.
Cleistostomata [sic!] nr. *dotilliforme* — Evans et al. 1973: 259, 276.

Type locality: Pakistan (Karachi).

Distribution

Persian Gulf: Iran (Stephensen 1946; Naderloo and Türkay 2012; Naderloo et al. 2013), Iraq (Ng et al. 2009), Kuwait (Jones and Clayton 1983; Jones 1986a; Snowden and Clayton 1995), Saudi Arabia (Basson et al. 1977; Apel 1994a, 1996b, 2001), Qatar (Al-Khayat and Jones 1996, 1999).

Gulf of Oman: Iran (Naderloo et al. 2015), UAE (Apel 2001), Oman (Hywel-Davies 1994).

World distribution: Northern Indian Ocean: Persian Gulf, Gulf of Oman and Pakistan.

Habitat: Intertidal, substrates muddy and mangroves.

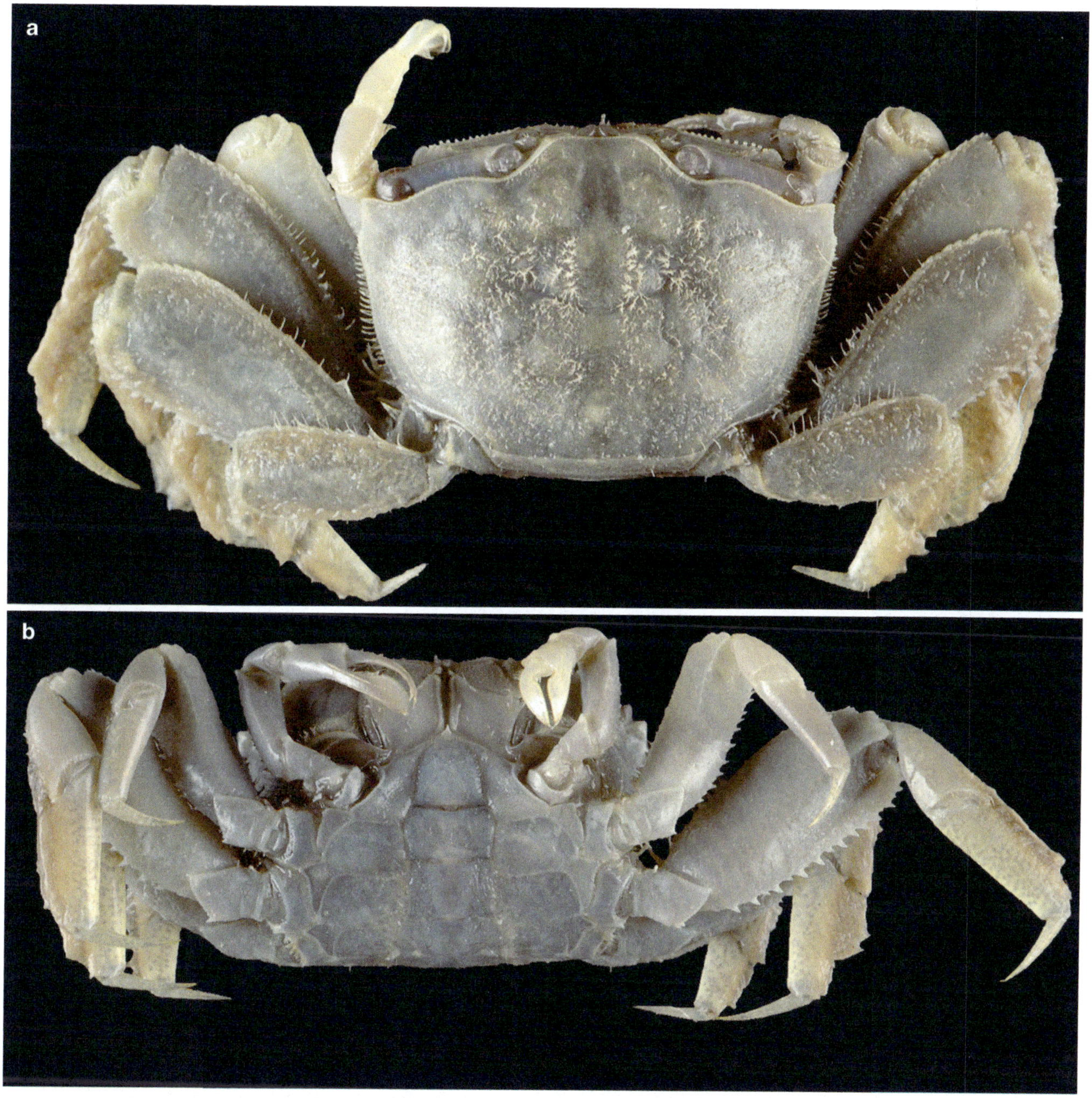

Fig. 34.5 *Nasima dotilliformis* (Alcock, 1900). Male (SMF 40534), CL = 7. 8 mm, CB = 11.5.8 mm, Mahshahr, Iran, Persian Gulf, 21.05.2008, coll. R. Naderloo and A. Kazemi, (**a**) dorsal view, (**b**) ventral view (Photos: S. Träkner)

Fig. 34.6 *Nasima dotilliformis* (Alcock, 1900). The crab in entrance of its burrow, Yadman-Shohadaye Valfajr, Arvandroud, Iran, 02.11.2012 (Photo: R. Naderloo)

Genus *Opusia* Ng, Rahayu and Naser, 2009

Opusia indica (Alcock, 1900)

(Figs. 34.7, 34.8, and 34.10d)

Tylodiplax indica Alcock 1900: 374, 375.

Tylodiplax indica — Alcock and Mcardle 1903: pl 64, fig 2; Stephensen 1946: 192–194, fig 58b–g; Clayton 1986: 87, 88, fig 3; Jones 1986a: 158, pl 46; Ismail and Ahmed 1993: 158; Snowden et al. 1994: 219–227; Apel and Türkay 1999: 136.

"*Tylodiplax*" *indica* — Ng et al. 2008: 234 (in list).

Serenella indica — Tirmizi and Ghani 1996: 95, 96, fig 36.

Opusia indica — Ng et al. 2009: 6–8, figs 1a, 2; Naderloo and Türkay 2012: 49; Naderloo et al. 2013: 450, tab 1, 456, tab 2.

Type locality: Pakistan (Karachi).

Distribution

Persian Gulf: Iran (Stephensen 1946; Ng et al. 2009; Naderloo and Türkay 2012; Naderloo et al. 2013), Iraq (Ng et al. 2009), Kuwait (Jones 1986a; Snowden et al. 1994), Saudi Arabia (Stephensen 1946), UAE (Apel 2001).

Gulf of Oman: Iran (Naderloo et al. 2015), UAE (Ismail and Ahmed 1993).

World distribution: Northern Indian Ocean: Persian Gulf, Gulf of Oman and Pakistan.

Habitat: Intertidal, substrates muddy and mangroves.

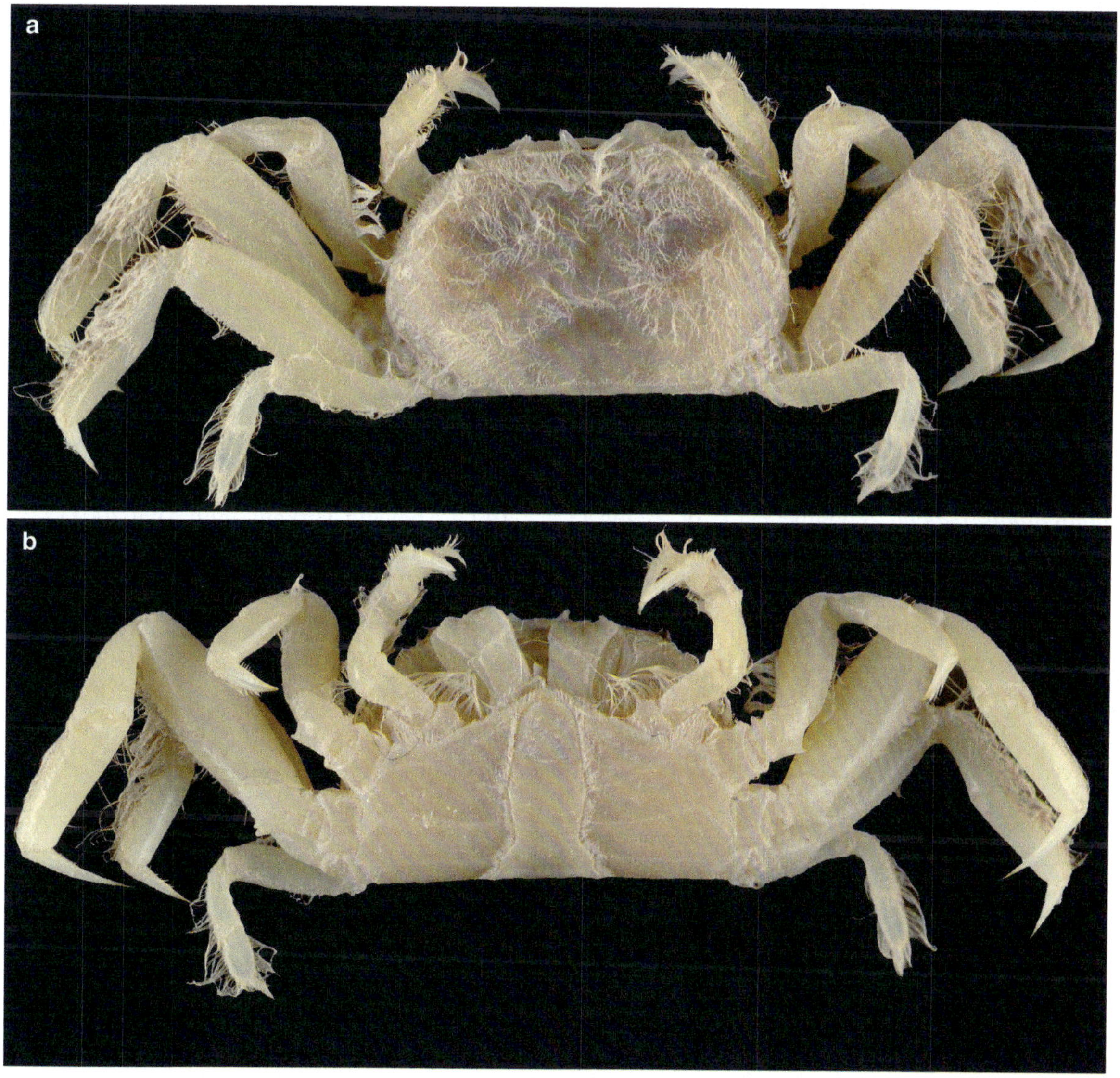

Fig. 34.7 *Opusia indica* (Alcock, 1900). Male (SMF 40558), CL = 5.60, CB = 10.20, Park Saheli, Bandar-Emam, Iran, Persian Gulf, 21.05.2008, coll. R. Naderloo and H. Salehi, (**a**) dorsal view, (**b**) ventral view (Photos: S. Träkner)

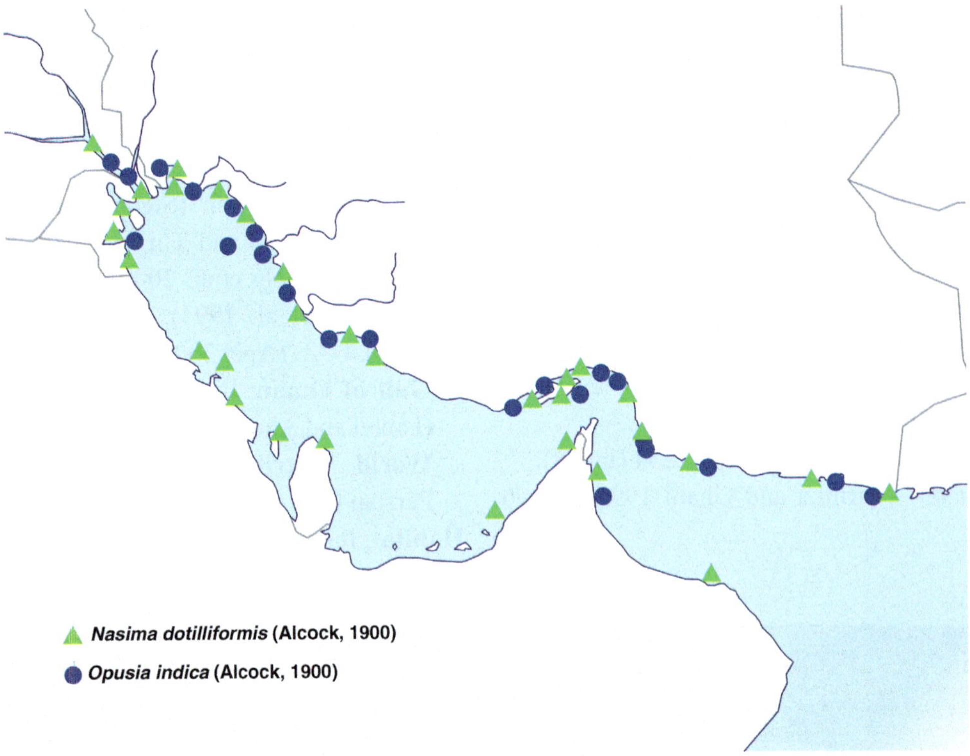

Fig. 34.8 Map showing the distribution of *Nasima dotilliformis* (Alcock, 1900), and *Opusia indica* (Alcock, 1900)

Genus Serenella Manning and Holthuis, 1981

Serenella leachii (Audouin, 1826)

(Figs. 34.4, 34.9, and 34.10e)

Macrophthalmus Leachii Savigny 1817: pl 2, fig 1; Audouin 1826: 257.

Serenella leachii — Manning and Holthuis 1981: 211, 212, fig 55a–d; Hywel-Davies 1994: 29, 37, 48; Al-Ghais and Cooper 1996: 422, 423, fig 9; Al-Khayat and Jones 1996: 797–799, 800–803, 812, figs 1, 2; Cooper 1997: 159–160; Al-Khayat and Jones 1999: 58, 61; Apel and Türkay 1999: 135, 136; Apel 2001: 113; Ng et al. 2008: 233.

Type locality: Red Sea (northern part).

Distribution

Persian Gulf: Qatar (Al-Khayat and Jones 1996, 1999), UAE (Al-Ghais and Cooper 1996; Cooper 1997; Apel and Türkay 1999; Apel 2001).

Gulf of Oman: UAE (Apel and Türkay 1999; Apel 2001).

World distribution: Northwestern Indian Ocean: Red Sea, Gulf of Aden, Persian Gulf and Gulf of Oman.

Habitat: Intertidal, substrates muddy and mangroves.

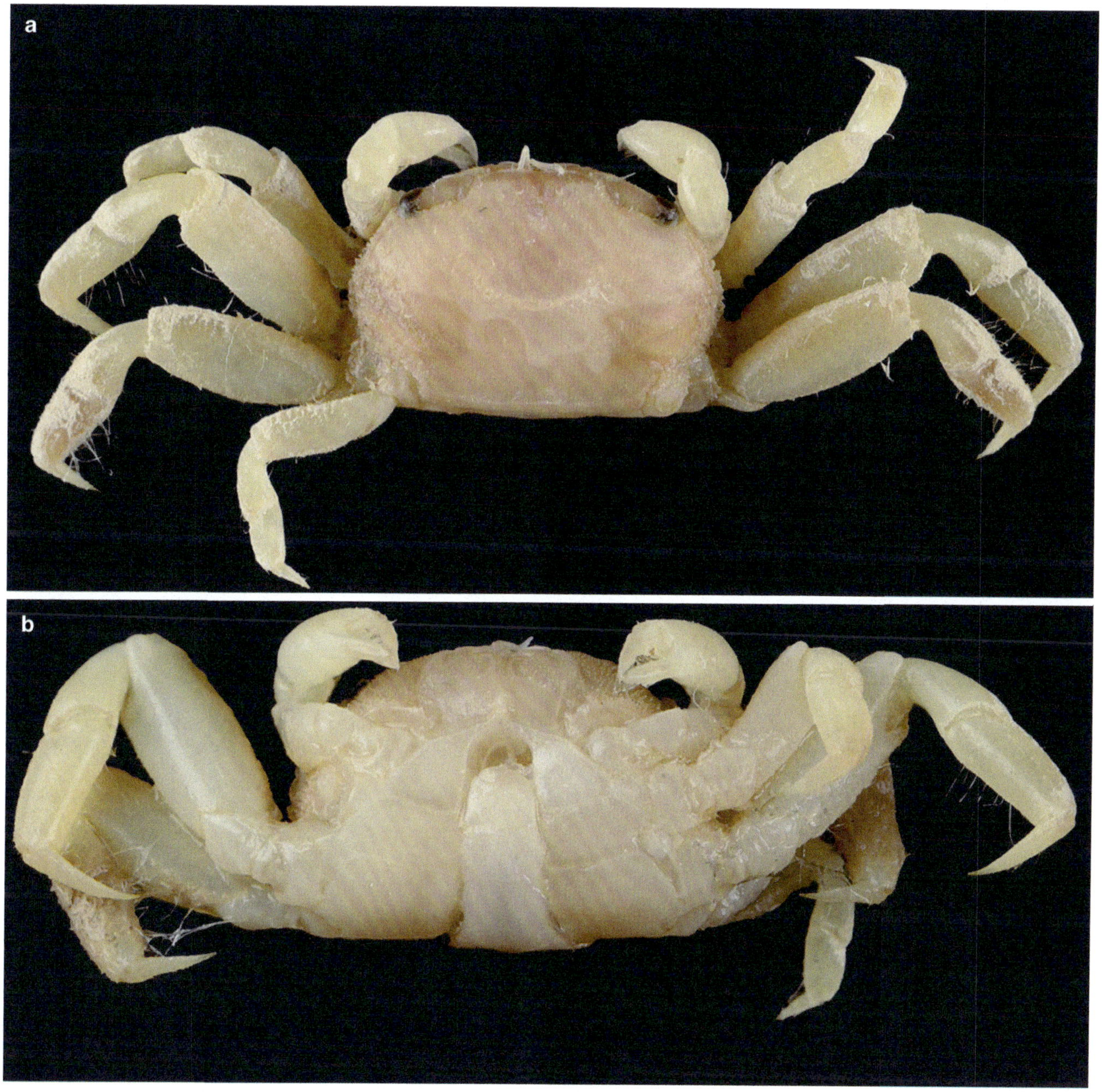

Fig. 34.9 *Serenella leachii* (Audouin, 1826). Male (SMF), CL = 3.7, CB = 6.8 mm, mangroves of Qurm, Muscat, Oman, Gulf of Oman, 07.1994, coll. A. Hywel-Davies, (**a**) dorsal view, (**b**) ventral view (Photos: S. Träkner)

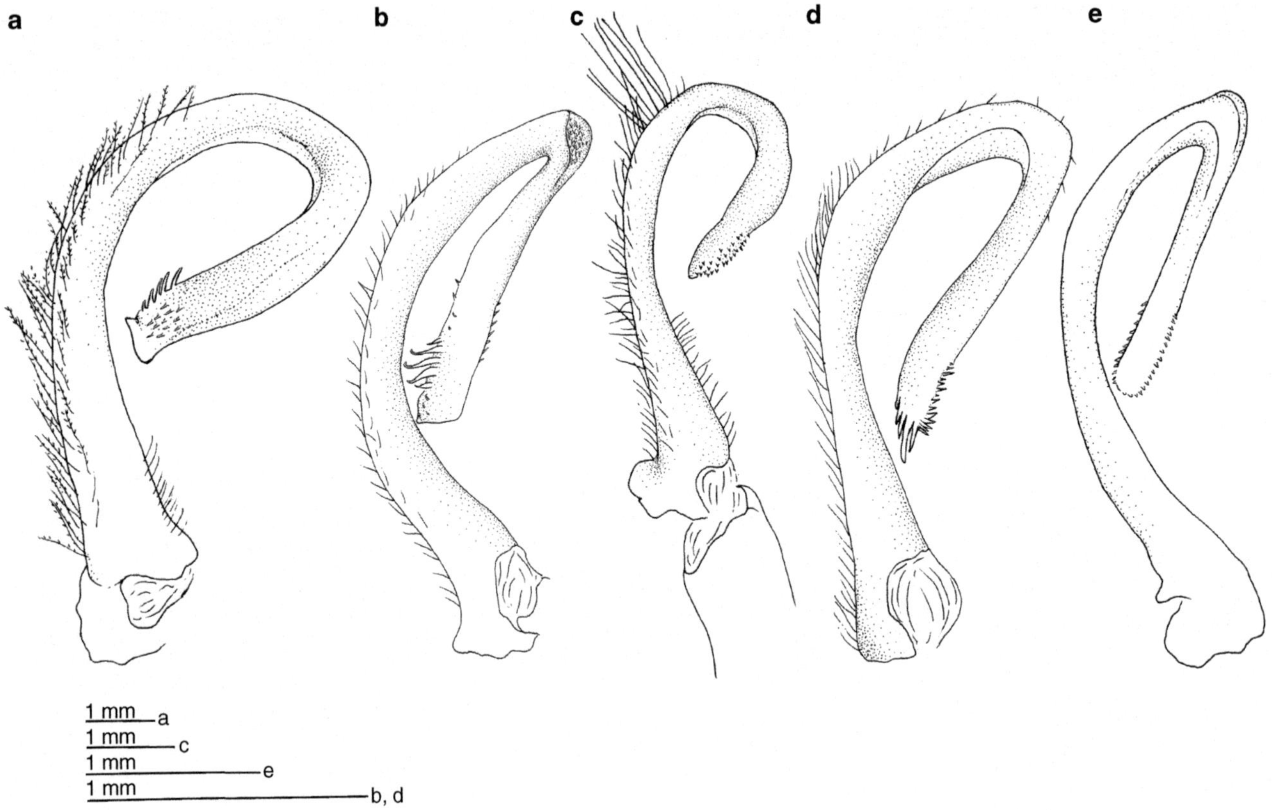

Fig. 34.10 Male G1, ventral view: (**a**) *Leptochryseus kuwaitensis* (Jones and Clayton, 1983) (SMF 40498); (**b**) *Manningis arabicum* (Jones and Clayton, 1983) (SMF 40510); (**c**) *Nasima dotilliformis* (Alcock, 1900) (SMF 40534); (**d**) *Opusia indica* (Alcock, 1900) (SMF 40558); (**e**) *Serenella leachii* (Audouin, 1826) (SMF: Muscat, Gulf of Oman)

Family Dotillidae Stimpson, 1858 (Soldier Crabs)

35

Diagnosis Carapace subglobular, squarish or transversely rectangular, always broader than long, relatively thick; dorsal surface usually flat, slightly convex, with special sculpture or without it; lateral margins entire, without any dentation, diverging posteriorly; no junction between anterolateral and posterolateral margins. Front distinctly narrow, simple. Orbits long, shallow, not covering eyestalks; eyestalks elongate, exposed. Third maxillipeds swollen, with narrow gap between, merus as large as or larger than ischium. Chelipeds markedly robust, equal to slightly subequal, palm and carpus distincly enlarged, fingers long, apically pointed. Walking legs relatively large, flat, setal tufts between first and second walking legs. Abdomen of males and females with freely moving somites; somites 4 and 5 markedly modified, narrowed or serrated with setae. G1 with apical long process, characteristically curved medially, apically serrated with long spines. Genital opening of males and females sternal.

Key to genera of the family Dotillidae in the region

1. Third maxillipeds medium-sized, not bulged anteriorly, merus longer than ischium; chelipeds of both males and females markedly stouter than walking legs, without tympanium on upper surface of merus ..... *Ilyoplax*
- Third maxillipeds markedly large, distinctly bulged anteriorly 2
2. Third maxillipeds with merus longer than ischium; male abdomen with somites 4 and 5 normal, anterior margin of somite 4 with setae .... *Dotilla*
- Third maxillipeds with merus shorter than ischium; male abdomen with somites 4 and 5 extraordinary narrow and long .... *Scopimera*

Genus *Dotilla* Stimpson, 1858

***Dotilla blanfordi* Alcock, 1900**
(Figs. 35.1, 35.3, and 35.7c)

Dotilla blanfordi Alcock 1900: 366.
Dotilla blanfordi Alcock and Mcardle 1903: pl 63, figs 3, 3a.
Dotilla blanfordi — Jones 1986a: 160, pl 45; Jones 1986b: 78; Titgen 1982: 290; Apel and Türkay 1999: 136; Tirmizi and Ghani 1996: 127–129, fig 48; Apel 2001: 106, 107; Ng et al. 2008: 235 (in list); Fatemi et al. 2011: 1747–1749, fig 2; Naderloo and Türkay 2012: 50; Naderloo et al. 2013: 450, tab 1.
Dotilla (*wichmanni* De Man?) — Stephensen 1946 (partly): 190, 191, fig 57a, b.
Dotilla sulcata — Ismail and Ahmed 1993: 158; Hornby 1997: 15.

Type locality: Pakistan.

Distribution

Persian Gulf: Iran [Stephensen (1946) as *Dotilla wichmanni* De Man, 1892; Naderloo and Türkay 2012; Naderloo et al. 2013], Kuwait (Jones 1986a, b).

Gulf of Oman: Iran (Fatemi et al. 2011; Naderloo and Türkay 2012; Naderloo et al. 2015).

World distribution: Northern Indian Ocean: Persian Gulf, Gulf of Oman, Pakistan and India.

Habitat: High intertidal, substrate sandy.

R. Naderloo, *Atlas of Crabs of the Persian Gulf*, DOI 10.1007/978-3-319-49374-9_35

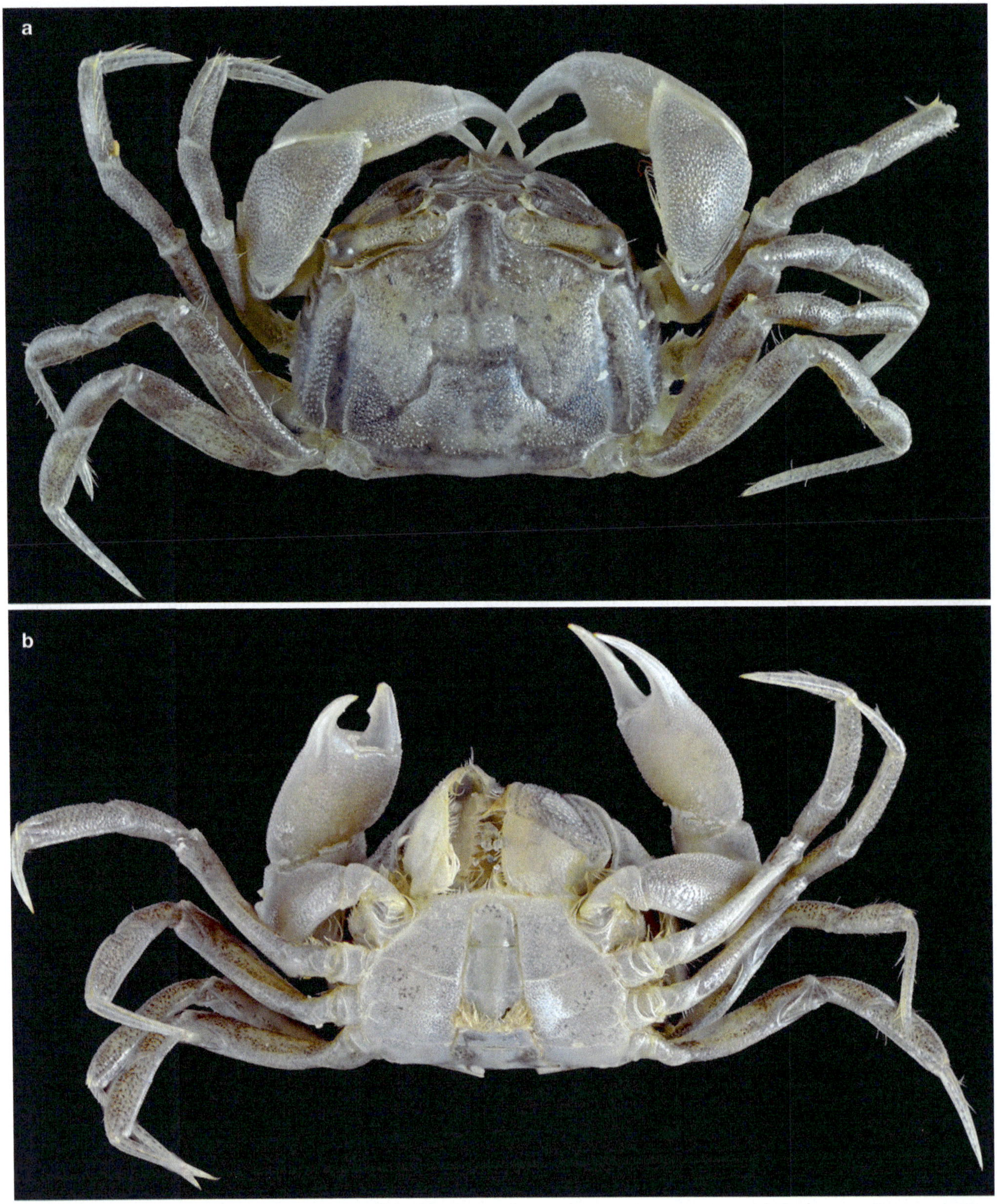

Fig. 35.1 *Dotilla blanfordi* Alcock, 1900. Male (ZUTC 5520), CL = 6.50, CB = 8.65 mm, Pozm, Sistan-Baluchestan, Iran, Persian Gulf, 22.05.2013, coll. R. Naderloo, (**a**) dorsal view, (**b**) ventral view (Photos: R. Abdollahi)

Dotilla sulcata (Forskål, 1775)

(Figs. 35.2, 35.3, and 35.7d)

Cancer sulcatus Forskål 1775: 92.
Dotilla affinis Alcock 1900: 364, 365.
Myctiris sulcatus — Savigny 1817: pl 1, fig 3i–iv.
Dotilla sulcata — De Man 1888b: 129–136, pl 9, fig 7; Titgen 1982: 146–148, 290 (in list); Clayton and Al-Kindi 1998: 197–212; Apel and Türkay 1999: 136; Apel 2001: 107; Ng et al. 2008: 235 (in list).
Dotilla affinis — Alcock and Mcardle 1903: pl 63, figs 1, 1a, b.
Dotilla blanfordi — Ismail and Ahmed 1993: 158.
Dotilla blandfordi [sic!] — Hornby 1997: 15.

Type locality: *Cancer sulcatus*: Red Sea (Suez). — *Dotilla affinis*: Gulf of Aden.

Distribution

Persian Gulf: UAE [Titgen 1982; Hornby (1997) as *D. blanfordi*].

Gulf of Oman: UAE [Ismail and Ahmed (1993) as *D. blanfordi*; Clayton and Al-Kindi 1998].

World distribution: Northwestern Indian Ocean: Somalia, Red Sea, Gulf of Aden, southeastern Persian Gulf and southwestern Gulf of Oman.

Habitat: High intertidal, substrate sandy.

Fig. 35.2 *Dotilla sulcata* (Forskål, 1775). Male (SMF 23897), CL = 7.5, CB = 8.7 mm, Khor Al-Hamra, Ras Al-Khaimah, UAE, Persian Gulf, coll. R.J. Hornby (Photo: S. Träkner)

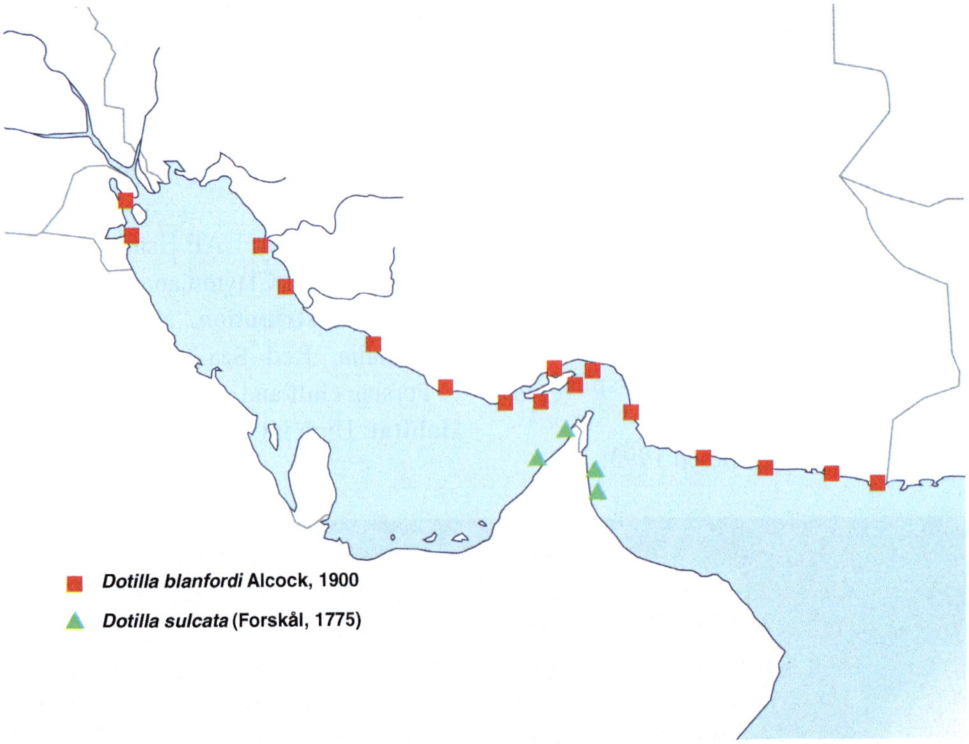

Fig. 35.3 Map showing the recorded localities of *Dotilla blanfordi* Alcock, 1900, and *Dotilla sulcata* (Forskål, 1775)

Genus *Ilyoplax* Stimpson, 1858

Key to species of the genus *Ilyoplax* in the region

- Carapace with lateral margins slightly converging anteriorly; front relatively broad, about 1/7 as long as breadth of carapace; G1 with apical part bent outward with nearly 45 degree .... *Ilyoplax frater*
- Carapace with lateral margins relatively striaght on anterior part; front narrow, about 1/11 as long as breadth of carapace; G1 with apical part gently bent outward .... *Ilyoplax stevensi*

***Ilyoplax frater* (Kemp, 1919)**
(Figs. 35.4, 35.7a, and 35.8)

Tympanomerus frater Kemp 1919: 342, 343, figs 16, 17.
Ilyoplax frater — Pretzmann 1971: 481, pl 7; Titgen 1982: 253 (in list); Clayton 1986: 87, 88, fig. 3; Jones 1986a: 159, pl 46; Jones 1986b: 77, fig 4; Vousden 1987: 37, pls 4, 7; Apel 1994a: 42, 45, figs 1, 2; Apel 1996b: 330–337, figs 1–3; Al-Khayat and Jones 1996: 797, 800, 809, 811–813, fig 8; Tirmizi and Ghani 1996: 133–136, fig 52; Cooper 1997: 163; Hornby 1997: 15; Al-Khayat and Jones 1999: 58, 61; Apel and Türkay 1999: 136; Apel 2001: 107; Ng et al. 2008: 235 (in list); Fatemi et al. 2011: 1749, 1750, fig 3; Naderloo and Türkay 2012: 50; Naderloo et al. 2013: 450, tab 1, 456, tab 2.

Type locality: Pakistan (Karachi).

Distribution

Persian Gulf: Iran (Pretzmann 1971; Naderloo and Türkay 2012; Naderloo et al. 2013), Kuwait (Jones 1986a, b; Clayton 1986), Saudi Arabia (Apel 2001), Bahrain (Vousden 1987; Apel 2001), Qatar (Al-Khayat and Jones 1999), UAE (Cooper 1997; Hornby 1997; Apel 2001).

Gulf of Oman: Iran (Fatemi et al. 2011; Naderloo et al. 2015).

World distribution: Northern Indian Ocean: Persian Gulf, Gulf of Oman and Pakistan.

Habitat: Intertidal, substrates muddy and mangroves.

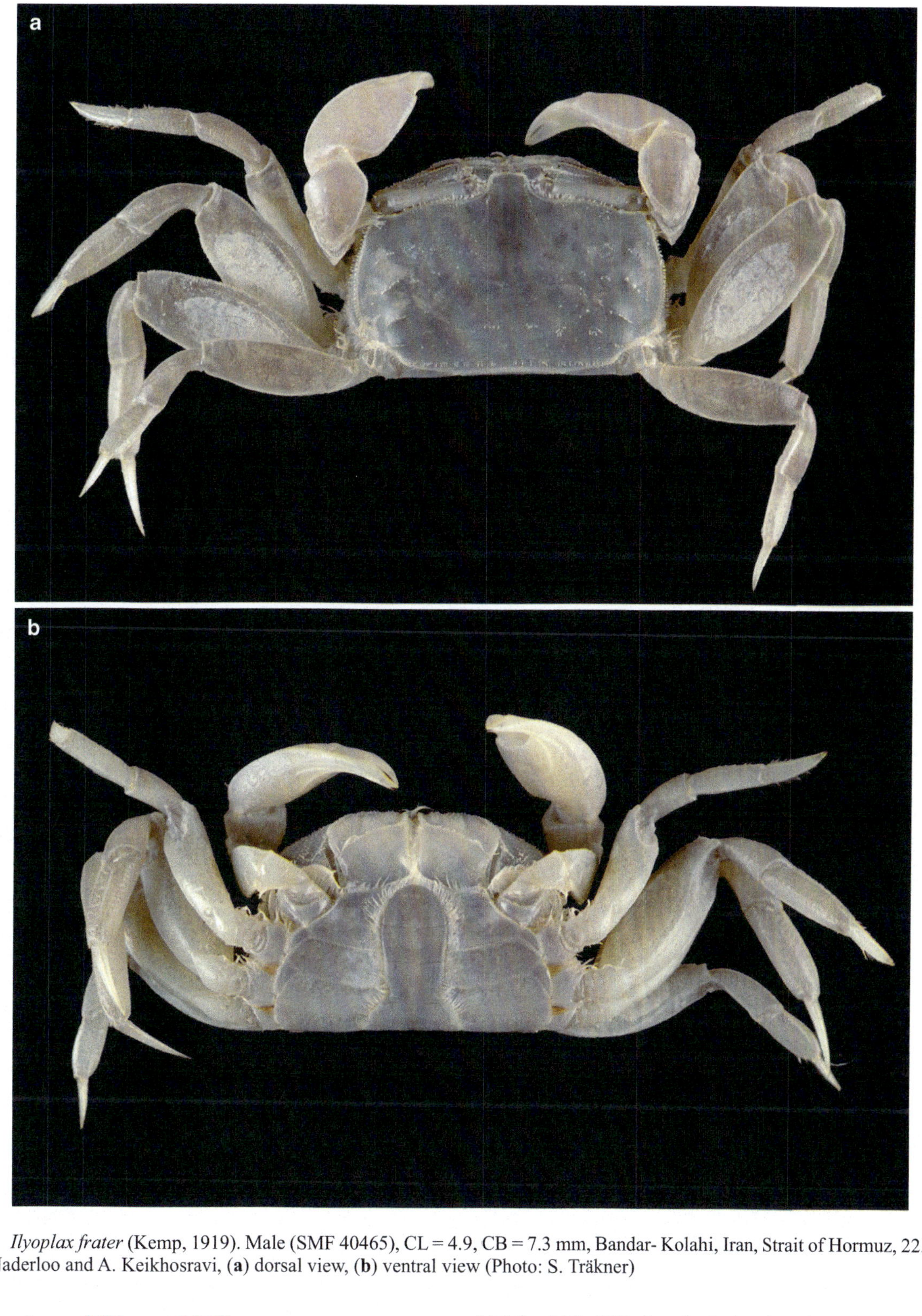

Fig. 35.4 *Ilyoplax frater* (Kemp, 1919). Male (SMF 40465), CL = 4.9, CB = 7.3 mm, Bandar- Kolahi, Iran, Strait of Hormuz, 22.04.2008, coll. R. Naderloo and A. Keikhosravi, (**a**) dorsal view, (**b**) ventral view (Photo: S. Träkner)

Ilyoplax stevensi (Kemp, 1919)
(Figs. 35.5, 35.7b, and 35.8)

Tympanomerus stevensi Kemp 1919: 339–342, figs 13–15, pl 23, fig 2.
Ilyoplax stevensi — Serène and Lundoer 1974: 9; Jones 1986a: 159, pl 46; Apel 1994a: 42–45, figs 1, 2; Apel 1996b: 330–337, figs 1–3; Apel and Türkay 1999: 136; Apel 2001: 107, 108; Ng et al. 2008: 235 (in list); Naderloo and Türkay 2012: 50; Naderloo et al. 2013: 450 tab 1, 456, tab 2.
Scopimera sp. — Basson et al. 1977: 56, 228.
Type locality: Pakistan (Karachi).

Distribution

Persian Gulf: Iran (Naderloo and Türkay 2012; Naderloo et al. 2013), Kuwait (Jones 1986a; Clayton 1986), Saudi Arabia (Apel 1994a, Apel 1996b, 2001).

Gulf of Oman: Iran (Fatemi et al. 2011; Naderloo et al. 2015).

World distribution: Northern Indian Ocean: Persian Gulf, Gulf of Oman and Pakistan.

Habitat: Intertidal, substrates muddy and mangroves.

Fig. 35.5 *Ilyoplax stevensi* (Kemp, 1919). Male (SMF 40485), CL = 4.7, CB = 7.1 mm, Hendijan, Iran, Persian Gulf, 21.05.2008, coll. R. Naderloo and H. Salehi, (**a**) dorsal view, (**b**) ventral view (Photos: S. Träkner)

Genus *Scopimera* De Haan, 1833

Scopimera crabicauda Alcock, 1900
(Figs. 35.6, 35.7e, and 35.8)

Scopimera crabicauda Alcock 1900: 370, 371.
Scopimera crabicauda — Alcock and Mcardle 1903: pl 63, fig 5a, b; Titgen 1982: 152, 153, 253 (in list); Jones 1986a: 160, pl 46; Vousden 1987: 34, 36, 37, tab 3, 6, 7; Apel and Türkay 1992: 191, 204, 205; Tirmizi and Ghani 1996: 124–126, fig 47; Cooper 1997: 160–163, figs 2, 8, 9; Hornby 1997: 15; Clayton and Al-Kindi 1998: 197–212; Al-Khayat and Jones 1999: 58, 61; Apel and Türkay 1999: 136; Apel 2001: 118; Ng et al. 2008: 235 (in list); Fatemi et al. 2011: 150–52, fig 4; Naderloo and Türkay 2012: 51; Naderloo et al. 2013: 450, tab 1.
Scopimera scabicauda [sic!] — Stephensen 1946: 189, 190, fig 56; Pretzmann 1971: 481, pl 6, figs 13–15; Basson et al. 1977: 26, 56, 59, 60, 63, 231, 235, fig 40; Clayton 1986: 87, 88, fig 3; Jones 1986b: 77, 78, fig 4.
Scopimera sp. — Evans et al. 1973: 275.
Type locality: Pakistan (Karachi).
Distribution
Persian Gulf: Iran (Stephensen 1946; Pretzmann 1971; Naderloo and Türkay 2012; Naderloo et al. 2013), Kuwait (Jones 1986a, b; Clayton 1986), Saudi Arabia (Basson et al. 1977; Apel 2001), Bahrain (Vousden 1987; Apel 2001), Qatar (Al-Khayat and Jones 1999), UAE (Cooper 1997; Hornby 1997; Apel 2001).
Gulf of Oman: Iran (Fatemi et al. 2011; Naderloo et al. 2015).
World distribution: Northern Indian Ocean: Persian Gulf, Gulf of Oman and Pakistan.
Habitat: High intertidal, substrate sandy.

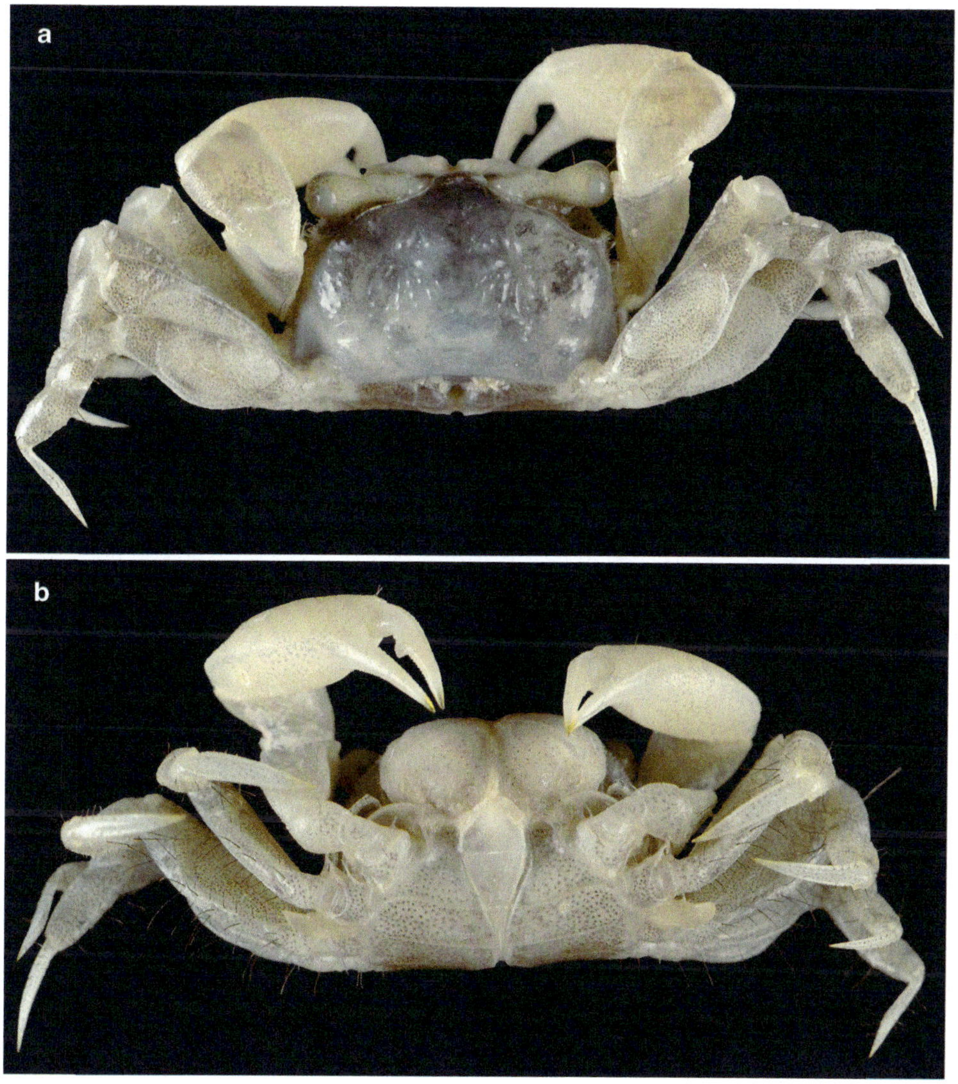

Fig. 35.6 *Scopimera crabicauda* Alcock, 1900. Male (SMF 40445), CL = 5.1, CB = 7.5 mm, Bandar-Rig, Iran, Persian Gulf, 24.04.2008, coll. R. Naderloo and A. Keikhosravi, (**a**) dorsal view, (**b**) ventral view (Photo: S. Träkner)

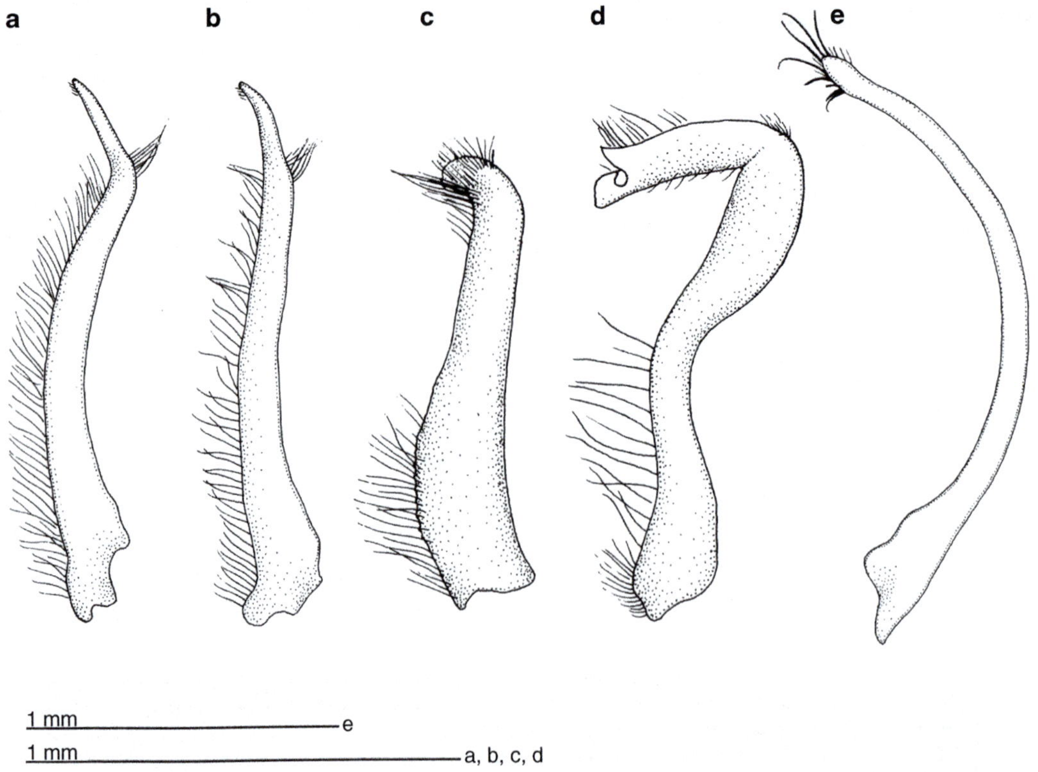

Fig. 35.7 Male G1, ventral view: (**a**) *Ilyoplax frater* (Kemp, 1919) (SMF 40470); (**b**) *Ilyoplax stevensi* (Kemp, 1919) (SMF 40477); (**c**) *Dotilla blanfordi* Alcock, 1900 (SMF 40434); (**d**) *Dotilla sulcata* (Forskål, 1775) (SMF 23897); (**e**) *Scopimera crabicauda* Alcock, 1900 (SMF 40450)

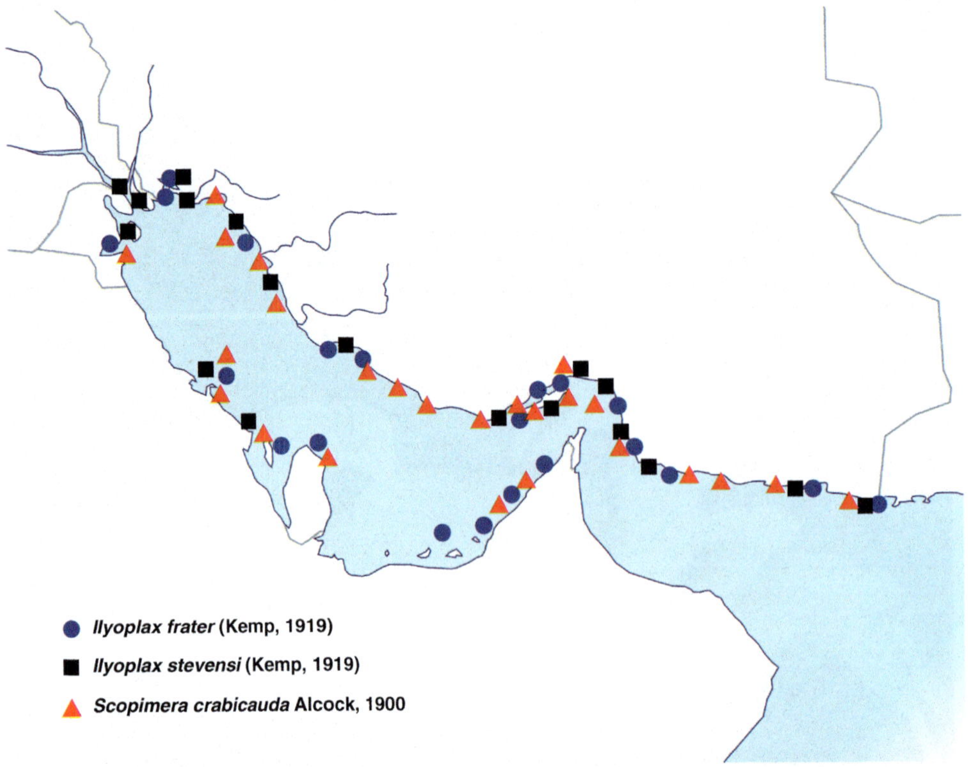

Fig. 35.8 Map showing the recorded localities of *Ilyoplax frater* (Kemp, 1919), *Ilyoplax stevensi* (Kemp, 1919), and *Scopimera crabicauda* Alcock, 1900

Family Macrophthalmidae Dana, 1851 (Sentinel Crabs) 36

Diagnosis Carapace subquadrate, broad, mostly much broader than long, dorsal surface usually flat, slightly convex, regions nearly well defined, lateral margins usually running parallel posteriorly, with one or more teeth. Front in various breadth, very narrow to relatively broad. Antennules folded transversely, more or less obliquely. Antennae with long flagellum. Orbits broad, eyestalks markedly long, sometimes extending exorbital angle. Third maxillipeds with gap between, but not rhomboidal, mandibles no visible when maxillipeds closed. Chelipeds subequal, sexually dimorphic, palm usually longer than fingers in adults, with patch of setae on inner surface. Walking legs robust, moderately flat, walking legs 2 and 3 normally longer than others. Abdomen of males and females with seven freely moving somites. G1 robust, long, usually tapering distally, apically with dense setae; G2 shorter than G1. Genital opening of males and females sternal.

Key to genera of the family Macrophthalmidae in the region

1. Carapace subquadrate, markedly broader than long, more than 1.5 times as broad as long; eyestalks narrow, markedly long; male chelipeds with or without chitinous plectrum on upper surface of merus; medium- to large-sized species, with carapace breadth normally more than 20 mm .... *Macrophthalmus*

- Carapace quadrate, slightly broader than long, around 1.2 times as broad as long; eyestalks stout, not markedly long; male chelipeds with chitinous plectrum on upper surface of merus; small-sized species, carapace hardly reaching to 10 mm of breadth 2

2. Carapace with three teeth on anterolateral margins, second tooth acute, shorter than third one; dense setal patch on inner surface of cheliped palm; front about 0.25 time as broad as carapace; living in rocky/cobble habitat .... *Chaenostoma*

- Carapace with four teeth on anterolateral margins, second tooth blunt, longer than third one; no setal patch on inner surface of cheliped palm; front about 0.4 time as broad as carapace; living in muddy habitat .... *Ilyograpsus*

R. Naderloo, *Atlas of Crabs of the Persian Gulf*, DOI 10.1007/978-3-319-49374-9_36

Genus *Chaenostoma* Stimpson, 1858

Chaenostoma sinuspersici (Naderloo and Türkay, 2011)
(Figs. 36.1, 36.2b, c and 36.3)

Macrophthalmus sinuspersici Naderloo and Türkay 2011: 508–513, figs 3c, d, 4c, d, 5a–g, 6a–g, 7a–c.
Macrophthalmus sinuspersici — Naderloo et al. 2011: 30–33, 38 (in key), figs 17e, f, 18a–i; Naderloo et al. 2013: 450, tab 1.
Macrophthalmus (*Chaenostoma*) *sinuspersici* — Naderloo and Türkay 2012: 51.
Chaenostoma sinuspersici — Naderloo 2013: 2 (in key); Teng et al. 2016: figs 1d–g, 2, tab1.
Macrophthalmus boscii — Crosnier 1965:134–136, figs 244–248; Tirmizi and Ghani 1996: 118–121, fig 45.
Type locality: Persian Gulf (Moghdan, Parsian, Iran, 27°02′N, 53°16′E).

Persian Gulf: Iran (Naderloo and Türkay 2011, 2012; Naderloo et al. 2011, 2013; Naderloo 2013).

Gulf of Oman: Iran (Naderloo and Türkay 2011; Naderloo 2013; Naderloo et al. 2015).

World distribution: Western Indian Ocean: East Africa, Madagascar, Gulf of Aden, Red Sea, Persian Gulf, Gulf of Oman and Pakistan.

Habitat: Intertidal, substrate rocky/cobble.

Fig. 36.1 *Chaenostoma sinuspersici* (Naderloo and Türkay, 2011). Male holotype (SMF 36904), CL = 7.33 CB = 9.09 mm, Moghdan, Parsian, Iran, Persian Gulf, 12.07.2006, coll. R. Naderloo (Photo: S. Träkner. Reproduced from Naderloo et al. (2011) by permission of Magnolia Press)

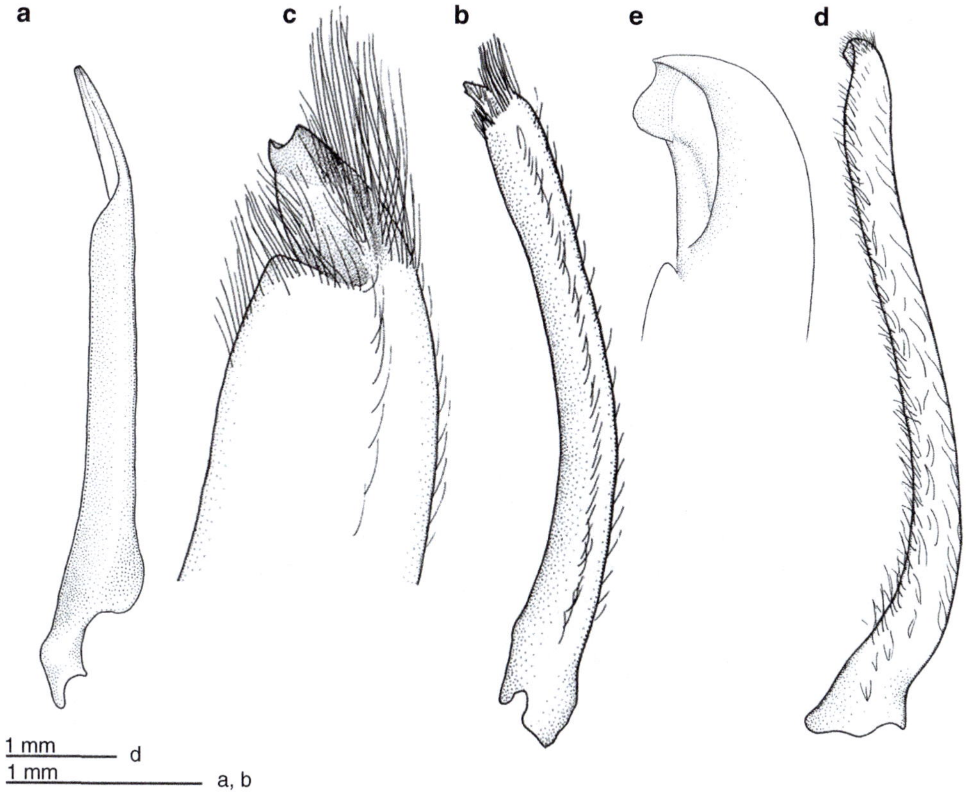

Fig. 36.2 Male G1, ventral view: (**a**) *Ilyograpsus rhizophorae* Barnard, 1955 (ZUTC Brach1261); (**b**) *Chaenostoma sinuspersici* (Naderloo and Türkay, 2011) (SMF 36904); (**c**) apical part of (**b**); (**d**) *Macrophthalmus* (*Macrophthalmus*) *grandidieri* A. Milne-Edwards, 1867 (SMF 26291); (**e**) apical part of (**d**)

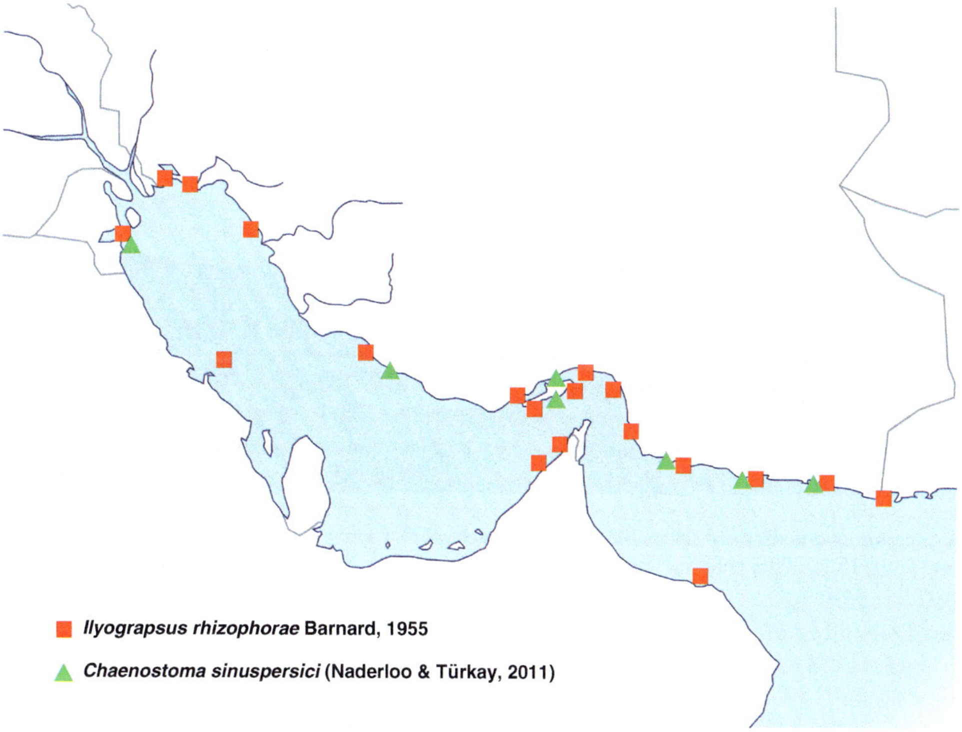

Fig. 36.3 Map showing the recorded localities of *Ilyograpsus rhizophorae* Barnard, 1955, and *Chaenostoma sinuspersici* (Naderloo and Türkay, 2011)

Genus *Ilyograpsus* Barnard, 1955

Ilyograpsus rhizophorae Barnard, 1955
(Figs. 36.2a, 36.3 and 36.4)

Ilyograpsus rhizophorae Barnard 1955: 26, fig 8.
Ilyograpsus paludicola — Crosnier 1965: 31–33, figs 36, 37, 38a, b, 39, 59; Fishelson 1971: 128; Basson et al. 1977: 228, 234; Titgen 1982: 254 (in list); Jones 1986a: 160, pl 46; Vousden 1987: tab 4; Ismail and Ahmed 1993: 158; Hywel-Davies 1994: 37, 48; Apel 1996a, b: 331 (in list); Al-Ghais and Cooper 1996: 415; Tirmizi and Ghani 1996: 143, fig 55; Hornby 1997: 14; Apel and Türkay 1999: 132; Apel 2001: 117.
Ilyograpsus rhizophorae — Komai and Wada 2008: 361, figs 2–5; Naderloo et al. 2011: 3–6, figs 1, 2a–j, 3a, b; Naderloo and Türkay 2012: 451; Naderloo et al. 2013: 450, tab 1, 456, tab 2.
Type locality: Mozambique (Inhambane).

Fig. 36.4 *Ilyograpsus rhizophorae* Barnard, 1955. Male (ZUTC Brach1261), CL = 4.6, CB = 5.9 mm, 2 km E Desalination Plants, Qeshm Island, Iran, Persian Gulf, 15.12.2008, coll. M. Türkay and R. Naderloo (Photo: S. Träkner. Reproduced from Naderloo et al. (2011) by permission of Magnolia Press)

Distribution

Persian Gulf: Iran (Naderloo et al. 2011; Naderloo and Türkay 2012; Naderloo et al. 2013), Kuwait [Jones (1986a) as *I. paludicola*], Saudi Arabia [Basson et al. (1977) and Apel (1996a, b, 2001) as *I. paludicola*], UAE [Al-Ghais and Cooper (1996) and Hornby (1997) as *I. paludicola*].

Gulf of Oman: Iran (Naderloo et al. 2011), Oman [Apel (2001) as *I. paludicola*], UAE [Hornby (1997) as *I. paludicola*].

World distribution: Western Indian Ocean: East Africa, Madagascar, Gulf of Aden, Red Sea, Persian Gulf, Gulf of Oman and Pakistan.

Habitat: Intertidal, substrates muddy and mangroves.

Key to species of the genus *Macrophthalmus* in the region

1. Male chelipeds with distinct chitinous plectrum on inner surface of merus; eyestalks narrow, not reaching to exorbital angle; lower orbital margin with long tubercles; palm without longitudinal ridge on outer surface, inner surface only with a longitudinal row of setae on upper potion; large-size species; occurring on muddy substrates .... *Macrophthalmus dentipes*

- Male chelipeds without chitinous plectrum on upper surface of merus 2

2. Eyestalks projecting beyond tip of exorbital angle 3

- Eyestalks never projecting beyond tip of exorbital angle 5

3. Eyestalks extremely long, extending beyond exorbital angle for half of its length; second anterolateral teeth of carapace nearly as long as exorbital angle .... *Macrophthalmus serenei*

- Eyestalks not extremely long, extending only slightly beyond exorbital angle 4

4. Apical portion of cornea with unique extension, about half of the cornea extending beyond exorbital angle; exorbital angle longer than second lateral tooth of carapace; inner surface of palm of male chelipeds without spine and patch of setae .... *Macrophthalmus indicus*

- Apical portion of cornea without extension, extending beyond exorbital angle; second lateral tooth of carapace larger than exorbital angle; inner surface of palm of male chelipeds with spines on proximal portion, patch of dense setae covering most part of inner surface, immovable finger strongly deflected .... *Macrophthalmus sulcatus*

5. Exorbital angle much smaller than second lateral tooth of carapace; eyestalks hardly reaching tip of exorbital angle; palm with longitudinal ridge on outer surface, inner surface of palm with spines on proximal portion .... *Macrophthalmus grandidieri*

- Exorbital angle not markedly smaller than second lateral tooth of carapace; eyestalks never reaching tip of exorbital angle 6

6. Walking legs with dense setae; palm of male chelipeds with inner surface fully covered with dense setae, without spines on proximal portion .... *Macrophthalmus depressus*

- Walking legs without dense setae; palm of male chelipeds without setae on inner surface, with spines on proximal portion .... *Macrophthalmus laevis*

Genus *Macrophthalmus* Desmarest, 1823

Macrophthalmus (*Macrophthalmus*) *grandidieri* A. Milne-Edwards, 1867

(Figs. 36.2d, e, 36.5, 36.6 and 36.12)

Macrophthalmus grandidieri A. Milne-Edwards 1867: 285.
Macrophthalmus grandidieri — Crosnier 1965: 127, figs 230, 235; Hywel-Davies 1994: 29, 37, 48; Apel 2001: 108, 109; Naderloo et al. 2011: 15–19, 38 (in key), figs 8a–f, 9a–d, 21a, b; Naderloo et al. 2013: 450, tab 1.
Macrophthalmus (*Macrophthalmus*) *grandidieri* — Barnes 1967: 23; Barnes 1970: 233, fig 6a–c; Barnes 1971: 40 (in key); Barnes 2010: 35 (in key), 40; Ng et al. 2008: 237 (in list); Naderloo and Türkay 2012: 52.
Macrophthalmus (*Macrophthalmus*) *sulcatus grandidieri* — Pretzmann 1974: 438, figs 1, 4.
Macrophthalmus brevis — Nobili 1906b: 318.

Type locality: Zanzibar.

Distribution

Persian Gulf: Iran (Naderloo et al. 2011; Naderloo and Türkay 2012; Naderloo et al. 2013).

Gulf of Oman: Oman (Hywel-Davies 1994; Naderloo et al. 2011; Naderloo et al. 2015).

World distribution: Western Indian Ocean: South Africa, East Africa, Red Sea, Persian Gulf and Gulf of Oman.

Habitat: Intertidal, substrate muddy-sand.

Macrophthalmus (*Macrophthalmus*) *indicus* Davie, 2012

(Figs. 36.7, 36.11a, b and 36.12)

Macrophthalmus (*Macrophthalmus*) *indicus* Davie 2012: 168, 169, figs 10d–f, 11a–c, 12a.
Macrophthalmus (*Macrophthalmus*) *graeffei* — Barnes 1970: 225; Barnes 1971: 13, 36 (in key), fig 3; Barnes 1977: 276 (in key), 279 (in list); Barnes 2010: 34 (in key), 39; Apel 2001: 108; Ng et al. 2008: 237 (in list); Naderloo et al. 2011: 19–23, fig 11a–e, 12a–d, 17a, b.
Macrophthalmus (*convexus,* Stimpson?) — Stephensen 1946: 191.

Type locality: Persian Gulf (Khark Island, Iran).

Distribution

PersianGulf: Iran [Stephensen (1946) as *Macrophthalmus* (*convexus,* Stimpson?); Naderloo et al. (2011) as *M. graeffei*; Davie 2012].

Gulf of Oman: No record.

World distribution: Indian Ocean: Red Sea, Persian Gulf and Indonesia.

Habitat: Subtidal, 5–74 m, substrates muddy and sandy with shells.

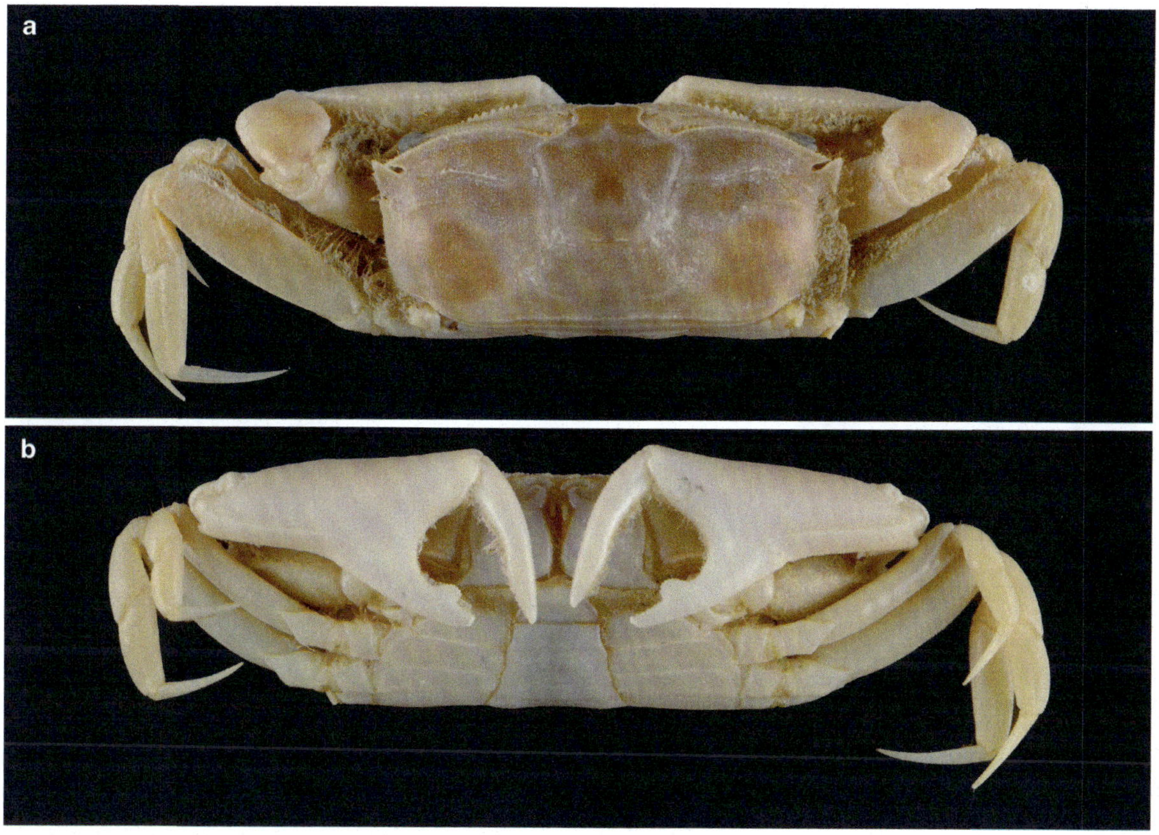

Fig. 36.5 *Macrophthalmus* (*Macrophthalmus*) *grandidieri* A. Milne-Edwards, 1867. Male (SMF 26291), CL = 10.54, CB = 30.24 mm, Muscat, Oman, Gulf of Oman, 07.1995, coll. A. Hywel-Davies; (**a**) dorsal view, (**b**) ventral view (Photos: S. Träkner. Reproduced from Naderloo et al. (2011) by permission of Magnolia Press)

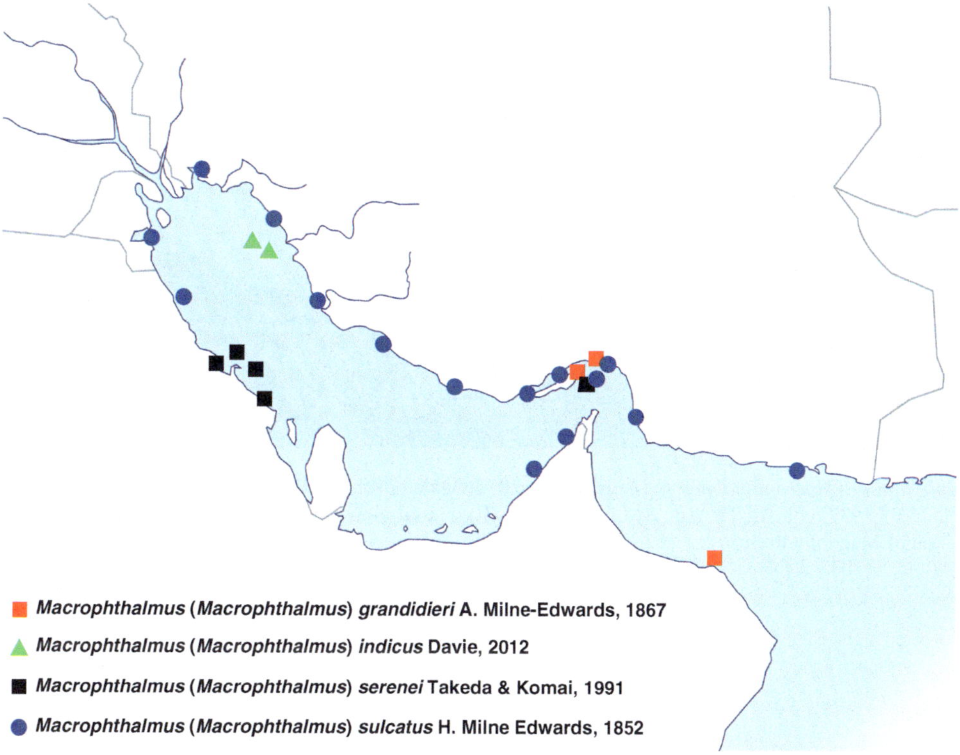

Fig. 36.6 Map showing the recorded localities of *Macrophthalmus* (*Macrophthalmus*) *grandidieri* A. Milne-Edwards, 1867, *Macrophthalmus* (*Macrophthalmus*) *indicus* Davie, 2012, *Macrophthalmus* (*Macrophthalmus*) *serenei* Takeda and Komai, 1991, and *Macrophthalmus* (*Macrophthalmus*) *sulcatus* H. Milne Edwards, 1852

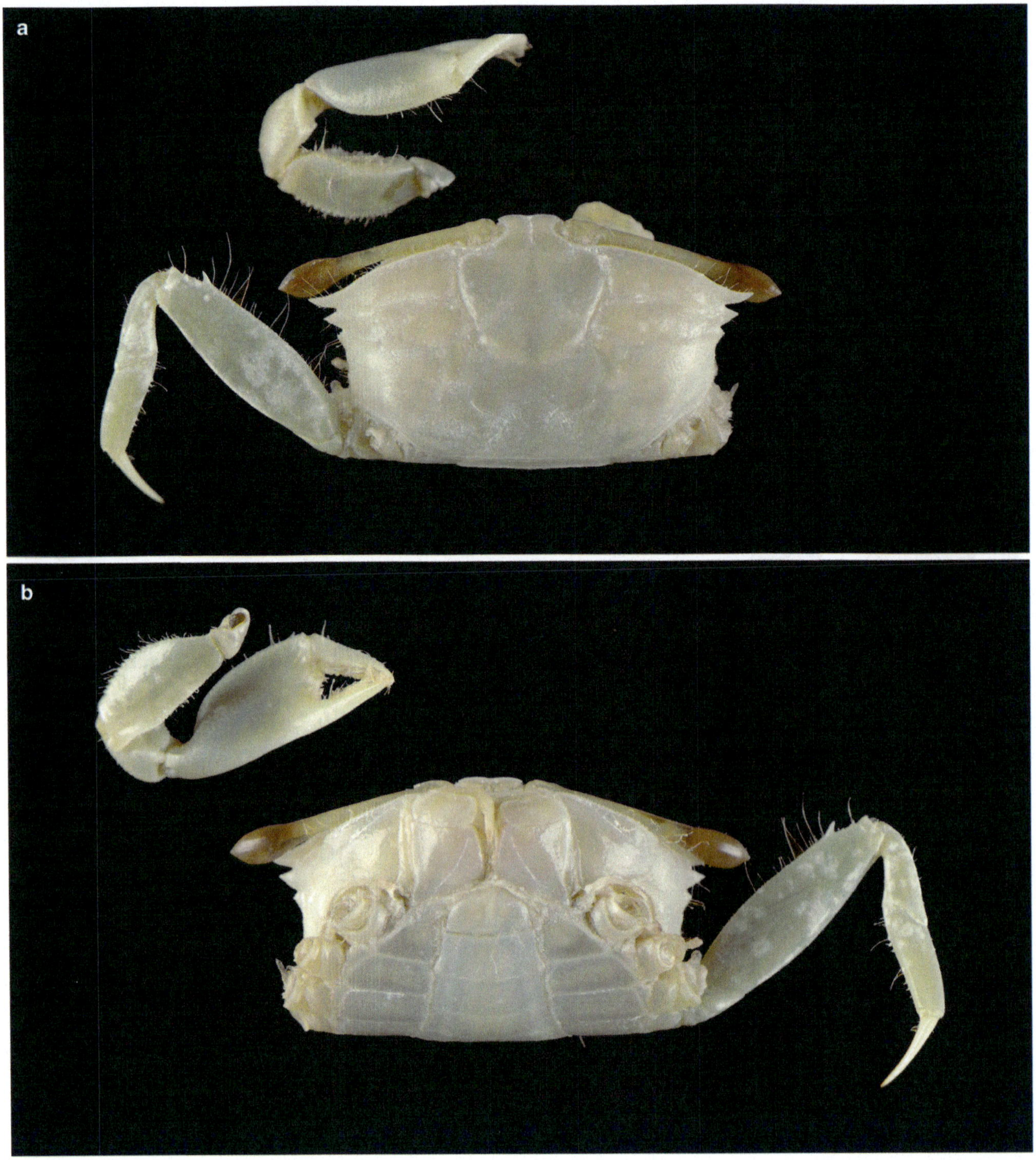

Fig. 36.7 *Macrophthalmus* (*Macrophthalmus*) *indicus* Davie, 2012. Male (ZMUC CRU1782), CL = 4.8, CB = 8.4 mm, Kharg Island, Iran, Persian Gulf, 15.03.1937, coll. G. Thorson; (**a**) dorsal view, (**b**) ventral view (Photos: S. Träkner. Reproduced from Naderloo et al. (2011) by permission of Magnolia Press)

Macrophthalmus (*Macrophthalmus*) *serenei* Takeda and Komai, 1991
(Figs. 36.6, 36.8, 36.9 and 36.10e, f)

Macrophthalmus serenei Takeda and Komai 1991: 168, fig 3a–e.
Macrophthalmus verreauxi — Nobili 1906b: 317.
Macrophthalmus telescopicus — Basson et al. 1977: 231, 256; Titgen 1982: 253 (in list); Vousden 1987: 35.
Macrophthalmus serenei — Naderloo et al. 2011: 27–30, 38 (in key), figs 15a–e, 16a–e, 17c, d.
Macrophthalmus (*Macrophthalmus*) *serenei* — Apel 2001: 109, 110; Ng et al. 2008: 237 (in list); Barnes 2010: 34 (in key), 40.

Type locality: Red Sea and Gulf of Aden.

Distribution

Persian Gulf: Iran (present study), Saudi Arabia [Basson et al. (1977) as *Macrophthalmus telescopicus*; Naderloo et al. 2011], Bahrain [Vousden (1987) as *Macrophthalmus telescopicus*].

Gulf of Oman: No record.

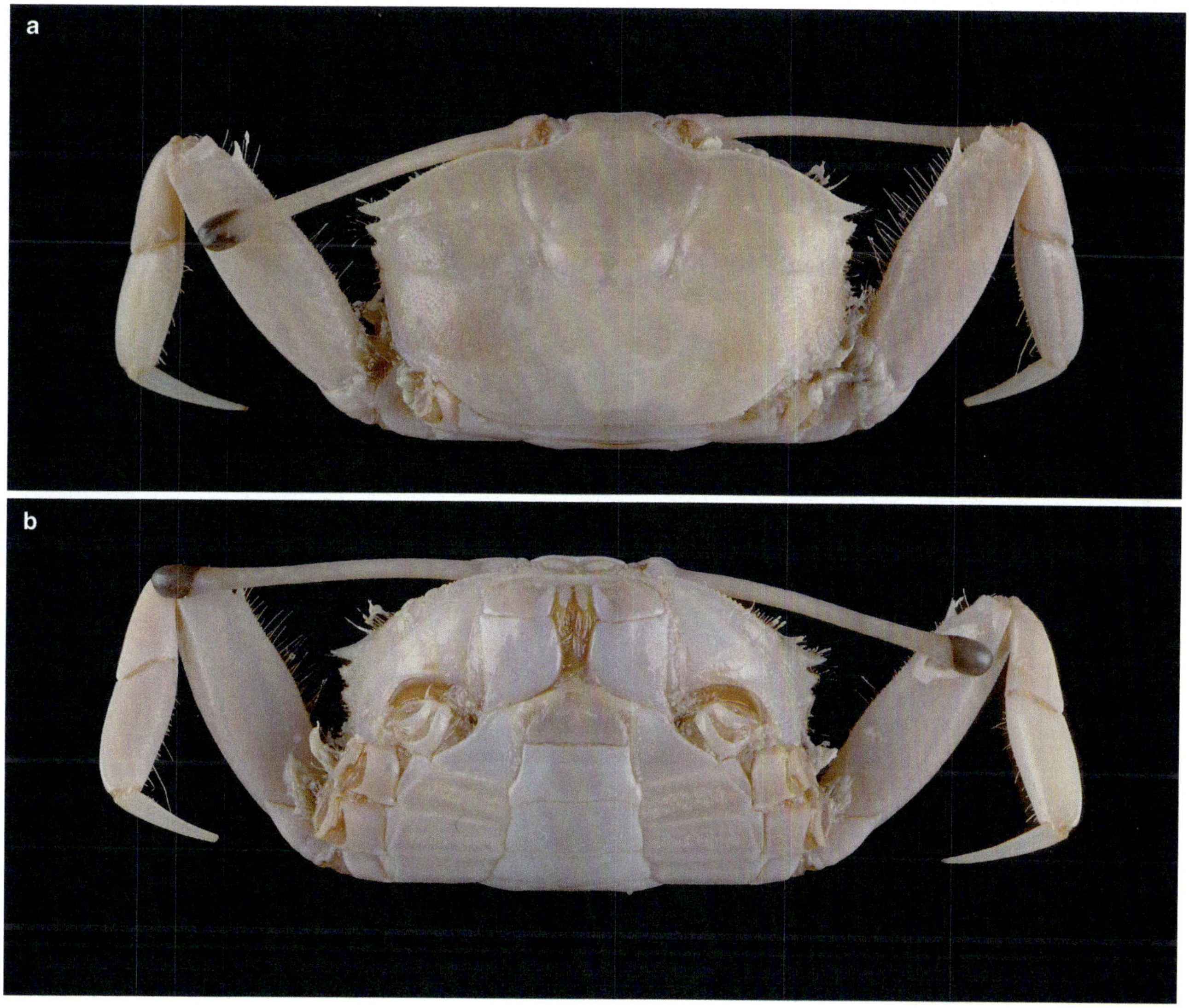

Fig. 36.8 *Macrophthalmus* (*Macrophthalmus*) *serenei* Takeda and Komai, 1991. Male (SMF 36098), CL = 17.8, CB = 11.7 mm, N Jubail, Abu Ali, Saudi Arabia, Persian Gulf, 08.04.1993, coll. M. Richmond; (**a**) dorsal view, (**b**) ventral view (Photos: S. Träkner. Reproduced from Naderloo et al. (2011) by permission of Magnolia Press)

Fig. 36.9 *Macrophthalmus* (*Macrophthalmus*) *serenei* Takeda and Komai, 1991. Burrow of the species, seawards sandy flat of mangroves in Hormuz Island, Iran, Persian Gulf, 03.01.2013 (Photo: R. Naderloo)

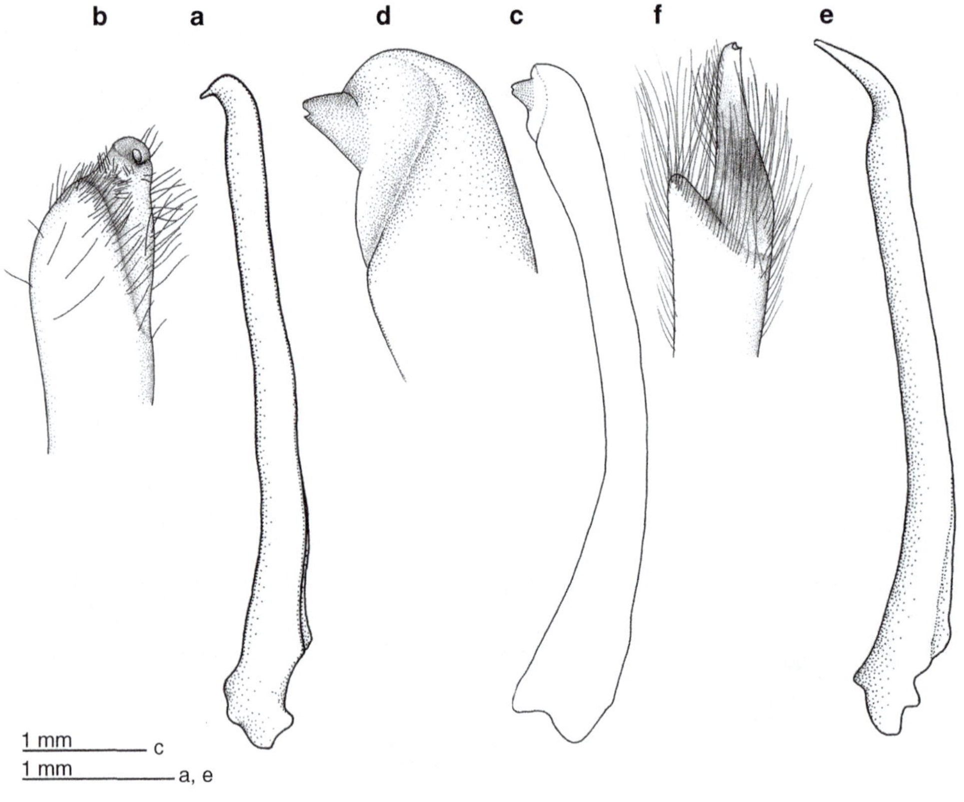

Fig. 36.10 Male G1, ventral view: (**a**) *Macrophthalmus* (*Macrophthalmus*) *indicus* Davie, 2012 (ZMUC CRU1782); (**b**) apical part of (**a**); (**c**) *Macrophthalmus* (*Mareotis*) *laevis* A. Milne-Edwards, 1867 (SMF 26293); (**d**) apical part of (**c**); (**e**) *Macrophthalmus* (*Macrophthalmus*) *serenei* Takeda and Komai, 1991 (SMF 36098); (**f**) apical part of (**e**)

World distribution: Indo-West Pacific: Madagascar, Red Sea, Gulf of Aden, Persian Gulf, Japan and Australia.

Habitat: Intertidal, substrate sandy.

Macrophthalmus (Macrophthalmus) sulcatus H. Milne Edwards, 1852

(Figs.36.6, 36.11, 36.12 and 36.13a, b)

Macrophthalmus sulcatus H. Milne Edwards 1852: 156.

Macrophthalmus sulcatus — Alcock 1900: 379; Naderloo et al. 2011: 33–37, 38 (in key), figs 19a–e, 20a–f, 21c–d; Naderloo et al. 2013: 450, tab 1.

Macrophthalmus (*Macrophthalmus*) *dilatatus sulcatus* — Barnes 1970: 216, fig 4; Titgen 1982: 149, 253 (in list); Tirmizi and Ghani 1996: 112, fig 42.

Macrophthalmus (*Macrophthalmus*) *sulcatus* — Pretzmann 1971: 482; Apel and Türkay 1999: 134; Apel 2001: 110; Ng et al. 2008: 237 (in list); Naderloo and Türkay 2012: 52.

Macrophthalmus sulcatus sulcatus — Pretzmann 1974: 438, figs 2, 3; Barnes 2010: 35 (in key), 40.

Macrophthalmus grandidieri — Basson et al. 1977: 60, 63, 228, 231, 235; Titgen 1982: 253 (in list); Clayton 1986: 87, fig 3; Jones 1986a: 159, pl 45; Vousden 1987: 35; Tirmizi and Ghani 1996: 114, fig 43; Al-Khayat and Jones 1999: 58, 61.

Macrophthalmus dilatatus sulcatus — Al-Ghais and Cooper 1996: 417, fig 6.

Type locality: Mauritius.

Distribution

Persian Gulf: Iran (Pretzmann 1971; Naderloo et al. 2011; Naderloo and Türkay 2012; Naderloo et al. 2013), Kuwait [Clayton (1986) and Jones (1986a) as

Fig. 36.11 *Macrophthalmus* (*Macrophthalmus*) *sulcatus* H. Milne Edwards, 1852. Male (SMF 36885), CL = 9.99, CB = 20.23 mm, Bandar-Rig, Bushehr, Iran, Persian Gulf, 16.10.2008, coll. R. Naderloo and A. Kazemi; (**a**) dorsal view, (**b**) ventral view (Photos: S. Träkner. Reproduced from Naderloo et al. (2011) by permission of Magnolia Press)

Fig. 36.12 *Macrophthalmus* (*Macrophthalmus*) *sulcatus* H. Milne Edwards, 1852. Burrow of the species, Bandar-Rig, Bushehr, Iran, Persian Gulf, 16.10.2008 (Photo: R. Naderloo)

M. grandidieri], Saudi Arabia [Basson et al. (1977) as *M. grandidieri*; Apel and Türkay 1999; Apel 2001; Naderloo et al. 2011], UAE [Al-Ghais and Cooper (1996) as *M. dilatatus sulcatus*; Apel and Türkay 1999; Apel 2001; Naderloo et al. 2011], Bahrain [Vousden (1987) as *M. grandidieri*].

Gulf of Oman: No record.

World distribution: Northern Indian Ocean: Mauritius, Persian Gulf, Pakistan, India and western Malaysia.

Habitat: Intertidal, substrate muddy-sand.

Macrophthalmus (*Mareotis*) *depressus* Rüppell, 1830

(Figs. 36.13c, d, 36.14 and 36.15)

Macrophthalmus depressus Rüppell 1830: 19, pl 4, fig 6.
Macrophthalmus depressus — Nobili 1906a: 155; Stephensen 1946: 191, 210, fig 58a; Basson et al. 1977: 56, 228, 235, fig 37; Clayton 1986: 87, fig 3; Vousden 1987: 35–37, tab 4, 7; Hogarth 1989: 117; Ismail and Ahmed 1993: 158; Apel 1994a, b: 42–45, pl 2; Apel 1996a, b: 331, 337, figs 1–3; Al-Ghais and Cooper 1996: 415, fig 5; Tirmizi and Ghani 1996: 116, fig 44; Cooper

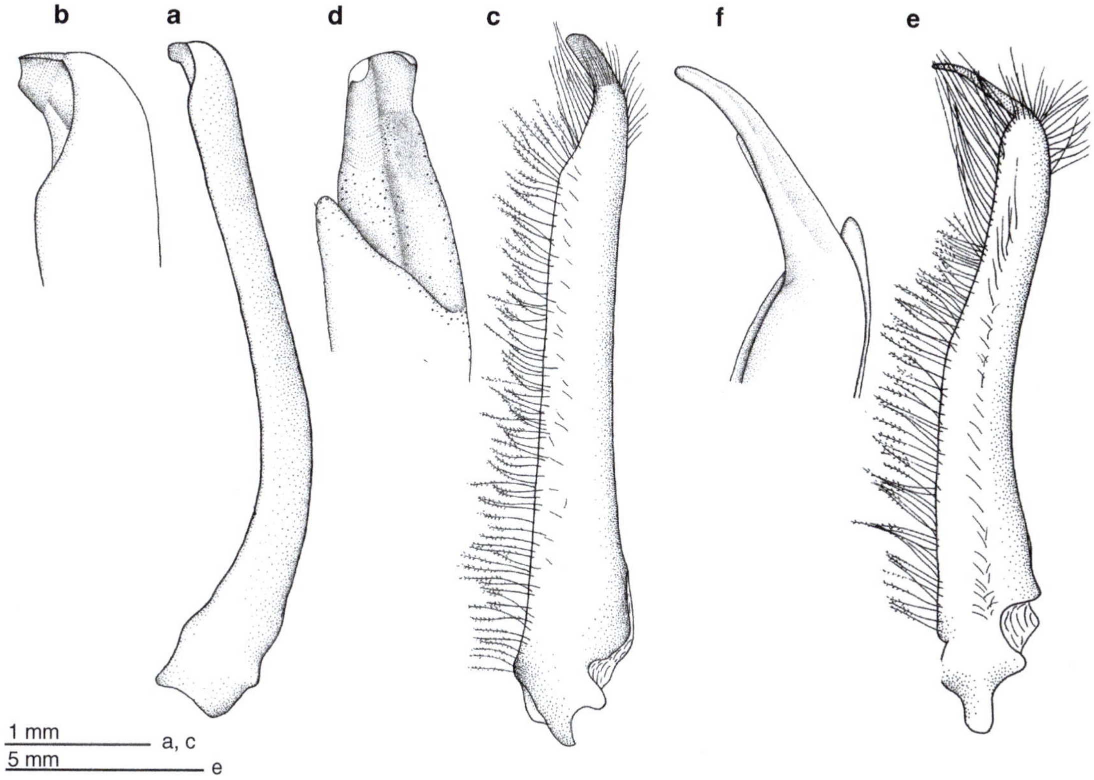

Fig. 36.13 Male G1, ventral view: (**a**) *Macrophthalmus* (*Macrophthalmus*) *sulcatus* H. Milne Edwards, 1852 (SMF 36901); (**b**) apical part of (**a**); (**c**) *Macrophthalmus* (*Mareotis*) *depressus* Rüppell, 1830 (36868); (**d**) apical part of (**c**); (**e**) *Macrophthalmus* (*Venitus*) *dentipes* Lucas, 1836 (SMF 26095); (**f**) apical part of (**e**)

Fig. 36.14 *Macrophthalmus* (*Mareotis*) *depressus* Rüppell, 1830. Male (SNMNH 101), CL = 18.69, CB =27.94 mm, Abu Ali, Jubail, Saudi Arabia, Persian Gulf, 27.08.1992, coll. M. Apel (Photos: S. Träkner. Reproduced from Naderloo et al. (2011) by permission of Magnolia Press)

1997: 158; Hornby 1997: 15; Al-Khayat and Jones 1999: 58, 61; Naderloo et al. 2011: 12–15, figs 6a–e, 7a–d, 10c, d; Naderloo et al. 2013: 450, tab 1.

Macrophthalmus (*Mareotis*) *depressus* — Barnes 1970: 206, 226, fig 7; Barnes 1977: 278 (in key), 279; Pretzmann 1971: 482, pl 9, fig 1; Pretzmann 1974: 440; Titgen 1982: 148, 253 (in list); Jones 1986a: 159, pl 47; Hywel-Davies 1994: 37, 48; Apel and Türkay 1999: 134; Apel 2001: 110; Ng et al. 2008: 237 (in list); Barnes 2010: 36 (in key); Naderloo and Türkay 2012: 52.

Type locality: Red Sea.

Distribution

Persian Gulf: Iran (Pretzmann 1971; Naderloo et al. 2011, 2013; Naderloo and Türkay 2012), Kuwait (Jones 1986a), Saudi Arabia (Basson et al. 1977; Apel 1994a, 1996a, b; Naderloo et al. 2011), Bahrain (Stephensen 1946; Naderloo et al. 2011), Qatar (Apel 2001), UAE (Titgen 1982; Al-Ghais and Cooper 1996; Cooper 1997; Hornby 1997; Apel 2001; Naderloo et al. 2011).

Gulf of Oman: Iran (Naderloo et al. 2015), Oman (Hywel-Davies 1994).

World distribution: Western Indian Ocean: South Africa, East Africa, Madagascar, Gulf of Aden, Red Sea, southern Oman, Persian Gulf, Gulf of Oman, Pakistan and west coast of India.

Habitat: Intertidal, substrates muddy and mangroves.

Macrophthalmus (*Mareotis*) *laevis* A. Milne-Edwards, 1867

(Figs. 36.10c, d, 36.15, 36.16 and 36.17)

Macrophthalmus laevis A. Milne-Edwards 1867: 287.

Macrophthalmus (*Macrophthalmus*) *ressli* Pretzmann 1971: 382, pl 9, fig 23.

Macrophthalmus laevis — Barnes 1976: 143, fig 6a–c; Titgen 1982: 150; Naderloo et al. 2011: 23–26, 38 (in key), figs 13a–f, 14a–e, 10e–f; Naderloo et al. 2013: 45, tab 1.

Macrophthalmus resseli [sic!] — Pretzmann 1974: 441.

Macrophthalmus (*Macrophthalmus*) *laevis* — Barnes 1977: 277 (in key), 280 (in list); Barnes 2010: 35 (in key), 40; Tirmizi and Ghani 1988b: 253, figs 1–11;

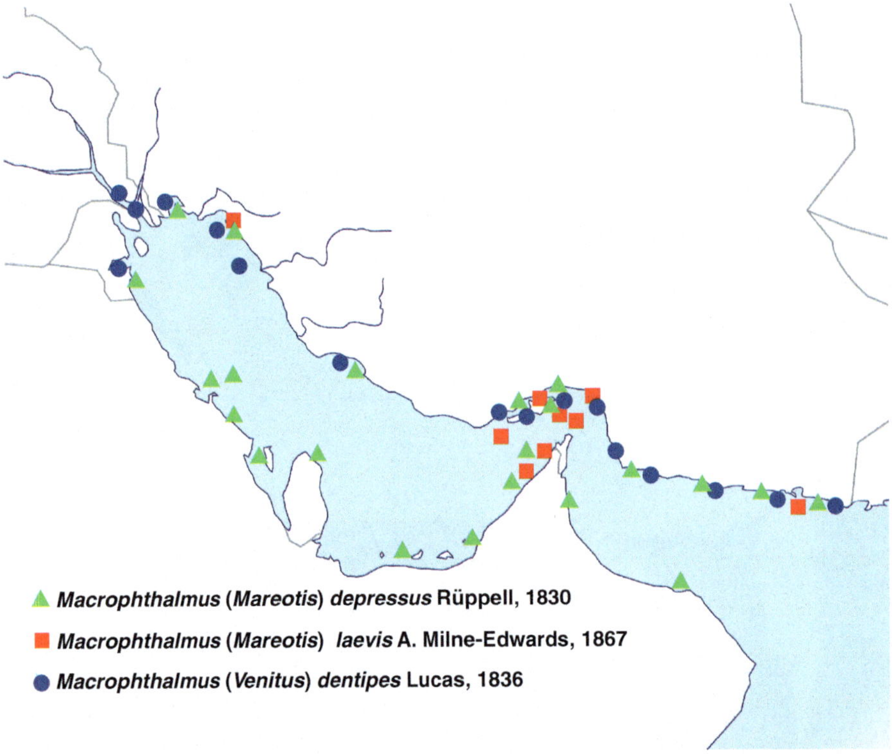

Fig. 36.15 Map showing the recorded localities of *Macrophthalmus* (*Mareotis*) *depressus* Rüppell, 1830, *Macrophthalmus* (*Mareotis*) *laevis* A. Milne-Edwards, 1867, and *Macrophthalmus* (*Venitus*) *dentipes* Lucas, 1836

Fig. 36.16 *Macrophthalmus* (*Mareotis*) *laevis* A. Milne-Edwards, 1867. Male (SMF 36869), CL = 20.54, CB = 12.07 mm, Kuwei, Qeshm Island, Iran, Persian Gulf, 06.05.2008, coll. R Naderloo and A. Kazemi; (**a**) dorsal view, (**b**) ventral view (Photos: S. Träkner. Reproduced from Naderloo et al. (2011) by permission of Magnolia Press)

Fig. 36.17 *Macrophthalmus* (*Mareotis*) *laevis* A. Milne-Edwards, 1867. Burrow of the species, mangroves of Hormuz Island, Iran, Persian Gulf, 03.01.2013 (Photo: R. Naderloo)

Tirmizi and Ghani 1996: 109, fig 41; Apel and Türkay 1999: 135; Apel 2001: 109; Ng et al. 2008: 237 (in list); Naderloo and Türkay 2012: 52.

Type locality: *Macrophthalmus laevis*: "Indian Seas". — *Macrophthalmus resseli*: Persian Gulf (Bandar-Abbas).

Distribution

Persian Gulf: Iran [Pretzmann (1971) as *Macrophthalmus* (*Macrophthalmus*) *ressli*; Naderloo et al. 2011, 2013; Naderloo and Türkay 2012], UAE (Titgen 1982; Apel and Türkay 1999; Apel 2001; Naderloo et al. 2011).

Gulf of Oman: Iran (Naderloo et al. 2015).

World distribution: Northern Indian Ocean: Persian Gulf, Gulf of Oman and Pakistan.

Habitat: Intertidal, substrates muddy-sand and sandy-mud.

Macrophthalmus (*Venitus*) *dentipes* Lucas, 1836

(Figs. 36.13e, f, 36.15, 36.18 and 36.19)

Macrophthalmus dentipes Lucas 1836: 551.

Macrophthalmus dentipes — Holthuis 1995: 401; Naderloo et al. 2011: figs 4a–e, 5a–f, 10a, b; Naderloo et al. 2013: 450, tab 1.

Macrophthalmus pectinipes — Guerin-Méneville 1838: 1, pl 23; Alcock 1900: 377; Chhapgar 1957b: 512; Barnes 1970: 237, fig 10; Pretzmann 1971: 31; 1974: 442; Titgen 1982: 253 (in list); Jones 1986a: 159, pl 45; Tirmizi and Ghani 1996: 121, fig 46.

Macrophthalmus (*Venitus*) *dentipes* — Apel and Türkay 1999: 135; Apel 2001: 110; Ng et al. 2008: 238 (in list); Barnes 2010: 33 (in key), 43; Naderloo and Türkay 2012: 53.

Fig. 36.18 *Macrophthalmus* (*Venitus*) *dentipes* Lucas, 1836. Male (SMF 26095) CL = 40. 00, CB = 64. 7 mm, mangroves of Barr-Al Hikman, southern Oman, 14.01.2001, coll. M. Apel (Photos: S. Träkner. Reproduced from Naderloo et al. (2011) by permission of Magnolia Press)

Fig. 36.19 *Macrophthalmus* (*Venitus*) *dentipes* Lucas, 1836. The species in its habitat, Tiab, Strait of Hormuz, 26.12.2013 (Photo: R. Naderloo)

Type locality: India (Mumbai).

Distribution

Persian Gulf: Iran [Pretzmann (1971) as *M. pectinipes*; Naderloo et al. 2011, 2013; Naderloo and Türkay 2012], Iraq [Barnes (1970) as *M. pectinipes*; Naderloo et al. 2011], Kuwait [Jones (1986a) as *M. pectinipes*; Naderloo et al. 2011].

Gulf of Oman: Iran (Naderloo et al. 2011, 2015).

World distribution: Northern Indian Ocean: southern Oman, northern and eastern Persian Gulf, Gulf of Oman, Pakistan and India.

Habitat: Intertidal, substrates muddy and mangroves.

Family Ocypodidae Rafinesque, 1815 (Ghost Crabs, Fiddler Crabs)

37

Diagnosis Carapace quadrate, transversely rectangular, pentagonal, markedly thick, dorsal surface moderately convex, smooth, regions weakly or never defined, lateral margins entire, without any dentation, junction between anterolateral and posterolateral margins not distinct. Front markedly narrow, straight, or very faintly bilobed, strongly deflexed. Orbits broad, occupying most part of anterior margin of carapace, eyestalks slender, characteristically long. Antennules folded obliquely to longitudinally, basal segment markedly enlarged, globose to ovate. Antennae with small basal segment and short flagellum. Third maxillipeds completely covering buccal cavern, with gap between, but never rhomboidal, exopod slender and sometimes very short, not easily visible when maxilliped closed. Chelipeds characteristically unequal, larger cheliped always longer than carapace, sexually dimorphic. Walking legs comparatively slender, long, characteristic pouch between bases of walking legs 2–3 or 3–4 serrated by setae, leading into buccal cavity, functioning as supplementary to breathing system. Abdomen of males and females with seven freely moving somites. Genital opening of males and females sternal.

Genus Ocypode Weber, 1795

Key to species of the genus *Ocypode* in the region

1. Eyestalks with prolonged stylus extending markedly beyond cornea 2
- Eyestalks without prolonged stylus 3
2. First and second walking legs without setal patch on anterior surface of propodus; stridulating ridge on inner surface of cheliped palm with 23–28 tubercles .... *Ocypode brevicornis*
- First walking legs with and second ones without setal patch on anterior surface of propodus; stridulating ridge on inner surface of cheliped palm with 10–15 tubercles .... *Ocypode rotundata*
3. Inner surface of palm of larger cheliped without stridulating ridge; propodus of first walking legs with setae along dorsal margin, transverse rows of setae on dorsal half, median row of setae on anterior surface; propodus of second walking legs with dense setal patch along dorsal margin .... *Ocypode cordimanus*
- Inner surface of palm of larger cheliped with stridulating ridge with 41–79 tubercles, getting to striae downwards; propodus of first walking legs with median row of setae on anterior surface; exorbital angle directed obliquely forward; dorsal surface of carapace with coarse granulations .... *Ocypode jousseaumei*

R. Naderloo, *Atlas of Crabs of the Persian Gulf*, DOI 10.1007/978-3-319-49374-9_37

Ocypode brevicornis H. Milne Edwards 1837
(Figs. 37.1, 37.8a, and 37.9)

Ocypode brevicornis H. Milne Edwards 1837a: 48.
Ocypode brevicornis — Apel 2001: 111; Türkay et al. 1996: 114 (in text); Ng et al. 2008: 240 (in list); Sakai and Türkay 2013: 682–685, figs 1c, 9, 31.
Ocypode platytarsis — Clayton 1996: 31, 32, pls 1 d, 2d, 3d.
Type locality: East coast of India.
Distribution
Persian Gulf: No record.
Gulf of Oman: Oman [Clayton (1996) as *O. platytarsis*; Sakai and Türkay 2013].
World distribution: Northern Indian Ocean: Gulf of Oman, India, Nicobar Island and Sri Lanka.
Habitat: High intertidal, substrate sandy.

Ocypode cordimanus Latreille, 1818
(Figs. 37.2, 37.8b, and 37.9)

Ocypode cordimana Latreille 1818: 198.
Ocypode cordimana — Nobili 1906b: 310.
Ocypode cordimanus — Sakai and Türkay 1977: 97–100, figs 1, 2; Clayton 1996: 32, pls 1e, 2e; Türkay et al. 1996: 102, figs 1–3; Apel 2001: 112; Ng et al. 2008: 240 (in list); Sakai and Türkay 2013: 696–702, figs 12, 34.
Type locality: "Indes Orientales". — Neotype: Papua New Guinea (Tarawai Island).
Distribution
Persian Gulf: No record.
Gulf of Oman: Oman (Clayton 1996; Sakai and Türkay 2013).
World distribution: Indo-West Pacific: South Africa, East Africa, Madagascar, Réunion, Mauritius, Seychelles, Red Sea, Socotra, southern Oman, Gulf of Oman, Pakistan, India, Sri Lanka, Maldives, Chagos Archipelago, Malaysia, Taiwan, China, Philippines, Japan, Indonesia, Cariline Islands, Mariana Islands, Palau Islands, Vanuatu, Solomon Islands, Papua New Guinea, Australia and New Caledonia.
Habitat: High intertidal, substrate sandy.

Ocypode jousseaumei (Nobili, 1905)
(Figs. 37.3, 37.4, 37.8c, and 37.9)

Ocypoda Jousseaumei Nobili 1905d: 233–235, fig 2.
Ocypoda Jousseaumei — Nobili 1906b: 310.
Ocypode jousseaumei — Türkay et al. 1996: 102–104, figs 4–6, 11; Clayton 1996: 31, pls 1c, 2c, 3c; Apel 2001: 112; Clayton 2001: 37; Clayton 2005: 53, figs 1–9; Ng et al. 2008: 238 (in list); Sakai and Türkay 2013: 712–714, figs 3a, 16, 38.
Type locality: Gulf of Aden (Obock, Djibouti).

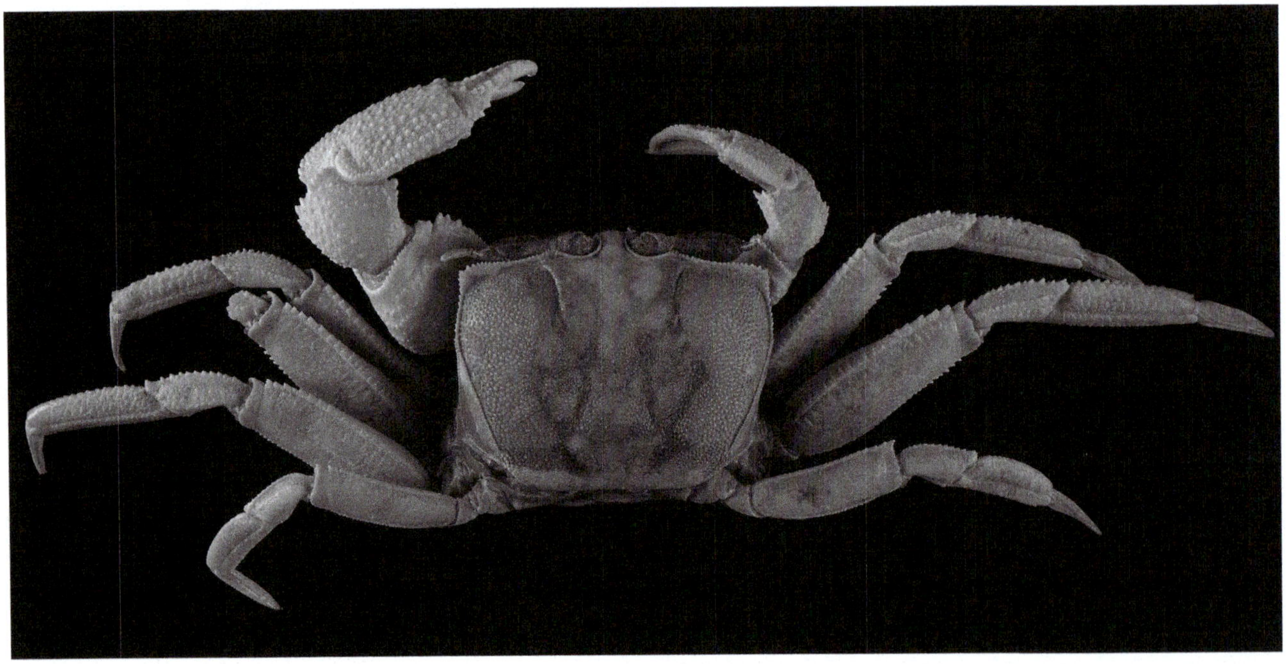

Fig. 37.1 *Ocypode brevicornis* H. Milne Edwards 1837. Male (SMF 24536), CL = 46.7, CB = 60.1 mm, Khor Al-Milh, Barr Al-Hikmann, Gulf of Masirah, Oman, Arabian Sea, 31.05.1995, coll. D. Clayton (Photo: S. Träkner)

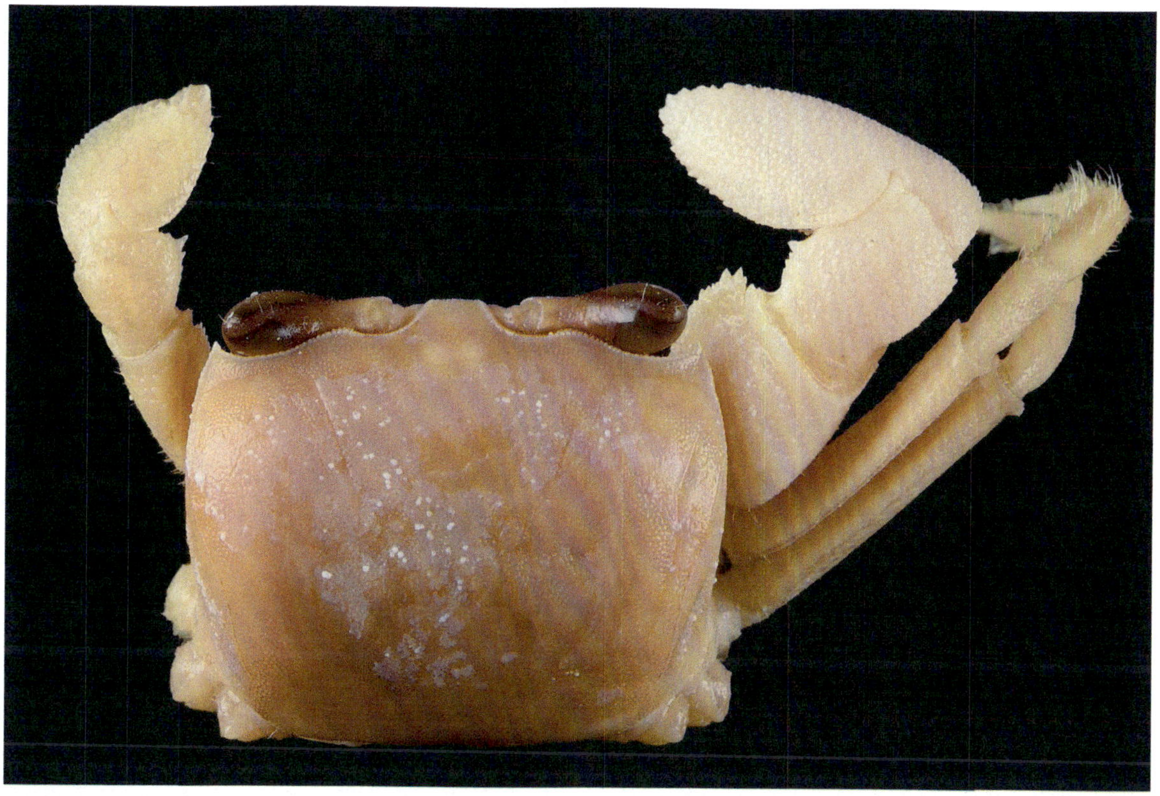

Fig. 37.2 *Ocypode cordimanus* Latreille, 1818. Male (NHCY 98), CL = 15.2, CB = 18.8 mm, N coast of Socotra Island, 08.10.1999, coll. N. Simoes (Photo: S. Träkner)

Fig. 37.3 *Ocypode jousseaumei* (Nobili, 1905). Male (ZUTC 5007), CL = 15.2, CB = 18.8 mm Sirik, Sistan-Baluchestan, Iran, Gulf of Oman, 17.04.2007, coll. R. Naderloo (Photo: R. Abdollahi)

Fig. 37.4 *Ocypode jousseaumei* (Nobili, 1905). The same specimen, frontal view (Photo: R. Abdollahi)

Distribution

Persian Gulf: No record.

Gulf of Oman: Iran (Naderloo et al. 2015), Oman (Clayton 1996).

World distribution: Northwestern Indian Ocean: Red Sea, Gulf of Aden, southern Oman and Gulf of Oman.

Habitat: High intertidal, substrate sandy.

Ocypode rotundata Miers, 1882

(Figs. 37.5, 37.6, 37.7, 37.8d, and 37.9)

Ocypode rotundata Miers 1882: 378, 382, pl 17 fig.
Ocypode rotundata var. *arabica* Nobili 1906a: 152, pl 5, fig 26.
Ocypode rotundata — Pretzmann 1971: 480, pl 4 figs 8–10; Titgen 1982: 152; Türkay et al. 1996: 104–107, 111, 113, figs 7, 8, 12, pls 1–3; Clayton 1996: 29–31, pls 1b, 2b, 3b; Tirmizi and Ghani 1996: 98–100, fig 37; Cooper 1997: 163–165, figs 10, 11; Hornby 1997: 15; Al-Khayat and Jones 1999: 58; Apel and Türkay 1999: 133; Apel 2001: 112; Ng et al. 2008: 238 (in list); Naderloo and Türkay 2012: 53; Naderloo et al. 2013: 450, tab 1, 458, tab 3; Sakai and Türkay 2013: 735–738, figs 5a, 25, 47.
Ocypode aegyptiaca — Stephensen 1946: 188, fig 55.
Ocypode aegyptica [sic!] — Evans et al. 1973: 252, 275.
Ocypode saratan — Basson et al. 1977: 38, 40, 56, 60, 126, 140, 145, 224, 231, figs 14; Jones 1986a: 157, pl 42; Jones 1986b: 77, 78, fig 4; Vousden 1987: 34, 35, 71, tabs 3, 4, 6.

Type locality Dukhan (probably Qatar) or western coast of India (see Sakai and Türkay 2013).

Distribution

Persian Gulf: Iran [Stephensen (1946) as *O. aegyptica*; Türkay et al. 1996; Hosseini 2009; Naderloo and Türkay 2012; Naderloo et al. 2013], Kuwait [Clayton 1986; Jones (1986a, b) as *O. saratan*], Saudi Arabia [Basson et al. (1977) as *O. saratan*; Türkay et al. 1996], Bahrain [Vousden (1987) as *O. saratan*], UAE [Evans et al. (1973) as *O. aegyptica*; Hornby 1997; Cooper 1997; Türkay et al. 1996].

Gulf of Oman: Iran (Naderloo et al. 2015).

World distribution: Northern Indian Ocean: southern Oman, Persian Gulf, Gulf of Oman, Pakistan and India.

Habitat: High intertidal, substrate sandy.

Fig. 37.5 *Ocypode rotundata* Miers, 1882. Female (ZUTC 5000), CL = 48.10, CB = 52.2 mm, Beris, Iran, Gulf of Oman, 2012, coll. N. Ghotbeddin (Photo: R. Abdollahi)

Fig. 37.6 *Ocypode rotundata* Miers, 1882. The ghost crab trapped in a fishing net, Jod Village, Iran, Gulf of Oman, 18.08.2007 (Photo: R. Naderloo)

Fig. 37.7 *Ocypode rotundata* Miers, 1882. Burrow and pyramid constructed by a male ghost crab in a breeding season, Jod Village, Iran, Gulf of Oman, 18.08.2007 (Photo: R. Naderloo)

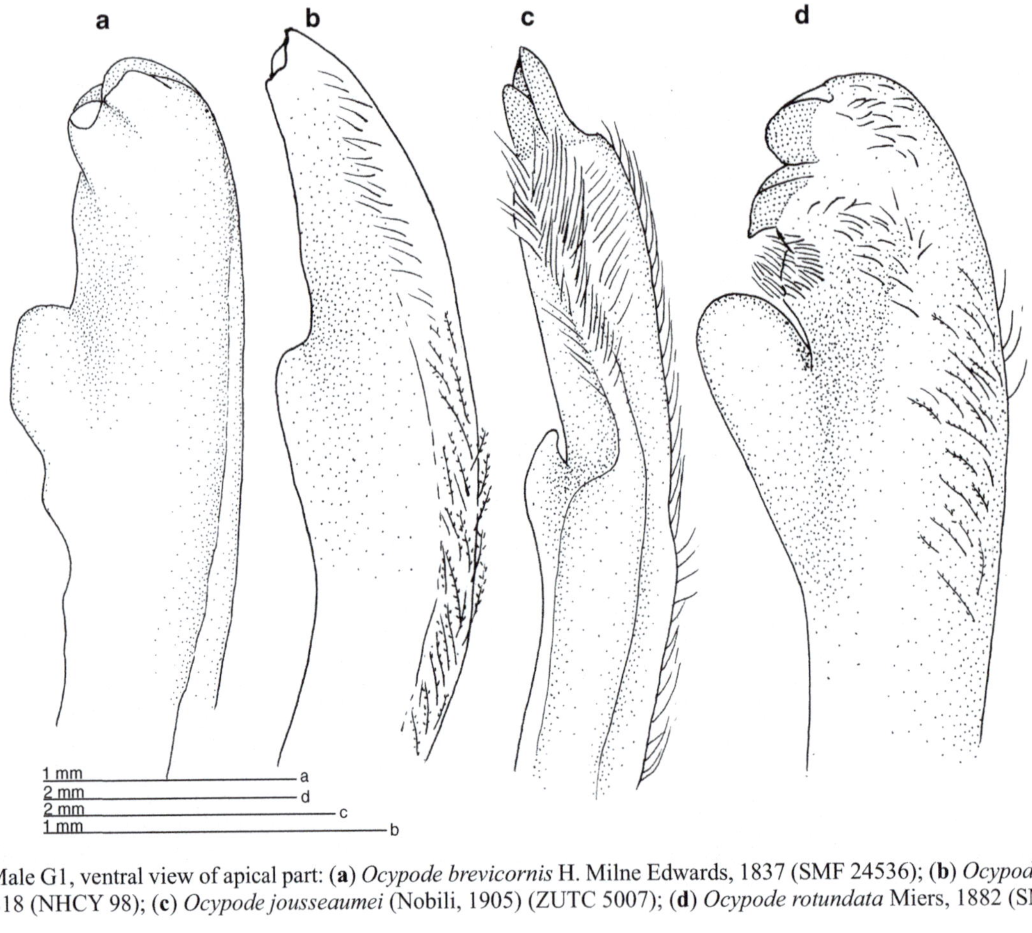

Fig. 37.8 Male G1, ventral view of apical part: (**a**) *Ocypode brevicornis* H. Milne Edwards, 1837 (SMF 24536); (**b**) *Ocypode cordimanus* Latreille, 1818 (NHCY 98); (**c**) *Ocypode jousseaumei* (Nobili, 1905) (ZUTC 5007); (**d**) *Ocypode rotundata* Miers, 1882 (SMF 23031)

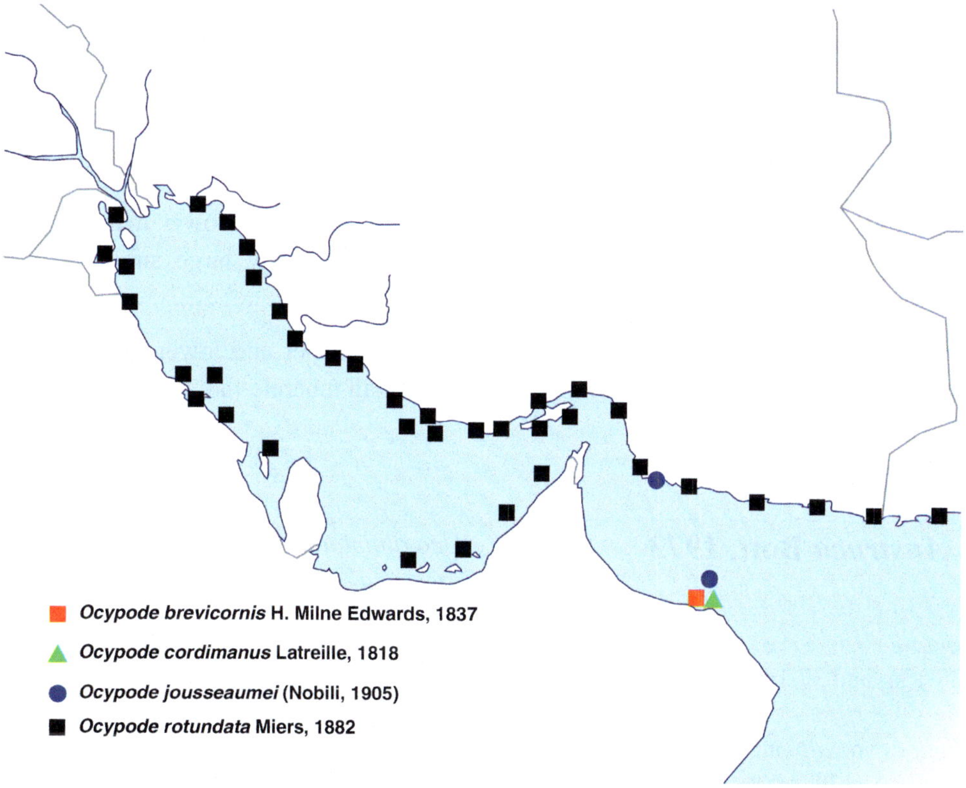

Fig. 37.9 Map showing the recorded localities of *Ocypode brevicornis* H. Milne Edwards, 1837, *Ocypode cordimanus* Latreille, 1818, *Ocypode jousseaumei* (Nobili, 1905), and *Ocypode rotundata* Miers, 1882

Key to species of fiddler crabs in the region

1. Front broad, around 1/6 to 1/8 of carapace breadth 2
- Front narrow, less than 1/10 of carapace breadth 5
2. Male large chelipeds with oblique granular ridge on inner surface of palm; immovable finger with distinct subdistal tooth on cutting edge; G1 with large subdistal process 3
- Male large chelipeds without oblique granular ridge on inner surface of palm; immovable finger without subdistal tooth on cutting edge; G1 without large subdistal process 4
3. Lateral margins of carapace nearly straight; movable finger of larger cheliped with median tooth; immovable finger with proximal, median and subdistal teeth; median tooth plate of gastric mill with six teeth, first four similar .... *Austruca iranica*
- Lateral margins of carapace slightly converging distally; proximal and median teeth on both fingers of larger cheliped; median tooth plate of gastric mill with eight teeth, first six similar ..... *Austruca albimana*

4. Movable finger of male larger cheliped with subdistal finger, looking apically bifurcate; male abdomen with locking apparatus; G1 without any subdistal process .... *Cranuca inversa*

- Movable finger of male large chelipeds without subdistal finger; male abdomen without locking apparatus; G1 with small subdistal process .... *Austruca sindensis*

5. Male larger cheliped with large tubercles on outer surface, upper and lower margins; immovable finger markedly arched downward medially, cutting edge with markedly large subdistal and large median elevation .... *Gelasimus hesperiae*

- Male larger cheliped with small tubercles on outer surfaces,upper and lower margins; immovable finger modertely arched downward medially, cutting edge with small tubercle-like teeth ..... *Gelasimus tetragonon*

The genus *Austruca* Bott, 1973

Austruca albimana (Kossmann, 1877)
(Figs. 37.10, 37.13a, and 37.14)

Gelasimus annulipes var. *albimana* Kossmann 1877: 53.
Uca annulipes — Nobili 1906a: 150, 151; Hornby 1997: 15.
Uca lactea — Hogarth 1989: 114, 115; Ismail and Ahmed 1993: 158.
Uca lactea annulipes — Hywel-Davies 1994: 37, 48.
Uca annulipes albimana — Apel and Türkay 1999: 133; Apel 2001: 114.
Uca (*Paraleptuca*) *albimana* — Ng et al. 2008: 241 (in list).
Uca albimana — Shih et al. 2009: 373–382.
Austruca albimana — Naderloo et al. 2010: 4 (in key), 4–7, figs 1a–l, 4a, 6a, b.
Austruca albimana — Shih et al. 2016: 153.

Type locality: Red Sea.

Distribution

Persian Gulf: UAE [Nobili (1906a) as *U. annulipes*; Apel and Türkay (1999) as *U. annulipes albimana*; Apel (2001) as *U. annulipes albimana*; Naderloo et al. 2010].

Gulf of Oman: UAE [(Apel (2001) as *U. annulipes albimana*; Naderloo et al. 2010].

World distribution: Northwestern Indian Ocean: Red Sea, Gulf of Aden, Socotra, Arabian Sea, southern Oman, southwestern Gulf of Oman and southeastern Persian Gulf.

Habitat: Intertidal, substrate sandy-mud.

Fig. 37.10 *Austruca albimana* (Kossmann, 1877). Socotra Island, Yeman: (**a**) frontal view; (**b**) dorsal view (Photo: Michael Apel. Reproduced from Naderloo et al. (2010) by permission of Magnolia Press)

Austruca iranica (Pretzmann, 1971)
(Figs. 37.11, 37.13b, and 37.14)

Uca annulipes iranica Pretzmann 1971: 481, pl 5, figs 11, 12.
Uca annulipes — Stephensen 1946: 189; Pretzmann 1971: 481; Al-Ghais and Cooper 1996: 419–421, figs 7, 8; Hornby 1997: 15 (?partly); Tirmizi and Ghani 1996: 105–107 (partly; not fig 40).
Uca annulipes iranica — Apel and Türkay 1999: 133, 134.
Uca lactea annulipes — Crane 1975: 299, 301, 611 (partly: material from the Persian Gulf); Collins et al. 1984: 318, 326, 327; Jones 1986a: 157, pl 42.
Austruca lactea annulipes — Clayton 1986: 86–88, fig 3.
Uca iranica — Shih et al. 2009: 377–382.
Austruca iranica — Naderloo et al. 2010: 14–19, figs 6c–f, 7a, 8a, 9a–k; Naderloo and Türkay 2012: 54.

Fig. 37.11 *Austruca iranica* (Pretzmann, 1971). Naqasheh, Qeshm Island, Iran, Persian Gulf: (**a**) frontal view; (**b**, **c**) dorsal view, with different colour patterns (Photo: R. Naderloo. Reproduced from Naderloo et al. (2010) by permission of Magnolia press)

Austruca albimana — Shih et al. 2016: 153, fig 8c.

Type locality: Persian Gulf (Bandar-Abbas, Iran).

Distribution

Persian Gulf: Iran [Pretzmann (1971) as *U. annulipes* and *U. annulipes iranica*; Mokhtari et al. (2008) as *U. lactea annulipes*; Naderloo et al. 2010; Mokhlesi et al. (2011) as *U. annulipes*; Naderloo and Türkay 2012], Kuwait [Clayton (1986) as *Austruca lactea annulipes*; Collins et al. (1984) and Jones (1986a) as *U. lactea annulipes*; Naderloo et al. 2010], UAE [Al-Ghais and Cooper (1996) and Hornby (1997) as *U. annulipes*; Apel and Türkay (1999) and Apel (2001) as *U. annulipes iranica*; Naderloo et al. 2010].

Gulf of Oman: Iran (Naderloo et al. 2010; Naderloo et al. 2015).

World distribution: Northern Indian Ocean: Persian Gulf, Gulf of Oman and Pakistan.

Habitat: Intertidal, substrates muddy-sand and mangroves.

Austruca sindensis (Alcock, 1900)

(Figs. 37.12, 37.13c, and 37.14)

Gelasimus inversus var. *sindensis* Alcock 1900: 356.

Uca (*Amphiuca*) *inversa sindensis* — Crane 1975: 108, pl 16e–h, figs 69j, 99.

Uca sindensis — Collins et al. 1984: 318–328, figs 1–4; Jones 1986a: 157, pl 42; Apel and Türkay 1999: 134; Apel 2001: 115; Mokhlesi et al. 2011: 245–249; Naderloo et al. 2013: 450, tab 1.

Uca (*Paraleptuca sindensis*) — Ng et al. 2008: 241 (in list); Naser et al. 2011: 1–3, fig 2; Naderloo and Türkay 2012: 54.

Austruca sindensis — Shih et al. 2016: 152, fig 8g.

Type locality: Pakistan (Karachi).

Distribution

Persian Gulf: Iran (Mokhlesi et al. 2011; Naderloo and Türkay 2012; Naderloo et al. 2013), Iraq (Naser et al. 2010), Kuwait (Collins et al. 1984; Jones 1986a).

Gulf of Oman: Iran (Naderloo et al. 2015).

World distribution: Northern Indian Ocean: Persian Gulf, Gulf of Oman and Pakistan.

Habitat: Intertidal, substrates muddy, sandy-mud and mangroves.

Fig. 37.12 *Austruca sindensis* (Alcock, 1900). Mangroves, behind Airport of Bandar-Abbas, Iran, Persian Gulf, 01.01.2008: (**a**) frontal view; (**b**) dorsal view (Photo: R. Naderloo)

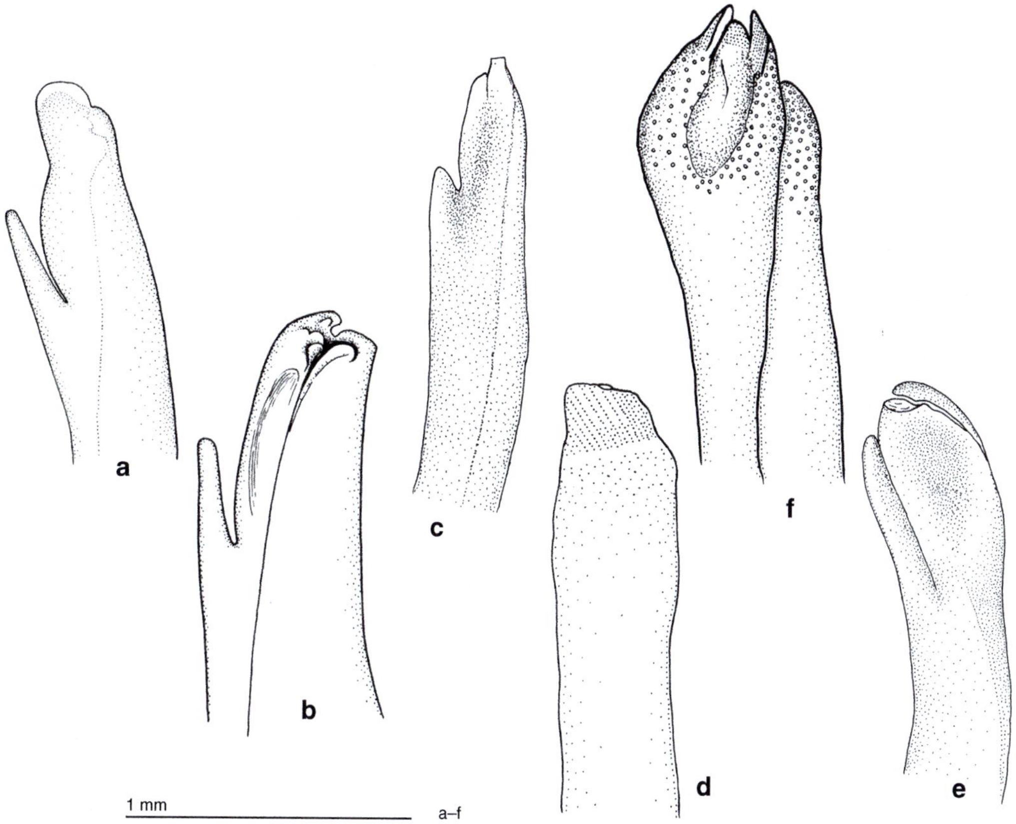

Fig. 37.13 Male G1, ventral view of apical part: (**a**) *Austruca albimana* (Kossmann, 1877); (**b**) *Austruca iranica* (Pretzmann, 1971); (**c**) *Austruca sindensis* (Alcock, 1900) (SMF 40575); (**d**) *Cranuca inversa* (Hoffman, 1874) (SMF 26045); (**e**) *Gelasimus tetragonon* (Herbst, 1790) (SMF: Red Sea); (**f**) *Gelasimus hesperiae* (Crane, 1975) (SMF 26035)

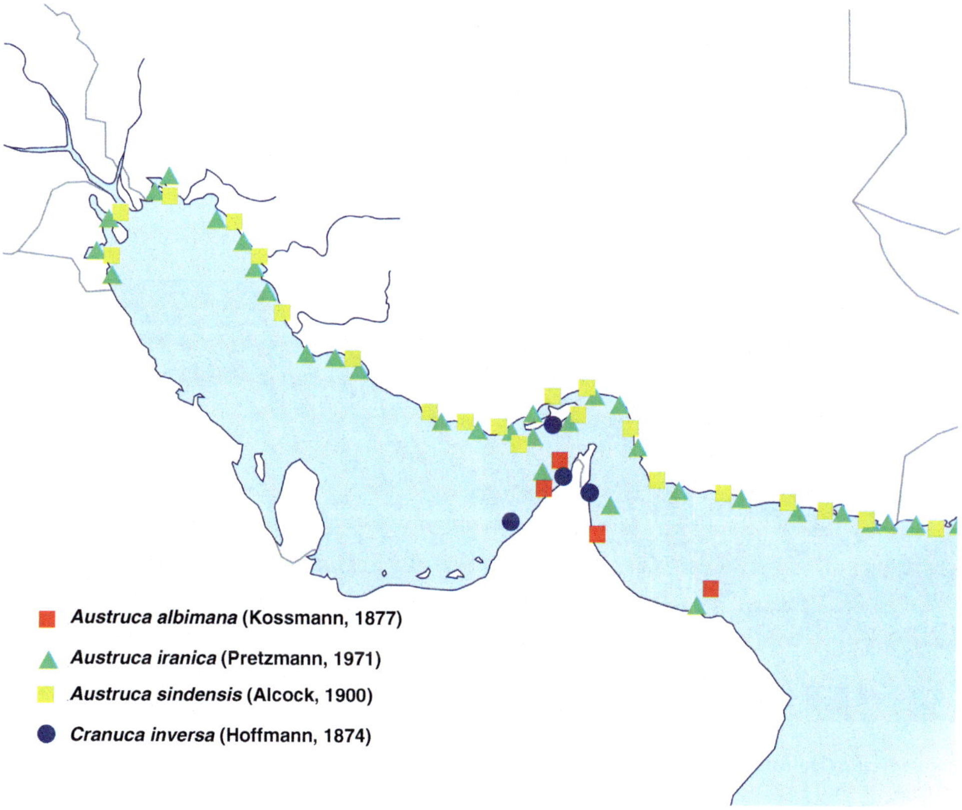

Fig. 37.14 Map showing the recorded localities of *Austruca albimana* (Kossmann, 1877), *Austruca iranica* (Pretzmann, 1971), *Austruca sindensis* (Alcock, 1900), *Cranuca inversa* (Hoffman, 1874)

The genus *Cranuca* Beinlich & von Hagen, 2006

Cranuca inversa (Hoffman, 1874)

(Figs. 37.13d, 37.14 and 37.15)

Gelasimus inversus Hoffmann 1874: 19, pl 4, figs 23–26.
Uca inversa — Collins et al. 1984: 325, 326, figs 3f, g, 4c–e, g; Hogarth 1989: 117; Al-Ghais and Cooper 1996: 421, 422; Hornby 1997: 15; Apel and Türkay 1999: 134; Apel 2001: 115; Naderloo et al. 2013: 450, tab 1.
Uca (*Amphiuca*) *inversa inversa* — Crane 1975: 105–108, pl 16, figs 39g, 69i, 99.
Uca sindensis — Ismail and Ahmed 1993: 158.
Uca inversa inversa — Hywel-Davies 1994: 37, 48.
Cranuca inversa — Beinlich and von Hagen 2006: 10, figs 5a, 9i; Ng et al. 2008: 240 (in list); Naderloo and Türkay 2012: 53.
Cranuca inversa — Shih et al. 2016: 153, fig 9a.
Type locality: Madagascar.
Distribution

Persian Gulf: Iran (Naderloo and Türkay 2012; Naderloo et al. 2013), UAE (Al-Ghais and Cooper 1996; Apel and Türkay 1999; Apel 2001).

Gulf of Oman: UAE [Ismail and Ahmed (1993) as *U. sindensis*; Hornby 1997; Apel 2001].

World distribution: Western Indian Ocean: South Africa, East Africa, Madagascar, Red Sea, Gulf of Aden, southern Oman, Persian Gulf, Gulf of Oman and Pakistan.

Habitat: Intertidal, substrates muddy and mangroves.

The genus *Gelasimus* Latreille, 1817

Gelasimus hesperiae (Crane, 1975)

(Figs. 37.13f, 37.16 and 37.17)

Uca (*Thalassuca*) *vocans hesperiae* Crane 1975: 87–89, 92, 93, 94 (partly).
Uca vocans vocans — Hywel-Davies 1994: 37, 48.
Uca vocans forma *excisa* — Al-Ghais and Cooper 1996: 421.
Uca hesperiae — Apel and Türkay 1999: 134; Apel 2001: 115.
Gelasimus hesperiae — Ng et al. 2008: 241 (in list).
Gelasimus hesperiae — Shih et al. 2016: 151, fig 7c.
Type locality: Tanzania (Zanzibar).
Distribution

Persian Gulf: UAE [Al-Ghais and Cooper (1996) as *Uca vocans* forma *excisa*; Apel 2001]

Gulf of Oman: UAE (Apel 2001).

World distribution: Indian Ocean: East Africa, Red Sea, Gulf of Oman, Persian Gulf, Pakistan, India, Sri Lanka, Burma, Malaysia and Singapore.

Habitat: Intertidal, substrates muddy and sandy/muddy.

Fig. 37.15 *Cranuca inversa* (Hoffman, 1874). Khor Ghirmah, N coast of Socotra Island, Yemen, 13.04.1999: (**a**) frontal view; (**b**) dorsal view (Photo: M. Apel)

Fig. 37.16 *Gelasimus hesperiae* (Crane, 1975). Al-Khawkhah, S Al-Hudaydah, Yemen, southern Red Sea, 03.04.2004. (**a**) Frontal view; (**b**) dorsal view (Photo: M. Apel)

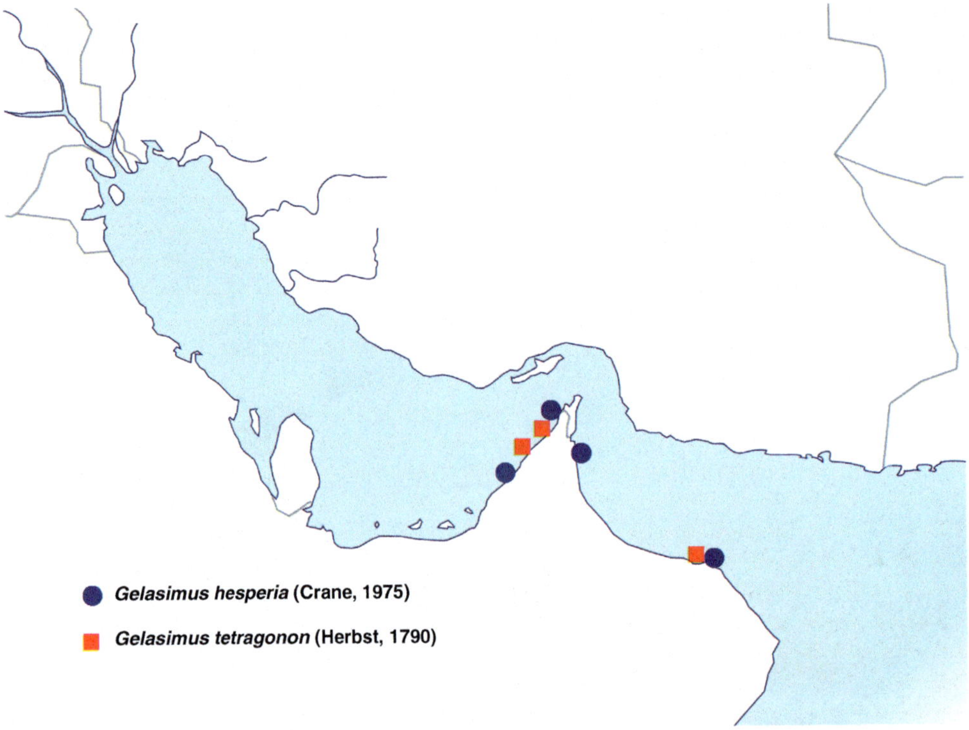

Fig. 37.17 Map showing the recorded localities of *Gelasimus hesperaie* (Crane, 1975), *Gelasimus tetragonon* (Herbst, 1790)

Gelasimus tetragonon (Herbst, 1790)

(Figs. 37.13e, 37.17, and 37.18)

Cancer tetragonon Herbst 1790: 257, 258, pl 20, fig 110.
Uca tetragonon — Nobili 1906a: 151, 152; Maccagno 1928: 22, 23, figs 6–8; Apel and Türkay 1999: 134; Apel 2001: 115.
Uca (*Thalassuca*) *tetragonon* — Crane 1975: 77–82, 596, 597, pl 13, figs 37d, 63a, b, 81f, 82e, 99; Sakai 1999: 38.
Gelasimus tetragonon — Ng et al. 2008: 241 (in list).
Gelasimus tetragonon — Shih et al. 2016: 151, fig 7f.
Type locality: Unknown.

Distribution

Persian Gulf: UAE (Apel and Türkay 1999; Apel 2001).

Gulf of Oman: Oman (Apel and Türkay 1999; Apel 2001).

World distribution: Indo-West Pacific: South Africa, East Africa, Red Sea, Persian Gulf, Gulf of Oman, Philippines, Australia and Tahiti.

Habitat: Intertidal, substrates muddy, sandy-mud and mangroves.

Fig. 37.18 *Gelasimus tetragonon* (Herbst, 1790). Al-Khawkhah, S Al-Hudaydah, Yemen, southern Red Sea, 03.04.2004: (**a**) frontal view; (**b**) dorsal view (Photo: M. Apel)

Family Pinnotheridae De Haan, 1833 (Pea Crabs) 38

Diagnosis Members of the family are small-sized crab usually living commensally with other marine species like bivalves. Sexual dimorphism is conspicuous in this family with females markedly larger than males. Females are mostly settling inside their host and are more frequently seen in comparison to males which are mostly free-living. Therefore most of the following characteristics are provided for females. Carapace usually circular to subcircular, soft in females, hard in males. Front narrow, usually entire. Orbits and eyes distinctly very small. Chelipeds medium-sized, fingers usually long, but shorter than manus, cutting edges with small rounded teeth. Walking legs slightly flattened. Third maxillipeds operculiform, merus completely fused to ischium. Abdomen of males and females with seven movable somites, abdomen of males narrow, abdomen of females markedly broad. Genital opening of males and females sternal.

Key to species of the family Pinnotheridae in the region

1. Dactylus of last two walking legs longer than dactylus of first two ones and in different form; third maxillipeds with merus broad, spatulate, dactylus attaching to medial part of merus .... *Arcotheres tivelae*

- Dactylus of walking legs equal or subequal, in similar form; third maxillipeds with merus slender or broad 2

2. Third maxillipeds without dactylus; propodus larger than carpus .... *Ostracotheres spondyli*

- Third maxillipeds with small dactylus attaching to proximal part of propodus; propodus smaller than carpus .... *Pinnotheres perezi*

Genus Arcotheres Manning, 1993

Arcotheres tivelae (Gordon, 1936)

(Figs. 38.1, 38.2, 38.5a, and 38.6)

Pinnotheres tivelae Gordon 1936: 167 (in key), 174, 175, fig 4.
Pinnotheres tivelae Saeedi and Ardalan 2010: 355–361.
Arcotheres placunae — Naderloo and Türkay 2012: 54; Naderloo et al. 2013: 450, tab 1.
Arcotheres tivelae — Ng et al. 2008: 248 (in list).
Type locality: Gulf of Oman (Muscat).
Distribution
Persian Gulf: Iran [Naderloo and Türkay (2012) as *A. placunae*; Naderloo et al. (2013) *A. placunae*).
Gulf of Oman: Iran [Naderloo et al. (2015) as *A. placunae*], Oman (Gordon 1936).

R. Naderloo, *Atlas of Crabs of the Persian Gulf*, DOI 10.1007/978-3-319-49374-9_38

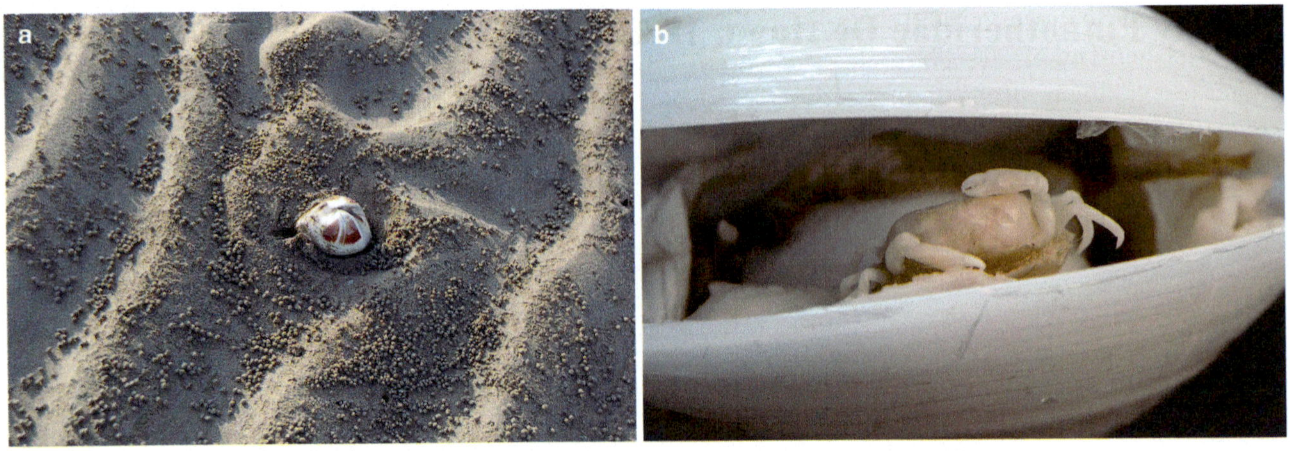

Fig. 38.1 *Arcotheres tivelae* (Gordon, 1936). (**a**) Host of the pea crab, namely *Callista umbonella* (Lamarck, 1818), which is the most common bivalve occuring on the sandy flats, Bandar-Abbas, Iran, Persian Gulf, photo: R. Naderloo; (**b**) the female pea crab in the mantle cavity of *Callista umbonella* (Photo: S. Shahbazi)

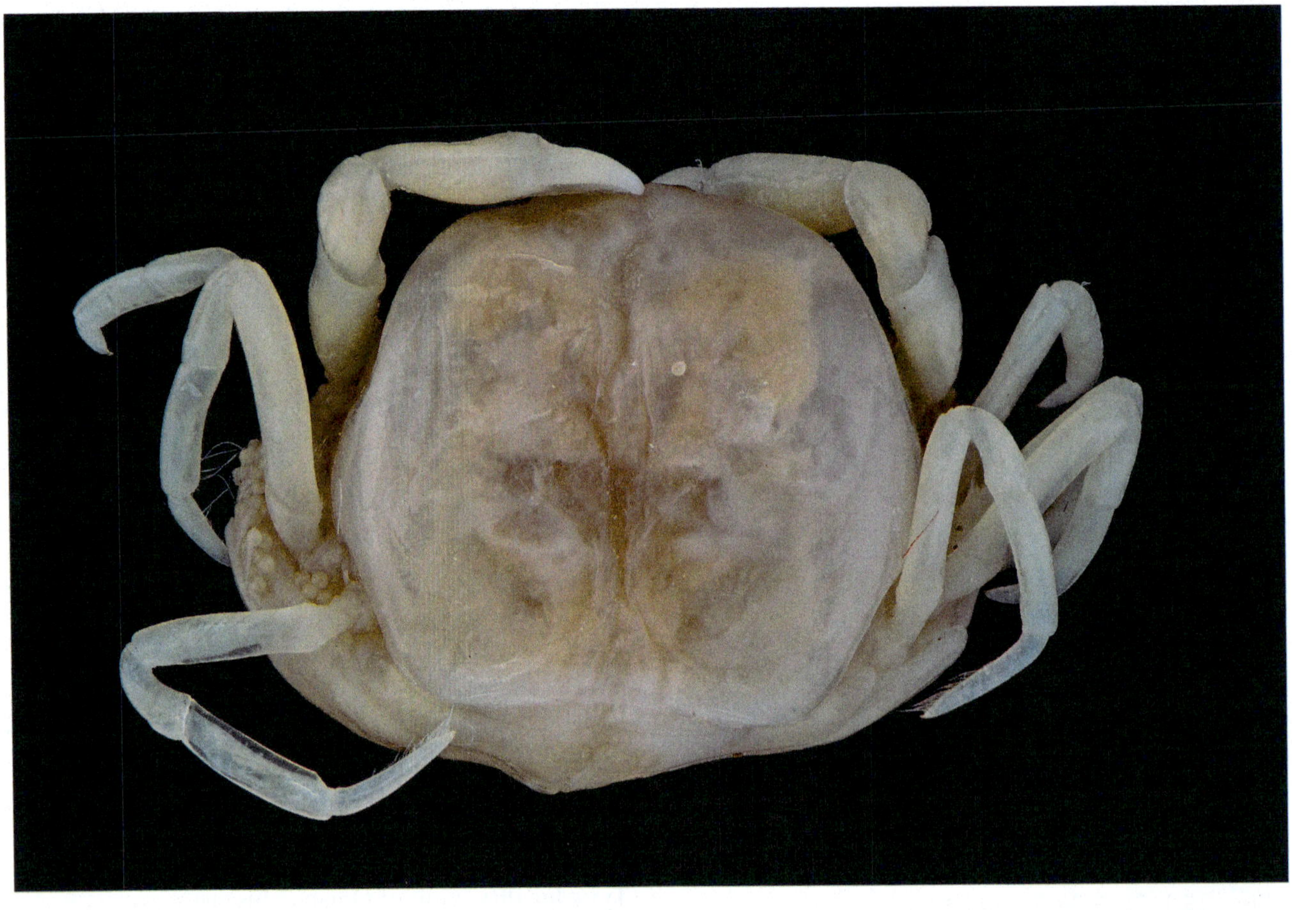

Fig. 38.2 *Arcotheres tivelae* (Gordon, 1936). Female (ZUTC 5931), CL = 8.4, CB = 9.5 mm, E Bandar-Abbas, 23.04.2008, coll. R. Naderloo (Photo: S. Shahbazi)

World distribution: Northern Indian Ocean: Persian Gulf, Gulf of Oman.

Habitat: Intertidal, substrates sandy and muddy-sand. This species occupies the most common bivalve, *Callista umbonella* (Lamarck, 1818) along the Iranian coast.

Remarks: This species was originally described from an intertidal clam, *Tivela ponderosa* (Koch in Philippi, 1844) in Muscat along the southwestern Gulf of Oman by Gordon (1936). *Arcotheres tivelae* has been frequenlty identified as *Arcotheres placunae*

(Hornel and Southwell (1909) from the region (e.g. Naderloo and Türkay 2012; Naderloo et al. 2013), the latter is a common species in the western Indian and Paksitani coast. I examined holotype of *Arcotheres tivelae* (Gordon, 1936) deposited in the Natural History Museum of London. Here examined specimens from the Persian Gulf and Gulf of Oman are completely identical with the female holotype of *A. tivelae*. Using the brief description and incomplete illustrations of the type material of *A. placunae* provided by Hornel and Southwell (1909) and regarding the georgraphical affinity, I decided to the synomymized *A. tivelae* with *A. placunae* with giving periority to the latter (see Naderloo & Türkay 2012). But, after discussion with Peter Ng (Lee Kong Chian Natural History Musuem, Singapore) and exmination of material from Karachi (Pakistan) deposited in ZRC, I was relatviely convinced that the two species are distinct and here exmined species belong to *A. tivilae*. However, further taxonomic works with comparing males, in particular G1 of males (no male of the either species has been described so far), are needed in order to establish the taxonommic position of these two species. Currently, *Arcotheres tivelae* is the only species of the genus occuring in the Persian Gulf and the Gulf of Oman, and this is more common than two other species of the family in the region. Whearas, *A. placunae* is apparently distributed on the Paksitani coast eastwards to western western Indian coast, but their exact distribution range is not still well documented.

Genus *Ostracotheres* H. Milne Edwards, 1853

Ostracotheres spondyli Nobili, 1905

(Figs. 38.3, 38.5c, and 38.6)

Ostracotheres spondyli Nobili 1905a: 164.
Ostracotheres spondyli — Nobili 1906a: 149, 150, pl 5, fig 24, pl 6, fig 33.

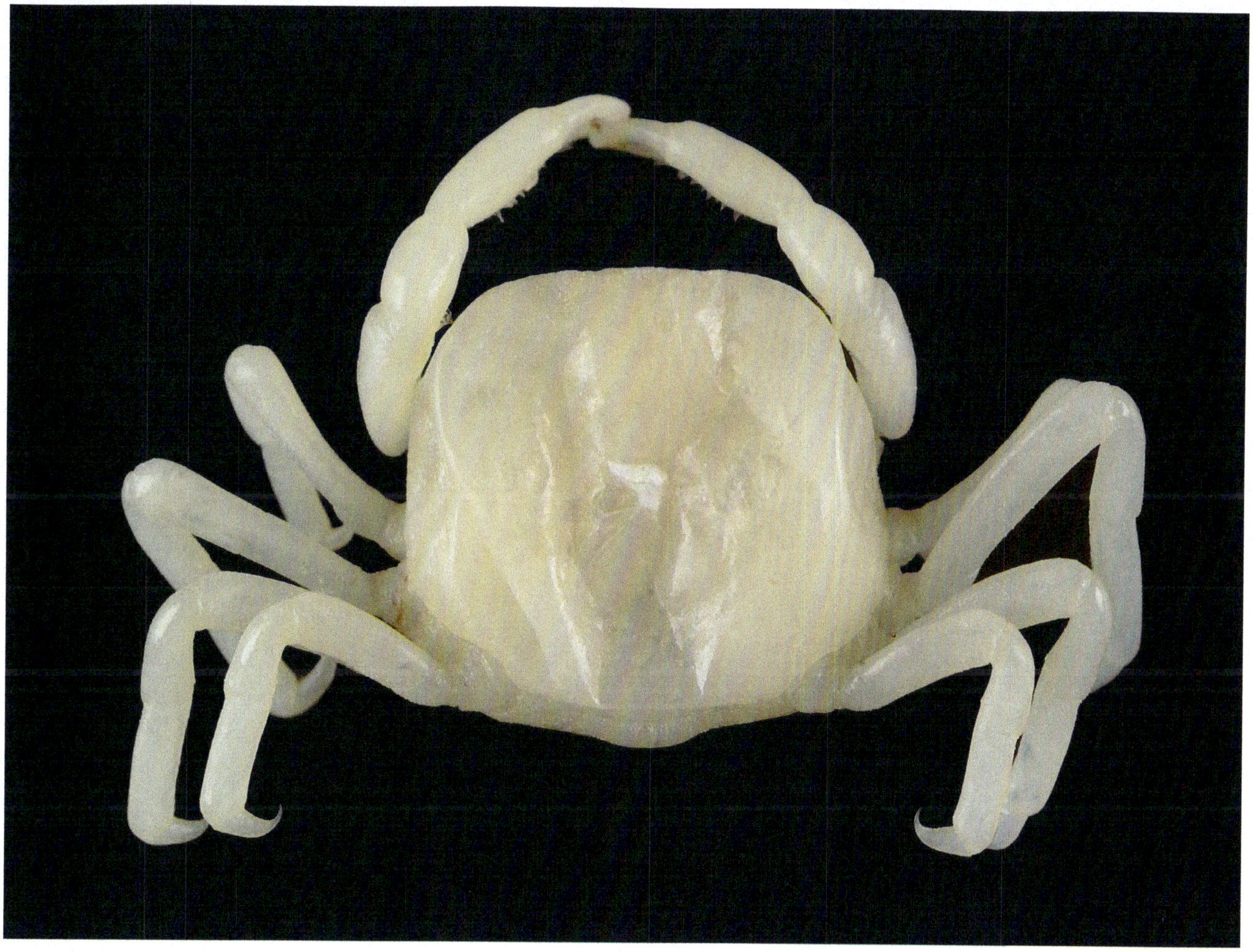

Fig. 38.3 *Ostracotheres spondyli* Nobili, 1905. Syntype female (MNHN-IU-2-14-8621), CL = 7.8, CB = 8.7 mm, Arzanah Island, Abu Dhabi, UAE, Persian Gulf (Photo: S. Träkner)

Ostracotheres spondyli—Tesch 1918: 262; Stephensen 1946: 185, 186; Titgen 1982: 253 (in list); Apel 2001: 106; Ng et al. 2008: 250 (in list).

Type locality: Persian Gulf (Arzanah Island, Abu Dhabi, UAE).

Distribution

Persian Gulf: Saudi Arabia (Apel 2001), UAE (Nobili 1905a, 1906a).

Gulf of Oman: No record.

World distribution: Endemic: Persian Gulf.

Habitat: Apel (2001) collected the species from a spiny oyster, *Spondylus* sp. in the Persian Gulf.

Genus *Pinnotheres* Bosc, 1802

Pinnotheres perezi Nobili, 1905

(Figs. 38.4, 38.5b, and 38.6)

Pinnoteres [sic!] *perezi* Nobili 1905a: 164.

Pinnoteres [sic!] *perezi* — Nobili 1906a: 147, 148, pl 5, fig 25.

Pinnotheres perezi — Tesch 1918: 250, 253, 287; Stephensen 1946: 185; Titgen 1982: 253 (in list); Apel 2001: 106; Ng et al. 2008: 250 (in list).

Type locality: Persian Gulf (Arzanah Island, Abu Dhabi, UAE).

Distribution

Persian Gulf: UAE (Nobili 1905a, 1906a).

Gulf of Oman: No record.

World distribution: Endemic: Persian Gulf.

Habitat: Unknown.

Fig. 38.4 *Pinnotheres perezi* Nobili, 1905. Holotype female (MNHN-IU-2014-6292), CL = 6.4, CB = 7.2 mm, Arzanah Island, Abu Dhabi, UAE, Persian Gulf (Photo: S. Träkner)

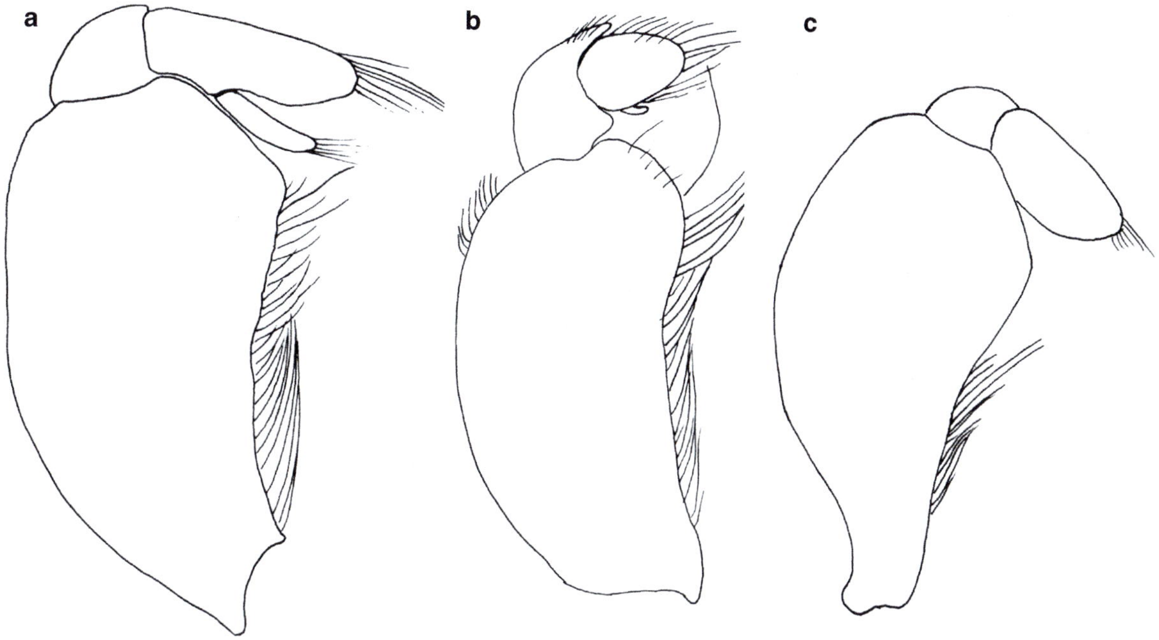

Fig. 38.5 Third maxillipeds of Pinnotheridae: (**a**) *Arcotheres tivelae* (Gordon, 1936) (ZUTC 5931); (**b**) *Pinnotheres perezi* Nobili, 1905 (MNHN-IU-2014-6292: holotype female); (**c**) *Ostracotheres spondyli* Nobili, 1905, female (SMF: Pole Reef)

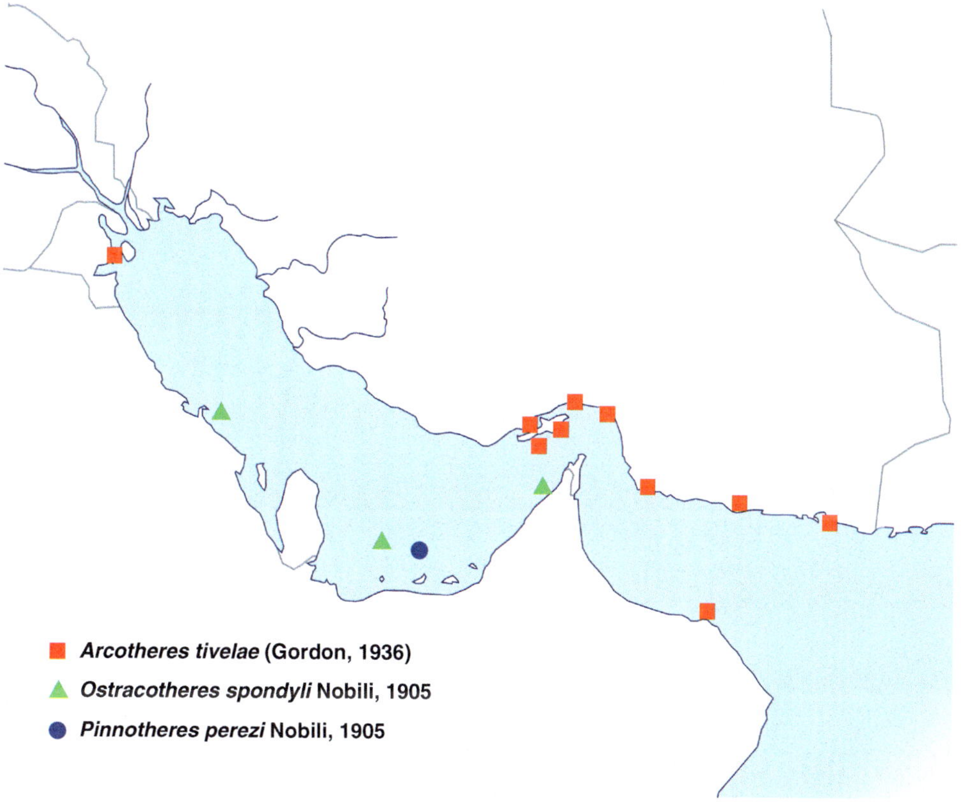

Fig. 38.6 Map showing the recorded localities of *Arcotheres tivelae* (Gordon, 1936), *Ostracotheres spondyli* Nobili, 1905, and *Pinnotheres perezi* Nobili, 1905

References

Abele LG (1972) Comparative habitat diversity and faunal relationships between the Pacific and Caribbean Panamian decapod Crustacea: a preliminary report, with some remarks on the crustacean fauna of Panama. Bull Biol Soc Wash 2:125–138

Adams A, White A (1849) Crustacea. In: Adams A (ed) The Zoology of the voyage of HMS "Samarang" , under command of Captain Sir Edward Belcher, during the years 1843–46. Reeve, Bentham & Reeve, London, pp 1–66, pls 61–13

Ahyong ST, Lai JC, Sharkey D, Colgan DJ, Ng PK (2007) Phylogenetics of the brachyuran crabs (Crustacea: Decapoda): the status of Podotremata based on small subunit nuclear ribosomal RNA. Mol Phylogenet Evol 45:576–586

Alcock A (1895) Material for a carcinological Fauna of India. No. 1. The Brachyura Oxyrhyncha. J Asiat Soc Bengal 64(2):157–291

Alcock A (1896) Material for a carcinological Fauna of India. No. 2: The Brachyura Oxystomata. J Asiat Soc Bengal 65(2):134–296

Alcock A (1898) Material for a carcinological fauna of India. No. 3: the Brachyura Cyclometopa. Part I. The family Xanthidae. J Asiat Soc Bengal 67:67–233

Alcock A (1899a) Material for a carcinological Fauna of India. No. 4: the Brachyura Cyclometopa. Part II: a revision of the Cyclometopa, with an account of the families Portunidae, Cancridae and Corystidae. J Asiat Soc Bengal 68:1–104

Alcock A (1899b) Material for a carcinological Fauna of India. No. 5: the Brachyura Primigenia or Dromiacea. J Asiat Soc Bengal 68:123–198

Alcock A (1899c) Illustrations of the Zoology of the Royal Indian Marine Survey Ship Investigator: Crustacea. Part VII, Plates XXXVI–XLV. Government Printing Press, Calcutta

Alcock A (1900) Material for a carcinological Fauna of India. No. 6: the Brachyura Catometopa, or Grapsoidea. J Asiat Soc Bengal 69:279–456

Alcock A, Anderson ARS (1898) Illustrations of the Zoology of the Royal Indian Marine Survey Ship Investigator: Crustacea, Part VI, Plates XXXIII-XXXV. Office of the Superintendent of Government Printing, Calcutta: pls. 33–35

Alcock A (1901) Catalogue of the Indian Decapod Crustacea in the collection of the Indian Museum. Part I. Brachyura. Fasciculus I. Introduction and Dromides or Dromiacea (Brachyura Primigenia). Indian Museum, Calcutta, pp 1–80, pls. 81–87

Alcock A (1905) Catalogue of the Indian Decapod Crustacea in the collection of the Indian Museum. Part II. Anomura. Fasciculus I. Pagurides. Indian Museum, Calcutta, pp 1–197, pls 191–116

Alcock A, Anderson ARS (1894a) An account of a recent collection of deep sea Crustacea from the Bay of Bengal and Laccadive Sea. Natural history notes from H.M. Indian Marine survey steamer "Investigator". Ser. II, No. 14. J Asiat Soc Bengal 69:141–185

Alcock A, Anderson ARS (1894b) List of the shore and shallow water Brachyura collected during the season 1893-94. Natural history notes from H.M. Indian Marine survey steamer "Investigator". Ser. II, No. 17. J Asiat Soc Bengal 63:197–209

Alcock A, Anderson ARS (1895) Illustrations of the zoology of the Royal Indian Marine Surveying Steamer Investigator: Crustacea, Part III, Plates IX–XV. Office of the Superintendent of Government Printing, Calcutta: pls 9–15

Alcock A, Anderson ARS (1896) Crustacea. Part IV Illustrations of the Zoology of the Royal Indian Marine Surveying Steamer Investigator: pls 16–27

Alcock A, Anderson ARS (1897) Illustrations of the zoology of the Royal Indian Marine Surveying Steamer Investigator: Crustacea, Part V, Plates XXVIII–XXXII. Office of the Superintendent of Government Printing, Calcutta

Alcock A, Anderson ARS (1899) Crustacea, Part VII. Illustrations of the zoology of the Royal Indian Marine Surveying Steamer "Investigator". Trustees of the Indian Museum, Calcutta

Alcock A, Annandale N (1907) Illustrations of the zoology of the Royal Indian Marine Survey Ship Investigator: Crustacea, Part XII, Plates LXXVII–LXXIX. Government Printing Press, Calcutta: pls 77–79

Alcock A, MacGilchrist AC (1905) Illustrations of the zoology of the Royal Indian Marine Survey Ship Investigator: Crustacea, Part XI, Plates LXVII–LXXVI. Fishes. Part VIII, Plates XXXVI–XXXVIII. Office of the Superintendent of Government Printing, Calcutta: pls 68–76, 36–38

Alcock A, McArdle AF (1900) llustrations of the zoology of the Royal Indian Marine Survey Steamer Ship Investigator, under the command of Commander T.H. Heming, R.N. Crustacea. Part VIII, Pls 46–48

Alcock A, McArdle AF (1903) Illustrations of the zoology of the Royal Indian Marine Survey Ship Investigator: Crustacea, Part X, Plates LVI–LXVII. Office of the Superintendent of Government Printing, Calcutta: pls 56–67

Al-Ghais S, Cooper R (1996) Brachyura (Grapsidae, Ocypodidae, Portunidae, Xanthidae and Leucosiidae) of Umm Al Quwain mangal, United Arab Emirates. Trop Zool 9:409–430

Al-Khayat JA, Jones DA (1996) Two new genera, *Manningis* and *Leptochryseus* (Decapoda: Camptandriinae), and descriptions of the first zoea of six brachyurans from the Arabian Gulf. J Crustac Biol 16:797–813

Al-Khayat JA, Jones DA (1999) A comparison of the macrofauna of natural and replanted mangroves in Qatar. Estuar Coast Shelf Sci 49:55–63

Al-Yamani F (2008) Importance of the freshwater influx from the Shatt-Al-Arab river on the Gulf marine environment. In: Abuzinada AH, Barth H-J, Krupp F, Böer B, Al-Abdessalaam TZ (eds) . Springer, Protecting the Gulf's marine ecosystems from pollution, Switzerland: pp 207–222

R. Naderloo, *Atlas of Crabs of the Persian Gulf*, DOI 10.1007/978-3-319-49374-9

Apel M (1994a) Effects of the 1991 Gulf War oil spill on the crab fauna of intertidal mudflats in the western Arabian Gulf. Cowier Forsch-Inst Senckenberg 166:40–46

Apel M (1994b) Biology, ecology and taxonomy of brachyuran and paguridean Crustacea. In: Krupp F, Abuzinada AH, Nader IA (eds) Establishment of a marine habitat and Wildlife Sanctuary for the Gulf region final report for phase II, Jubail/Frankfurt, Frankfurt a. Main: pp 406–437

Apel M (1996) Ecological observations on crab communities (Crustacea: Decapoda: Brachyura) on intertidal mudflats in the western Arabian Gulf and the effect of the 1991 oil spill. In: Krupp F, Abuzinada AH, Nader IA (eds) A marine wildlife sanctuaryfor the Arabian Gulf: environmental research and conservation following the 1991 Gulfwar oil spill. European Commission, Brussels National Commission for Wildlife Conservation and Development, Riyadh orschungsinstitut Senckenberg, Frankfurt a.M., Brussels/Riyadh/Frankfurt a.M, pp 327–340

Apel M (2001) Taxonomie und Zoogeographie der Brachyura, Paguridea und Porcellanidae (Crustacea: Decapoda) des Persisch-Arabischen Golfes. Fachbereich Biologie, Frankfurt am Main

Apel M, Spiridonov VA (1998) Taxonomy and zoogeography of the portunid crabs (Crustacea: Decapoda: Brachyura: Portunidae) of the Arabian Gulf and adjacent waters. Fauna Arab 17:159–331

Apel M, Türkay M (1992) The intertidal crabs and hermit crabs (Crustacea: Decapoda: Brachyura and Paguridea) in the study area and their condition after the oil spill. Final report for phase I CEC-Project: establishment of a Marine Habitat and Wildlife Sanctuary for the Gulf Region, p 187–205

Apel M, Türkay M (1999) Taxonomic composition, distribution and zoogeographic of the Grapsid and Ocypodid crab Fauna of intertidal soft bottoms in the Arabian Gulf. Estuar Coast Shelf Sci 49:131–142

Apel M, Bishop JM (2006) First record of the non-indigenous Chinese mitten crab *Eriocheir hepuensis* Dai, 1991 from the Arabian Gulf with remarks on potential impacts. – Presentation at the 1st International AEHMS Conference, on the state of the Gulf-Ecosystem: Future and Threats, Al-Ain, UAE, March 5–7

Audouin V (1826) Explication sommaire des planches de Crustacés de l'Égypte et de la Syrie, publiées par J.C. Savigny; offrant un exposé de caractères naturels des genre avec la distinction des espèces. Description de l'Égypte. [...] Description de l'Égypte, ou Recueil des Observations et des Recherches [...]. Histoire naturelle 1:77–98

Bahmani M (1997) A systematic study of crabs in the inter-tidal zone of hormozgan province. Iran Fish Sci J 6:1–16

Balss H (1922) Diagnosen neuer japanischer Decapoden. Zool Anz 54(1–2):1–6

Balss H (1933) Beiträge zur Kenntnis der Gattung Pilumnus (Crustacea, Decapoda) und verwandter Gattungen. Capita Zoologica Scripta 4:1–47, pls 41–17

Barnard KH (1955) Additions to the fauna-list of South African Crustacea and Pycnogonida. Ann S Afr Mus 43:1–107

Barnes RSK (1967) The Macrophthalminae of Australasia: with a review of the evolution and morphological diversity of the type genus *Macrophthalmus* (Crustacea: Brachyura). Trans Zool Soc Lond 31:195–262

Barnes RSK (1970) The species of *Macrophthalmus* (Crustacea: Brachyura) in the collections of the British Museum (Natural History). Bull Br Mus Nat Hist (Zool) 20:205–251

Barnes RSK (1971) Biological results of the Snellius Expedition XXIII. The genus Macrophthalmus (Crustacea, Brachyura). ZoologischeVerhandelingen, Leiden 115: 3–40

Barnes RSK (1976) Contributions towards a revision of *Macrophthalmus* (Crustacea: Brachyura), VIII. A re-examination of the *M. telescopicus* (Owen) complex; The status of *M. Laevis* A. Milne Edwards; And the affinities of *M. holthuisi* Serène. Zool Meded 50:133–151

Barnes RSK (1977) Concluding contribution towards a revision of, and a key to, the genus *Macrophthalmus* (Crustacea: Brachyura). J Zool 182:267–280

Barnes RSK (2010) A review of the sentinel and allied crabs (Crustacea: Brachyura: Macrophthalmidae), with particular reference to the genus *Macrophthalmus*. Raffles Bull Zool 58:31–49

Barth H-J, Khan NY (2008) Biogeophysical setting of the Gulf. In: Abuzinada AH, Barth H-J, Krupp F, Böer B, Al-Abdessalaam TZ (eds) . Springer, Protecting the Gulf's marine ecosystems from pollution, Switzerland: pp 1–21

Basson PW, Burchard JA, Hardy JT, Price ARG (1977) Biotopes of the Western Arabian Gulf: marine life and environments of Saudi Arabia. ARAMCO, Department of Loss Prevention and Environmental Affairs, Dhahran

Beinlich B, von Hagen HO (2006) Materials for a more stable subdivision of the genus *Uca* Leach. Zool Meded Leiden 80(4):9–32

Bell T (1855a) Horae carcinologicae, or notices of Crustacea. I A monograph of the Leucosiadae. Ann Mag Nat Hist 16:361–367

Bell T (1855b) Horae carcinologicae, or notices of Crustacea. I. A monograph of the Leucosiadae. Trans Linn Soc Lond 21:277–314, pls 230–234

Bianconi G (1851) Specimina zoologica Mosambicana. Memorie della Accademia delle Scienze dell'Istituto di Bologna 3

Bianconi JJ (1869) Specimina Zoologica Mosambicana, Fasciculus XVII memorie della Accademia delle Scienze dell'Istituto di Bologna 2:199–222

Borradaile LA (1902) Marine crustaceans. III. The Xanthidae and some other crabs. In: Gardiner JS (ed) The fauna and geography of the Maldive and Laccadive Archipelagoes; Being the account of the work carried on and of the collections made by an expedition during the years 1899 and 1900. University Press, Cambridge, pp 237–271

Brito Capello F (1871) Algunas especies novas ou pouco conhecidas de Crustaceos pertencentes aos generos *Calappa* e *Telphusa*. Jornal de sciencias mathematicas, physicas e naturaes, Lisboa 3:128–134, pl 122, figs 121–120

Brösing A, Richter S, Scholtz G (2007) Phylogenetic analysis of the Brachyura (Crustacea, Decapoda) based on characters of the foregut with establishment of a new taxon. J Zoolog Syst Evol Res 45(1):20–32

Brösing A, Al-Aidaroos AM, Türkay M (2014) The Red Sea species of *Cymo* de Haan, 1833 (Decapoda, Brachyura, Xanthidae), associates of scleractinian corals. Zootaxa 3779:195–214

Buitendijk AM (1939) Biological results of the Snellius expedition V. The Dromiacea, Oxystomata, and Oxyrhyncha of the Snellius expedition. Temminckia 4:223–276

Calman WT (1920) A new crab of the Genus *Sesarma* from Basra. Ann Mag Nat Hist 5:62–65

Campbell BM (1969) The genus *Eucrate* (Crustacea: Goneplacidae) in Eastern Australia and the Indo-West Pacific. Mem Qld Mus 15:117–140

Campbell BM, Stephenson W (1970) The sublittoral Brachyura (Crustacea: Decapoda) of Moreton Bay. Mem Qld Mus 15:235–301

Carpenter KE, Krupp F, Jones DA, Zajonz U (1997) The living marine resources of Kuwait, Eastern Saudi Arabia, Bahrain, Qatar, and the United Arab Emirates. FAO Species Identification Field Guide For Fishery Purposes, Rome

Castro P (1999) Trapeziid crabs (Crustacea, Brachyura, Xanthoidea, Trapeziidae) of the Indian Ocean and the Red Sea. Zoosystema 21:93–120

Castro P (2000a) Crustacea Decapoda: a revision of the Indo-West Pacific species of palicid crabs (Brachyura: Palicidae). Résultats des campagnes MUSORSTOM 21:437–610

Castro P (2000b) Crustacea Decapoda: a revision of the Indo-West Pacific species of palicid crabs (Brachyura Palicidae). In: Crosnier A (ed) Résultats des Campagnes MUSORSTOM, vol 20. Paris, pp 437–610

Castro P, Ng PKL (2010) Revision of the family Euryplacidae Stimpson, 1871 (Crustacea: Decapoda: Brachyura: Goneplacoidea). Zootaxa 2375:1–130

Castro P, Ng PKL, Ahyong ST (2004) Phylogeny and systematics of the Trapeziidae Miers, 1886 (Crustacea: Brachyura), with the description of a new family. Zootaxa 643:1–70

Chen H (1982) On the genus *Nursilia* (Crustacea, Decapoda: Leucosiidae) of Chinese waters. Oceanologia et Limnologia Sinica 13:267–274

Chen H (1984) A study of the genus *Carcinoplax* (Crustacea, Decapoda, Goneplacidae) of Chinese water. Oceanologica et Limnologica Sinica 15:188–201

Chen H (1989) Leucosiidae (Crustacea: Brachyura). In: Forest J (ed) Résultats des Campagne MUSORSTOM, Mémoires du Muséum National d'Histoire Naturalle, vol 5. Paris Éd. du Muséum, pp 181–263

Chhapgar BF (1957a) On the marine crabs (Decapoda, Brachyura) of Bombay State, Part I. J Bombay Nat Hist Soc 54:399–439

Chhapgar BF (1957b) On the marine crabs (Decapoda: Brachyura) of Bombay State Part II. J Bombay Nat Hist Soc 54:503–549

Chhapgar BF (1969) More additions to the crab fauna of Bombay State. J Bombay Nat Hist Soc 65:608–617

Chiong WL, Ng PKL (1998) A revision of the buckler crabs of the genus *Cryptopodia* H. Milne Edwards, 1834 (Crustacea: Decapoda: Brachyura: Parthenopidae). Raffles Bull Zool 46:157–216

Chopra B (1934) Further notes on Crustacea Decapoda in the Indian Museum. III On the decapod Crustacea collected by the Bengal Pilot Service off the mouth of the river Hughli Dromiacea and Oxystomata. Rec Indian Mus 35:25–52

Chopra B (1935) Further notes on Crustacea Decapoda in the Indian Museum. VIII. On the decapod Crustacea collected by the Bengal Pilot Service off the mouth of the river Hooghly. Brachygnatha (Oxyrhyncha and Brachyrhyncha). Rec Indian Mus Calcutta 37:463–514

Chopra B, Das K (1930) Further notes of Crustacea Decapoda in the Indian Museum. On two new species of hymenosomatid crabs, with notes on some other species. Rec Indian Mus 32:413–429

Chopra B, Das KN (1937) Further notes on Crustacea Decapoda in the Indian Museum. IX On three collections of crabs from Tavoy and Mergui Archipelago. Rec Indian Mus 39:377–434

Chu KH, Ho HY, Li CP, Chan TY (2003) Molecular phylogenetics of the mitten crab species in *Eriocheir*, sensu lato (Brachyura: Grapsidae). J Crustac Biol 23:738–746

Clark PF, Galil B (1993) A revision of the xanthid genus *Pilodius* Dana, 1851 (Crustacea: Brachyura: Xanthoidea). Journal of Natural History 27: 1119–1206

Clark PF, Abdul-Sahib I, Al-Asadi M (2006) The first record of *Eriocheir sinensis* H. Milne Edwards, 1853 (Crustacea: Brachyura: Varunidae) from the Basrah Area of Southern Iraq. Aquat Invasions 2:1–4

Clark (2006) is actually Clark et al. (2006) which is already in the list … must be corrected in the text as well.

Clayton DA (1986) Ecology of mudflats with particular reference to those of the northern Arabian Gulf. In: Halwagy R, Clayton DA, Behbehani M (eds) Marine environment and pollution First Arabian Gulf Congress on Environment and Pollution. University of Kuwait, Kuwait, pp 83–96

Clayton DA (1996) Ghost crabs of Oman (Crustacea: Brachyura: Ocypodidae). Sultan Quaboos Univ J Sci 1:27–35

Clayton DA (2001) Acoustic calling in four species of ghost crabs: *Ocypode jousseaumei*, *O. platytarsus*, *O. rotundata* and *O. saratan* (Brachyura: Ocypodidae). Bioacoustics 12:37–55

Clayton DA (2005) Substrate (acoustic/vibrational) communication and ecology of the ghost crab *Ocypode jousseaumei* (Brachyura: Ocypodidae). Mar Freshw Behav Physiol 38:53–70

Clayton DA, Al-Kindi A (1998) Population structure and dynamics of two scopimerine sand crabs *Scopimera crabricauda* Alcock, 1900 and *Dotilla sulcata* (Forskåll, 1775) in an estuarine habitat in Oman. Trop Zool 11:197–215

Collins MJ, Jones DA, Clayton DA (1984) Redescription of *Uca sindensis* (Alcock, 1900) (Brachyura: Ocypodidae) with notes on the ecology of a population from Kuwait. J Crustac Biol 4:318–328

Cooper RT (1997) Mangal-associated Brachyura (Ocypodidae, Grapsidae, Portunidae, Majidae, Xanthidae and Leucosiidae) from the north-eastern coastal Islands of Abu Dhabi, United Arab Emirates. Crustaceana 70:155–179

Crane J (1975) Fiddler crabs of the world (Ocypodidae: genus *Uca*). Princeton University Press, Princeton

Creasey S, Rogers AD, Tyler P, Young C, Gage J (1997) The population biology and genetics of the deep-sea spider crab, *Encephaloides armstrongi* Wood-Mason, 1891 (Decapoda: Majidae). Philos Trans R Soc B 352:365–379

Crosnier A (1965) Crustacés Décapodes, Grapsidae et Ocypodidae. Faune de Madagascar 18:1–153

Dai AY (1991) Studies on the subspecies differentiation of the genus *Eriocheir* (Decapoda: Brachyura). Sci Treatise Syst Evol Zool 1:61–71

Dana JD (1851) Conspectus Crustaceorum quæ in Orbis Terrarum circumnavigatione, Carolo Wilkes e Classe Reipublicæ Fœderatæ Duce, lexit et descripsit. Proc Acad Natl Sci Phila 5:247–254

Dana JD (1852a) Conspectus Crustaceorum, & Conspectus of the Crustacea of the Exploring Expedition under Capt. Wilkes, U.S.N., including the Crustacea Cancroidea Corystoidea. Proc Acad Natl Sci Phila 6:73–86

Dana JD (1852b) Crustacea. Part I. C. Sherman, Philadelphia

Dana JD (1855) Crustacea. In: United States exploring expedition during the years 1838 to 1842, under the command of Charles Wilkes, U.S.N. C. Sherman, Philadelphia

Daniel A, Chakrapany S (1983) Observations on the swarming, breeding habits and some larval stages of the deep-sea portunid crab *Charybdis* (*Goniohellenus*) *edwardsi* Leene et Buitenijk, 1949 in the northern Arabian Sea in January–February 1974 and off the Madras coast during January to March 1976 to 1979. Rec Zool Surv India 81:101–108

Davie PJF (1989) A re-appaisal of *Heteropanope* Stimpson, and *Pilumnopeus* A. Milne Edwards (Crustacea: Decapoda: Pilumnidae) with description of new spesies and genera. Mem Qld Mus 27:129–156

Davie PJF (2002) Crustacea: Malacostraca: Eucarida (Part 2): Decapoda — Anomura, Brachyura. Zoological Catalogue of Australia. Wells A, Houston WWK (eds) Vol 193B, Melbourne, CSIRO Publishing. pp 641

Davie PJF (2012) A review of *Macrophthalmus sensu lato* (Crustacea: Decapoda: Macrophthalmidae) from Australia including two new species and new records. Mem Qld Mus Nat 56:149–219

Davie PJF, Ghani NA (1993) A new species of *Pilumnopeus* (Crustacea: Decapoda: Pilumnidae) from Pakistan. Raffles Bull Zool 41:61–65

Davie PFJ, Ng PKL (2007) Two new subfamilies of Varunidae (Crustacea: Brachyura), with description of two new genera. Raffles Bull Zool 16:257–272

De Grave S, Pentcheff D, Ahyong ST (2009) A classification of living and fossil genera of decapod crustaceans. Raffles Bull Zool Suppl 21:1–109

De Man JG (1879) On some new or imperfectly known Podophthalmous Crustacea of the Leyden museum. Notes from the Leyden Museum 1

De Man JG (1887a) Übersicht der Indo-pacifischen Arten der Gattung *Sesarma* Say nebst einer Kritik der von W. Hess und E. Nauck in den Jahren 1865 und 1880 beschriebenen Decapoden. Zoologischen Jahrbüchen: Abtheilung für Systematik Geographie und Biologie der Thiere 2:639–722

De Man JG (1887b) Report on the podophthalmous Crustacea of the Mergui Archipelago, collected for the Trustees of the Indian Museum, Calcutta, by Dr. John Anderson, F.R.S., Superintendent of the Museum. J Linn Soc Lond Zool 22:1–64, pls 1–3 (pt 1, Nov 1887), 65–128, pls 4–8 (pt 2, Dec 1887), 129–176, pls 9–12 (pt 3, Jan 1888), 177–240, pls 13–15 (pt 4, Apr 1888), 241–312, pls 16–19 (pt 5, Jul 1888). J Linn Soc Lond Zool 22:1–64, pls 61–63 (pt 61, Nov 1887), 1865–1128, pls 1884–1888 (pt 1882, Dec 1887), 1129–1176, pls 1889–1812 (pt 1883, Jan 1888), 1177–1240, pls 1813–1815 (pt 1884, Apr 1888), 1241–1312, pls 1816–1819 (pt 1885, Jul 1888)

De Man JG (1888a) Bericht über die im indischen Archipel von Dr. J Brock gesammelten Decapoden und Stomatopoden Archiv für Naturgeschichte 53:289–600

De Man JG (1888b) Report on the podophthalmous Crustacea of the Mergui archipelago, collected for the trustees of the Indian Museum, Calcutta, by Dr. John Anderson, F.R.S., Superintendent of the Museum. J Linn Soc Lond Zool 22:1–305, pls 301–319

De Man JG (1895) Bericht über die von Herrn Schiffscapitän Storm zu Atjeh, an den westlichen Küsten von Malakka, Borneo und Celebes sowie in der Java-See gesammelten Decapoden und Stomatopoden. Zoologische Jahrbücher Abteilung für Systematik, Geographie und Biologie der Thiere 8:485–609

De Man JG (1896) Bericht über die von Herrn Schiffskapitän STORM zu Atjeh, an den westlichen Küsten von Malakka, Borneo und Celebes sowie in der Java-See gesammelten Decapoden und Stomatopoden. Dritter Theil. Zoologische Jahrbücher, Abtheilung für Systematik, Geographie und Biologie der Thiere 9:339–386, figs 340–349

de Saint Laurent M (1980a) Sur la classification et la phylogénie des Crustacés Décapodes Brachyoures. I. Podotremata Guinot, 1977 et Eubrachyura sect. nov. Comptes rendus hebdomadaires des séances de l'Académie des sciences, Paris, (D) 290:1265–1268

de Saint Laurent M (1980b) Sur la classification et la phylogénie des Crustacés Décapodes Brachyoures. II. Heterotremata et Thoracotremata Guinot, 1977. Comptes rendus hebdomadaires des séances de l'Académie des sciences, Paris, (D) 290:11317–1320

Evans G, Murry JW, Biggs HEJ, Bate R, Bush PR (1973) The oceanography, ecology, sedimentology and geomorphology of parts of the Trucial Coast Barrier Island Complex, Persian Gulf. Springer Verlag, Berlin

Eydoux F, Souleyet LFA (1842) Crustacés. In: Eydoux F, Souleyet LFA (eds) Voyage autour du monde exécuté pendant les années 1836 et 1837 sur la corvette la Bonite, commandée par M Vaillant, Capitaine de Vaisseau, publié par odre du Roi, sous les auspices du départment de la marine, Zoologie. Firmin Didot Frères, Paris, pp 219–272

Fabricius JC (1775) Systema Entomologiae, sistens Insectorum Classes, Ordines, Genera, Species, adjectis Sysnonymis, Locis, Descriptionibus, Observationibus. Officina Libraria Kortii, Flensburgi et Lipsiae

Fabricius JC (1781) Species Insectorum exhibentes eorum differentias specificas, synonyma auctorum, loca natalia, metamorphosia adectis observationibus, descriptionibus. 2. Impensis Carol. Ernest. Bohnii, Hamburg Kilonii

Fabricius JC (1787) Mantissa Insectorum sistens eorum species nuper detectas adjectis characteribus genericis, differentiis specificis, emendationibus, observationibus

Fabricius JC (1793) Entomologia systematica emendata et aucta. Secundum. Classes, ordines, genera, species adjectis synonymis, locis, observationibus, descriptionibus (Tom. II)

Fabricius JC (1798) Supplementum Entomologiae Systematicae

Fatemi SMR, Vosoughi G, Valinasab T, Savari A, Ghorbeddin N (2011) First report of dotilid crabs (Decapoda, Brachyura) from the northern Gulf of Oman, Iran. Crustaceana 84(14):1745–1753

Fatemi SMR, Valinasab T, Ng PKL, Ghotbeddin N (2012) First record of *Cryptopodia echinosa* (Crustacea: Decapoda: Brachyura: Parthenopidae) from the Oman Sea. Mar Biodivers Rec 5:e52

Fauvel P (1911) Annélides polychètes du Golfe Persique recueillis par M.N. Bogoyawlensky. Archives de Zoologie Expérimentale et Générale 4:353–439

Fishelson L (1971) Ecology and distribution of the benthic fauna in the shallow waters of the Red Sea. Mar Biol 10:113–133

Flipse HJ (1930) Decapoda F. (Decapoda Brachyura continued) Oxyrhyncha: Parthenopidae. Siboga Expéditie 39:1–96

Forskål P (1775) Descriptiones animalium, avium, amphibiorum, insectorum, vermium, quae in itinere orientali observavit Petrus Forskål. Post Mortem Auctoris edidit Carsten Niebuhr. Adjuncta est materia Medica Kahirina. Hafniae

Galil B (1986) *Quadrella* (Brachyura: Xanthoidea: Trapeziidae) – review and revision. J Crustac Biol 6:275–293

Galil B (1987) Trapeziidae (Decapoda: Brachyura: Xanthoidea) of the Red Sea. Isr J Zool 34:159–182

Galil B (1997) Crustacea Decapoda: a revision of the Indo-Pacific species of genus *Calappa* Weber, 1795 (Calappidae). In: Crosnier A (ed) Résultats des Campagnes MUSORSTOM, 18, Mémoires du Muséum national d'Histoire naturelle, Paris, pp 171–335

Galil BS (2000) *Halimede tyche* (Herbst, 1801)(Decapoda; Xanthoidea; Halimedinae) from the Mediterranean coast of Israel, with a review of the genus. J Crustac Biol 20(5):325–331

Galil B (2001) A revision of the genus *Arcania* Leach, 1817 (Crustacea: Decapoda: Leucosioidea). Zool Meded 75:169–206

Galil BS (2003) Contribution to the knowledge of Leucosiidae II. *Euclosia* gen. nov.(Crustacea: Brachyura). Zool Meded 77:331–347

Galil B (2005a) Contributions to the knowledge of Leucosiidae IV. *Seulocia* gen. nov. (Crustacea: Brachyura). Zool Meded 79(2):41–59

Galil BS (2005b) Contributions to the knowledge of Leucosiidae III. *Urnalana* gen. nov. (Crustacea: Brachyura). Zool Meded Leiden 79:9–40

Galil BS (2006) Contributions to the knowledge of Leucosiidae V. *Coleusia* gen. nov.(Crustacea: Brachyura). Zool Meded 80:55–69

Galil BS (2009) An examination of the genus *Philyra* Leach, 1817 (Crustacea, Decapoda, Leucosiidae) with descriptions of seven new genera and six new species. Zoosystema 31:279–320

Galil BS, Clark PF (1988) On a collection of *Acropora*-inhabiting trapeziids (Crustacea Brachyura Xanthoidea) from East Africa. Trop Zool 1(1):137–151

Galil BS, Clark PF (1994) A revision of the genus *Matuta* Weber, 1795 (Crustacea: Brachyura: Calappidae). Zool Verh 294:1–55, figs 51–57, pls 51–14

Galil B, Vannini M (1990) Research on the coast of Somalia. Xanthidae Trapeziidae Carpiliidae Menippidae (Crustacea Brachyura). Trop Zool 3:21–56

Galil BS, Ashelby CW, Clark PF (2012) New records of four leucosiid species (Crustacea: Decapoda: Brachyura: Leucosiidae) from Qatar, Persian Gulf. Mar Biodivers Rec 5:e84

Gerstaecker A (1856) Carcinologische Beitrage. Archiv für Naturgeschichte 22:101–169

Ghani N, Tirmizi NM (1992) Occurrence of four Xanthid (Brachyura) crabs in Karachi waters (northern Arabian Sea). Pak J Mar Sci 1:37–47

Ghani N, Tirmizi NM (1995a) New Leucosid crab from Pakistan waters of the Northern Arabian Sea. In: Thompson MF, Tirmizi NM (eds) The Arabian Sea – living marine resources and the environment. Oxford & IBH Publishing Co., New Delhi, pp 143–147

Ghani N, Tirmizi NM (1995b) *Xenophthalmus wolffi* Takeda & Miyake, a genus and species of free living Pinnotherid crab new to Pakistan waters (Decapoda, Brachyura). Crustaceana 68(8):653–655

Ghory F, Ng PKL, Kazmi QB (2013) On the identities of *Pilumnopeus convexus* Maccagno, 1936, and *P. salomonensis* Ward, 1942, from the Indian Ocean, with a note on *P. riui* Takeda, 2001 (Decapoda, Brachyura, Pilumnidae). Crustaceana 86(3):301–312

Ghotbeddin N, Fatemi SMR, Valinasab T (2012a) Intertidal crabs of Chabahar Bay (Northeast of Oman Sea): new collections and biogeographic considerations. Iran J Fish Sci 12:440–451

Ghotbeddin N, Fatemi R, Valinassab T (2012b) Identification of Iranian subtidal portunid crabs (Crustacea: Decapoda: Brachyura) of the Oman Sea with first record of 5 species. Oceanography 3:1

Ghotbeddin N, Naderloo R (2014) Confirming of the occurrence of three intertidal xanthid crabs (Crustacea: Decapoda: Brachyura: Xanthidae) in the Gulf of Oman. Mar Biodivers Rec 7:e98

Gillikin DP, Schubart CD (2004) Ecology and systematics of mangrove crabs of the genus *Perisesarma* (Crustacea: Brachyura: Sesarmidae) from East Africa. Zool J Linn Soc 141:435–445

Girard C (1859) Note monographique sur les genres Crabe et Platycarcin, avec indication d'espèces nouvelles. (Famille des Cancérides. Ordre des Décapodes. Section des Brachyoures. Classee des Crustacés). Annales de la Societe entomologique de France 3:143–162

Gordon I (1930) Seven new species of Brachyura from the coasts of China. Ann Mag Natl Hist 10(6):519–525

Gordon I (1931) Brachyura from the coasts of China. J Linn Soc Lond Zool 37:525–558

Gordon I (1936) On a few Indo-Pacific species of *Pinnotheres*, with special reference to asymmetry of the walking legs. Zool J Linn Soc 40:163–180

Gordon I (1950) Crustacea: Dromiacea. Part I. Systematic account of the Dromiacea collected by the "John Murray" Expedition. Part II. The morphology of the spermatheca in certain Dromiacea. Sci Rep John Murray Exped 1933–34 London Br Mus Nat Hist 9:201–253

Gravier C (1920) Sur une collection de crustacés recueillis à Madagascar par M. le Lieutenant Decary. Bulletin du Muséum d'Histoire Naturelle Paris 26:376–383

Griffin DJG (1968) *Hyastenus hilgendorfi* De Man, a majid spider crab new to Australia. Aust Zool 15:103–106, pl.101

Griffin DJG (1973) A revision of the spider crabs of the genus *Phalangipus* (Crustacea, Brachyura, Majidae). J Nat Hist 7:165–207

Griffin DJG (1974) Spider crabs (Crustacea: Brachyura: Majidae) from the International Indian Ocean expedition, 1963–1964. Smithson Contrib Zool 182:1–35

Griffin DJG, Tranter HA (1974) Spider crabs of the family Majidae (Crustacea: Decapoda: Brachyura) from the Red Sea. Isr J Zool 23(1974):162–198

Griffin DJG, Tranter HA (1986) The Decapoda Brachyura of the Siboga expedition. Part VIII: Majidae. Siboga Expéd 39:1–335

Guérin-Méneville FE (1838) Crustacés, Arachnides et Insectes. In: Duperrey LI (ed) Voyage autour de monde, execute par ordre du Roi, sur la corvette de sa Majeste, la Coquille, pendant les annees 1822, 1823, 1824 et 1825, Paris, p 319

Guinot D (1960) Les especes Indo-Pacifiques du genre *Globopilumnus* (Crustacea Brachyura Xanthidae). Mémoires de l'Institut scientifique de Madagascar F 3:97–119, figs 111–114

Guinot D (1964) Crustacés Décapodes Brachyoures (Xanthidae) des Campagnes de la Calypso en Mer Rouge (1952), dans la Golfe Persique et a l'ile Aldabra (1954). Mémoires du Muséum National d'Histoire Naturelle, (A, Zool) 32:1–108

Guinot D (1967a) La faune carcinologique (Crustacea, Brachyura) de l'Ocean Indien occidental et de la Mer Rouge. Catalogue remarques biogéographiques et bibliographie. Mémoires de l'Institut fondamental d'Afrique noire 77:235–352

Guinot D (1967b) Recherches préliminaires sur les groupements naturels chez les Crustacés Décapodes Brachyoures. II. Les anciens genres *Micropanope* Stimpson et Medaeus Dana. Bulletin du Muséum national d'Histoire naturelle, Paris, 2e série 39:345–374

Guinot D (1968) Recherches préliminares sur les groupments naturels chez les Crustacés Décapodes Brachyures. V. Établissement d'un caratères, évolutif: l'articulation ischiomérale des chélipdès. Bulletin du Muséum National d'Histoire Naturelle, Paris, (2) 40(1):149–166

Guinot D (1969) Sur divers Xanthidae, notamment sur *Actaea* de Haan et *Paractaea* gen. nov. (Crustacea Decapoda Brachyura). Cahiers du Pacifique 13: 223–267

Guinot D (1971) Recherches préliminaries sur les groupements naturels chez les Crustacés Décapodes Brachyoures, VIII Synthèse et bibliographie. Bulletin du Muséum national d'Histoire naturelle (Paris) 42:1063–1090

Guinot D (1976) Constitution de quelques groupes naturels chez les Crustacés Décapodes Brachyoures: I. La superfamille des Bellioidea et trois sous-familles de Xanthidae (Polydectinae Dana, Trichiinae de Haan, Actaeinae Alcock). Mémoires du Muséum national d'Histoire naturelle Nouvelle Série Série A, Zoologie 97:1–308

Guinot D (1979) Données nouvelles sur la morphologie, la phylogenèse et la taxonomie des Crustacés Décapodes Brachyoures. Mémoires du Muséum national d'Histoire naturelle. Nouvelle Série. Série A, Zoologie 112:3–354

Guinot D (1989) Le genre *Carcinoplax* H. Milne Edwards, 1852 (Crustacea, Brachyura, Goneplacidae). In: Forest J (ed) Résultats des Campagnes MUSORSTOM, 5, Mémoires du Muséum National d'Histoire Naturelle, (A, Zool). Éd. du Muséum, Paris, pp 265–345

Guinot D, Tavares M (2003) A new subfamilial arrangement for the Dromiidae de Haan, 1833, with diagnoses and descriptions of new genera and species (Crustacea, Decapoda, Brachyura). Zoosystema 25:43–130

Guinot-Dumortier D (1960) Révision des genres *Euxanthus* Dana et *Hypocolpus* Rathbun (Crustacea, Decapoda, Brachyura). Mémoires du Muséum National d'Histoire Naturelle, (A, Zool) 20(2):153–218, pls 1–12

Guo JY, Ng NK, Dai A, Ng PKL (1997) The taxonomy of three commercially important species of mitten crabs of the genus *Eriocheir* De Haan, 1835 (Crustacea: Decapoda: Brachyura: Grapsidae). Raffles Bull Zool 45:445–476

Haan WD (1833–1850) Crustacea. In: Siebold PF (ed) Fauna japonica. Ex Officin. Lithogr. Auctoris et Typis J.G. La Lau, Lugduni-Batavorum, p 244

Hartnoll RG (1975) The Grapsidae and Ocypodidae (Decapoda: Brachyura) of Tanzania. J Zool 177:305–328

Haswell WA (1882) Catalogue of the Australian stalk- and sessile-eyed Crustacea. Australian Museum, Sydney

Heller C (1861a) Beiträge zur Crustaceenfauna des Rothen Meeres. I Theil. Sitzungsberichte der mathematisch-naturwissenschaftlichen Klasse der kaiserlichen Akademie der Wissenschaften Wien 43:297–374

Heller C (1861b) Beiträge zur Crustaceen-Fauna des Rothen Meeres. II. Theil. Sitzungsberichte der mathematisch-naturwissenschaftlichen Klasse der kaiserlichen Akademie der Wissenschaften Wien 44:241–295, pls 241–243

Henderson JR (1893) A contribution to Indian carcinology. Trans Linn Soc Lond 5:325–458, pls 336–340

Herbst GFW (1782–1804) Versuch einer Naturgeschichte der Krabben und Krebse nebst einer systematischen Beschreibung ihrer verschiedenen Arten. Berlin & Stralsund: (Band I: 1782–1790, 1274 pp. Band II: 1796, 1226 pp. Band III: 1799–1804, 1216 pp)

Herklots JA (1852) Notice carcinologique. Bijdragen tot de Dierkunde 5:35–37

Hess W (1865) Beiträge zur Kenntnis der Decapoden-Krebse Ost-Australiens. Archiv für Naturgeschichte 31:127–173

Hilgendorf F (1878) Die von Herrn. Peters in Moçambique gesammelten Crustaceen Monatsbericht de Königlich Preussischen Akademie der Wissenschaften zu Berlin 1878:782–851

Hilgendorf F (1892) Eine neue *Brachynotus* Art aus Aden. Sitzungsberichte der Gesellschaft Naturforschender Freunde zu Berlin 4:37–40

Hoffmann CK (1874) Crustacés et Echinodermes de Madagascar et de l'île de la Réunion. In: Polen FPL, Van Dam DC (eds) Recherches sur la faune de Madagascar et ses dépendances, pp 1–58

Hogarth PJ (1988) Anomuran Crustacea (Paguridea, Porcellanidae, and Hippidea) from Oman, principally from Dhofar province, southern Oman. J Nat Hist 22:1095–1110

Hogarth PJ (1989) The marine Crustacea of Dhofar, Southern Oman. J Oman Stud 10:99–124

Hogarth PJ (1994) Brachyuran crabs (Xanthoidea: Xanthidae, Pilumnidae, Menippidae and Trapeziidae) of southern Oman. Trop Zool 7:93–108

Hogarth P, Beech M (2001) A frist modern record of Mangrove Crab *Scylla serrata* in U.A.E and south-eastern Arabian Gulf. Tribulus 11:30

Holthuis L B (1958) Crustacea Decapoda from the northern Red Sea (Gulf of Aqaba and Sinai Peninsula) II. Hippidea and Brachyura (Dromiacea, Oxystomata, and Grapsoidea). Contributions to the knowledge of the Red Sea, 9. Bulletin of the Sea Fisheries Research Station, Israel, 17: 41–54, figs 1–4

Holthuis LB (1977) The Grapsidae, Gecarcinidae and Palicidae (Crustacea: Decapoda: Brachyura) of the Red Sea. Isr J Zool 26:141–192

Holthuis LB (1995) The identities of *Macrophthalmus rouxii* Lucas, 1836, and *M. dentipes* Lucas, 1836, and the substitution of the latter name for *M. pectinipes* Guérin, 1838 (Decapoda, Brachyura, Ocypodidae). Crustaceana 68:401–404

Holthuis LB, Manning RB (1990) Crabs of the subfamily Dorippinae MacLeay, 1838, from the Indo-West Pacific Region (Crustacea: Decapoda: Dorippidae). Res Crust Tokyo 3:1–151, figs 151–158

Hornby R (1997) A survey of the habitats, invertebrate fauna and environmental sensitivity of the mainland coast of the UAE, with information on status and distribution of Crustaceans. TRIBULUS Bull Emirates Nat Hist Group 7:11–17

Hornell J, Southwell T (1909) Description of a new species of *Pinnotheres* from *Placuna placenta*, with a note on the genus. In: Hornell J (ed) Report to the government of Baroda on the marine zoology of Okhamandal in Kattiawar Part I. Williams and Norgate, London, pp 99–103

Hosseini SH (2009) The intertidal Decapods of Bushehr, northern part of the Persian Gulf. Iran Fish Scientific Journal 8:37–46

Hsueh P-W, Huang J-F, Ng PKL (2009). On a new genus and new species of pilumnid crab from Taiwan, and the generic placements of *Heteropanope changensis* (Rathbun, 1909) and *Pilumnopeus pereiodontus* Davie and Ghani, 1993 (Crustacea: Decapoda: Brachyura). Journal of Natural History, 43:323–334

Hywel-Davies A (1994) A quantitative analysis of the horizontal and vertical zonation of Brachyura and Mollusca aasociated with the Qurm mangal, Muscat, Sultanate of Oman. School of Ocean Sciences, Bangor

Ihle JEW (1918) Die Decapoda Brachyura der Siboga-Expedition. III. Oxystomata, Calappidae, Leucosiidae, Raninidae. Siboga Expeditie Monographie 39b2:159–322

Ismail NS, Ahmed MA (1993) Macrobenthic invertebrates of mangrove, *Avicennia marina* (Forskåll), and of intertidal flats of Khor Kalba, UAE, Gulf of Oman towards the rational use of high salinity tolerant plants. Springer, pp 155–161

Jones DA (1986a) A field guide to the sea shores of Kuwait and the Arabian Gulf. University of Kuwait, Distributed by Blandford Press, Poole

Jones DA (1986b) Ecology of the rocky and sandy shores of Kuwait. In: Halwagy R, Clayton DA, Behbehani M (eds) Marine environment and pollution, proceedings of the 1st Arabian Gulf Conference on Environment and Pollution. University of Kuwait, Kuwait, pp 69–81

Jones D, Clayton D (1983) The systematics and ecology of crabs belonging to the genera *Cleistostoma* De Haan and *Paracleistostoma* De Man on Kuwait mudflats. Crustaceana 45:183–199

Kam Y, Schlöder C, Roche D, Torchin M (2011) The Iraqi crab, *Elamenopsis kempi* in the Panama Canal: distribution, abundance and interactions with the exotic North American crab, *Rhithropanopeus harrisii*. Aquat Invasions 6:339–345

Kazmi QB, Kazmi NM (1987) Crustacean fauna (Stomatopoda and Decapoda) of the Manora Island, Karachi coast. Zool Pak 1:45–63

Kazmi QB, Tirmizi NM (1990) Report of three leucosiids from Pakistan (Decapoda, Brachyura, Leucosiidae). Crustaceana 58:17–26

Kazmi QB, Tirmizi NM (1995) A new species of *Cyphocarcinus* and a new record of *Macropodia formosa* Rathbun (Crustacea: Decapoda, Majidae) from Pakistan. Pak J Mar Sci 4:59–65

Kazmi QB, Tirmizi NM (1999) Two new species and a new record of spider crabs with a note on *Stilbognathus curvicornis* (Herbst, 1803) from Pakistan waters of the Northern Arabian Sea. Crustaceana 72:369–382

Kemp S (1919) Notes on Crustacea Decapoda in the Indian Museum. No XII Scopimerinae. Rec Ind Mus 16:305–348

Khvorov SA, Al-Kharusi LH, Al-Mosharfi MN, Al-Sinawi MN (2012) Catalogue of brachyuran crabs (Decapoda) of the Sultanate of Oman: part I. The Arabian Sea coast (Dhofar, Al-Wusta, southern part of Sharqiya). Crustaceana 85:1735–1744

Klunzinger CB (1906) Über einige Ergebnisse aus meiner soeben erschienenen Arbeit über die Spitz- und Spitzmundkrabben des Roten Meeres. Verhandlungen der Deutschen Zoologischen Gesellschaft 1906:229–232

Klunzinger CB (1913) Die Rundkrabben (Cyclometopa) des Roten Meeres. Abhandlungen der kaiserlich Leop-Carol Deutschen Akademie der Naturforscher Halle 2:97–402

Koh SK, Ng PKL (2008) A revison of the shore crabs of the genus *Eriphia* (Crustacea: Brachyura: Eriphiidae). Raffles Bull Zool 56:325–355

Komai T, Wada K (2008) A revision of the estuarine crab genus *Ilyograpsus* Barnard, 1955 (Crustacea: Decapoda: Brachyura: Macrophthalmidae), with descriptions of a new genus and one new species. Raffles Bull Zool 56:357–384

Kossmann R (1877) III. Malacostraca (1. Theil: Brachyura). In: Kossmann R (ed) Zoologische Ergebnisse einer im Auftrage der Königlichen Academie der Wissenschaften zu Berlin ausgeführten Reise in die Küstengebiete des Roten Meeres. Wilhelm Engelmann, Leipzig, pp 1–66

Kossmann R (1878) Kurze Notizen über einige neue Crustaceen, sowie über neue Fundorte einiger bereits beschriebener. Archiv für Naturgeschichte 44:251–258

Kossmann R (1880) Zoologische Ergebnisse einer im Auftrage der königlichen Akademie der Wissenschaften zu Berlin ausgeführten Reise in die Küstengebiete des Rothen Meeres. III. Malacostraca. 2. Theil: Anomura. Verlag W Engelmann, Leipzig, pp 67–79

Krauss F (1843) Die südafrikanischen Crustaceen. Eine Zusammenstellung aller bekannten Malacostraca. Bemerkungen über deren Lebensweise und geographische Verbreitung, nebst Beschreibung und Abbildung mehrerer neuer Arten, Stuttgart

Kuronuma K (1974) Arabian Gulf fishery-oceanography survey by the Umitaka-Maru, training-research vessel, Tokyo University of Fisheries with collaboration of Kuwait Institute for Scientific Research, December 1968. Trans Tokyo Univ Fish 1:1–118

Lai JC, Ng PK, Davie PJ (2010) A revision of the *Portunus pelagicus* (Linnaeus, 1758) species complex (Crustacea: Brachyura: Portunidae), with the recognition of four species. Raffles Bull Zool 58:199–237

Lamarck JBPA (1801) Système des animaux sans vertèbres, ou tableau général des classes, des ordres et des genres de ces animaux; présentant leurs caractères essentiels et leur distribution, d'après la considération de leurs rapports naturels et de leur organisation, et suivant l'arrangement établis dans les galeries du Muséum d'Histoire Naturelle, parmi leur dépouilles conservées; précédé du discours d'ouverture du cours de zoologie, donné dans le Muséum National d'Histoire Naturelle l'an 8 de la République. Déterville, Paris

Lamarck JBPAD (1818) Histoire naturelle des animaux sans vertèbres 5:1–612

Latreille PA (1812) Crustacés et insectes. Paris

Latreille PA (1818) Crustacés. In: Crustacés, Arachnides et Insectes. Atlas. Tableau Encyclopédique et Méthodique des trois Règnes de la Nature 86:1–39, pls 268–397

Latreille PA (1825) Familles naturelles du règne animal, exposées succinctement et dans un ordre analytique, avec l'indication de leurs genres. J.B. Baillière, Paris

Lattemann S, Höpner T (2008) Environmental impact and impact assessment of seawater desalination. Desalination 220:1–15

Laurie RD (1906) Report on the Brachyura collected by Professor Herdman at Ceylon, in 1902. In: Henderman WA (ed) Report to the Government of Ceylon on the Pearl Oyster Fisheries of the Gulf of Manaar Part v. Supplementary Report 40, pp 349–432, pls 341, 342

Laurie RD (1915) Reports on the marine biology of the Sudanese Red Sea. XXI. On the Brachyura. J Linn Soc Lond Zool 31:407–475, figs 401–405, pls 442–445

Leach WE (1815) *Podophthalmus*. In: The zoological miscellany – being descriptions of new or interesting animals, vol 2, pp 147–150

Lenz H (1905) Ostafrikanische Dekapoden und Stomatopoden Gesammelt von Herrn Prof. Dr. A. Voeltzkow. Abhandlungen herausgegeben von der Senckenbergischen Naturforschenden Gesellschaft 27: 341–392, Pls 47–48

Leene JE (1938) The Decapoda Brachyura of the Siboga-Expedition. VII. Brachygnatha: Portunidae. Siboga Expéd 39:1–156

Leene JE, Buitendijk AM (1949) Note on *Charybdis ihlei* nov. spec., *Charybdis beauforti* nov. spec., and *Charybdis edwardsi* nom. nov., from the collections of the British Museum (Natural History), London. Bijdragen tot de Dierkunde 28:291–298

Lewinsohn C (1977) Die Dromiidae des Roten Meeres (Crustacea, Decapoda, Brachyura). Zoologische Verhandelingen 151:3–41

Linnaeus C (1758) Systema naturae per regna tria naturae, secundum classes, ordines, genera, species, cum characteribus, differentiis, synonymis, locis. Tomus I. Edit. decima, reformata. Salvius, Stockholm, Stockholm

Linnaeus C (1764) Museum S.R.M. Ludovicae Ulricae Reginae Suecorum, Gothorum, Vandalorumque. In quo animalia rariora, exotica, imprimis Insecta and Conchilia describuntur and determinantur. Literis & impensis Direct. Laur. Salvii

Lucas H (1836) Macrophthalme, *Macrophthalmus*.(Crust.). FE Guérin-Méneville, Dictionnaire pittoresque d'Histoire Naturelle 4:551

Maccagno T (1928) Crostacei Decapodi. Le specie del genere Uca Leach conservate nel Regio Museo Zoologico di Torino Bolletino dei Musei di Zoologia e di Anatomia comparata della R Università di Torino 41:1–52

Maccagno T (1936) Crostacei di Assab: Decapodi Stomatopodi Anfipodi. Spedizione del Barone Raimondo Franchetti in Dancalia (1928–29). Annali del Museo Civico di Storia Naturale di Genova 59:171–186

MacGilchrist AC (1905) Natural history notes from the R.I.M.S. 'Investigator', Capt. T.H.Heming, R. N. (retired), commanding.- Series III., No. 6. An account of the new and some of the rarer decapod Crustacea obtained during the surveying seasons 1901–1904. The Annals and Magazine of Natural History, series 7(15):233–268

MacLeay WS (1838) On the Brachyurous Decapod Crustacea brought from the Cape by Dr. Smith. In: Smith A (ed) llustrations of the zoology of South Africa. Smith, Elder & Co., London, pp 53–71, pls 52–53

Manning RB (1982) A redescription of *Cancer sexpes* Fabricius, 1798 (Crustacea, Decapoda, Hexapodidae). Steenstrupia 8:157–161

Manning RB (1991) The status of *Cleistostoma dotilliforme* Alcock, 1900 (Crustacea: Decapoda: Ocypodidae), with the description of a new genus. Raffles Bull Zool 39:299–306

Manning RB, Holthuis LB (1981) West African Brachyuran crabs. Smithson Contrib Zool 306:1–379

Manning RB, Holthuis LB (1986) Preliminary descriptions of four new species of dorippid crabs from the Indo-West pacific region (Crustacea: Decapoda: Brachyura). Proc Biol Soc Wash 99:363–365

McLay CL (1989) The Occurrence of *Percnon affine* (H. Milne Edwards, 1853) on Niue Island (Decapoda, Brachyura, Grapsidae). Crustaceana 56:106–108

McLay CL (1993) Crustacea Decapoda: the sponge crabs (Dromiidae) of New Caledonia and the Philippines with a review of the genera. In: Crosnier A (ed) Résultats des Campagnes MUSORSTOM. Mémoires du Muséum National d'Histoire Naturelle, pp 111–252

Mendosa JCE, Chong VC, Ng PKL (2009) A new xanthid crab of the genus *Medaeops* Guinot, 1967, from Peninsular Malaysia, with a note on *Leptodius granulosus* Haswell, 1882 (Crustacea: Decapoda: Brachyura: Xanthidae). Zootaxa 2297:44–54

Miers EJ (1877) Notes upon the oxystomatous Crustacea. Trans Linn Soc Lond 2 Zool 1:235–249

Miers EJ (1879a) On a collection of Crustacea made by Capt. H. C. St. John, R. N., in the Corean and Japanese Seas. Part I. Podophthalmia. With an appendix by Capt. H. C. St. John. Proc Sci Meet Zool Soc Lond 1879:18–61

Miers EJ (1879b) Descriptions of new or little-known species of maioid Crustacea (Oxyrhyncha) in the collection of the British Museum. Ann Mag Nat Hist 4(Series 5):1–28

Miers EJ (1882) On the species of Ocypoda in the collection of the British Museum. Ann Mag Nat Hist 5:376–388

Miers EJ (1884a) Crustacea. In: Report on the zoological collections made in the Indo-Pacific Ocean during the voyage of H.M.S. "Alert" 1881–82. Part I. The collections from Melanesia. The collections from Melanesia. British Museum, London, pp 178–322, pls 118–132

Miers EJ (1884b) Crustacea. Report of the zoological collections made in the Indo–Pacific Ocean during the voyage of HMS 'Alert', 1881–1882: 178–331, Part 172:513–575

Miers EJ (1886) Report on the Brachyura collected by H.M.S. Challenger during the years 1873–1876. In: Murray J (ed) Zoology. Neill and Company, Edinburgh, pp 1–362

Miers EJ (1887) Notes upon the oxystomatous Crustacea. Trans Linn Soc Lond Zool 2:235–249, pls 238–240

Milne Edwards H (1834) Histoire naturelle des Crustacés, comprenant l'Anatomie, la Physiologie et la Classification de ces animaux. I. Paris

Milne Edwards H (1837a) Histoire naturelle des Crustacés, comprenant l'Anatomie, la Physiologie et la classification de ces animaux. II Paris 2:1–531

Milne Edwards H (1837b) Histoire naturelle des Crustacés (Atlas). Paris, 1837:1–32, pls 1–42

Milne Edwards H (1852) De la famille des ocypodides (Ocypodidae). Second Mémoire. Observations sur les affinités zoologiques et la classification naturelle des crustacés. Annales des Sciences Naturelles, 3e série 18:128–166

Milne Edwards H (1853) Mémoire sur la famille des Ocypodides. Suite (1). Deuxiéme Tribu Principale. Annales des Sciences Naturelles, 3e série 20:163–228

Milne Edwards H (1854) Notes sur queques Crustacés nouveaux ou peu connus conservés dans la collection du Muséum d'histoire Naturelle. Archives du Muséum d'Histoire naturelle, Paris 7:145–192

Milne-Edwards A (1861) Etudes zoologiques sur les Crustacés récents de la famille des Portuniens. Gide, Libraire-Editeur

Milne-Edwards A (1862) Faune carcinologique de l'île de la Réunion. In: Maillard L (ed) Notes sur l'île de la Réunion. Paris, pp 1–16

Milne-Edwards A (1865a) Descriptions de quelques Crustacés nouveaux appartenant a la tribu des Maiens. Annales de la Société entomologique de France, 4e série 5:133–147

Milne-Edwards A (1865b) Études zoologiques sur les Crustacés récents de la famille des Cancériens. Nouvelles Archives du Muséum d'Histoire naturelle, Paris 1:179–308

Milne-Edwards A (1867) Descriptions de quelques espéces nouvelles de Crustacés Brachyures. Annales de la Société entomologique de France, 4e série 7:263–288

Milne-Edwards A (1869) Note sur quelques nouvelles espèces du genre *Sesarma* (Say). Bulletin des Nouvelles Archives du Muséum 5:25–31

Milne-Edwards A (1872) Recherches sur la faune carcinologique de la Nouvelle-Calédonie. Nouvelles Archives du Muséum d'Histoire naturelle, Paris 8:229–267

Milne-Edwards A (1873) Recherches sur la faune carcinologique de la Nouvelle-Calédonie, II. Nouvelles Archives du Muséum d'Histoire naturelle, Paris 9:155–332

Milne-Edwards A (1874) Recherches sur la faune carcinologique de la Nouvelle-Calédonie troisième partie. Nouvelles Archives du Muséum d'Histoire naturelle, Paris 10:39–58

Mohammed S, Al-Ssadh S (1996) Coral reef grounds and its associated biota in the western side of the Arabian Gulf (ROPME Sea Area) with respect to 1991 Gulf war oil spill. Indian J Mar Sci 25:35–40

Mokhlesi A, Kamrani E, Backwell P, Sajjadi M (2011) Study on the behaviour of two fiddler crabs, *Uca sindensis* and *Uca annulipes* (Decapoda: Ocypodidae), in Bandar Abbas, Iran. J Mar Biol Assoc U K 91:245–249

Mokhtari M, Savari A, Rezai H, Kochanian P, Bitaab A (2008) Population ecology of fiddler crab, *Uca lactea annupiles* (Decapoda: Ocypodidae) in Sirik mangrove estuary, Iran. Estuarine, Coastal and Shelf Science, 76: 273–281

Moosa K, Serène R (1981) Observations on the Indo-West-Pacific Palicidae (Crustacea: Decapoda) with descriptions of two new subfamilies, four new genera and six new species. Mar Res Indones 22:21–66

Müller H (1887) Zur Crustaceen fauna von Trincomali. Verhandlungen der Naturforschenden Gesellschaft in Basel 8:470–485, pls 474–475

Naderloo R (2005) Taxonomic study of the subtidal crabs of the Persian Gulf, Iran. Unpublished M.Sc. thesis, University of Tehran, Iran: 131 pp

Naderloo R (2011) Grapsoid crabs (Decapoda: Brachyura: Thoracotremata) of the Persian Gulf and the Gulf of Oman. Zootaxa 3048:1–43

Naderloo R (2013) The sentinel crabs of the genus *Chaenostoma* (Stimpson, 1858) (Crustacea: Brachyura: Macrophthalmidae), with description of a new species and new records. J Nat Hist 47:2835–2848

Naderloo R (2014) Invasive Hepu mitten crab, *Eriocheir hepuensis* (Crustacea: Decapoda: Brachyura: Varunidae) from the Iranian marshland in the northern Persian Gulf estuarine system. Mar Biodivers Rec 7:e23

Naderloo R (2015a) A new species of decorator crabs, genus *Menaethiops* Alcock, 1895 (Crustacea: Decapoda: Brachyura: Majoidea: Epialthidae), from Abu Musa Island, Persian Gulf, Iran. Zootaxa 3925:431–437

Naderloo R (2015b) Two new species of *Palapedia* Ng, 1993 (Crustacea: Decapoda, Brachyura: Xanthidae) from the Persian Gulf. Zootaxa 3994(2):265–274

Naderloo R, Apel M (2012) Leucosiid Crabs of the Genus *Hiplyra* Galil, 2009 (Crustacea: Brachyura: Leucosiidae) from the Persian Gulf and Gulf of Oman, with Description of a New Species. Zool Stud 51:248–258

Naderloo R, Ebrahimnejad S, Sari A (2015) Annotated checklist of the decapod crustaceans of the Gulf of Oman, northwestern Indian Ocean. Zootaxa 4028:397–412

Naderloo R, Ebrahimnejad S, Doostali AH, Mahdian M (2016) Five Brachyuran crabs of the families Xanthidae and Pilumnidae (Crustacea: Decapoda: Brachyura) from Abu Musa Island, Iran; new records for the Persian Gulf. Mar Biodivers Rec 9:19

Naderloo R, Fatemi Y (2015) New record of the rare alien grapsoid crab, *Percnon affine* from the Gulf of Oman. Mar Biodivers Rec 8:e92

Naderloo R, Ng PKL (2011) Rediscovery of the rare pilumnid crab, *Actumnus simplex* Rathbun, 1911 (Decapoda, Brachyura) and a new record for the Persian Gulf. Crustaceana, 84(12–13):593–604

Naderloo R, Sari A (2005a) Iranian subtidal leucosiid crabs (Crustacea: Deacapoda: Brachyura) of the Persian Gulf: taxonomy and zoogeography. Iran J Anim Biosystematics 1:28–43

Naderloo R, Sari A (2005b) New record of a trapeziid crab, *Quadrella reticulata* Alcock, 1898 (Brachyura: Trapeziidae) from the Persian Gulf. Iran J Anim Biosystematics 1:67–73

Naderloo R, Sari A (2007a) Subtidal crabs of the Iranian coast of the Persian Gulf: new collections and biogeographic considerations. Aquat Ecosyst Health Manage 10:341–349

Naderloo R, Sari A (2007b) New record of *Hyastenus inermis* (Rathbun, 1911) (Brachyura, Majidae) from the Persian Gulf, Iran. Crustaceana 80:1261–1264

Naderloo R, Schubart CD (2009) Redescription and mitochondrial identification of *Chiromantes boulengeri* (Calman, 1920) (Decapoda: Brachyura: Sesarmidae) based on fresh material from the Persian Gulf, Iran. Zootaxa 2128:61–68

Naderloo R, Schubart CD (2010) Description of a new species of *Parasesarma* (Crustacea; Decapoda; Brachyura; Sesarmidae) from the Persian Gulf, based on morphological and genetic characteristics. Zool Anz 249:33–43

Naderloo R, Türkay M (2009) A new species of the genus *Nanosesarma* (Crustacea: Decapoda: Brachyura: Sesarmidae), and redescription of *Nanosesarma jousseaumei* (Nobili, 1906), including new records from the Persian Gulf. J Nat Hist 43:2911–2923

Naderloo R, Türkay M (2011) A new species of the *Macrophthalmus boscii*-group (Decapoda: Brachyura: Macrophthalmidae) from the Persian Gulf with designation of a Neotype for *M. boscii* Audouin (1826). Mar Biodivers 41:503–515

Naderloo R, Türkay M (2012) Decapod crustaceans of the littoral and shallow sublittoral Iranian coast of the Persian Gulf: faunistics, biodiversity and zoogeography. Zootaxa 3374:1–67

Naderloo R, Türkay M (2015) On a new species of decorator crabs of the genus *Paratymolus* Miers, 1879 (Crustacea: Decapoda: Brachyura: Majoidea: Inachidae) from the Persian Gulf. Trop Zool 28:66–70

Naderloo R, Türkay M, Chen H-L (2010) Taxonomic revision of the wide-front fiddler crabs of the *Uca lactea* group (Crustacea: Decapoda: Brachyura: Ocypodidae) in the Indo-West Pacific. Zootaxa 2500:1–38

Naderloo R, Türkay M, Apel M (2011) Brachyuran crabs of the family Macrophthalmidae Dana, 1851 (Decapoda: Brachyura: Macrophthalmidae) of the Persian Gulf. Zootaxa 2911:1–42

Naderloo R, Türkay M, Sari A (2013) Intertidal habitats and decapod (Crustacea) diversity of Qeshm Island, a biodiversity hotspot within the Persian Gulf. Mar Biodivers 43:445–462

Naser MD (2011) The Sesarmid crab *Parasesarma persicum* Naderloo and Schubart, 2010 (Crustacea: Decapoda: Brachyura: Sesarmidae), new to the Iraqi coastal waters of Khor Al-Zubair and Shatt Al-Basrah Canal, Basrah, Iraq. Jordan J Biol Sci 4:185–190

Naser MD, Ali MH, Yasser AG (2010) New record of the fiddler crab *Uca* (*Paraleptuca*) *sindensis* (Crustacea: Brachyura: Ocypodidae) from Khor Al-Zubair, Basrah, Iraq. Mar Biodivers Rec 3:e87

Naser MD, Son MO, Yasser AG (2011) Assessing the risks of invasions of aquatic invertebrates in the Shatt Al-Arab River. Russ J Biol Invasions 2:120–125

Naser MD, Page T, Ng N, Apel M, Yasser A, Bishop J, Ng P, Clark P (2012) Invasive records of *Eriocheir hepuensis* Dai, 1991 (Crustacea: Brachyura: Grapsoidea: Varunidae): implications and taxonomic considerations. BioInvasions Rec 1:71–86

Neumann V, Piridonov VA (1999) Shallow water crabs from the western Indian Ocean: Portunoidea and Xanthoidea excluding Pilumnidae (Crustacea Decapoda Brachyura). Trop Zool 12:9–66

Ng NK, Dai AY, Guo J, Ng PKL (1998) The complete larval development of the southern Chinese mitten crab, *Eriocheir hepuensis* Dai, 1991 (Decapoda, Brachyura, Grapsidae) reared under laboratory conditions. Crustaceana, 71: 493–517

Ng NK, Guo J, Ng PKL (1999) On the generic affinities of *Eriocheir leptognathus* Rathbun, 1913, and *Eriocheir formosa* Chan, Hung and Yu, 1995, with description of a new genus (Crustacea: Grapsidae: Varuninae). Journal of Crustacean Biology, 19: 154–170

Ng PKL (1998) Crabs. In: Carpenter KE, Niem VH (eds) FAO species identification guide for fishery purposes the living marine resources of the Western Central Pacific. FAO, Food and Agriculture Organisation, Rome, pp 1045–1155

Ng PKL, Clark PF (2005) The Indo-West Pacific Pilumnidae XVIII: On the taxonomy of *Pilumnus scabriusculus* Adams and White, 1849, and *P. sluiteri* De Man, 1892, with a note on *Cancer incanus* Forskål, 1775 (Brachyura: Xanthoidea). Zootaxa 841:1–14

Ng PKL, Davie P (2002) A checklist of the brachyuran crabs of Phuket and Western Thailand. In: Bruce NL, Berggren M, Bussawarit S (eds) Biodiversity of Crustacea of the Andaman Sea, Proceedings of the International Workshop on the Biodiversity of Crustacea in the Andaman Sea, Phuket Marine Biological Center, 29 November – 20 December 1998, Phuket Marine Biological Center Special Publication, vol 23, pp 369–384

Ng PKL, Davie PJF (2007) On the identity of *Atergatis floridus* (Linnaeus, 1767) and recognition of *Atergatis ocyroe* (Herbst, 1901) as avalid species from the Indian Ocean (Crustacea: Brachyura: Xanthidae). Raffles Bull Zool 16:169–175

Ng PKL, Wang CH (1994) Notes on the enigmatic genus *Pseudozius* Dana, 1851 (Crustacea, Decapoda, Brachyura). J Taiwan Mus 47:83–99

Ng PKL, Guinot D, Davie PJF (2008) Systema Brachyurorum, part I. An annotated checklist of extant brachyuran crabs of the world. Raffles Bull Zool Suppl 17:1–286

Ng PKL, Rahayu D, Naser MD (2009) The Camptandriidae of Iraq, with description of a new genus and notes on *Leptochryseus* Al-Khayat & Jones, 1996 (Crustacea: Decapoda: Brachyura). Zootaxa 2312:1–26

Ng PKL, Lai JCY, Ghotbeddin N, Safaei M (2011a) *Calappa exanthematosa* Alcock & Anderson, 1894, as a valid species of box crab from the Indian Ocean (Crustacea: Decapoda: Calappidae). Zootaxa 3042:1–14

Ng PKL, Safaie M, Ghotbeddin N (2011b) The Box Crab *Calappa Dumortieri* Guinot, 1964 (Decapoda, Brachyura, Calappidae) in the Gulf of Oman. Crustaceana 84:1781–1785

Ng PKL, Trivedi JN, Vachhrajani KD (2015) *Dentoxanthus iranicus* Stephensen, 1946 (Crustacea: Brachyura: Galenidae): a new record from Gujarat, India, with systematic notes. Mar Biodivers Rec 8:e132

Nobili G (1905a) Décapodes nouveaux des cotes d'Arabie et du Golfe Persique (Diagnoses préliminaires). Bulletin du Muséum National d'Histoire Naturelle, Paris 11

Nobili G (1905b) Quatre décapodes nouveaux du Golfe Persique. Bulletin du Muséum National d'Histoire Naturelle, Paris 11:238–239

Nobili G (1905c) Diagnoses prèliminaires de 34 espéces et varietés nouvelles et de 2 genres nouveaux de Décapodes de la Mer Rouge. Bulletin du Muséum d'Histoire Naturelle Paris 11:393–411

Nobili G (1905d) Note sur *Ocypoda fabricii* Milne Edwards. Bulletin du Muséum d'Histoire Naturelle Paris 11:229–235

Nobili G (1906a) Crustacés décapodes et stomatopodes. In: Mission J. Bonnier et Ch. Perèz (Golfe Persique, 1901). Bulletin Scientifique de la France et de la Belgique 40:13–159

Nobili G (1906b) Faune carcinologique de la Mer Rouge. Décapodes et Stomatopodes. Annales des Sciences naturelles, (Zoologie) 9:1–347

Nobili G (1907) Ricerche sui Crostacei della Polinesia. Decapodi, Stomatopodi, Anisopodi e Isopodi. Memorie della Reale Accademia della Scienze di Torino, Serie 2(57):351–430

Odhner T (1925) Monographierte Gattungen der Krabbenfamilie Xanthidae. I. Göteborgs Kungliga Vetenskaps—och Vitterhets-Sämhalles Handlingar 29(1): 3–92, Pls 1–5

Owen R (1839) Crustacea the zoology of Captain Beechey's voyage to the Pacific Ocean and Bering's Strait, performed in HMS Blossom in the years 1825–1828. London, pp 144

Paul'son OM (1875) Izsledovaniya rakoobraznykh Krasnago Morya s zametkami otnositel'no rakoobraznykh drugikh morei. Chast' I. Podophthalmata i Edriophthalmata (Cumacea), Kiew

Piontkovski SA, Al-Gheilani HM, Jupp BP, Al-Azri AR, Al-Hashmi KA (2012) Interannual changes in the Sea of Oman ecosystem. Open Mar Biol J 6:38–52

Pretzmann G (1971) Ergebnisse einiger Sammelreisen nach Vorderasien. 2. Teil: Marine Brachyura. Annalen des Naturhistorischen Museums Wien 75:477–487

Pretzmann G (1974) Die Unterfamilie Macrophthalminae Dana im Wiener Naturhistorischen Museum. Annalen des Naturhistorischen Museums in Wien 78:437–444

Price ARG, Izsak C (2005) Is the Arabian Gulf really such a lowspot of biodiversity?: scaling effects and management implications. Aquat Ecosyst Health Manage 8:363–366

Raffaelli D, Hawkins SJ (1996) Intertidal ecology. Kluwer Academic Publishers, Dordrecht

Rajkumar M, Vasagam KK (2009) First record of marine crab, *Eucrate alcocki* Serène, in Serène and Lohavanijaya, 1973 (Crustacea, Decapoda, Brachyura, Euryplacidae) from India. Chinese J Oceanol Limnol 27:832–834

Randall JW (1840) Catalogue of the Crustacea brought by Thomas Nuttall and J. K. Townsend, from the west coast of North America and the Sandwich Islands, with descriptions of such species as are apparently new, among which are included several species of different localities, previously existing in the collection of the Academy. Journal of the Academy of Natural Sciences of Philadelphia 8 (1): 106–147, Pls 3–7

Rathbun MJ (1910) The Danish Expedition to Siam, 1899–1900. 5. The Brachyura. Det Kongelige Danske Videnskabernes Selskabs Skrifter 7:303–367

Rathbun MJ (1911) The Percy Sladen trust expedition to the Indian Ocean in 1905, under the leadership of Mr. J. Stanley Gardiner. Volume III. No. XI. Marine Brachyura. Trans Linn Soc Lond 2 Zool 14:191–261

Rathbun MJ (1923) An analysis of "*Dromia dormia* (Linnaeus)". Proc Biol Soc Wash 36:65–70

Reynolds RM (1993) Physical oceanography of the Gulf, Strait of Hormuz, and the Gulf of Oman—Results from the Mt Mitchell expedition. Mar Pollut Bull 27:35–59

Rezaie-Atagolipour M, Naderloo R, Kamrani E, Savari R (2013) Preliminiary biological information of *Scylla serrata* (Forskål, 1775) (Brachyura, portunidae) in the Persian Gulf and Gulf of Oman: a conservation priority. Crustaceana 86(93):322–325

Riegl BM, Purkis SJ (2012) Coral reefs of the Gulf: adaptation to climatic extremes in the world's hottest sea. In: Riegl BM, Purkis SJ (eds) Coral reefs of the Gulf: adaptation to climatic extremes. Springer, Dordrecht/New York, pp 1–4

Rumphius GE (1741) D'Amboinsche Rariteitkammer, behelzende eene Beschryvinge van allerhande zoo weeke als harde Schaalvisschen, te weeten raare Krabben, Kreeften, en diergelyke Zeedieren, als mede allerhande Hoorntjesen Schulpen, die men in d'Amboinsche Zee vindt: 1–340, pls 341–360, 343 e cycle, Océanologie, Univ. Aix-Marseille II:341–167, figs 341–328

Rüppell E (1830) Beschreibung und Abbildung von 24 Arten kurzschwänzigen Krabben, als Beitrag zur Naturgeschichte des rothen Meeres. H.L. Brönner, Frankfurt a. M.

Saeedi H, Ardalan AA (2010) Incidence and biology of *Arcotheres tivelae* (Crustacea: Decapoda) in *Amiantis umbonella* (Bivalvia: Veneridae) on the northern coast of the Persian Gulf, Iran. J Mar Biol Assoc U K 90:655–661

Sakai T (1938) Studies on the crabs of Japan III. Brachygnatha, Oxyrhyncha. Yokendo Co., Tokyo

Sakai K (1999) J.F.W. Herbst-Collection of Decapod Crustacea of the Berlin Zoological Museum, with remarks on certain species. Naturalists, Publications of Tokushima Biological Laboratory, 6, Shikoku University, Tokushima, pp 1–45, pls 41–21

Sakai K, Türkay M (1977) Die Gattung *Ocypode* in der Sammlung des GenferNaturhistorischen Museums (Crustacea: Decapoda). Revue Suisse De Zoologie 84(1):177–180

Sakai K, Türkay M (2013) Revision of the genus *Ocypode* with the description of a new genus, *Hoplocypode* (Crustacea: Decapoda: Brachyura). Mem Qld Mus 56:665–794

Sakai K, Türkay M, Yang S-l (2006) Revision of the *Helice/Chasmagnathus* complex (Crustacea: Decapoda: Brachyura). Schweizerbart, Stuttgart

Sale PF, Feary DA, Burt JA, Bauman AG, Cavalcante GH, Drouillard KG, Kjerfve B, Marquis E, Trick CG, Usseglio P, Van Lavieren H (2010) The growing need for sustainable ecological management of marine communities of the Persian Gulf. Ambio 40:4–17

Sankarankutty C (1975) On a new species of *Hexapus* De Haan (Decapoda, Goneplacidae) from Cochin. Crustaceana 28(1):1–6

Sankarankutty C, Subramanian S (1976) Taxonomic notes on Crustacea Decapoda collected by Deep-Sea trawling off Dar es Salaam. Univ Sci J (Dar Univ) 2:17–24

Savigny JCD (1817) Crustacés. Atlas Description de l'Égypte, ou recueil des observations et des recherches qui ont été faites en Égypte pendant l'Expédition de l'Armée Française, publié par les orders de Sa Majesté l'Empereur Napoléon le Grand: Histoire naturelle, pls 1–13

Schubart C, Ng PKL (2000) On the identities of the rafting crabs *Cancer depressus* Frabricius, 1775, *Cancer squamosus* Herbst, 1790, *Plagusia immaculata* Lamarck, 1818, and *Plagusia tuberculata* Lamarck, 1818 (Crustacea: Decapoda: Brachyura: Plagusiidae). Raffles Bull Zool 48:325–336

Serène R (1964) Goneplacidae et Pinnotheridae. Papers from Dr Th Mortensen's Pacific Expedition 1914–1916 80:181–282

Serène R (1968) The Brachyura of the Indo Pacific region. In: Prodromus for a check list of the non-planctonic marine fauna of South East Asia special, vol 1. Publication of the Singapore National Academy of Science, Singapore, pp 33–120

Serène R (1973) Notes on Indo-West Pacific species of *Macrophthalmus* (Crustacea, Brachyura). Zool Meded Leiden 46:99–116, pls 111–114

Serène R (1984) Crustacés Décapodes Brachyoures de l'Ocean Indien Occidental et de la Mer Rouge. Xanthoidea: Xanthidae et Trapeziidae. Avec un addendum par A. Crosnier: Carpiliidae et Menippidae. Faune Trop 14:1–400

Serène R, Lundoer S (1974) Observations on the male pleopod of the species of *Ilyoplax* Stimpson with a key to the identification of the species. Phuket Mar Biol Cent Res Bulletin 3:3–10

Serène R, Soh CL (1976) Brachyura collected during the Thai-Danish expedition (1966). Phuket Mar Biol Cent Res Bull 12:1–37

Sheppard C, Price A, Roberts C (1992) Marine ecology of the Arabian region: patterns and processes in extreme tropical environments. Academic Press, London

Sheppard C, Al-Husiani M, Al-Jamali F, Al-Yamani F, Baldwin R, Bishop J, Benzoni F, Dutrieux E, Dulvy NK, Durvasula SR, Jones DA, Loughland R, Medio D, Nithyanandan M, Pilling GM, Polikarpov I, Price AR, Purkis S, Riegl B, Saburova M, Namin KS, Taylor O, Wilson S, Zainal K (2010) The Gulf: a young sea in decline. Mar Pollut Bull 60:13–38

Shih H-T, Kamrani E, Davie PJF, Liu M-Y (2009) Genetic evidence for the recognition of two fiddler crabs, *Uca iranica* and *U. albimana* (Crustacea: Brachyura: Ocypodidae), from the northwestern Indian Ocean, with notes on the *U. lactea* species-complex. Hydrobiologia 635:373–382

Shih Hsi-Te, Ng PKL, Davie PJF, Schubart CD, Türkay M, Naderloo R, Jones D, Liu Min-Y (2016) Systematics of the family Ocypodidae Rafinesque, 1815 (Crustacea: Brachyura), based on genetic relationships, with a reorganization of subfamily rankings and a review of the taxonomic status of *Uca* Leach, 1814, sensu lato and its subgenera. Raffles Bull Zool 64:139–175

Silas EG, Alagarswami K (1967) On an instance of parasitisation by the pea-crab (*Pinnotheres* sp.) on the backwater clam *Meretrix casta* (Chemnitz) from India, with a review of the work on the systematics, ecology biology and ethology of pea crabs of the genus *Pinnotheres* latreille The symposium on Crustacea. Marine Biological Association of India, Emakalum, pp 1161–1227

Simoes N, Apel M, Jones DA (2001) Intertidal habitats and decapod faunal assemblages (Crustacea: Decapoda) of Socotra Island, Republic of Yemen. Hydrobiologia 449:81–97

Snowden RJ, Clayton DA (1995) Aspects of the life history and seasonal ecology of *Nasima dotilliformis* (Alcock) (Brachyura: Ocypodidae) on a Kuwait mudflat. J Univ Kuwait (Sci) 22:84–94

Snowden RJ, Clayton DA, Al-Taher EY (1991) Population biology of *Ilyoplax stevensi* (Brachyura: Ocypodidae) on a Kuwait mudflat. Mar Ecol Prog Ser 71:219–225

Snowden RJ, Clayton DA, Al-Taher EY (1994) Aspects of the ecology of *Tylodiplax indica* Alcock (Brachyura: Ocypodidae) on a Kuwait mudflat. J Univ Kuwait (Sci) 21:219–228

Spiridonov VA (1994) The swimming crabs (Crustacea, Brachyura, Portunidae) of submerged rises and insular shelves of the Atlantic and Indian Oceans. Bottom fauna of seamounts NAUKA–Transactions of the P P Shirshov Institute of Oceanology 129:126–152

Spiridonov VA, Apel M (2007) A new species and new records of deep-water Calappidae (Crustacea: Decapoda) from the Indian Ocean with a key to the *Mursia* Desmarest, 1823 species of the region. J Nat Hist 41:2851–2890

Spiridonov V, Türkay M (2001) Deep sea swimming crabs of the *Charybdis miles* species group in the western Indian Ocean (Crustacea: Decapoda: Portunidae). J Nat Hist 35:439–469

Stephensen K (1946) The Brachyura of the Iranian Gulf. Danish scientific investigations in Iran, Part IV. E. Munksgaard, Copenhagen

Stephenson W (1972a) An annotated check list and key to the Indo-West-Pacific swimming crabs (Crustacea: Decapoda: Portunidae). Bull R Soc N Z 10:1–64

Stephenson W (1972b) Portunid crabs from the Indo-West-Pacific and Western America in the Zoological Museum, Copenhagen (Decapoda, Brachyura, Portunidae). Steenstrupia 2:127–156

Stephenson W, Rees M (1961) Sur deux nouveaux crustacés (Portunidae) Indo-pacifique. Bulletin du Muséum national d'Histoire naturelle, Paris, 2e série 33:421–427

Stephenson W, Rees M (1967a) Some portunid crabs from the Pacific and Indian Oceans in the collections of the Smithsonian Institution. Proc US Natl Mus 120:1–114

Stephenson W, Rees M (1967b) Portunid crabs (Crustacea: Decapoda: Brachyura) collected by the "Discovery" in the Indian Ocean. J Nat Hist 1:285–288

Stephenson W, Rees M (1968) A revision of the *Charybdis miles* "group" of species (Crustacea: Portunidae), with description of a new species from Queensland waters. Mem Qld Mus 15:91–109

Števčić Z (2005) The reclassification of Brachyuran Crabs (Crustacea: Decapoda: Brachyura). Fauna Croat 14(1):1–159

Stevcic Z, Ng PKL (1988) The systematic position of the genus *Dentoxanthus* Stephenson, (1945) (Crustacea, Decapoda, Brachyura, Pilumnldae). Steenstrupla, 14(1):1–5

Stimpson W (1857) Prodromus descriptionis animalium evertebratorum, quae in Expeditione ad Oceanum Pacificum Septentrionalem, a Republica Federata missa, Cadwaladaro Ringgold et Johanne Rodgers Ducibus, observavit et descripsit. Pars III. Crustacea Maioidea. Proc Acad Natl Sci Philadelphia 9:216–221

Stimpson W (1858a) Prodromus descriptionis animalium evertebratorum, quae in expeditione ad Oceanum Pacificum Septentrionalem, a Republica Federata missa, C. Ringoold et J. Rogers ducibus, observavit et descripsit. Pars IV. Crustacea Cancroidea et Corystoidea. Proc Acad Natl Sci Philadelphia 10:31–40

Stimpson W (1858b) Prodromus descriptionis animalium evertebratorum, quae in Expeditione ad Oceanum Pacificum Septentrionalem, a Republica Federata missa, C. RINGGOLD et J. RODGERS ducibus, observavit et descripsit. Pars V. Crustacea Ocypodoidea. Proc Acad Natl Sci Philadelphia 10:93–110

Stimpson W (1858c) Prodromus descriptionis animalium evertebratorum, quae in Expeditione ad Oceanum Pacificum Septentrionalem, a Republica Federata missa, C. Ringgold et J. Rodgers ducibus, observavit et descripsit. Pars VII. Proc Acad Natl Sci Philadelphia 10:225–252

Stimpson W (1907) Report on the Crustacea (Brachyura and Anomura) collected by the North Pacific Exploring Expedition, 1853–56. Smithson Misc Collections 49:1–240

Sun H, Zhou K, Yang X (2003) Phylogenetic relationships of the mitten crabs inferred from mitochondrial 16S rDNA partial sequences (Crustacean, Decapoda). Acta Zoollogica Sinica, 49:592–599

Takeda M, Komai T (1991) Japanese species of the *Macrophthalmus telescopicus* complex (Crustacea: Decapoda: Brachyura: Ocypodidae). Bull Natl Sci Mus Tokyo A (Zool) 17:165–171

Takeda M, Miyake S (1968) Pilumnid crabs of the family Xanthidae from the West Pacific. I. Twenty-three species of the genus Pilumnus, with description of four new species. Zool Lab Fac Agric Kyushu Univ 1:1–60

Takeda M, Miyake S (1969) Pilumnid crabs of the family Xanthidae from the West Pacific. II. Twenty-one species of four genera, with descriptions of four new species. OHMU Occas Pap Zool Lab Fac Agric Kyushu Univ 2:93–156

Takeda M, Miyake S (1970) A new free-living pinnotherid crab (Crustacea, Brachyura) from the Iranian Gulf. 3 (2). OHMU Occas Pap Zool Lab Fac Agric Kyushu Univ 3:11–18

Tan CGS (1996) Leucosiidae of the Albatross expedition to the Philippines, 1907–1910 (Crustacea: Brachyura: Decapoda). J Nat Hist 30:1021–1058

Tan CGS, Ng PKL (1995) A revision of the Indo-Pacific genus *Oreophorus* Rüppell, 1830 (Crustacea: Decapoda: Brachyura: Leucosiidae). In: Richer de Forges B (ed) Les fonds meubles des lagons de Nouvelle-Calédonie (Sédimentologie, benthos). ORSTOM, Paris, pp 101–189

Tan S, Ng PK (2007) Descriptions of new genera from the subfamily Parthenopinae (Crustacea: Decapoda: Brachyura: Parthenopidae). Raffles Bull Zool 16:95–119

Tan SH, Huang JF, Ng PKL (1999) Crabs of the family Parthenopidae (Crustacea: Decapoda: Brachyura) from Taiwan. Zool Stud 38:196–206

Tang B, Zhou K, Song D, Yang G, Dai A (2003) Molecular systematics of the Asian mitten crabs, genus *Eriocheir* (Crustacea: Brachyura). Mol Phylogenet Evol 29:309–316

Teng Shao-J, Shih Hsi-Te, Naderloo R, Corbari L (2016) A study on the *Chaenostoma boscii-complex* group (Decapoda: Brachyura: Macrophthalmidae) from the Indo-West Pacific. Crustacean Research 45:15–27

Tesch JJ (1918) Decapoda F. (Decapoda Brachyura continued) Hymenosomatidae, Retropilumnidae, Ocypodidae, Grapsidae and Gecarcinidae. Siboga Expéd 39:1–148

Tirmizi NM (1970) *Ixa holthuisi* n. sp., a new species of crab from the Northern Arabian Sea (Decapoda, Brachyura, Oxystomata). Crustaceana 18:312–314

Tirmizi NM, Ghani N (1982) Three brachyrhynchan crabs new to the coastal waters of Pakistan (Decapoda, Brachyura). Crustaceana 43:105–110

Tirmizi NM, Ghani N (1988a) Occurrence of two interesting brachyrhunchan crabs in Karachi waters. In: Thompson MF, Tirmizi NM (eds) Marine Science of the Arabian Sea. American Institute of Biological Sciences, Washington DC, pp 137–142

Tirmizi NM, Ghani N (1988b) The rediscovery of *Macrophthalmus* (*Macrophthalmus*) *laevis* A. Milne-Edwards, 1867, in the Arabian Sea (Decapoda Brachyura). Crustaceana 55:253–256

Tirmizi NM, Ghani N (1992) A new distributional record of *Zozymodes cavipes* (Dana, 1852)(Crustacea; Decapoda) from the coastal waters of Pakistan. Pak J Mar Sci 1:135–138

Tirmizi NM, Ghani N (1996) Marine fauna of Pakistan: 5: Crustacea: Brachyura, Brachyrhyncha, part 1: Xanthidae, Goneplacidae, Pinnotheridae, Ocypodidae, Grapsidae. Center of Excellence, University of Karachi, Karachi

Tirmizi NM, Kazmi QB (1982) Range extension of *Harrovia elegans* De Man, 1887, with a note on the male of *Dentoxanthus iranicus* Stephensen, 1945, from the northern Arabian Sea (Decapoda, Brachyura, Eumedoninae). Crustaceana 43(3):308–313

Tirmizi NM, Kazmi QB (1986) Marine fauna of Pakistan. 4. Crustacea: Brachyura (Dromiacea, Archaebrachyura, Oxystomata, Oxyrhyncha). Center of Excellence, University of Karachi

Tirmizi NM, Kazmi QB (1993) Carcinological studies in Pakistan, with remarks on species of the Red Sea and the Mediterranean. Bull Inst Oceanogr Fish 9:347–380

Tirmizi NM, Kazmi QB (1995) A new species of *Schizophrys* from Pakistan with a note on *Schizophrys aspera* (Crustacea: Decapoda: Majidae). Oxford & IBH Publishing Co., New Delhi

Tirmizi NM, Serène R (1971) The rediscovery of two species of crabs (Decapoda, Brachyura) with observations on three species from Pakistan. Crustaceana 21:1–32

Tirmizi NM, Ghani N, Khan K (1986) Mangrove crabs of Karachi. Pakistan Agricultural Research Council, Islamabad

Titgen RH (1982) The systematics and ecology of the Decapods of Dubai, and their zoogeographic relationships to the Arabian Gulf and the Western Indian Ocean. D Phil thesis, Texas A&M University, p 269

Türkay M (1986) Crustacea Decapoda Reptantia der Tiefsee des Roten Meeres. Senckenberg Marit 18:123–185

Türkay M, Sakai K, Apel M (1996) The *Ocypode* ghost crabs (Crustacea: Decapoda: Brachyura) of the Arabian Peninsula and adjacent regions. Fauna Saudi Arab 15:99–117

Tweedie MWF (1950) The fauna of Cocos (Keeling) Islands, Brachyura and Stomatopoda. Bull Raffles Mus 22:105–148

Tyndale-Biscoe M, George RW (1962) The Oxystomata and Gymnopleura (Crustacea, Brachyura) of Western Australia with descriptions of two new species from Western Australia and one from India. J R Soc West Aust 45:65–96

Valinasab T, Fatemi SMR, Ghotbeddin N (2012) First record of *Galene bispinosa* Herbst, 1783 [Decapoda; Brachyura; Galenidae] from the Gulf of Oman, Iran. World Appl Sci J 17:439–441

Van Couwelaar M, Angel MV, Madin LP (1997) The distribution and biology of the swimming crab *Charybdis smithii* McLeay, 1838 (Crustacea; Brachyura; Portunidae) in the NW Indian Ocean. Deep Sea Res II Top Stud Oceanogr 44:1251–1280

Vannini M, Valmori P (1981) Researches on the coast of Somalia. The shore and the dune of Sar Uanle. 30. Grapsidae (Decapoda, Brachyura). MZIS 14:57–101

Vousden DH (1987) The Bahrain Marine Habitat Survey. A study of the marine habitats in the waters of Bahrain and their relationship to physical, chemical, biological and anthropogenic influences Appendices. ROPME, Kuwait

Wagner HP (1986) A revision of the genus *Doclea* Leach, 1815 (Crustacea, Brachyura, Majidae). Bulletin du Muséum national d'histoire naturelle, Paris, 4e Série 8:893–953

Weber F (1795) Nomenclator entomologicus secundum entomologiam systematicum III. Fabricii adjectis speciebus recens detectis et varietabilis. CE Bohn, Kiel & Hamburg 1–172

White A (1847a) List of the specimens of Crustacea in the collection of the British Museum. Trustees, London

White A (1847b) List of the specimens in the collection of the British Museum. British Museum, London

White A (1847c) No. VIII. Descriptions of a new genus and five new species of Crustacea. Appendix. In: Jukes JB (ed) Narrative of the surveying voyage of HMS Fly, commanded by Captain FP Blackwood, RN in Torres Strait, New Guinea, and other islands of the Eastern Archipelago, during the years 1842–1846; together with an excursion into the interior of the eastern part of Java. T. & W. Boone, London, pp 335–338

Wilson EO (1988) The current state of biological diversity. Biodivers 521:3–18

Wood-Mason J (1877) Description of a new species of Portunidae from the Bay of Bengal. Ann Mag Nat Hist 4:422

Wood-Mason J, Alcock A (1891) Natural history notes from H.M. Indian Marine Survey Steamer 'Investigator,' Commander R. F. Hoskyn, R. N., commanding. No. 21. Note on the results of the last season's deep-sea dredging. Ann Mag Nat Hist 6 7:1–19, 186–202, 258–272

Yang S, Sun X (1979) New species of *Eucrate* [sic] from China. Mem Peking Nat Hist Mus 8:1–11

Yazdani M, Taheri M, Sanjani MS, Seyfabadi J (2008) New observation of two species, *Thalamita admete* and *Thalamita prymna* (Decapoda: Portunidae), from Chabahar Bay (Gulf of Oman). Mar Biodivers Rec 2:e56

Zaouali J, Ben Souissi J, Galil BS, d'Udekem d'Acoz C, Ben Abdallah A (2007) Grapsoid crabs (Crustacea: Decapoda: Brachyura) new to the Sirte Basin, southern Mediterranean Sea—the roles of vessel traffic and climate change. Mar Biodivers Rec 1:e73

Zarenkov NA (1969) Crabs of the family Leucosiidae (subfamilies Ebaliinae and Iliinae) collected in tropical waters of Indian and Pacific Oceans. Nauchnye Doklady Vysshei Shkoly, biologicheskie Nauki, SSSR 12:16–26

Zarenkov NA (1971) Species composition and ecology of Crustacea Decapoda of the Red Sea. In: Vodianitzky VA (ed) Benthos of the continental shelf of the Red Sea (In Russian). Editions Naukova Dumk, Kiev, pp 155–203

Index

R. Naderloo, *Atlas of Crabs of the Persian Gulf*, DOI 10.1007/978-3-319-49374-9

X

Z

The manufacturer's authorised representative in the EU is Springer Nature Customer Service Centre GmbH, Europaplatz 3, 69115 Heidelberg, Germany. If you have any concerns regarding our products, please contact ProductSafety@springernature.com

Printed and bound by CPI Group (UK) Ltd, Croydon, CR0 4YY
30/09/2026
02225476-0005